W9-DGI-528

Shands'
Handbook of Orthopaedic Surgery

Eighth edition of
HANDBOOK OF ORTHOPAEDIC SURGERY

Shands' Handbook of Orthopaedic Surgery

R. BEVERLY RANEY, Sr., M.D.
Clinical Professor of Orthopaedic Surgery and Chairman Emeritus of Division of Orthopaedic Surgery, University of North Carolina School of Medicine and North Carolina Memorial Hospital, Chapel Hill, N. C.

H. ROBERT BRASHEAR, Jr., M.D.
Professor of Orthopaedic Surgery, University of North Carolina School of Medicine; Attending Orthopaedic Surgeon, North Carolina Memorial Hospital, Chapel Hill, N. C.

With the collaboration of

ALFRED R. SHANDS, Jr., M.D.
Medical Director Emeritus, Alfred I. duPont Institute of the Nemours Foundation, Wilmington, Del.; Visiting Professor of Orthopaedic Surgery (Emeritus), University of Pennsylvania School of Medicine, Philadelphia, Pa.

With 419 illustrations and a bibliography of 1466 titles

Eighth edition of HANDBOOK OF ORTHOPAEDIC SURGERY

Saint Louis
THE C. V. MOSBY COMPANY
1971

Eighth edition of HANDBOOK OF ORTHOPAEDIC SURGERY

Printed in the United States of America

Library of Congress Catalog Card Number 70-166318

Standard Book Number 8016-4081-4

Distributed in Great Britain by Henry Kimpton, London

CB/CB/B 9 8 7 6 5 4 3 2

This edition of
HANDBOOK OF ORTHOPAEDIC SURGERY
is dedicated to

Polly

Gay

Carolyn

Preface

The eighth edition of the *Handbook of Orthopaedic Surgery* incorporates changes in title, authorship, and content. The title of this edition honors Dr. A. R. Shands, Jr., who developed the concept of this comprehensive but brief introductory text while Associate Professor of Orthopaedic Surgery at Duke University and produced its first edition in 1937 with the collaboration of his resident and associate, Dr. R. B. Raney; after many editions as author or co-author, Dr. Shands continues his constructive interest in the *Handbook* as collaborator. The new authorship of the eighth edition is the third in the history of the *Handbook.*

The content of the eighth edition of the *Handbook* continues its primary objective: presentation of the fundamental facts and principles of orthopaedic surgery concisely but in sufficient detail to convey a well-rounded knowledge of the subject. Chapter lengths have been modified to conform to the relative importance of their subject material rather than to fit uniform lecture schedules. Modifications in the text reflect the changing orthopaedic scene. A multitude of small alterations have been made in an effort to add what is new, pertinent, and reliable; to retain what is fundamental; and to delete what is disproved, misleading, or inconsequential.

A similar policy has been followed with respect to the illustrations and the bibliography. A few of the older illustrations have been deleted, and a number of new ones added. Many older bibliographic references have been replaced by newer ones, preference having been given to articles of a review nature published in the English language in easily available journals and texts. Some timeless references, often eponymic, have been retained.

We acknowledge with gratitude the generous assistance of Dr. Frank C. Wilson, Chairman of the Orthopaedic Division at the University of North Carolina School of Medicine, and that provided by other associates of the North Carolina School of Medicine faculty, the orthopaedic resident staff of the North Carolina Memorial Hospital, and Mr. Charles Wright of the Department of Medical Illustration. We are also indebted to Mrs. Martha Thomas for her efficient secretarial assistance and to Mr. Felton Parker for his many contributions to bibliographic research and proofreading.

R. Beverly Raney, Sr.
H. Robert Brashear, Jr.
Chapel Hill, N. C.

Contents

Shands'
Handbook of Orthopaedic Surgery

Eighth edition of
HANDBOOK OF ORTHOPAEDIC SURGERY

CHAPTER 1 Introduction to orthopaedics

The development of orthopaedic surgery as a specialized division of medical practice has been a long and gradual process. It was early recognized that problems peculiar to this field of work should be grouped together as one subject for study and that such problems are best handled by individuals especially trained and experienced in their diagnosis and treatment. This principle can be observed in the works accredited to Hippocrates in the fifth century B.C., which contain many excellent descriptions of affections of the bones and joints. The first work devoted exclusively to the subject of orthopaedics, Andry's *L'Orthopédie,* was published in 1741. It provided impetus for the frank separation of orthopaedics as a specialized branch of medical science. With the accelerated development of surgical technic one hundred years later, orthopaedic surgery finally became separated from the general field of surgery and established as a specialty. With the introduction of anesthesia and asepsis, rapid progress was made in the development of surgery of the bones and joints. Reconstructive operations could be performed, which in earlier years had been impossible because of the extreme suffering of the patients and the severe infection that often followed the opening of joints and the exposure of bones. The development and widespread use of roentgenography in visualizing bone and joint lesions increased greatly the accuracy of diagnosis and the effectiveness of surgical treatment. With the advent of modern industrial machinery and the automobile and airplane, the incidence of traumatic orthopaedic problems increased rapidly. The care of large numbers of men crippled by injuries during the first and second world wars led to an increased development of orthopaedic surgery. In recent decades attention has been focused on the importance of the care and treatment of the crippled child, and as a result extensive state and national programs have been instituted for the medical care of the crippled child and his restoration to normal living. Still more recently, special projects for orthopaedic rehabilitation of the crippled adult have been inaugurated and expanded. Because of their effectiveness in reducing physical disability, orthopaedic surgery and orthopaedic rehabilitation are playing an important role in state and federal health programs.

In definitions of orthopaedic surgery emphasis is placed equally upon the prevention and the correction of deformity and disability. Certain phases of orthopaedics can be interpreted as representing a mechanical aspect of preventive medicine—for example, the deformity that might follow a crippling disease can often be prevented by the orthopaedic measure of using braces, splints, traction, or similar devices. This concept of prevention dates back to the original orthopaedic textbook, written by Nicholas Andry of the University of Paris. The term *orthopaedic* has been adopted from the title of Andry's work; he originated it by combining two Greek words: *orthos,* meaning straight, and *pais,* child. Andry stated that the purpose of his book was "to teach the different methods of preventing and correcting the deformities of children." From this definition has been expanded the modern interpretation of orthopaedic surgery as applying to patients of all ages.

The subject matter of orthopaedic surgery includes the injuries, diseases, and deformities of bones and joints and of their related structures—the muscles, tendons, ligaments, and nerves. Of the many definitions of orthopaedic surgery, that adopted by the American Academy of Orthopaedic Surgeons is most descriptive: "Orthopaedic Surgery is the medical specialty that includes the investigation, preservation, restoration, and development of the form and function of the extremities, spine, and associated structures by medical, surgical, and physical methods."

In this textbook of orthopaedic surgery, etiologically related entities have, insofar as possible, been placed together. Affections that cannot be satisfactorily classified in this manner have been grouped under headings that indicate the anatomic regions involved. It is hoped that this arrangement will enable the physician and the student unfamiliar with orthopaedic surgery to obtain a clear understanding of the relationships between its different entities.

GENERAL CONSIDERATIONS OF BONE AND JOINT AFFECTIONS

Embryology, anatomy, and physiology. Bone is derived from the mesenchyme, or primitive connective tissue. In its development it passes through either a membranous or a cartilaginous phase. In membranous bone formation there is gradual replacement of primitive connective tissue by osteoid matrix, which calcifies promptly to become bone. Membranous bone formation occurs in the bones of the cranial vault and the face. In the cartilaginous or enchondral type of development, hyaline cartilage models of the bones are formed in the mesenchymal tissue during embryonic life; the cartilage model is later invaded by blood vessels, and the cartilage cells are destroyed and replaced by bone. The long bones, spine, scapulae, ribs, sternum, and pelvis develop in this manner. In either membranous or enchondral bone formation the bone tissue is formed by osteogenic cells known as osteoblasts. In time all bone that originally replaced cartilage is in turn replaced by more mature bone, and in adults none of the existing bone is of enchondral origin.

Grossly, bones are of three shapes: flat, irregular, and long. Except for the haversian systems contained in the compact portion of long bones, all mature bone is histologically similar. The flat and the irregular bones, such as the scapula and the ilium, consist of an inner and an outer plate of compact bone, between which is situated a cancellous or spongy portion. Each long bone comprises, from within outward: an elongated medullary canal; a fine layer of connective tissue called the endosteum; the cancellous or spongy portion at either end, which in children is the epiphysis; the compact layers, with their numerous haversian systems, which form the diaphysis, or shaft; and an outer fibrous covering called the periosteum. The periosteum is firmly bound to the compact layer of bone, or cortex, by anchoring fibers called Sharpey's fibers.

Bone formation in the diaphyses, from primary centers of ossification, is well developed by the time of birth. Ossification of the epiphyses, which proceeds from

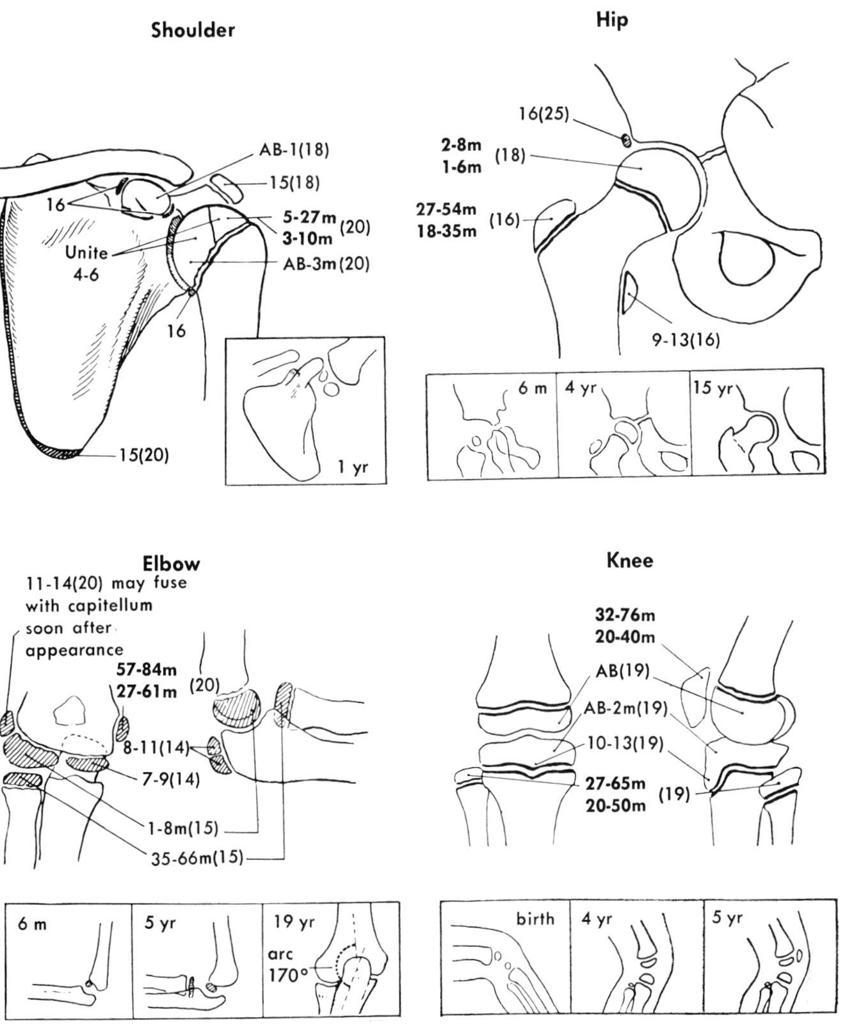

Fig. 1-1. Epiphyseal development at several major joints, showing the age at which these secondary ossification centers appear in roentgenograms and the age, in parentheses, at which union occurs. Age is expressed in months, **m,** or in years; when two sets of age figures appear, that in upper line refers to males and that in lower line to females; **AB** signifies at birth. The inset diagrams show average development at sample ages. (From Pugh, D. C.: The roentgenologic diagnosis of diseases of bones. In Golden, R., editor: Diagnostic roentgenology, Baltimore, 1959, The Williams & Wilkins Co.)

secondary centers, is a much slower process. At birth, ossification centers are not usually visible roentgenographically in any epiphyses except those of the lower end of the femurs and, less frequently, the upper end of the tibiae. Thereafter, ossification appears in the various epiphyses in orderly chronologic sequence (Fig. 1-1). Between epiphysis and diaphysis is the epiphyseal cartilaginous plate, which is silhouetted in roentgenograms as the epiphyseal line. Long bones increase in length by growth at the epiphyseal plates; their shafts thicken by appositional growth beneath the periosteum. As local skeletal maturity is reached, the thinned epiphyseal plate is replaced by fusion between diaphysis and epiphysis. Certain of the vertebral epiphyses are last to fuse, doing so at about 25 years of age.

Bone is a highly specialized form of connective tissue composed of branching cells in an organic calcified matrix. The chief components of the matrix are collagen fibers and a ground substance. The hardness of bone results when hydroxyapatite, a complex mineral crystal composed of calcium, phosphate, and carbonate, is deposited in the soft matrix. The ground substance contains mucopolysaccharides, which include hyaluronic acid and chondroitin sulfates; it is the extracellular and interfibrillar component that permits the exchange of inorganic calcium, phosphorus, and other substances between the blood and the bone. The collagen fibers that make up the greatest portion of the organic matrix of bone are similar to those found elsewhere in the body.

Osteoblasts, osteocytes, and osteoclasts are the cellular components of bone. Each has a specific function. The osteoblast is associated with the formation of new bone tissue and is found in great numbers on the surfaces of actively growing bone. Its function is the manufacture of organic bone matrix or its components. As new bone tissue is formed, the osteoblast is incorporated within the matrix to become an osteocyte. The osteocyte is the living element of bone tissue. Osteoclasts are multinucleated cells associated with bone resorption. They are often found in indentations in the bone surface called Howship's lacunae. During active bone growth and the healing of fractures, transformations of these cells from one to another probably occur, and each cell retains characteristics common to all three.

Compact or cortical bone contains haversian systems, or *osteones* (Fig. 1-2), which can be considered as long, patent columns, irregularly parallel to the long axis of the shaft of the bone. The central space of the column is known as the haversian canal; surrounding it are concentric layers of calcified intercellular substance, called lamellae. Within the lamellae are spaces called lacunae, occupied by the bone cells, or osteocytes. The lacunae are connected with one another and with the haversian canal by tiny irregular channels, termed canaliculi, which contain the processes of the osteocytes. Volkmann's canals are transverse or oblique channels that connect haversian canals and transmit the vascular supply from the periosteum to the osteones and thence to the medullary canal.

Cancellous or spongy bone, found in the metaphyseal and epiphyseal areas (Fig. 1-3), consists of a network of fine trabeculae oriented in the bone in a manner best to withstand the stresses placed upon it. Cancellous bone also has a lamellar pattern, but the lamellae are not arranged concentrically about a central canal as in the osteones of haversian bone. All normal adult bone, cancellous or cortical, is lamellar. Nonlamellar bone is found where bone has been formed rapidly in re-

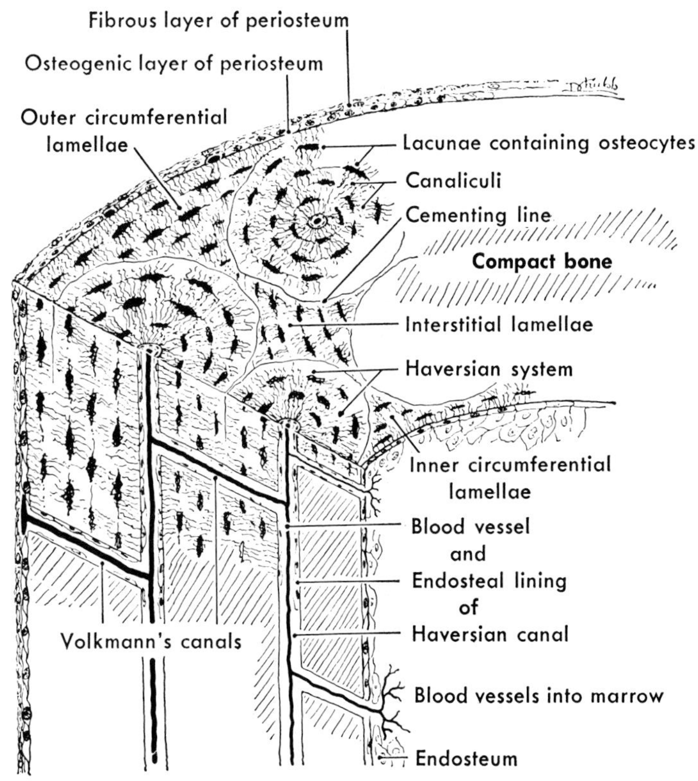

Fig. 1-2. Schematic drawing of a section of the cortex of a long bone. Note especially the haversian systems, or osteones, the several types of lamellae, and the several kinds of vascular channels. (From Ham, A. W.: Histology, ed. 6, Philadelphia, 1969, J. B. Lippincott Co.)

sponse to injury or disease. It is called fiber bone, or woven bone, because its collagen bundles, having been laid down in an irregular manner, give it a fibrillar appearance.

The bone receives its nourishment from blood circulating through the nutrient arteries and the periosteal vessels. The nutrient arteries of the long bones enter the shafts, usually obliquely, through nutrient foramina situated in most cases near the middle of the shaft; they send branches to the ends of the diaphysis where they form an abundant capillary bed close to the epiphyseal plate. Vessels from the nutrient arteries also supply the marrow and endosteum. The periosteum is supplied from the outside, and from its dense arterial network numerous small vessels pass into minute orifices in the compact bone and run through the haversian canals. Other vessels pass from the periosteum through orifices in the compact bone to supply its spongy portion. The epiphysis is nourished by branches of the anastomotic vessels surrounding the joint. Lymphatic vessels are present in the periosteum; although they have been traced into the bone substance, bone essentially has no lymphatic circulation. Nerves are distributed freely to the periosteum; nonmedullated fibers accompany the nutrient arteries into the interior of the bone.

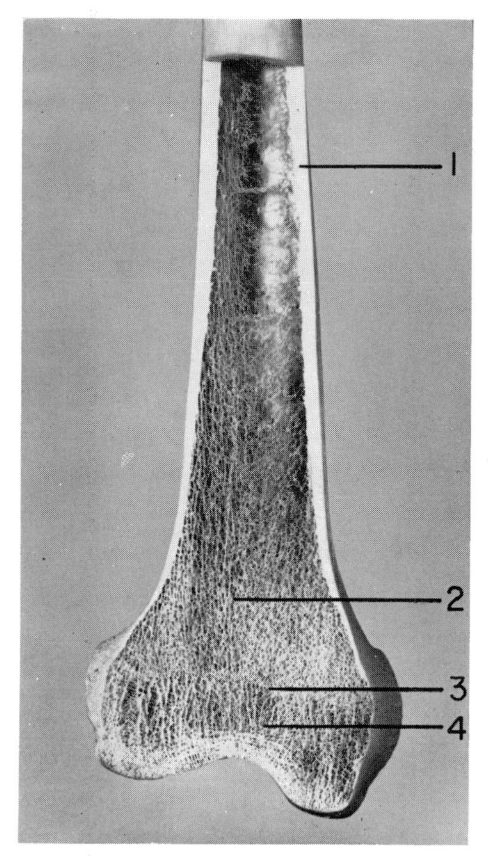

Fig. 1-3. Frontal section of distal portion of adult femur: **1,** cortical (haversian) bone of diaphysis; **2,** cancellous (spongy) bone of metaphysis; **3,** remains of epiphyseal line; **4,** cancellous bone of epiphysis.

A joint is made up of the contacting articular surfaces of two or more bones, surrounded by a capsule and held in place by ligaments and muscles. The cancellous ends of the long bones are covered by a thin layer of compact subchondral bone and a layer of hyaline cartilage. In such cartilage the cells, or chondrocytes, lie rather sparsely within a hyperhydrated matrix that includes collagen fibers and protein polysaccharides. The cartilage contains no blood vessels and is nourished through the joint fluid, arteries of the subjacent bone, and small vessels in the region of the attachment of the synovial membrane. For articular cartilage to maintain its normal state it is necessary that the joint function normally.

The joint cavity has a delicate lining called the synovial membrane. Within the joint is the synovial fluid, a clear, light yellow, viscid liquid consisting of a plasma dialysate plus other substances, chiefly mucin, which are added by the synovial membrane. The lubricating properties of the synovial fluid depend chiefly upon its hyaluronic acid content. Nerve fibers are absent in articular cartilage but are contained in the capsule and, to a lesser extent, the synovial membrane and there mediate the pain of joint reactions.

The synovial membrane is surrounded by a strong fibrous capsule. Flexible but inelastic ligaments thicken and reinforce the capsule and, in turn, are partly covered by muscles and muscle attachments. In addition, certain joints, such as the knee and the jaw, possess intra-articular fibrocartilages that decrease shock, facilitate joint motion, and provide increased stability. Fibrocartilaginous disks between the bodies of adjacent vertebrae perform a similar function. The motion in joints is brought about by muscle activity. In order to function properly, a joint requires normal tone, strength, and correlated action in the muscles by which it is moved.

About the joints are bursae, which are closed sacs lined by specialized connective tissue and containing synovial fluid. Bursae are usually found over bony prominences, especially where a muscle or tendon moves over a projection of bone. The function of bursae is to facilitate gliding movements by diminishing friction.

Etiology. The most common causes of pathologic changes in the structures of the musculoskeletal system are congenital anomalies, trauma, infection, metabolic disorders, endocrine disorders, tumors, circulatory disorders, neurologic disorders, and psychologic disorders.

Congenital anomalies. Abnormalities in the development of the limbs or spine, present at birth, are frequent and make up a large part of children's orthopaedics. Congenital deformities may be caused by genetic abnormalities, by changes in the environment of the developing fetus, or by a combination of these two influences.

Trauma. Mechanical injury may be acute, such as a sudden blow or wrench, or chronic, such as the repeated stresses on the foot occasioned by weight bearing.

Infection. Pathogenic organisms may enter a bone or joint by way of the bloodstream, directly through a lacerating wound, or by direct extension from a neighboring focus. Common sources of bloodstream infection are pustules, boils, and infected abrasions. In bone, bacteria may cause osteomyelitis; in a joint, they may cause pyogenic arthritis.

Metabolic disorders. Numerous changes take place in and about the bones and joints as a result of disturbances of metabolism. Gouty arthritis, occasioned by a disturbance of purine metabolism, is an example.

Endocrine disorders. Extensive changes in the bones may take place as a result of abnormalities of the endocrine glands. An example is the resorption of bone associated with excessive production of parathyroid hormone in hyperparathyroidism.

Tumors. Benign bone tumors are common and amenable to treatment; malignant bone tumors have a lower incidence and a far more serious prognosis. Tumors of joints, muscles, and tendons are uncommon.

Circulatory disorders. Disturbances that decrease the blood supply of certain epiphyses are believed to cause profound changes in these growing areas. Lesions of this type have been termed aseptic, ischemic, or avascular necrosis. Changes sometimes observed in the head of the femur, tarsal navicular, distal end of the second metatarsal, lunate, and certain other bones are examples of avascular necrosis. Disturbances that increase the blood supply to an epiphysis, such as a healing fracture of the femoral shaft, may cause an increase in bone length.

Neurologic disorders. Neurologic disorders, a large and varied group, make up a considerable part of orthopaedic practice. Lesions located in the brain may cause cerebral palsy; lesions in the spinal cord may cause affections such as paraplegia, poliomyelitis, progressive muscular atrophy, and neuropathic joint disease; and lesions in the peripheral nerves may cause various types of localized paralysis.

Psychologic disorders. Psychiatric disorders sometimes lead to joint lesions, such as contractures. Most orthopaedic affections may be aggravated by psychologic disorders. Neuromuscular manifestations of hysteria often simulate primary orthopaedic affections.

Physical diagnosis.* Every student of orthopaedic surgery must gain an accurate understanding of the physical diagnosis of orthopaedic affections. Such knowledge is best acquired by careful study of the history and the examination of each patient.

History. When a history is being taken, the age, sex, occupation, racial background, and economic status of the patient should be recorded. Then a careful analysis of the presenting complaint is made. It is necessary to determine whether the complaint concerns a new symptom or the recurrence of an old one. It is important to gain an accurate understanding of the time and manner of onset and to determine whether the onset was (1) gradual or sudden, (2) associated with an injury or strain, and (3) accompanied by constitutional symptoms such as chills or fever. A gradual onset may indicate a static disability such as that which may accompany a postural strain; the sudden onset of disability suggests a traumatic lesion or, if accompanied by febrile symptoms, an acute bone or joint infection.

In orthopaedic patients the presenting complaint usually concerns pain, deformity, or paralysis. In analysis of the character and type of *pain,* determination of the following points is important: (1) the severity of the pain and whether it is aching or sharp, (2) whether the pain is becoming progressively worse or is diminishing, (3) whether it is less severe in the morning after rest and worse at night, (4) whether activity and cold or damp weather increase its severity, (5) whether radiation of the pain occurs (and in what course), and (6) whether any pain is present in other parts of the body. The extent of disability produced by the pain should be ascertained, as well as the character and result of previous treatment. Pain, although always an important symptom, is variable and, as far as possible, must be analyzed with full appreciation of the patient's psychic stability and tolerance of discomfort. Sharp pain may indicate bone injury with muscle spasm, or pressure on a sensory nerve; increasing pain may indicate progression of an infectious process or malignant tumor; pain that becomes worse with activity during the day may indicate joint strain; pain that is worse in bad weather and is felt in more than one part of the body may be associated with chronic arthritis; and radiating pain frequently accompanies rupture of an intervertebral disk and pressure on a nerve root.

If *deformity* is present, the following points should be ascertained: (1) the patient's concept of the deformity, (2) when and by whom it was first noted, (3) whether its onset was associated with known injury or disease, (4) whether the deformity appears to be increasing, and (5) the degree of disability experienced by the patient. Deformity, unless immediately associated with trauma, is seldom recognized first by the patient. Lateral curvature of the spine, for example, is often first called to the attention of the child's parents during the fitting of a dress.

---

*For more detailed information on orthopaedic physical diagnosis, the reader is referred to the *Manual of Orthopaedic Surgery* (published by the American Orthopaedic Association and available at 430 N. Michigan Ave., Chicago, Ill. 60611).

If there has been *paralysis,* the physician should note these factors: (1) the time and mode of onset, (2) the distribution and degree of paralysis, (3) the improvement or the increase of symptoms, (4) the presence of sensory disturbance, (5) the presence of trophic changes, and (6) any disturbance in the control of bladder or bowel.

When reviewing the past history, the physician should pay careful attention to whether there have been previous orthopaedic disabilities such as the following: a short leg; a limp when growing up, which might indicate a coxa plana or mild slipping of the upper femoral epiphysis; and sprains, fractures, or dislocations. It should also be noted whether the patient has had diabetes, a venereal disease, a series of furuncles, or a chest complaint that might indicate pulmonary tuberculosis. It is essential that the physician always be on the lookout for psychic disorders, misleading subjective exaggeration of the symptoms, and malingering, especially in patients with low back pain. The family history is of significance, particularly in reference to hemophilia, malignancy, and congenital anomalies. The review of systems is particularly important in the medical history of an orthopaedic patient. An almost forgotten episode of mild hemoptysis may be the only hint that the patient's low back pain is caused by metastatic cancer of the lung or by tuberculosis. A history of renal stones may help to indicate that the patient's skeletal problem is a result of hyperparathyroidism.

Examination. Whenever possible, a thorough physical examination should be done initially, in addition to a careful study of the area of local complaint. A comprehensive physical examination is always indicated in the search for contributing factors, such as visceral neoplasms that may have metastasized to bone. In order to obtain the confidence of the patient, which is so important, especially with children, the examiner should proceed quietly and gently. In addition to the routine physical examination, there should be careful observation of the patient as he stands and walks, since abnormalities of posture and gait are often helpful in suggesting the diagnosis. Whether the patient is of slender or stocky build and whether his muscles are overdeveloped, normal, or underdeveloped should be noted. In a child, overdevelopment of certain muscles may indicate pseudohypertrophic muscular dystrophy. Underdeveloped muscles and faulty posture may underlie backache and foot pain.

Often a tentative diagnosis can be made on the basis of gait alone, such as the duck waddle associated with congenitally dislocated hips, coxa vara, and progressive muscular dystrophy, or the unilateral sway associated with a weak gluteus medius after poliomyelitis.

Careful inspection of the affected region should be made, and any dissimilarity as compared with a corresponding normal area should be observed. Any redness, swelling, atrophy, or other visible abnormality should be noted. Muscular atrophy is frequently a valuable confirmatory sign of disuse and local disability. Following inspection, a careful palpation of the affected region should be performed in an attempt to determine the presence of increased local heat, tenderness, crepitation, changes in consistency, or abnormal masses.

Both active and passive motion of the joint or joints should be observed carefully. It is important to compare the range of motion of the pathologic joint with that of the corresponding normal joint. The range of motion should be measured

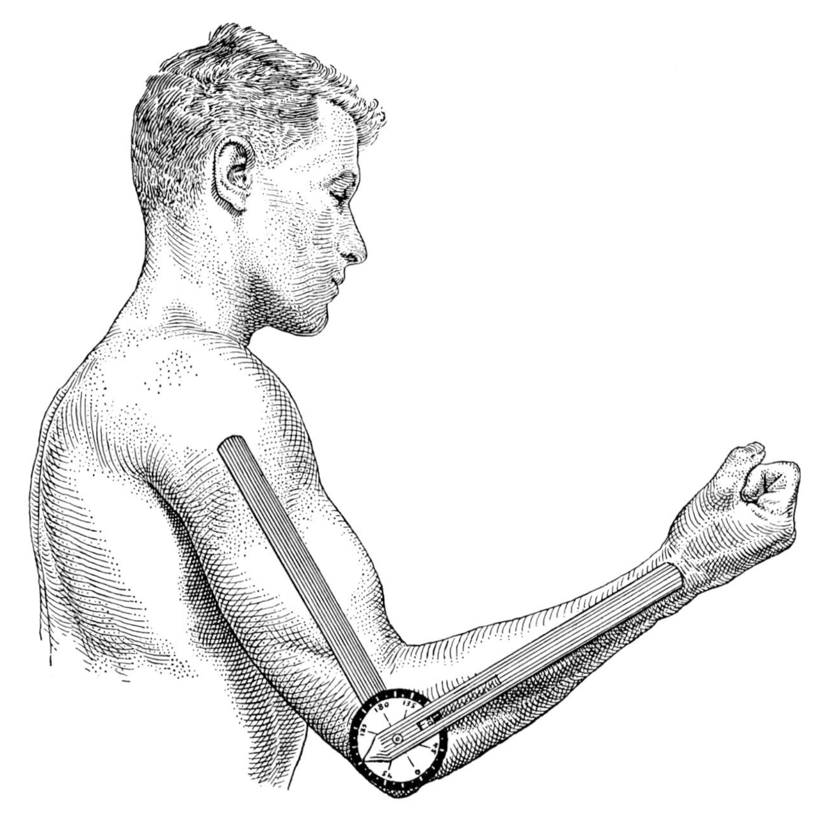

Fig. 1-4. Goniometer for measuring the range of joint motion.

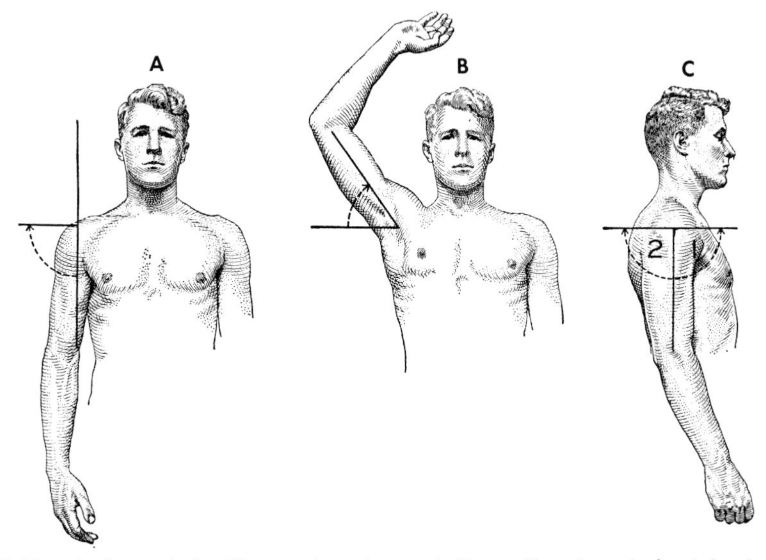

Fig. 1-5. Terminology of shoulder motion. Arrows indicate direction of: **A,** abduction; **B,** elevation; **C,** flexion, **1,** and extension, **2.** (After Cave and Roberts.)

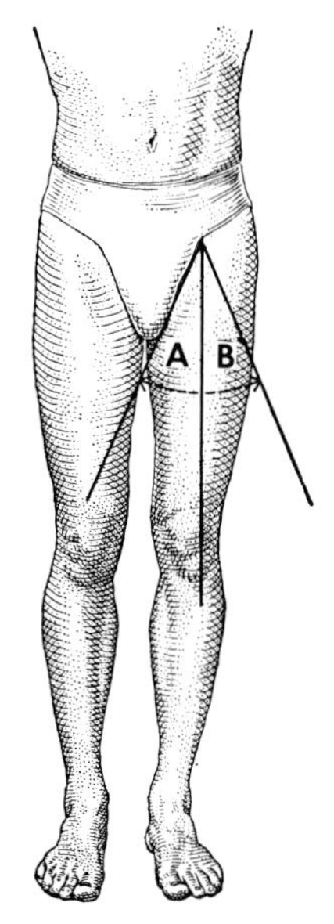

Fig. 1-6. Terminology of hip motion. Arrows indicate direction of **A,** adduction, and **B,** abduction. (After Cave and Roberts.)

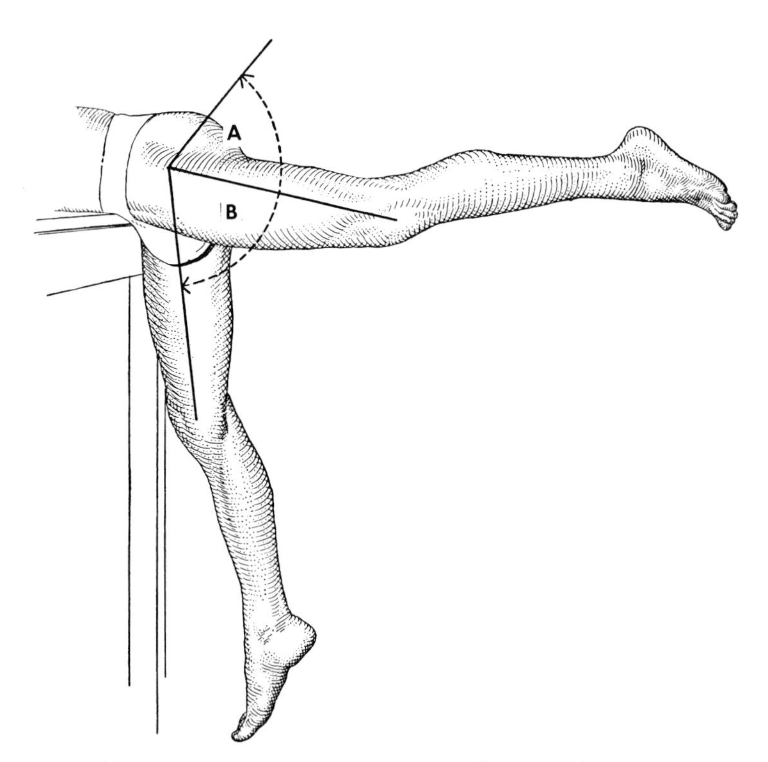

Fig. 1-7. Terminology of hip motion. Arrows indicate direction of **A,** hyperextension, and **B,** flexion. (After Cave and Roberts.)

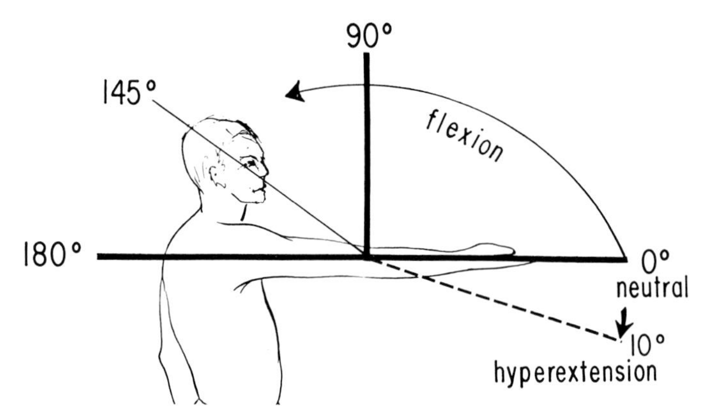

Fig. 1-8. Terminology of elbow motion. Range of motion is expressed in number of degrees of flexion and of hyperextension from the straight, or neutral zero, position. (After Rowe.)

with a *goniometer* wherever possible (Fig. 1-4). If such an instrument is not available, an approximation of the range of joint motion should be made.

The terminology of joint motions has not been standardized, but progress toward this goal is being made. At the shoulder, for example, the terminology long in use (Fig. 1-5) describes certain end positions adequately but is confusing when applied to intermediate positions. An approved method of measuring and recording joint motion, published by the American Academy of Orthopaedic Surgeons,* expresses mobility in degrees of deviation from a defined neutral zero position for each joint. The neutral zero position of the elbow, wrist, fingers, hip (Figs. 1-6 and 1-7), knee, and toes, for example, is that of the extended extremity or anatomic position. Motion from the neutral position (Fig. 1-8) and limitation of motion (Fig. 1-9) are expressed in degrees. The medical student should learn this terminology and use it for describing joint mobility in the records of his patients.

During the examination of joint motion, the presence of muscle spasm and of crepitus should be noted. A joint that has a normal range of smooth, painless, active motion can be presumed to be free of any advanced lesion. It is often desirable to measure the length of the extremities and their circumference at corresponding levels. In determining the length of the lower limbs, one should measure the distance between the lower margin of the anterior superior iliac spine and the tibial malleolus. Such measurements approach accuracy only when made with the patient lying relaxed on a table, with his pelvis level, his hips and knees fully extended, and both hips equally abducted or adducted. To determine the length of the upper limbs measurements should be taken from the tip of the acromion to the tip of the middle finger while the shoulder is adducted and the other joints are in neutral zero posi-

*Available in booklet form from the American Academy of Orthopaedic Surgeons, 430 N. Michigan Ave., Chicago, Ill. 60611, and also printed in the *Manual of Orthopaedic Surgery* (published by the American Orthopaedic Association and available at 430 N. Michigan Ave., Chicago, Ill. 60611).

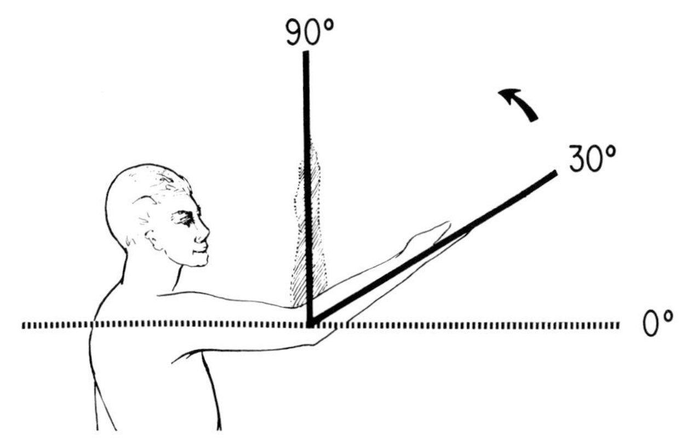

Fig. 1-9. Terminology of limited motion in the elbow joint. In this instance motion is present from 30 degrees to 90 degrees. This may be expressed also as "a flexion deformity of 30 degrees with further flexion to 90 degrees." (After Rowe.)

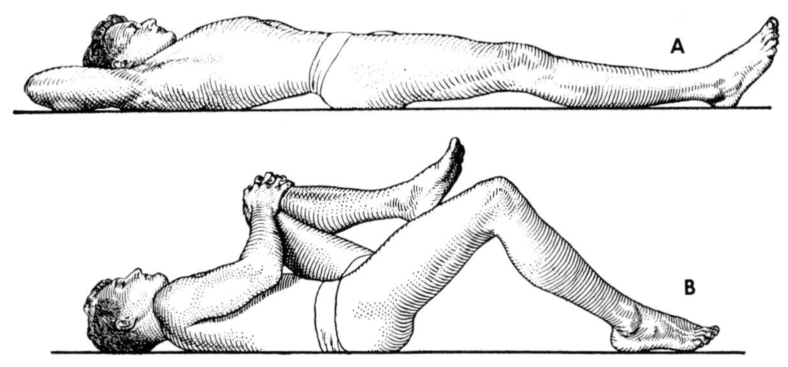

Fig. 1-10. Thomas test for flexion contracture of the right hip. **A,** With patient supine, flexion of the right hip is concealed by arching of the lumbar spine. **B,** With normal left hip held in extreme flexion to flatten the lumbar lordosis, flexion contracture of the right hip is evident.

tion. The student should become acquainted with conventional methods of expressing deformity and the degree of limitation of motion. The following will serve as an example. In deformities of the hip, routine examination sometimes fails to disclose the degree of hip flexion contracture in the presence of a lordosis or hyperextension of the lumbar spine (Fig. 1-10). In such cases the examiner should flex the patient's normal hip fully on the abdomen. This procedure, described in 1875 by Hugh Owen Thomas, will flatten the lumbar spine against the examining table, and an accurate determination of the amount of flexion contracture of the affected hip can then be made. At times it is difficult to ascertain abnormal limitation of motion of the spine. There is great variation in the flexibility of the spinal column in different individuals. Abnormal limitation of motion usually accompanies tenderness and muscle spasm; stiffness of long standing may be present in the absence of pain, tenderness, and spasm.

A careful neurologic examination is necessary when a neuromuscular disturbance is suspected. The type and distribution of motor paralysis should be determined, the character of the reflexes should be recorded, any change in sensation should be charted, and the presence of atrophy should be noted. In cases of paralysis it is most important to estimate the power of each muscle or muscle group. Muscle testing should be done with great care, and the findings should be recorded in detail. It is convenient to use the following scale:

100%	N	Normal	Complete range of motion against gravity with full resistance
75%	G	Good	Complete range of motion against gravity with some resistance
50%	F	Fair	Complete range of motion against gravity
25%	P	Poor	Complete range of motion with gravity eliminated
10%	T	Trace	Evidence of slight contractility but no joint motion
0	0	Zero	No evidence of contractility

Congenital abnormalities should be recorded. Static deformities should be noted, as well as abnormalities of posture, and especially those that may be associated with unequal lengths of the legs, flatfeet, or knock-knees. As a rule such postural abnormalities can be recognized without difficulty.

Roentgenographic diagnosis. Roentgenographic examination of the bones and joints is frequently necessary to clarify or confirm the clinical diagnosis. The roentgen-ray findings may include bone atrophy or hypertrophy, erosion or increased density, change in bone relationships, change in joint contour, or change in the soft tissues. For interpretation of roentgenograms the student or practitioner should not rely on the roentgenologist's report but should himself examine and interpret every film in the light of the clinical findings. Comparison of roentgenograms of the involved structure with films of the opposite, normal side is often essential, particularly in children, because of normal variations of the epiphyses incident to growth. In children, roentgenograms of the carpal bones may be indicated to determine bone or physiologic age. Certain cases may require, in addition to routine anteroposterior and lateral roentgenograms, films made in oblique positions, stereoscopic films, or laminagrams. Roentgenograms made after the injection of contrast media often provide useful diagnostic data by outlining joint cavities, the spinal subarachnoid space, blood vessels, or sinus tracts. Scintillation scanning after intravenous injection of a bone-seeking radioactive isotope may reveal the local increase of osteoblastic activity that is characteristic of metastatic bone lesions or osteomyelitis.

Laboratory diagnosis. The clinical examination of the patient should be supplemented, whenever indicated, by laboratory studies of the blood, urine, synovial fluid, spinal fluid, aspirates, or other material, which often reveal facts essential to the diagnosis and treatment. In many bone diseases it is important to have a determination of the serum proteins, calcium, phosphorus, and phosphatase. Microscopic examination of pathologic tissue is especially indicated when the presence of infection or tumor is suspected. Cultures of joint fluid are often essential in establishing the diagnosis. The intracutaneous tuberculin test is of considerable diagnostic value, especially in children. Electromyography and nerve conduction studies are frequently helpful in the differential diagnosis of neuromuscular affections.

Treatment. Orthopaedic treatment may be divided into nonoperative and operative types. Nonoperative treatment comprises procedures such as the following: rest and support, secured by strapping, braces, splints, traction, and plaster casts; physical therapy, including the use of selected exercises, heat, and massage; occupational therapy; and medical treatment, such as the administration of drugs and the prescription of diets. Operative treatment may be classified as closed or open. The closed operations consist of manipulative procedures, such as the reduction of dislocations and fractures and the stretching of contractures. The most frequently performed open operations are tenotomy, osteotomy, arthrotomy, arthrodesis, arthroplasty, and the open reduction of fractures. Tendon and muscle transference, nerve transference or resection, and bone grafting are indicated less frequently. The object of all operative treatment is to improve function. Operation should never be performed at the risk of decreasing function, lessening needed stability, or leading to pain.

The treatment of orthopaedic patients frequently requires consultation with other specialists. The consultants' advice should be carefully considered with the orthopaedic findings before the final decision concerning treatment is made. The orthopaedic surgeon who fails to obtain a consultation when it is indicated may be embarrassed by finding, after weeks or months of unsuccessful treatment, that the diagnosis is not primarily orthopaedic.

Lesions of the bones, joints, and allied structures often require a longer time for healing than do those of other tissues of the body; hence the convalescence of the average orthopaedic patient is notably slow. Because of the long period often necessary for treatment, the patient may become disheartened, lose morale, and develop a mental attitude that acts as a psychologic barrier to normal convalescence. To offset this, patience and optimism are required of the physician, and a well-rounded program of rehabilitation should be started early in the course of treatment. The rehabilitation program should include intelligently planned physical and mental activities. Too quick a recovery cannot be expected and should not be promised. Because of the slowness of the improvement often encountered in orthopaedic patients, they may change from doctor to doctor and from one form of treatment to another and may ultimately find themselves in the hands of an unethical practitioner or cultist. For the best understanding between patient and doctor, a frank and honest statement should be made by the doctor to the patient, to the family, or to both, explaining in detail as much as is known about the cause of the complaint, the treatment, and the prognosis. Consultation with other physicians should be encouraged if the diagnosis is uncertain or the results of treatment are slow.

Rehabilitation. In recent years increased emphasis has been placed on the physician's part in restoration of the patient to normal living after his illness or injury. This phase of medicine, which has always been a part of good, comprehensive medical care, is termed rehabilitation. Programs of rehabilitation are designed (1) to separate the patient from bed and hospital as soon as possible, (2) to restore the patient to maximum functional activity as quickly as possible, and (3) to facilitate the patient's early adjustment to work, home, and community.

The orthopaedic patients most often requiring long-term rehabilitation are those with hemiplegia, paraplegia, cerebral palsy, poliomyelitis, low back pain, arthritis, fresh fractures, fracture deformities, and amputations. The orthopaedic surgeon, who deals predominantly with chronic disabilities, has always included rehabilitation as an essential part of orthopaedic treatment. Effective rehabilitation in orthopaedic cases often requires a team composed of representatives from various disciplines and including, in addition to the orthopaedic surgeon, a physiatrist, physical therapist, occupational therapist, vocational counselor, and social worker. It is important to the patient's welfare that his attending physician—who in orthopaedic cases is usually the orthopaedic surgeon—plan, direct, and interpret the rehabilitative procedures.

In hospital, rehabilitation should start when the patient is admitted. The orthopaedist should explain to the adult patient the nature of his illness or injury, what is to be his treatment, and what is to be his own part in getting well. A physical conditioning program, occupational therapy, planned recreational activity, and vocational counseling may be indicated. If on discharge from hospital the patient will obviously be unable to resume his former occupation, the orthopaedist should consider sending him to a rehabilitation facility or sheltered workshop where, after an evaluation of his interests, aptitudes, and capacities, the patient can learn a suitable occupation. The orthopaedic surgeon should work closely with the state rehabilitation counselor in obtaining vocational training for his handicapped patients and as a rule should follow all of his patients at intervals until their occupational and social readjustments have been achieved.

NOTE: Anyone further interested in these subjects should make use of the bibilography (p. 445) for more extended reading. The bibliography is by no means exhaustive; an attempt has been made, however, to select as references the most authoritative orthopaedic textbooks and the most informative articles written in English.

CHAPTER 2

Congenital deformities

Congenital deformities are abnormalities of development present at birth. They are frequently observed in all orthopaedic clinics, particularly on the crippled children's services. There are many different types of anomalies and numerous minor variations; they may involve any bone or joint structure. Congenital deformities vary in significance from very minor abnormalities, such as webbing of the toes, to serious and disabling defects, such as absence of the major portion of an extremity.

Much progress has been made in study of the causes of congenital anomalies. They are believed to result from abnormal genetic factors, alterations in the environment of the developing fetus, or both. Individuals affected with certain congenital disorders, such as Down's syndrome, or mongolism, possess an abnormal number or type of chromosomes. Genetically determined defects may result from such chromosomal aberrations or from mutations of genes in recent or distant ancestors. New hereditary defects appear as the result of either small point mutations or disorders involving larger segments of the chromosomes. Animal experimentation has shown that mutations can be brought about by exposing the germ cells to ionizing radiation or to any of a great variety of chemical agents. This fact is of considerable importance in an age when human exposure to radiation and chemicals in the environment is of increasing occurrence. According to Fraser, however, only about 10% of all human congenital deformities can be definitely classed as hereditary in origin.

Harmful influences in the environment of the developing fetus possibly account for the majority of congenital defects. Factors shown to have an adverse effect on the human fetus include heavy irradiation, thalidomide, the folic acid antagonist aminopterin, rubella, toxoplasmosis, and certain androgenic hormones. Probably many other agents can injure the human fetus. All of these produce damage during the first trimester of pregnancy. The incidence of congenital defects increases as maternal age increases. Maternal diabetes also has an unfavorable influence on the fetus.

Congenital defects have been produced experimentally in many animals, including mammals. Defects of the nervous system, cleft palate, clubfoot, polydactyly, syndactyly, and many other deformities have been produced in the laboratory by a variety of means. Agents capable of inducing congenital abnormalities include deficient diets, roentgen rays, hypothermia, hypoxia, certain viruses, and many chemicals. Hormones such as androgens, insulin, and cortisone, when injected into animals during early pregnancy, have also produced a variety of defects. These teratogenic agents are not specific in their action. The type of congenital defect depends greatly on the stage of fetal development at which the agent is applied.

Largely because of their uncertain etiology, congenital deformities are generally not preventable, but it is possible that improved care of the expectant mother's health and protection from environmental poisons may lessen their incidence. Needed treatment should in most instances be started quite early in infancy, before abnormal changes in the affected tissues become more advanced and fixed by increasing age and trauma and before they lead to deforming secondary changes in adjacent structures. The effectiveness of treatment in reducing disability and disfigurement varies widely with the type of anomaly and with the promptness with which the treatment is instituted.

Of common congenital anomalies involving the musculoskeletal system, the two most important are dislocation of the hip and clubfoot. In the present chapter these and a number of other congenital deformities are described. Several additional types, including cervical rib, torticollis, and lumbosacral variations, are discussed in subsequent chapters that deal with individual regions of the body.

CONGENITAL DYSPLASIA OF THE HIP

Congenital dysplasia is a malformation of the hip joint, with which a subluxation or dislocation of the hip is often associated at the time of birth or shortly thereafter. Three degrees of congenital dysplasia have been described: acetabular dysplasia, subluxation, and dislocation. Acetabular dysplasia, or preluxation, is the mildest form of hip dysplasia; in it there is neither subluxation nor dislocation. The femoral head remains seated in its socket, but the acetabulum is shallow. This phase merges imperceptibly into the second, that of subluxation, in which the femoral head, although in contact with the acetabulum, is no longer well seated but begins to ride upward and outward. The third and most advanced phase is that of complete dislocation, in which the femoral head is no longer in contact with the acetabulum. The difference between subluxation and dislocation may be dependent upon the position of the fibrocartilaginous labrum, or *limbus*. If the femoral head lies within the limbus, it is in contact with the acetabulum, and only subluxation exists. In some cases this difference can be determined only by arthrograms or at open surgery.

Incidence. Dysplasia of the hip is one of the more common congenital deformities. It is seen with great frequency in certain areas of the world, particularly in the Latin races. The incidence is high, for example, in northern Italy, and also in Japan. A study carried out in Sweden showed the incidence of dislocation to be 1.7 cases per 1,000 live births. Hip dysplasia is seldom observed in Negroes but is quite common in Navajo Indians. It is six to eight times as common in girls as in boys. Complete dislocation is more frequently unilateral than bilateral.

Etiology. The exact cause of hip dysplasia is unknown. Evidence from the study of twins suggests that both genetic and environmental factors play a part in causing dislocation. Its higher incidence in girls may result from anatomic differences in female and male pelves, or from joint laxity secondary to hormonal changes in girls at birth. An important environmental factor may be the position in which the hips are held during infancy. Babies wrapped in swaddling clothes with the hips extended and adducted, as are the Navajo infants, are susceptible to hip dislocation. Elongation of the joint capsule or a primary developmental defect causing imperfect formation of the posterosuperior margin of the acetabulum may be an etiologic factor. Dislocation may also be the result of some teratogenic factor that adversely affects the embryo at a critical stage in the development of the hip; this view is supported by the fact that other congenital defects are frequently associated with dysplasia of the hip.

ACETABULAR DYSPLASIA AND SUBLUXATION OF THE HIP

Clinical picture. Acetabular dysplasia and subluxation of the hip are frequently discovered shortly after birth by the examining physician. A most important finding is limitation of abduction of the flexed hip or hips (Fig. 2-1). Another commonly noted sign is asymmetry of the gluteal folds. Neither of these findings is sufficient to establish the diagnosis, but either should alert the doctor to look more closely. Looking for these signs takes only a moment or two and should be part of the routine examination of every newborn child. Another useful test in congenital subluxation of the hip is the snapping sign *(Ortolani's sign)*. If traction is applied to the subluxated hip while abduction is being tested, a definite click may be felt if the femoral head slips into its socket. It may be possible to produce a jerk of exit as well as a jerk of entry. The presence of this sign indicates that the subluxation may progress to dislocation if untreated. Obvious shortening and instability of the hip as demonstrated by the telescoping test are not found in subluxation but are indicative of dislocation. When dysplasia in the newborn infant remains untreated, the mother may observe in succeeding months that the hip is abnormal, that she has difficulty spreading the infant's thighs to put on diapers, or that the thigh creases are unequal.

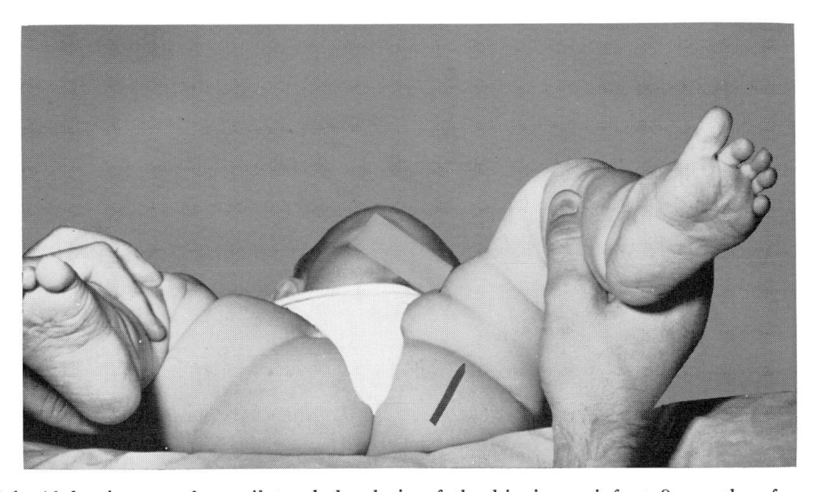

Fig. 2-1. Abduction test for unilateral dysplasia of the hip in an infant 8 months of age who has subluxation on the left side. When the test is positive, the thigh on the dysplastic side cannot be abducted so far as the thigh on the side with a normal hip. Note deepening of the proximal fold in the left thigh, which also suggests dysplasia.

Roentgenographic picture. The roentgenographic findings of acetabular dysplasia may not be obvious at birth. The most readily observable change is an increase in the obliquity of the acetabular roof. A slight upward and lateral displacement of the femur from the acetabulum may also be seen in early subluxation. At 2 to 6 months of age, delay in the appearance of the ossification center of the femoral head will become evident. As pointed out by Caffey, however, some of these radiologic signs can be seen in children with normal hips. The diagnosis must therefore be based on a combination of physical and roentgenographic findings.

Treatment. The treatment of congenital subluxation of the hip is to maintain the head of the femur as deeply as possible within the acetabulum until there is roentgenographic evidence of an adequate, well-developed socket. This is accomplished by maintaining the hips in a widely abducted position by means of a pillow splint or brace (Fig. 2-8). Forced abduction is to be avoided. The treatment may require a few months to more than a year. With early treatment the prognosis is excellent. Some of the milder degrees of dysplasia improve without treatment. Since there is no method of determining which subluxations will persist or progress to cause degenerative hip changes in later years, it is far better to institute the rather simple treatment than to risk the development of serious sequelae.

CONGENITAL DISLOCATION OF THE HIP

In 1826 Dupuytren first described accurately the pathologic changes in congenital dislocation of the hip, but not until 1888 did Paci suggest reduction of the dislocation as a means of treatment. Until then such deformities of the hip had been considered incurable. The suggestions made by Paci were first popularized in 1895 by Lorenz. It is now generally agreed that if a congenitally dislocated hip can be reduced early, and if the corrected position can be maintained for an adequate period of time by means of a plaster cast or brace, a stable joint with satisfactory function will often result.

Pathology. The pathologic changes are characteristic and vary with age, being minimal in the early case and increasing with the duration of the dislocation. The acetabulum is usually shallow with an oblique superior and posterior surface. It may be filled with fat and fibrous tissue. The head of the femur is nearly always displaced upward and backward upon the ilium. Anterior dislocation is uncommon. When the ossification center of the femoral head appears in roentgenograms, it is smaller than that of a normal hip; this is believed to be due to decreased functional stimulation. With continued dislocation the neck of the femur may show an increase in its angle with the shaft (coxa valga). The whole upper portion of the femur usually develops a structural alteration, as a result of which the neck or shaft and neck undergo torsion, assuming a more anterior position in relation to the lower end of the femur (antetorsion or anteversion). The capsule over the head, which on weight bearing acts as a suspensory ligament, becomes elongated, thick, and fibrous; it may become constricted in the middle and assume an hourglass shape. The ligamentum teres may be extremely thin, ribbonlike, and atrophic and may even be absent; in some cases, however, it is broad and thick. The muscles about the joint, especially the adductor group, become shortened and contracted.

With continued weight bearing, a shallow secondary acetabulum may develop on

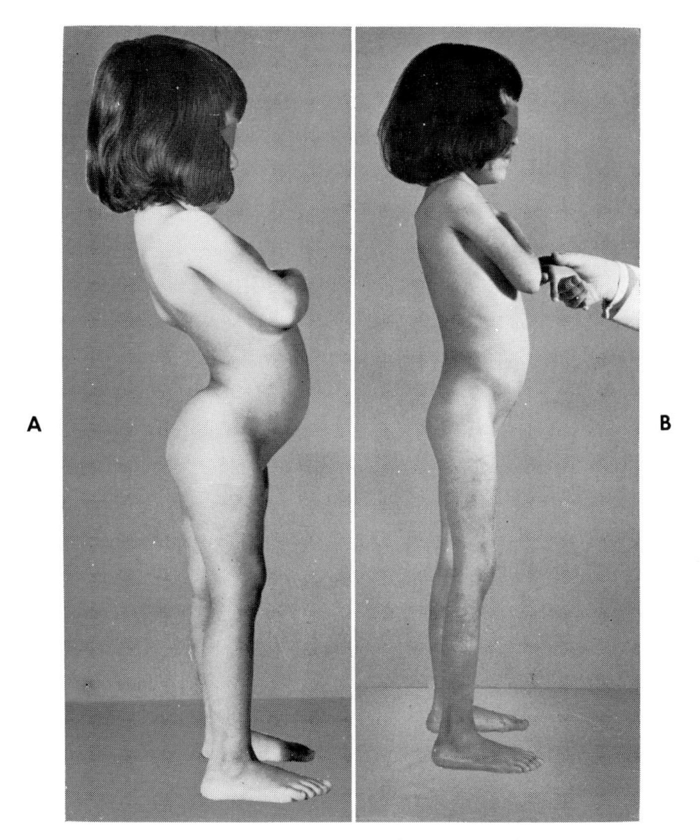

Fig. 2-2. Bilateral congenital dislocation of the hip in a girl 3½ years of age. **A,** Note extreme lumbar lordosis and protrusion of the abdomen. **B,** One year following reduction. Note normal curve of lumbar spine and improvement of abdominal protrusion.

the wing of the ilium. The pelvis is underdeveloped on the affected side, and a postural deviation of the lumbar spine toward this side takes place. Forward tilting of the pelvis and increase of the lumbar lordosis develop to a slight degree in unilateral dislocation and to a much greater degree in bilateral dislocation (Fig. 2-2).

Clinical picture. It may be difficult to recognize dislocation of the hip in early infancy. In a unilateral case the mother may notice that the affected hip is prominent laterally and that the extremity is short. Ortolani's sign (p. 19) may be elicitable. By placing the child supine on a hard examining table and then flexing the child's hips and knees, the examiner can see that on the affected side the knee is lower than on the normal side (Fig. 2-3). Then by abducting the flexed hip as far as possible the examiner will note restriction of motion on the side of the dislocation or subluxation (Fig. 2-1); the restriction is caused primarily by adductor tightness. If the extended thigh is first pushed toward the patient's head and then pulled distally, the greater trochanter and head of the femur can be felt to move up and down in the buttock. This is commonly called telescoping or piston mobility. The instability may sometimes be recognized more easily by testing with the hip and knee flexed to

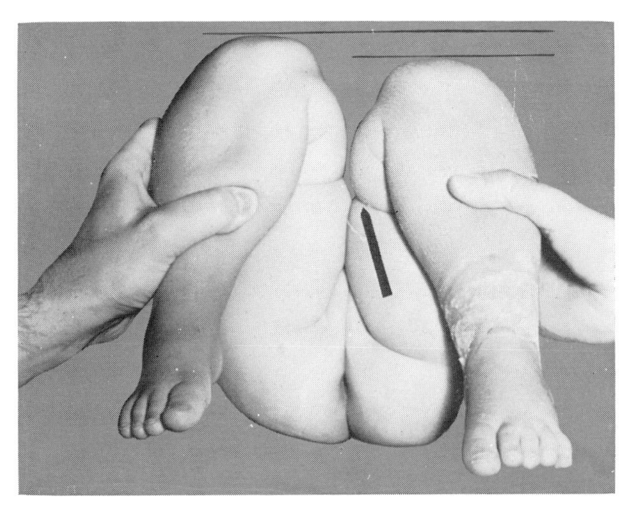

Fig. 2-3. Knee level test for unilateral congenital dislocation of the hip in a girl 2½ years of age. With infant supine on a hard surface, hips and knees are flexed. When the test is positive, the knee on the dislocated side is lower. Note increased depth of the left thigh fold, which also suggests dislocation. (Same patient as shown in Fig. 2-6.)

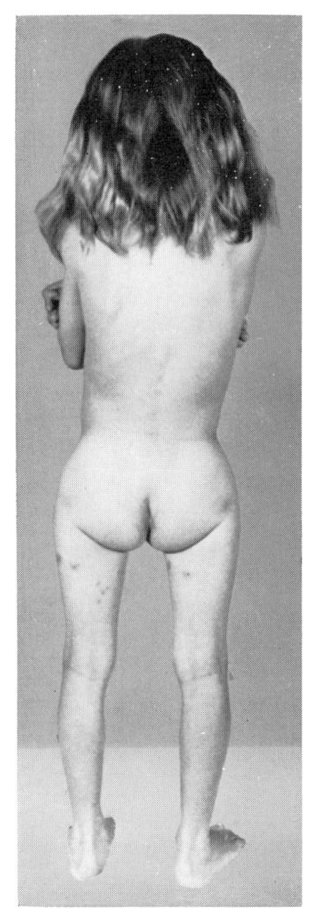

Fig. 2-4. Bilateral congenital dislocation of the hip in a girl 5 years of age. Note widening of the perineum and broadening of the buttocks.

90 degrees. In the unilateral case the skin folds of the thighs are usually asymmetric (Figs. 2-1 and 2-3), and the trochanter is more prominent than normal. Instability of the hip on weight bearing delays the child in learning to stand and walk and causes a characteristic limp. In young children there are no other complaints, and function is usually good. In older children fatigue may be present on exertion, and pain may develop after activity. Sometimes, even in adult patients, there is no complaint of pain.

In bilateral dislocation the gait is especially characteristic: the patient, because of instability of the hips, sways from side to side, exhibiting a duck waddle gait. A similar, less awkward gait is present in unilateral dislocation. In bilateral dislocation the perineum is wide, the buttocks are broad, and the transverse gluteal folds are altered (Fig. 2-4). The normal lordosis of the lumbar spine is increased, and there is protrusion of the abdomen (Fig. 2-2, *A*). In both unilateral and bilateral cases the greater trochanter is prominent and appears above Nélaton's line (the line from the anterior superior iliac spine to the ischial tuberosity). When the patient stands, bearing weight on the affected hip, the pelvis is tilted downward on the normal side (positive Trendelenburg sign) instead of tilting upward as it would if the hip being

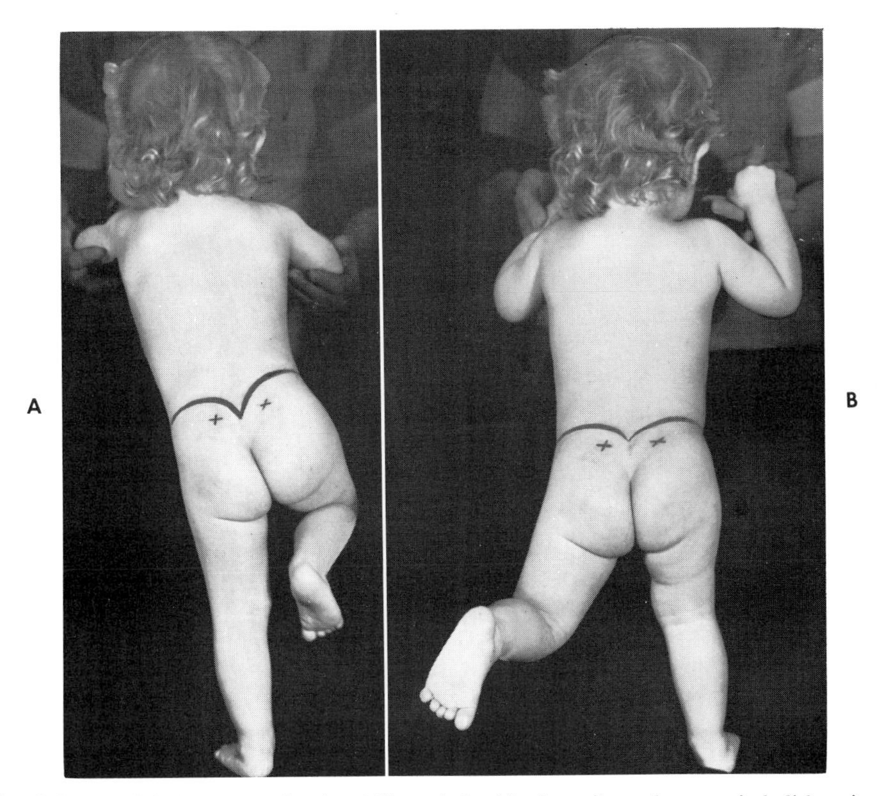

Fig. 2-5. Trendelenburg test for instability of the hip in unilateral congenital dislocation in a child 2½ years of age. **A,** When child bears weight on the normal left hip, the opposite side of the pelvis is elevated to maintain balance (negative test). **B,** When weight is borne on the dislocated right hip, the opposite side of the pelvis cannot be elevated normally (positive test).

tested possessed normal stability (Fig. 2-5). The femoral head can usually be felt outside of the acetabulum and forward on external rotation. The range of abduction and of external rotation may be decreased. In older children a flexion deformity of the hip can be demonstrated on flattening the lumbar spine on the examining table (Thomas test, p. 13).

Roentgenographic picture. In the infant it may be impossible to make a diagnosis with certainty until roentgenograms of the hip have been examined. Three classic signs of dislocation are delayed growth of the ossification center of the capital epiphysis, upward and outward displacement of the femoral head, and increased obliquity of the acetabular roof (Fig. 2-6). The acetabular angle is increased, and Shenton's line is disrupted. Arthrograms, made with the injection of radiopaque dye into the hip joint and carefully interpreted, may be of the greatest value in demonstrating structural details not otherwise visualized. In older children the upward displacement of the femur before treatment is greater (Fig. 2-7, *A*), and after treatment the obliquity of the acetabular roof tends to persist (Fig. 2-7, *B*).

Diagnosis. Usually the diagnosis is made on the clinical findings and confirmed by the roentgenographic changes. The waddling gait of unreduced congenital dislocation of the hip may be simulated by that of congenital coxa vara, gluteus medius weakness from neurologic diseases such as progressive muscular dystrophy or poliomyelitis, and pathologic dislocation following paralysis or sepsis.

Treatment. Treatment should be begun as soon as the dislocation is recognized, since the earlier it is started the more favorable should be the ultimate result. The treatment varies with the age of the patient and may be considered in four age categories: (1) under 1 year, (2) between 1 and 3 years, (3) between 3 and 6 years, and (4) over 6 years.

First age period—under 1 year. In the infant under 12 months of age it has been shown that if the dislocated hip is brought gently into abduction and held in this position, there is a definite tendency for the dislocation to become reduced and for a normal acetabulum to form. The treatment is often begun with from 1 to 3 weeks of skin traction in increasing abduction. When the hip has been reduced in the abducted and flexed position, this position is maintained by a splint (Fig. 2-8, *A*), brace (Fig. 2-8, *B*), or cast (Fig. 2-9). It is important that extreme or forced abduction be avoided, since it may result in avascular necrosis of the femoral head. Gradual resumption of hip adduction is allowed after clinical and roentgenographic examination shows sufficient stability to prevent redislocation.

Second age period—1 to 3 years. Often a congenital dislocation of the hip is not recognized until the child begins to walk at about 1 year of age. At this age it may still be possible to reduce the dislocation by skeletal traction in gradually increasing abduction. Reduction by traction alone is preferable to that by manipulation or surgery because it is less likely to injure the femoral head or its blood supply.

When dislocation persists despite a thorough trial of traction, an attempt to effect reduction by manipulation is in order. This should be done with the child under general anesthesia and supine on a hard surface. Gentleness is essential. If the adductor muscles have remained tight despite the period of preliminary traction, adductor tenotomy should be done. The hip and knee are then flexed to 90 degrees and the hip is externally rotated slightly; then gradually and slowly the thigh is

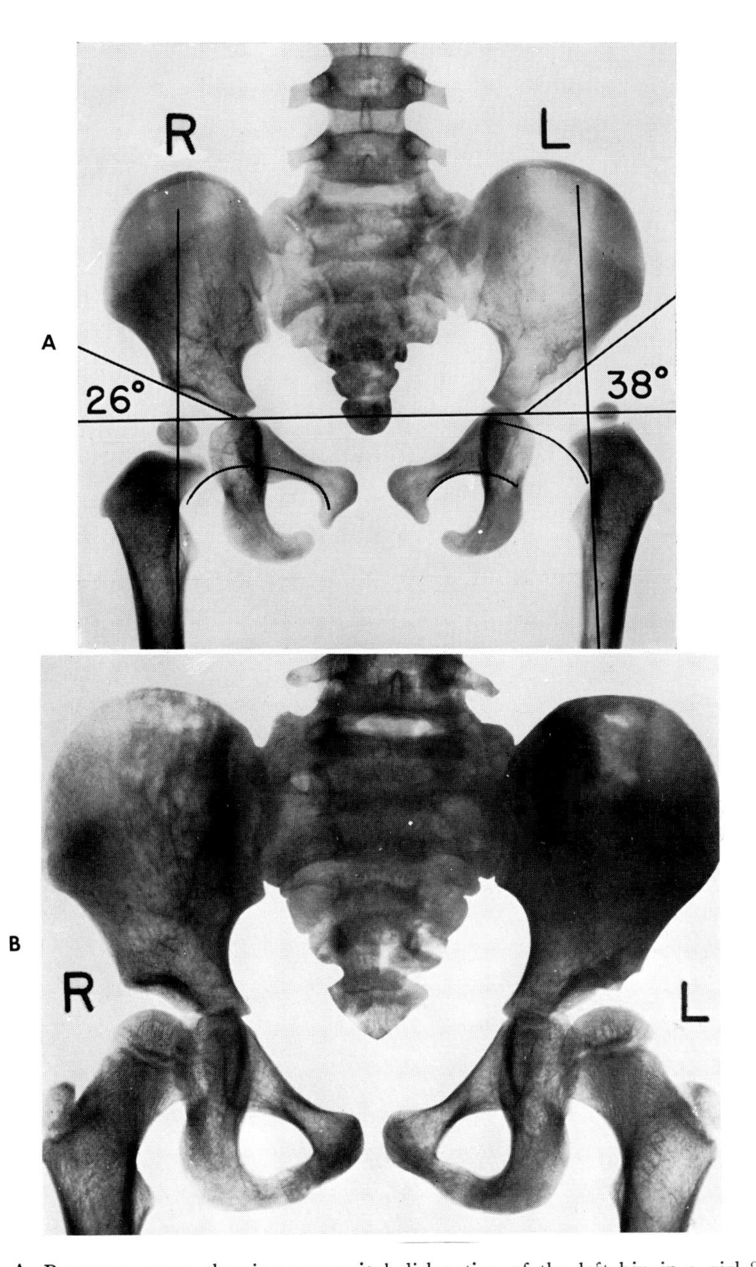

Fig. 2-6. A, Roentgenogram showing congenital dislocation of the left hip in a girl 2½ years of age. Note upward and lateral displacement of the left femoral head, underdevelopment of the left capital epiphysis, increased obliquity of the superior margin of the left acetabulum (increased acetabular angle), and disruption of the left obturator-coxofemoral line (Shenton's line). (Same patient as shown in Fig. 2-3.) **B,** Roentgenogram of the same patient as shown in **A,** 3 years after closed reduction. Note that the left hip shows well-rounded, normally trabeculated capital epiphysis and well-deveoped acetabulum, and that Shenton's line is intact.

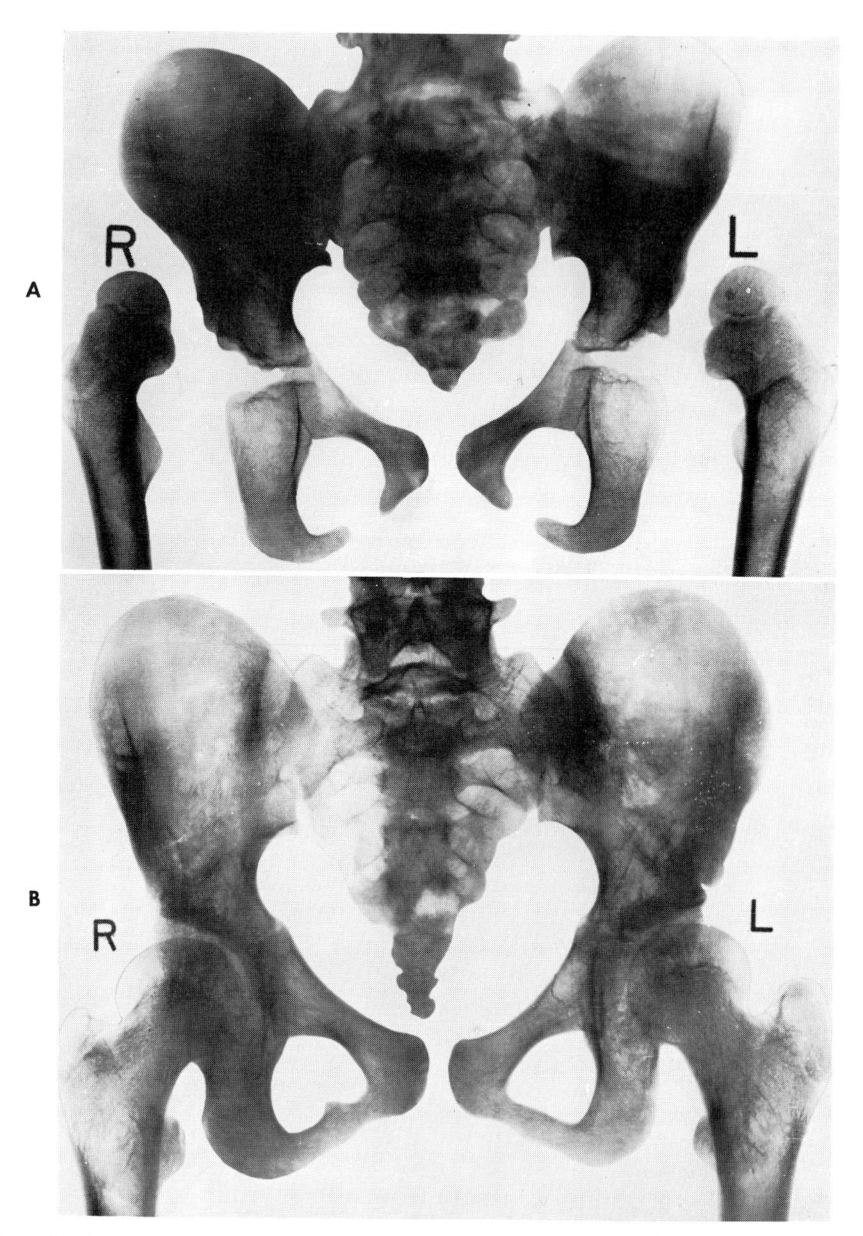

Fig. 2-7. A, Roentgenogram of bilateral congenital dislocation of the hip in the 5-year-old girl shown in Fig. 2-4. Note upward displacement of the proximal end of the femurs and marked obliquity of acetabular roofs. **B,** Roentgenogram of same patient as shown in **A,** 7 years after closed reductions. Now 12 years old, the patient has a perfect functional result. Note that femoral heads are well rounded, acetabula are shallow, and considerable obliquity of the acetabular roofs persists.

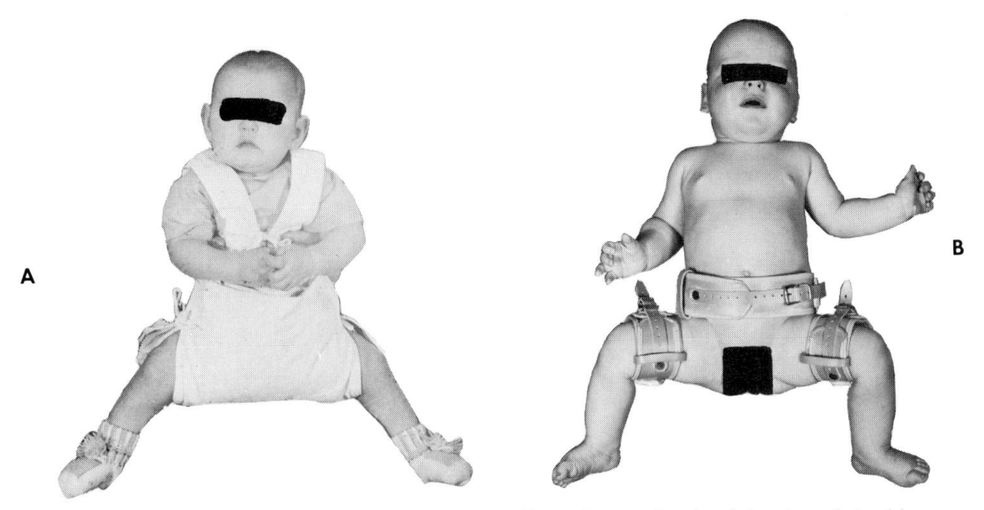

Fig. 2-8. Pillow splint (Frejka), **A,** and brace, **B,** used to maintain abduction of the hips.

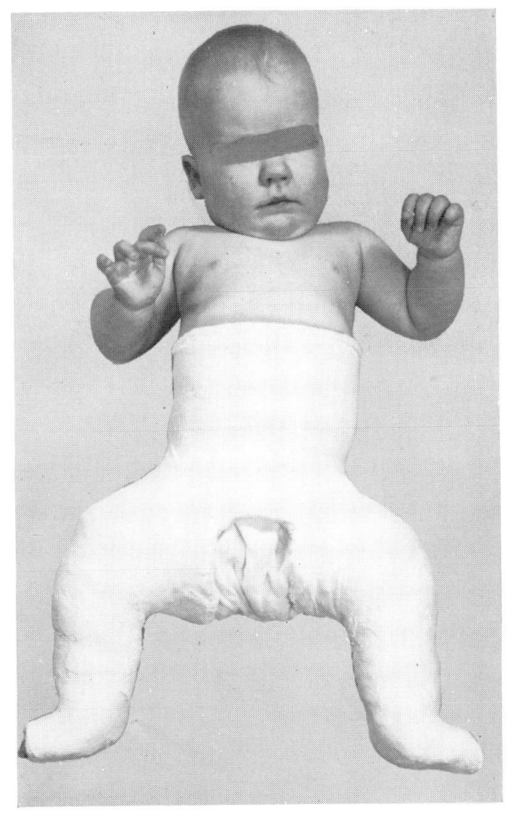

Fig. 2-9. Bilateral hip spica cast applied in moderate abduction, 90 degrees of external rotation, and 90 degrees of flexion of the hips, for congenital dislocation of the hip in a girl 6 months of age.

abducted and further externally rotated, while a forward pressure is exerted on the head and greater trochanter of the femur through the buttock. As the dislocation is reduced, the head of the femur can be felt to slip over the posterior acetabular rim into the acetabulum, usually with a click. In a successful reduction the head of the femur will be well placed in the acetabulum when the hip is flexed 90 degrees and in moderate abduction.

After reduction a bilateral hip spica cast (Fig. 2-9) should be applied to support the lower extremities with the hips in moderate abduction, 90 degrees of external rotation, and 90 degrees of flexion, the knees in 90 degrees of flexion, and the ankles in neutral zero position. The hips should not be forced into full abduction because this may cause avascular necrosis from excessive pressure on the femoral head and hip capsule. A double hip spica should be applied for the unilateral case as well as for the bilateral case. The infant's mother should be instructed concerning methods of keeping the cast clean and dry.

If the hip, when first reduced, seems unstable in abduction and external rotation, the cast should be applied with the hips in abduction and internal rotation (Fig. 2-10).

In different clinics the length of time in the initial cast varies from 3 to 9 months. The immobilization period varies also according to the stability of the individual patient's hip; 6 months is usually advisable. After this a bilateral hip spica is applied with the hips brought down into 45 degrees of abduction, 45 degrees of external rotation, and 45 degrees of flexion; this cast is left on for 2 to 3 months. After removal of the cast the child is allowed to move around in bed for about 1 week. The lower limbs should be examined frequently for joint mobility, since mild contractures may require passive exercises. Walking is allowed as soon as the legs will return to a nor-

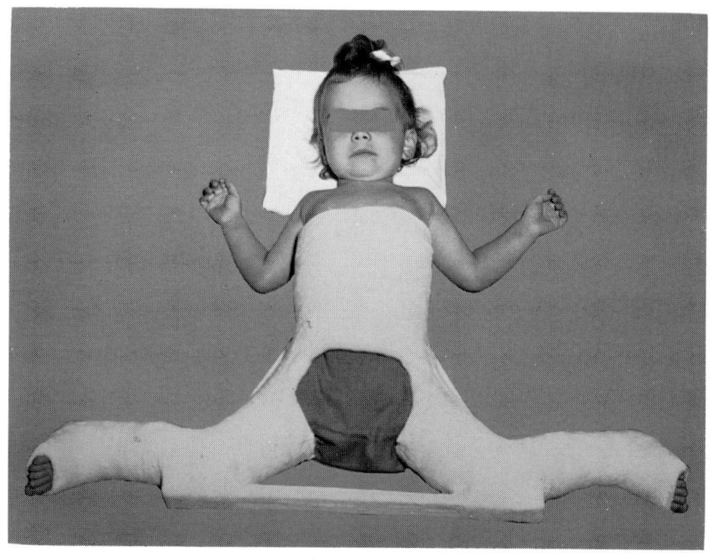

Fig. 2-10. Bilateral hip spica cast applied in a position of abduction and internal rotation. This girl, aged 1 year, 9 months, had bilateral dislocation of the hip.

mal standing position; this is usually several weeks after removal of the plaster. Roentgenograms of the hips should be made at long intervals while the child is in plaster, since redislocation sometimes occurs in spite of all precautions. The cause of redislocation may be the persistence of soft tissues, such as an inverted acetabular labrum, between femoral head and acetabulum, failure of the acetabular roof to develop, or excessive anteversion that allows the head to dislocate anteriorly when the thigh rotates externally.

Similarly the interposition of soft tissues or the presence of bony maldevelopment may make the initial attempt at closed reduction fail. After unsuccessful closed reduction, as well as after redislocation following closed reduction, open reduction is indicated. This usually consists of approaching the hip joint anteriorly, incising the capsule, removing any soft tissue obstacle, and placing the femoral head in the acetabulum; if the reduction seems insecure and the patient is more than 1½ years old, the acetabular socket should be redirected by osteotomy of the innominate bone (*Salter operation,* Fig. 2-11, *A*) or by pericapsular osteotomy of the ilium (*Pemberton operation,* Fig. 2-11, *B*).

After open reduction and a period of immobilization in a plaster cast, physical therapy, including pool therapy, may be helpful. A bilateral, long leg brace that has a waistband to include the pelvis and that holds the hips in abduction is sometimes used after either closed or open reduction. It is believed that walking in an

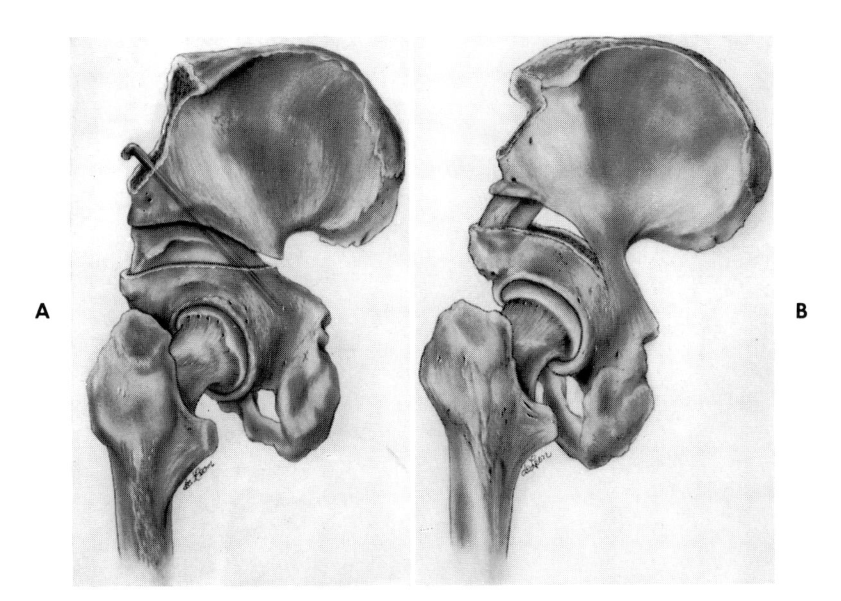

Fig. 2-11. A, Salter osteotomy of the innominate bone to decrease the obliquity of the acetabulum and provide increased coverage of the femoral head. A wedge of bone removed from the anterior part of the iliac crest is held in the opened osteotomy site by a stout Kirschner wire. The wire is removed 6 weeks later. **B,** Pemberton pericapsular osteotomy of the ilium. This procedure also decreases the obliquity of the acetabulum and provides increased coverage of the femoral head. A block of bone from the iliac crest is wedged into the open end of the curved osteotomy line.

abduction brace tends to improve and deepen the acetabulum. For a year or more after removal of the plaster cast it is generally desirable for the patient to use a night splint that maintains the hips in wide abduction by means of a long bar between the shoes.

An increase in the normal anteversion of the femoral neck is present in nearly all cases of congenital dislocation of the hip. The anteversion may measure from 30 to 90 degrees. This is an important factor in hip dislocations; if uncorrected, it may lead to unsatisfactory therapeutic results. In order to correct excessive anteversion some surgeons advocate a derotation femoral osteotomy either below the greater trochanter or above the condyles. In children under 3 years of age much of the excessive anteversion may undergo spontaneous correction, and derotation osteotomy may not be necessary. When osteotomy is to be done, the amount of anteversion should be determined before operation as accurately as possible by one of various roentgenographic technics. A method of measuring anteversion developed at the Alfred I. duPont Institute *(Dunlap-Shands method)* is believed to have less than a 10% error.

The percentage of satisfactory results after manipulative treatment of congenital dislocation of the hip in patients under 3 years of age has been reported to be from 60% to 80%.

Third age period—3 to 6 years. After skeletal traction for at least 3 weeks, during which the head of the femur has been pulled down to a point opposite or below the center of the acetabulum, a closed manipulative reduction may be attempted if the hip has not been reduced with abduction, internal rotation, and traction alone. The size of the child and the tightness of the structures about the hip should be taken into consideration in making a decision concerning a manipulative reduction. Excessive force in closed reduction undoubtedly will injure the epiphysis of the femur and its blood supply and is always contraindicated. An adductor tenotomy is usually indicated when a closed reduction is to be done in this age group. If an attempt at gentle closed reduction proves unsuccessful, open reduction and either innominate osteotomy (Fig. 2-11, *A*) or pericapsular iliac osteotomy (Fig. 2-11, *B*) is indicated. Surgical treatment is indicated much more often for patients in this age group than for younger children. In patients from 3 to 8 years of age, Colonna reported successful results following an arthroplastic procedure consisting of enlarging or deepening the acetabulum, covering the head of the femur with the elongated capsule, and placing it deeply in the acetabulum. The hip is then immobilized in plaster in abduction and internal rotation, and several weeks later a derotation osteotomy is performed.

Fourth age period—over 6 years of age. In this age group reduction of the dislocation is usually impossible or inadvisable because of severe shortening and contracture of muscles and fasciae as well as deformity of the femoral head and the acetabulum.

The patient's disability may often be improved, however, by a palliative operation, of which several types are available. A roof or buttress can be constructed by turning down a flap of iliac bone over the femoral head as near as possible to the acetabulum *(shelf operation)*. Increased support of the pelvis can be supplied by an abduction osteotomy of the femur below the lesser trochanter. In adults, interposition or replacement arthroplasty (Fig. 5-19, *A* and *B*) is sometimes indicated. For permanent relief of hip instability and pain, arthrodesis may be advisable.

In older untreated cases a high cork sole under the shortened leg will sometimes relieve the strain. When back pain due to strain is present, a corset or snugly fitting back brace should be applied.

Complications. Avascular necrosis of the epiphysis of the head of the femur is a common complication and affects the end result of treatment. Reports have shown the incidence of aseptic necrosis to vary between 18% and 64%. Such changes are much more common and have a worse prognosis in older patients than in younger ones. Avascular necrosis is probably caused by circulatory changes incident to the reduction; some observers believe it may be the result of vascular occlusion associated with immobilization in wide abduction and external rotation. When the epiphysis begins to show evidence of disintegration, it is often best to put the hip at rest by discontinuing weight bearing. This avascular necrosis is in many ways similar to that seen in coxa plana. In addition to the epiphyseal changes there is often associated coxa vara; in these cases transtrochanteric osteotomies are sometimes recommended to increase the neck-shaft angle. Avascular necrosis of the head of the femur may be associated with excessive anteversion. Hips showing avascular necrosis usually develop some flattening and irregularity of the femoral head and, in later life, osteoarthritis.

Gill and others showed that disability from subluxation or redislocation of the hip frequently develops years after an apparently successful closed or open reduction.

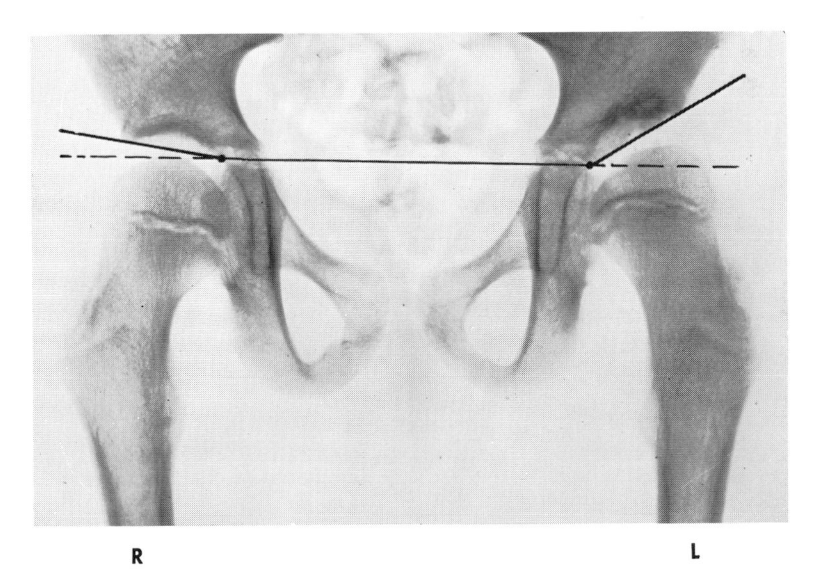

Fig. 2-12. Roentgenogram showing dysplastic changes in the left hip of a 6-year-old boy, 4 years after closed reduction of congenital dislocation. Note density changes and increased obliquity of the superior margin of the left acetabulum. Left acetabular angle is 30 degrees; normal right acetabular angle is 10 degrees. The superior portion of the left acetabulum does not cover the femoral head as in the normal hip. Note bilateral coxa valga (155 degrees).

Patients with reduced dislocations should be kept under observation for many years.

Prognosis. In congenital dislocation of the hip the prognosis for good joint function is excellent if treatment is started during the first year of life, and good if started during the second or third year and if late redislocation or subluxation is prevented. In older patients good results are much less certain. Circulatory changes in the femoral head sometimes lead to local deformity and impairment of function. Roentgenographic changes (Fig. 2-12) may persist for many years. The patient with an incomplete reduction is likely to complain of weakness and early fatigue and, later in life, of pain on weight bearing. Secondary hypertrophic bone changes that gradually develop in the hip region and the lumbar spine may cause considerable pain of a chronic type. A completely unreduced dislocation causes deformity, limping, and hip and low back discomfort, all of which tend to increase as the patient grows older.

CONGENITAL DISLOCATION OF THE KNEE

Two types of congenital luxation of the knee are encountered. One is a hyperextension of the knee, or congenital genu recurvatum (Fig. 2-13); this is more common than true dislocation. The second type exhibits a complete or incomplete posterior displacement of the condyles of the femur on those of the tibia, constituting an anterior dislocation of the knee joint. The cause of these conditions may be an abnormal position of the knee in utero; when the deformity is corrected, the knee usually develops in normal fashion.

Clinical picture. In true congenital dislocation of the knee the leg can be brought to a straight line but will not flex beyond this point because of contracture of the

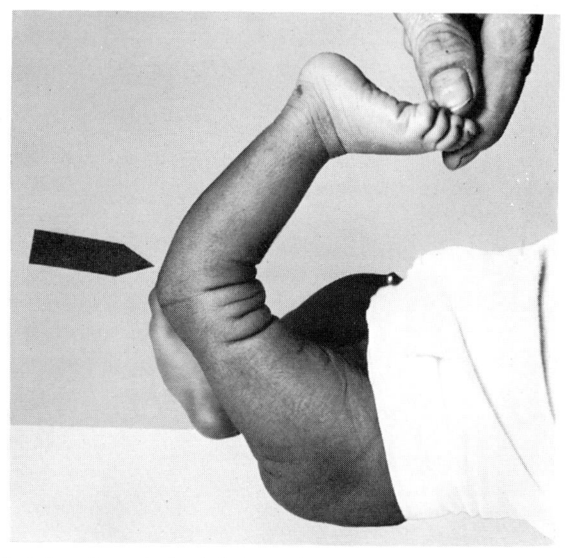

Fig. 2-13. Congenital luxation of the left knee in an infant. Note extreme hyperextension of the knee (genu recurvatum) when the leg is held up.

quadriceps muscle and the patellar ligament. Often the patella is absent, or it may be very small. There is usually a wrinkling of the skin over the patella, and there may be an associated varus or valgus deformity of the knee. Lateral instability is often present. Congenital dislocation of the knee may be associated with congenital dislocation of the hip and with arthrogryposis.

Treatment. As soon as dislocation of the knee is recognized, an attempt should be made to correct the displacement. This can be done usually either by gently manipulating the joint and stretching the quadriceps tendon or by wedged casts. After reduction the knee should be kept in a flexed position by means of a posterior splint. This position should be maintained until the flexors of the knee are strong enough to prevent a recurrence of the deformity. In resistant cases surgical reduction is sometimes required. This usually involves capsulotomy, mobilization of the collateral ligaments, and lengthening of the quadriceps muscle.

CONGENITAL TALIPES

The term *talipes* is used in connection with many foot deformities, whether congenital or acquired. It is derived from the Latin *talus,* meaning ankle, and *pes,* meaning foot, and was originally used to designate a foot deformity that caused the patient to walk on the ankle. Deformities of the foot and ankle are conventionally described according to the position of the foot. The four cardinal positions (Fig. 2-14) are (1) *varus,* or inversion, (2) *valgus,* or eversion, (3) *equinus,* or plantar flexion, and (4) *calcaneus,* or dorsiflexion. Of the congenital deformities of the foot, the combination of equinus and varus is the most common, and that of calcaneus and valgus next. The equinovarus foot has a clublike appearance and is the classic type of congenital clubfoot.

Talipes equinovarus (congenital clubfoot)

Etiology. The cause of congenital clubfoot is unknown. A hereditary factor is often concerned, its presence being observed in from 5% to 22% of the cases in

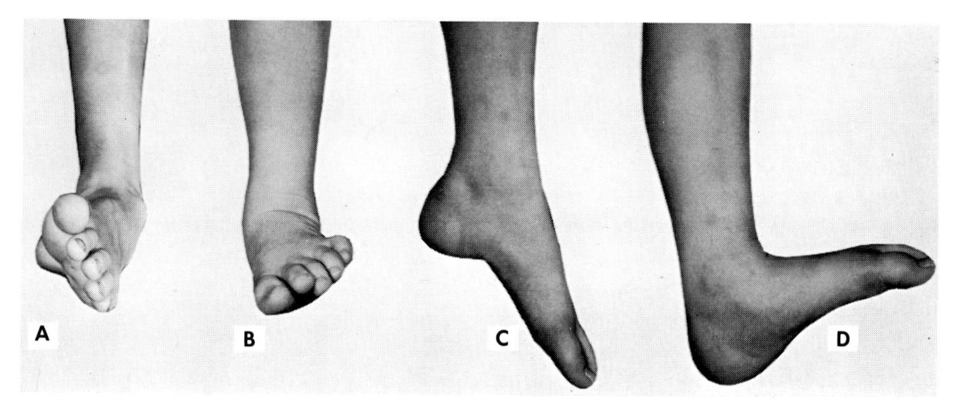

Fig. 2-14. Four cardinal positions of foot deformity: **A,** varus; **B,** valgus; **C,** equinus; and **D,** calcaneus.

reported series. Of numerous hypotheses of the pathogenesis of clubfoot, the most widely accepted are the following:

1. Arrested or anomalous development of this particular part of the embryo in the first trimester of pregnancy may be the cause. This genetic theory receives indirect support from the statistical fact that between 4% and 12% of the patients with congenital clubfoot in reported series have other congenital anomalies as well.

2. At about the third month of intrauterine life the foot occupies normally an equinovarus position. As the fetus develops, inward rotation of the leg normally takes place, and by the seventh month the foot is no longer in the equinovarus position. If inward rotation fails to occur, however, the foot remains in equinovarus, and the child is born with a clubfoot. By many embryologists, however, this hypothesis is not accepted.

3. Investigations of pathologic changes in the soft tissues in congenital clubfoot have shown abnormalities in the relative maturity and length of the muscles, as well as variations in their tendon insertions. These findings suggest that muscle imbalance may play a considerable part in the causation of congenital clubfoot.

Incidence. Talipes equinovarus is one of the most common congenital anomalies of the foot. According to Duthie the incidence of talipes equinovarus is one in 800 births. However, when a child has this condition, the chance of a sibling's having it is increased to one in thirty-five births. Congenital clubfoot is twice as common in boys as in girls and may involve either one or both feet. Occasionally there is talipes equinovarus on one side, with metatarsus varus or talipes calcaneovalgus on the other.

Pathology. All degrees of talipes equinovarus, from a very mild deformity to one in which the toes touch the medial side of the lower leg, are found. The Achilles tendon is always shortened. The anterior and posterior tibial tendons are contracted in proportion to the degree of varus deformity. In some cases degenerative changes in the fibers of certain muscles and anomalous insertions of tendons in the foot have been demonstrated. The ligaments and joint capsules on the medial side of the foot are thickened and contracted.

At birth the talus is greatly deformed. Its cartilaginous neck is shortened and deviated medially and plantarward. The subtalar surfaces are tilted into varus and equinus and medially rotated. The calcaneus is slightly shortened and widened. As the child grows, adaptive changes take place in these bones. The talus may become wedge shaped, with only its posterior surface apposed to the tibia. A small portion of the navicular bone articulates with the inner border of the head of the talus. The calcaneus points downward and is tilted in such fashion that its medial process approaches the tibial malleolus. The anterior extremity of the calcaneus is pointed medially and follows the direction of the neck of the talus. There is frequently an associated external torsion of the tibia or external rotation of the talus in the ankle mortise. Structural changes may be found in all other bones of the foot but are less extreme in degree. The anatomic changes vary according to the degree of the deformity; in the mild clubfoot bony changes may be minimal.

In older, untreated patients proliferative bone changes due to weight bearing take place about the edges of the articulating surfaces. The articular cartilage undergoes atrophy. New bone is almost always formed around the greatly thickened and

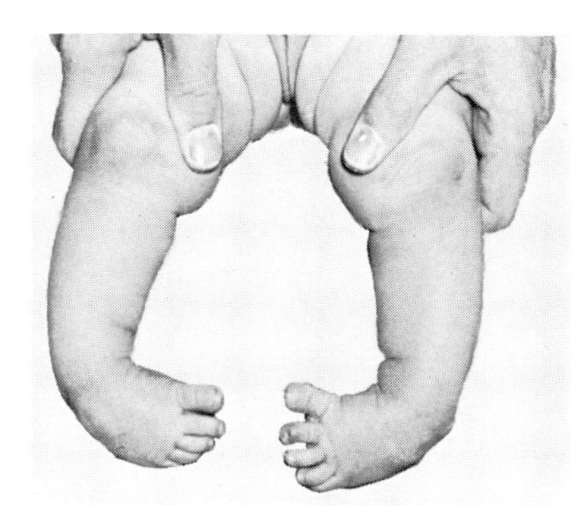

Fig. 2-15. Bilateral congenital talipes equinovarus (congenital clubfeet) in a 2-month-old infant.

broadened proximal end of the fifth metatarsal bone. A large bursa forms over the weight-bearing surface on the lateral portion of the dorsum of the foot and may contain a semisolid gelatinous material.

Clinical picture. The heel is drawn up, the entire foot below the talus is inverted, and the anterior half of the foot is adducted (Fig. 2-15). A variable degree of stiffness prevents complete manual correction of the deformity. The medial border of the foot is concave, the lateral border is convex, and there is a transverse crease across the sole at the level of the mediotarsal joint. When the infant starts to walk, he suffers a tremendous handicap because of his inability to bear weight normally. The muscles of the leg quickly become fatigued and soon show marked atrophy. Pain is experienced only by the adult patient in whom arthritic changes have developed.

Diagnosis. In infants the diagnosis of congenital clubfoot is made with ease, but in older children and adults it is sometimes difficult to exclude paralysis as a cause of the deformity. Clubfoot is seen frequently in association with paralytic changes in the lower extremities in spina bifida with meningomyelocele, in cerebral palsy, and in poliomyelitis. Often the first sign of the peroneal type of progressive muscular atrophy is the development of slight equinovarus. Occasionally in progressive muscular dystrophy the foot has a similar appearance. Old injury of the lower tibial epiphysis, osteomyelitis, and fracture in the region of the ankle joint are other causes of equinovarus deformity.

Prognosis. The earlier the treatment is started, the better the result. When treatment is begun within the first weeks of life, the deformity may be completely corrected by nonoperative procedures in a relatively short time. The course of treatment is sometimes long, however, requiring several years. According to Kite, from 10% to 15% of the patients tend later to relapse and require a second period of treatment. In patients treated late the prognosis for normal appearance and function of the foot is poor.

Treatment. Treatment should be started as soon as the deformity is recognized. If treatment is begun immediately after birth, the problem of correcting the deformity is simple as contrasted with the difficulty encountered in older infants or children.

The treatment is divided into three stages: (1) correction of the deformity, (2) maintenance of correction until normal muscle balance has been regained, and (3) observation for several years in order to forestall any recurrence of the deformity.

It is most important that the first stage of treatment be continued until a position of marked overcorrection has been reached. Anything short of this is inadequate. The position of dorsiflexion and eversion of the foot and abduction of the forefoot is the goal of every form of treatment.

It is generally agreed that the functional result is far better if the final position can be obtained without the trauma of forceful manipulation and without radical surgery, either of which may be followed by interference with bone growth and by stiffness of the joints of the foot. Lengthening of the Achilles tendon and posterior capsulotomy of the ankle joint, however, usually produce no unfavorable effects if unaccompanied by forceful manipulation.

The ease with which correction may be obtained and the choice of method in the *first stage of treatment* vary according to the age of the patient. In young infants a series of plaster casts, applied without manipulation other than holding the foot in a position of as much correction as can be obtained without the use of force, is usually effective (Fig. 2-16). Painting the skin with a nonirritating adhesive liquid helps to prevent slipping of the cast and loss of the corrected position. The cast should be changed at intervals of from 3 to 14 days. Successive casts provide an opportunity for the tight structures on the medial side of the foot to stretch and for the lax structures on the lateral side of the foot to contract. In this way a gradual correction of the deformity is obtained without application of excessive force. The casts should be padded, particularly over bony prominences, and usually should extend above the flexed knee to allow effective correction of the rotation deformity. It is often convenient to use the technic of wedging the cast instead of changing the entire cast. A wedge of plaster is taken out over that aspect of the foot toward which correction is desired, the sides of the wedge are brought together, and plaster is applied in circular fashion over the closed-in wedge.

The adduction deformity of the forefoot should be corrected first; this phase of

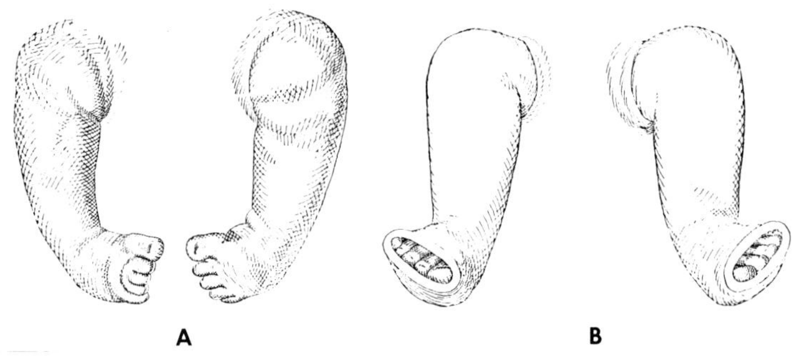

Fig. 2-16. Bilateral congenital talipes equinovarus in an infant. **A,** Before correction. **B,** Undergoing correction in plaster casts.

the treatment must be continued until the navicular bone is brought directly in front of the head of the talus. Next, the inversion deformity is corrected; care must be taken to continue this until there is no inversion of the calcaneus. Correction of the relationship between talus and calcaneus may be verified by an anteroposterior roentgenogram of the foot showing anterior divergence of their long axes. Finally, the plantar flexion or equinus is corrected by gradually increased dorsiflexion of the ankle. Attempts to correct simultaneously all three elements of the deformity may produce a misshapen, rocker-bottom foot. If necessary, correction of the equinus may be aided by (1) lengthening of the Achilles tendon and (2) posterior capsulotomy of the ankle joint. The final cast is often left on for a period of 4 to 6 weeks to allow thorough accommodation to the new position. It may be necessary, for many months, to use a night splint to hold the foot in the overcorrected position.

An alternative method of correcting the deformity of congenital clubfoot is use of the Denis Browne splint. This splint (Fig. 2-17) consists of two padded metal plates that are securely fastened with adhesive tape to the infant's feet and are then connected by a crossbar. The stretching of contracted structures is accelerated by the infant's kicking. The footplates can be so rotated outward on the crossbar that progressive correction of the adduction and later of the equinus may be obtained.

After full overcorrection of the deformity in all planes has been secured by means of casts or the Denis Browne splint, measures are taken to maintain the correction and to encourage the development of normal muscle balance. A clubfoot shoe to hold the corrected position should be used. This shoe has a lateral deviation of its anterior half and a slight raise (1/8 to 1/4 inch) of its entire lateral border (Fig. 2-18). When special clubfoot shoes cannot be obtained, ordinary shoes can be altered to aid in preventing a recurrence of the deformity. Corrective night shoes and splints are also useful. The Denis Browne splint may be applied by means of shoes and used to maintain correction (Fig. 2-19). Walking should be encouraged

Fig. 2-17. Denis Browne splint for the correction of clubfoot. Felt-padded plates are strapped to the feet in corrected position with adhesive tape. The control of rotation, eversion, and dorsiflexion is adjustable.

in order to strengthen the weakened muscles. If the patient is old enough to co-operate in muscle training, exercises to strengthen the muscles of abduction, eversion, and dorsiflexion of the foot and ankle may be helpful.

In certain of the more difficult cases, casts alone may prove inadequate for correcting the deformity within a reasonable length of time. This is particularly true in older children and in cases of recurrent clubfoot. For such patients surgical treatment should be considered. If the child is less than 8 to 10 years old, operation should usually be limited to the soft tissues. The simplest and most frequently used

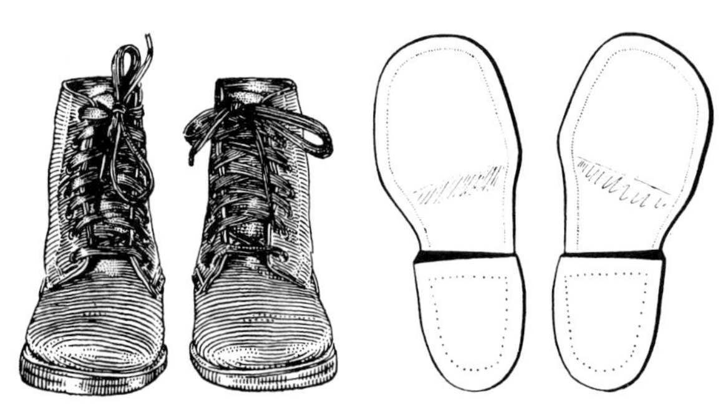

Fig. 2-18. Clubfoot shoes. The lateral side of the soles and heels is raised to maintain correction of varus deformity, and the front half of the shoes is turned out to maintain correction of fore-foot adduction.

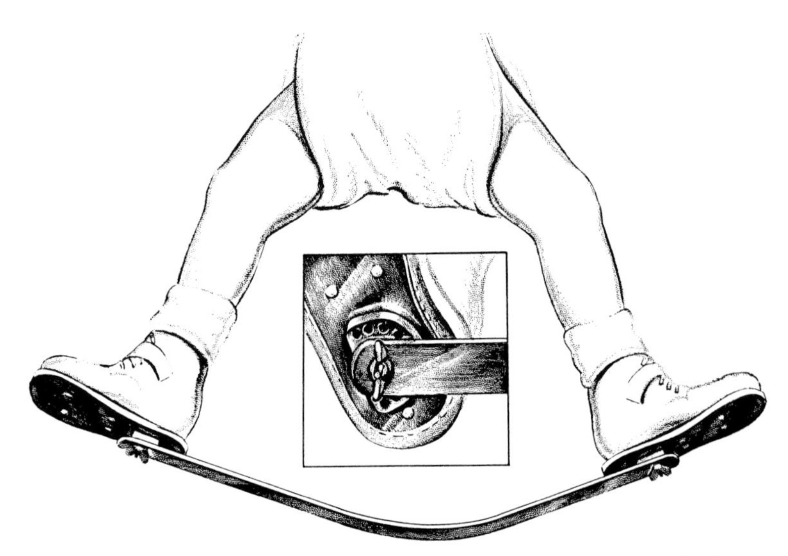

Fig. 2-19. Denis Browne splint worn day and night, or at night only, to assist in maintaining corrected position after full overcorrection of clubfoot deformity has been accomplished by earlier treatment. Note dorsiflexion of ankles and external rotation. Attachment of footplate to crossbar, shown in inset, allows adjustment of rotation. Bending the crossbar increases dorsi-flexion.

operation is lengthening of the Achilles tendon to lessen the equinus deformity. Often this should be accompanied by section of the posterior capsule of the ankle joint, which may be tightly contracted. Surgical correction of the varus and forefoot adduction is more difficult. Various soft tissue releases may be used. They usually involve cutting the tendinous and ligamentous structures on the medial side of the foot. An osteotomy of the calcaneus to lessen inversion of the heel, as described by Dwyer, may be helpful. Lateral transference of the anterior tibial tendon, advocated by Garceau, may help to prevent a recurrence of the varus and adduction deformity. Surgical release of the talonavicular joint associated with resection of the calcaneocuboid joint, as described by Evans, has been helpful in the treatment of recurrent clubfoot in children 4 to 9 years old. Such surgical procedures should usually be followed by further attempts at overcorrection by plaster casts.

The correction of a clubfoot deformity in a child more than 10 years old or in an adult usually requires operation upon the tarsal bones. A *wedge osteotomy,* the base of the wedge being on the dorsal and lateral aspect of the foot, may be indicated. As much bone as is necessary to correct the deformity completely is removed from the region of the transverse tarsal joint. At the same time the subtalar joint is usually fused. Although such operations stiffen one or more of the abnormal tarsal joints, they usually improve the function, endurance, and appearance of the foot.

After either nonoperative or surgical treatment the corrected clubfoot should be observed until the child has reached adolescence, or in the case of older patients for at least several years. During this period any tendency toward recurrence of the deformity should be promptly and thoroughly treated. A recurrent deformity is usually more difficult to treat than the original clubfoot.

Talipes calcaneovalgus

Talipes calcaneovalgus, a congenital deformity characterized by eversion of the foot, increased dorsiflexion of the ankle, and apparent lengthening of the Achilles tendon, is seen frequently. It is often of mild degree and, unlike congenital equinovarus, is not fixed but can be easily overcorrected by stretching. It is most noticeable immediately after birth. As the child begins to use his muscles and kick his legs, the condition becomes less obvious and often seems to disappear completely. When the infant later begins to stand, however, persistent changes in the affected foot or feet may be evidenced by excessive pronation.

Treatment. Mild degrees of calcaneovalgus require no treatment. Moderate ones may be improved by gentle stretching exercises applied by the mother. Feet severely deformed at birth should be held in light corrective casts for at least several weeks. A Denis Browne splint holding the feet in an equinovarus position for several months is sometimes indicated. Every foot that has shown an early calcaneovalgus deformity should be fitted with a firm shoe with a ⅛-inch raise of its medial border when the infant begins to stand and walk; usually the shoe modification should be continued for several years.

Talipes valgus

At times a valgus deformity is seen without calcaneus. The treatment is similar to that of talipes calcaneovalgus.

Talipes calcaneus

Simple calcaneus deformity is treated in the manner outlined for talipes calcaneovalgus.

Talipes varus

Talipes varus is usually an incomplete form of congenital equinovarus deformity; it sometimes resembles a relapsed equinovarus foot. The treatment is like that described for talipes equinovarus.

Talipes equinus

Congenital talipes equinus is about half as common as talipes varus. It should be treated at birth or soon thereafter by corrective casts. In the older child it is often necessary to lengthen the Achilles tendon and to perform a posterior capsulotomy of the ankle joint before the equinus can be fully corrected. Operation should be followed by a plaster cast for from 4 to 6 weeks.

Metatarsus varus (metatarsus adductus)

The deformity known as metatarsus varus or metatarsus adductus is quite common, being seen more frequently than talipes equinovarus. It consists of adduction of the forefoot at the tarsometatarsal joints (Fig. 2-20). Sometimes supination of the foot is present; it is not fixed, however, and corrects itself spontaneously in the weight-bearing position. Metatarsus varus has been described as one third of a clubfoot (the forefoot adduction component), but this is not precisely correct because in true clubfoot much of the adduction takes place at the talonavicular joint, which is not the case in metatarsus varus. Adduction of the first metatarsal bone may be greater than that of the remaining metatarsals. When this is true, widening between

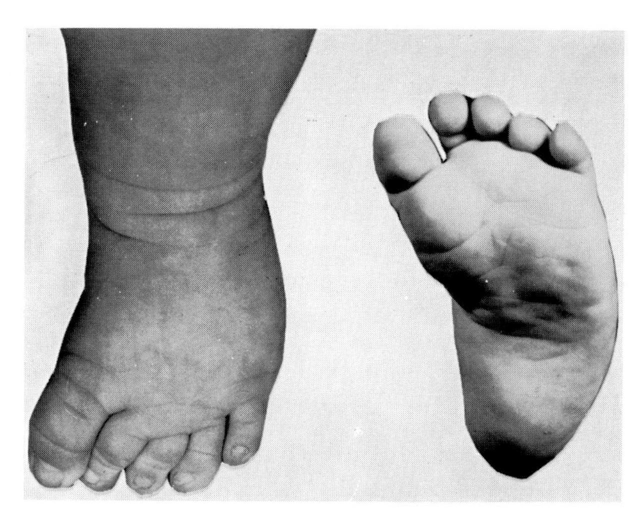

Fig. 2-20. Left metatarsus varus in an infant 3 months of age. Note turning-in of anterior part of the foot.

the first and second toes will be noted. This is sometimes called metatarsus primus adductus. Many observers, among them Kite, report that the incidence of metatarsus varus is increasing. It may involve one or both feet; when unilateral, it is sometimes associated with clubfoot of the opposite extremity. At times it may be a residual deformity resulting from an old congenital talipes equinovarus. Metatarsus varus may also be of acquired type, in which case it is usually combined with valgus deformity of the posterior portion of the foot. The deformity is obvious on inspection. If it is untreated until after walking has begun, the mother will usually complain that the child toes in, walks clumsily, and trips over his feet. Metatarsus varus is one of the causes of pigeon-toe. The intoeing may be accentuated by an accompanying internal tibial torsion.

Treatment. In very mild cases the treatment may involve only stretching exercises carried out by the mother five or six times a day. The exercise is simple, consisting of stretching the forefoot gently into abduction. If the deformity is more than minimal, cast treatment is indicated. The earlier such treatment is started, the better the end result. A bootee cast extending from the toes to just above the ankle is sufficient. As the plaster begins to set, the forefoot is molded into a position of abduction, with care to prevent valgus of the heel that might cause a flatfoot deformity. The cast is changed at intervals of a week or two. Overcorrection of the deformity is desired. If the treatment is early and adequate, the results are usually quite satisfactory. After the correction, special shoes with outswung soles are sometimes prescribed. In severe or late cases complete correction may not be obtained. Severe metatarsus varus in older children can be greatly improved by surgical treatment.

Congenital vertical talus (congenital convex pes valgus)

An uncommon and severe form of flatfoot, congenital vertical talus is characterized by excessive plantar flexion of the talus, to such a degree that sometimes its

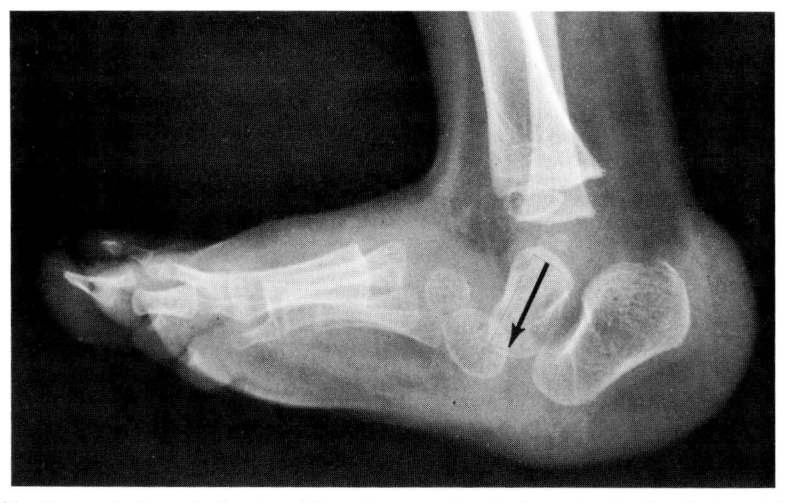

Fig. 2-21. Congenital vertical talus. Note downward pointing of talar head, dorsal luxation of talonavicular joint, equinus position of calcaneus, and reversal of longitudinal arch.

long axis almost parallels that of the tibia. The calcaneus is also in equinus, but since the forefoot is dorsiflexed the diagnosis of calcaneovalgus may be mistakenly made. The navicular bone is displaced dorsally onto the neck of the talus, and the calcaneocuboid joint also suffers dorsal subluxation (Fig. 2-21). The long arch of the foot is thereby reversed, and the prominent head of the talus presents in the sole. The appearance of the foot simulates the rocker-bottom deformity of an inadequately treated clubfoot. The deformity is rigidly fixed by contracted ligaments and tight tendons. The posterior tibial tendon may even pass dorsal to the head of the talus to its insertion on the navicular bone.

Treatment. The treatment of this rigid foot is most difficult. In infants, alignment of the forefoot with the plantar-flexed talus by a series of casts may be attempted. Early surgical treatment is usually necessary to reduce the talonavicular joint and to correct the hindfoot equinus. If the deformity is not corrected, the foot may later become painful. It may then be necessary to reconstruct the bony architecture surgically or to remodel and stabilize it by subtalar arthrodesis after the age of 10 years.

CONGENITAL CLUBHAND

Clubhand is an uncommon congenital malformation usually associated with complete or partial absence of the radius. It is sometimes hereditary and is bilateral in almost half the cases. The deformity is quite obvious and is characterized by marked radial deviation of the hand and shortening of the forearm (Fig. 2-22). The thumb ray is usually underdeveloped or absent, as are the radial carpal bones, usually the navicular and the trapezium. In this type of deformity the ulna is almost always bowed, with its concavity directed toward the radial side (Fig. 2-23). The hand and arm are small, and the shoulder girdle is underdeveloped. Muscles, particularly those controlling the thumb, are frequently absent, as are the radial nerve and artery. The median nerve often takes over the sensory function of the radial nerve. The elbow is often abnormal, and its motion may be quite limited. The right side is affected twice as often as the left. Clubhand with radial deviation is often associated with the presence of a cervical rib and other anomalies. Despite the deformity, the hand is ordinarily quite useful.

Treatment. If the patient is in infancy or early childhood, an attempt should be made to improve the deformity by means of a series of corrective casts. A brace or a dynamic splint may then be used. Several operative procedures for correction of the deformity have been devised. Centralization of the carpus over the end of the ulna may be accomplished surgically after preliminary stretching of the soft tissues by serial plaster casts. The corrected position may be maintained by means of a Kirschner wire inserted down the third metacarpal, across the center of the distal ulnar epiphysis, and into the medullary canal of the ulna. Although functional improvement may be obtained, it is often not great. Careful analysis of the function of the entire limb should be made before surgery is advised. This is especially true of bilateral cases.

Clubhand due to congenital absence of the ulna, which is much rarer than that due to absence of the radius, is an analogous anomaly and is treated in a similar manner.

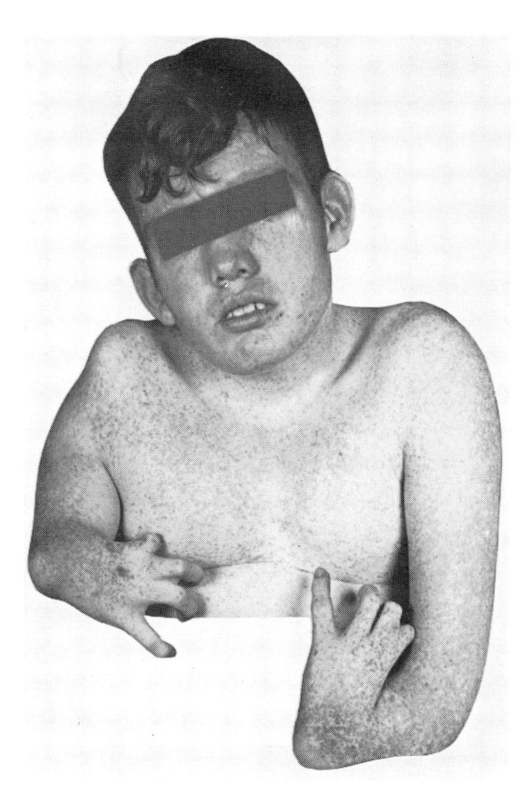

Fig. 2-22. Bilateral congenital clubhand in a boy 14 years of age. Note turning-in of hands and shortness of forearms. This patient showed bilateral complete absence of the radius, bilateral absence of the thumb, and absence of the proximal two thirds of the right humerus.

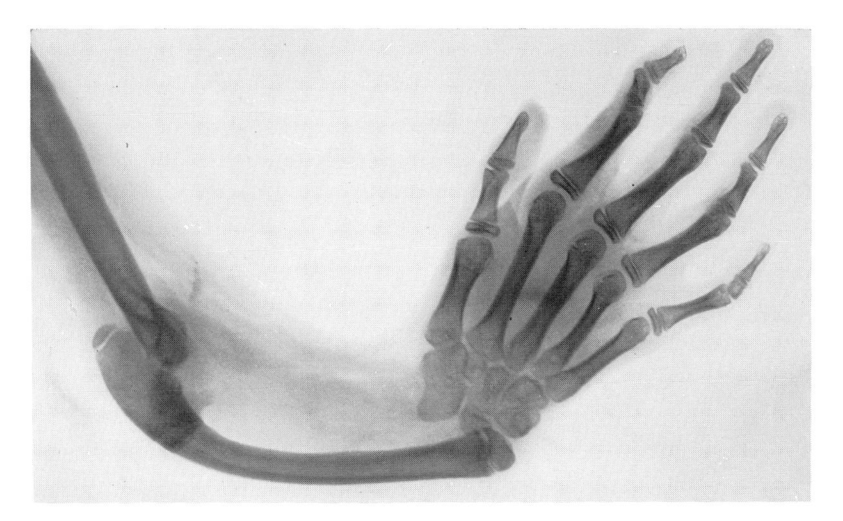

Fig. 2-23. Roentgenogram of congenital clubhand in a girl 10 years of age. Note turning-in of hand, absence of radius, bowing of ulnar shaft, and prominence of distal end of ulna.

CONGENITAL DEFECTS OF INDIVIDUAL BONES

Congenital partial absence of a long bone is seen more frequently than total absence. Defects in the bones of the upper extremity are more often bilateral than those occurring in the lower extremity. Complete absence of a middle or intercalary segment of a limb is called *phocomelia* (Fig. 2-24). Congenital bone defects are often associated with other anomalies. In many instances they form a most difficult therapeutic problem.

Humerus, radius, and ulna

Humeral defects associated with other anomalies in the upper limb are uncommon; isolated defects of the humerus are rare. Defects of the radius and ulna are described under clubhand (p. 42).

Sacrum and coccyx

Congenital absence of the sacrum is rare. In total absence, with which agenesis of the lumbar spine may be associated, the lower limbs show motor and sensory paralysis of varying degree, atrophy, and contractures; the treatment is similar to that of paraplegia due to a myelomeningocele. Absence of the lower coccygeal segments is of no clinical importance.

Femur

Underdevelopment or partial absence of the femur, especially of its upper third (proximal focal femoral deficiency), is much more common than complete absence.

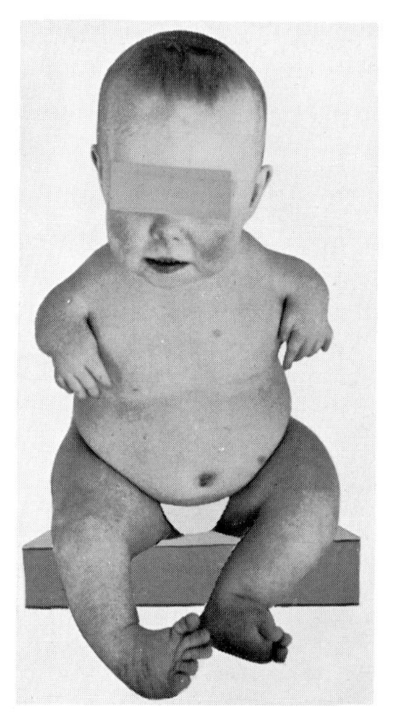

Fig. 2-24. Phocomelia of upper limbs and right clubfoot in an infant 11 months of age.

It presents a very complicated problem for orthopaedic management. The aim of treatment is to make the shortened limb as suitable as possible for prosthetic fitting. Extensive reconstructive surgery may be required. With partial absence of the femur are often associated complete absence of the patella, absence of the fibula, and anomalies of the pelvis on the affected side.

Tibia

Absence of the tibia is rare. The limb exhibits a characteristic deformity. The thigh is rotated externally and adducted, the knee is flexed, and the upper end of the fibula is displaced laterally and backward from the femur. The fibula is bowed, with its concavity directed medially, and the foot is in a varus or equinovarus position. It is possible to correct the deformity of the foot surgically and to fuse the upper end of the fibula to the femur. The end results, however, have not always been satisfactory, and amputation at or just above the knee is usually preferable.

Fibula

Congenital absence of the fibula is more common than that of any other long bone. Total absence (Fig. 2-25) is observed more frequently than partial absence.

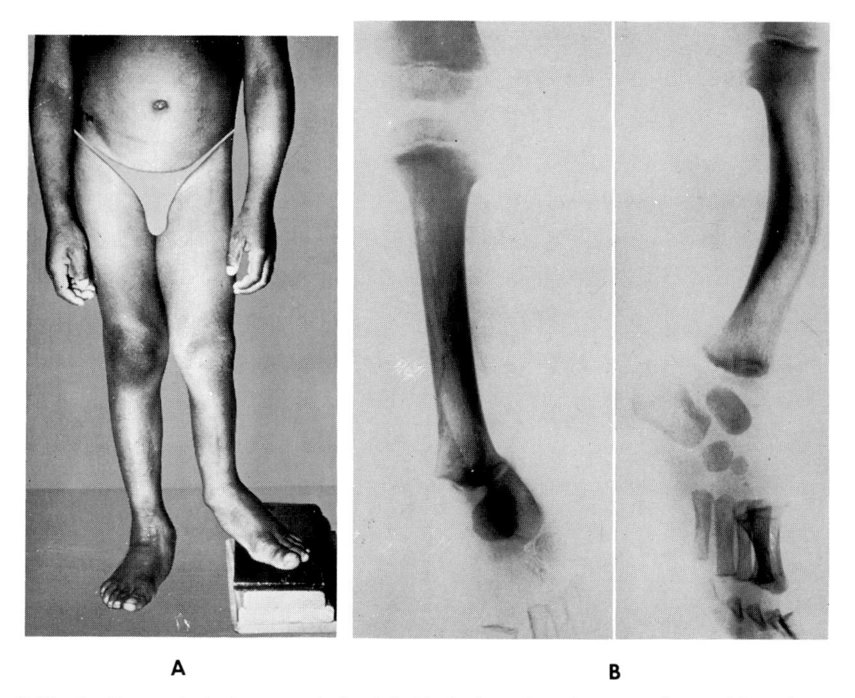

A B

Fig. 2-25. A, Congenital absence of the left fibula in a boy 4 years of age. Note shortness of the left lower limb, external rotation of the leg and foot, and valgus deformity of the foot. The shortness measured 3 inches. **B,** Anteroposterior and lateral roentgenograms of the left lower limb of the patient shown in **A.** Note absence of the fibula, bowing of the short and thick tibia, lateral displacement of the foot, and equinovalgus deformity.

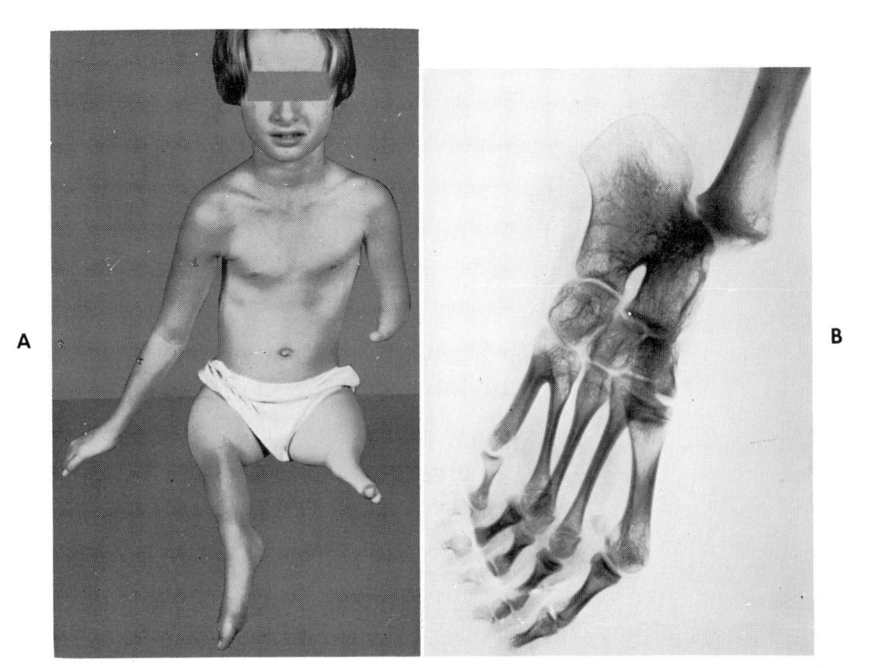

Fig. 2-26. A, Multiple congenital deformities in a girl 9 years of age. Note absence of major parts of the left upper and lower limbs, apparent absence of the right knee, and severe equinus of the right foot. **B,** Roentgenogram of the right foot of patient in **A,** showing absence of the fibula, extreme equinus, and synostosis of calcaneus, talus, and navicular bone.

In complete or almost complete absence the tibia is usually bowed anteriorly; the foot is short, occupies an equinovalgus position, and may show deformities of the toes. The affected limb is short. Congenital absence of the fibula may occur in association with other congenital anomalies (Fig. 2-26). The treatment of congenital absence of the fibula involves correction of the foot deformity and of the tibial bowing. A strong fibrous band may be present in the region of the missing fibula; excision of this band may be helpful. Osteotomy of the tibia is sometimes necessary. When the shortening is not too great, an elevated shoe and brace may be used to equalize the leg length. In some instances the shortening is severe, and amputation may be indicated to permit the fitting of a conventional prosthesis.

Patella

The development of the patella may be delayed or imperfect, or the bone may be completely absent. Occasionally an ossification center in the upper and lateral segment fails to fuse with the remainder of the bone, producing a *bipartite patella.* This condition is usually bilateral and ordinarily causes no symptoms. Partial or complete absence of the patella is usually associated with underdevelopment of the quadriceps tendon and occasionally with congenital dislocation of the knee. Hypoplastic or absent patellae are a characteristic of the uncommon inherited disorder known as onycho-osteodysplasia or nail-patella syndrome. Clubfoot, dislocation of the hip, and other congenital anomalies are sometimes associated with absence of

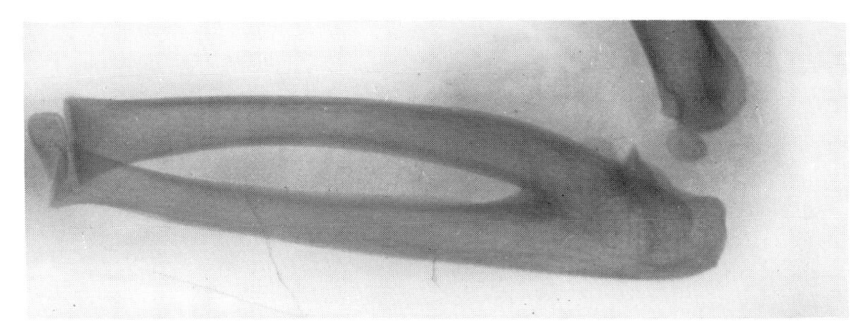

Fig. 2-27. Roentgenogram showing congenital synostosis of the radius and the ulna. Note that their proximal ends have failed to develop separately.

the patella. Deformity and weakness of the knee depend more upon the associated anomalies than upon the defect of the patella.

CONGENITAL RADIOULNAR SYNOSTOSIS

An infrequent congenital anomaly, usually bilateral, is radioulnar synostosis. It occurs as a rule at the proximal end of the radius and ulna (Fig. 2-27). Sometimes there is only a small bridge of connecting bone. There is usually a fibrous union between the bones in the lower third of the forearm. The head of the radius is sometimes dislocated. Occasionally there is fusion of the radius, ulna, and humerus, with absence of the elbow joint. The forearm may be fixed in a position of pronation or in one that is midway between pronation and supination. Extension of the elbow is limited. Since the patient learns to compensate for his loss of forearm rotation by increased use of the shoulder, the disability is slight.

Treatment. The whole upper end of the radius may be resected, or only the bone between the radius and the ulna, and the space so formed may be filled with fat, muscle, or fascia. Rarely, however, do these operative procedures meet with success. The disappointing postoperative results have been ascribed to persistence of fibrous bands between the bones in the distal part of the forearm and to congenital absence of muscles that rotate the forearm.

CONGENITAL CONTRACTURES

Congenital contractures may involve any of the joints. When present at the ankle or the wrist, they may produce clubfoot or clubhand, respectively. They are usually associated with other congenital malformations. Flexion contracture of the little finger, or camptodactyly, is common and frequently bilateral; it is inherited as an autosomal dominant trait and may be associated with shortening of the flexor digitorum sublimis tendon. Congenital contracture of the elbow is usually associated with a flexion deformity of the wrist. The shoulder often suffers a limitation of abduction. Congenital hip flexion contracture is rare; it may be associated with flexion of the knee. When the knee is flexed, there may be redundant skin in the popliteal space. It is thought that these contractures of the knee and of the hip may result from a long-continued position of flexion in utero.

Treatment. Congenital contractures are treated by (1) gradual manipulation into the corrected position, (2) exercises to develop the weak muscles, usually the extensors, and (3) the use of retention splints. Occasionally it may be desirable to incise the tight skin and underlying resistant bands; it may be necessary to graft skin over resulting denuded areas.

CONGENITAL ABNORMALITIES OF FINGERS AND TOES

Deformities of the hand or foot occur approximately once in every 600 live births; most of them affect only the fingers or toes.

Syndactyly (webbed fingers or toes)

Syndactyly (Fig. 2-28, *A*) occurs twice as often in boys as in girls and more often in the hand than in the foot. In the hand the fingers on the ulnar side are more often affected than those on the radial side, and the thumb is seldom involved. The union between affected digits may consist only of skin and connective tissue or may include bone. Webbing is sometimes associated with polydactyly, and both of these conditions may be hereditary. Syndactyly is often associated with multiple congenital deformities of the hands and feet. An inherited disorder characterized by syndactyly of the hands and sometimes of the feet, associated with a pointing of the head in the region of the anterior fontanel, was described in 1906 by Apert and is called Apert's syndrome (acrocephalosyndactyly).

Treatment. Separation of the involved fingers is accomplished by dividing the soft tissues and bony structures and should be done between the second and the fifth years. Many plastic procedures have been devised; the results are usually satisfactory. Rarely does webbing of the toes interfere with function.

Macrodactyly

Macrodactyly is an overdevelopment of one or more fingers or toes (Fig. 2-28, *B*). Presumably it results from an abnormal growth capacity in the embryonic tissue forming this part. In some cases lesions of local nerve trunks have been found, suggesting neurofibromatosis. The usual treatment is amputation, but occasionally a plastic operation to effect a reduction in size is preferable.

Polydactyly

Supernumerary digits are the commonest deformity of the hand (Fig. 2-28, *C*). They occur in the hand and foot with about equal frequency. The condition is often bilateral and may be combined with syndactyly. Amputation of the extra finger or toe is indicated.

Cleft hand and cleft foot

This anomaly consists of a cleft or division in the middle of the hand or foot, which results in the formation of large digits or parts (Fig. 2-29). It is spoken of as a lobster-claw hand or foot and is sometimes associated with syndactyly or polydactyly. These hands and feet are unsightly but often function satisfactorily. They are sometimes treated surgically for cosmetic improvement.

CONGENITAL AMPUTATIONS AND CONSTRICTING BANDS

Congenital amputations (Fig. 2-29) may result from genetic defects or amniogenic constrictions. Often there is a deep circular constriction about the arm or leg

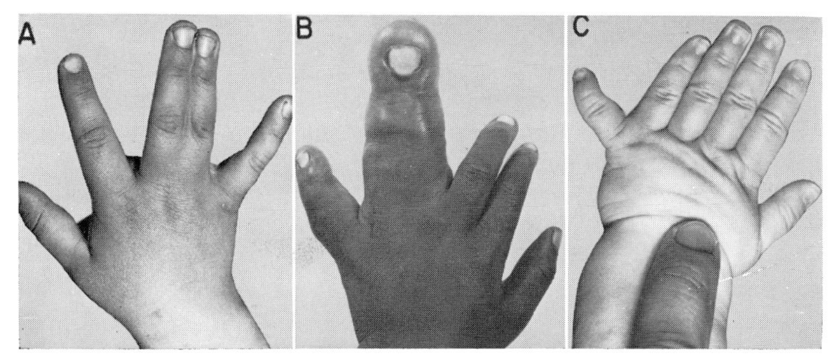

Fig. 2-28. Congenital deformities of the hand: **A,** syndactyly; **B,** macrodactyly; **C,** polydactyly.

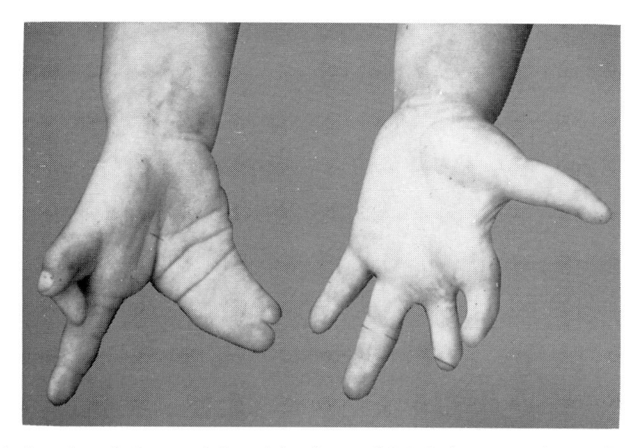

Fig. 2-29. Cleft hand and finger deformities in a child 2½ years of age. In the right hand, note midpalmar cleft, syndactyly of the ring and little fingers, and deformity of the index finger and thumb. In the left hand, note abnormally long thumb and congenital amputations of the distal phalanges of the index and middle fingers.

without amputation. The constriction may occur at any level of the arm or leg. Distal to it the extremity may be enlarged.

Treatment. The treatment of congenital amputations is discussed in Chapter 9. When a deep indented band is associated with disturbance of the distal circulation, it may be necessary to dissect out the band through a Z-plasty incision. The operation may be done in two or three stages to avoid increasing the circulatory embarrassment. Occasionally it may be desirable to complete the amputation.

ASYMMETRIC DEVELOPMENT (CONGENITAL HEMIHYPERTROPHY)

In the normal individual there are often slight differences in the development of the two lateral halves of the body. This condition, however, is sometimes severe and of congenital origin. It may be caused by an increased blood supply to one side of the body, resulting from a primary developmental abnormality of the circulatory system. Occasionally an arteriovenous fistula may be the cause. Neurofibromatosis may produce a similar appearance. Hemihypertrophy may at times resemble a unilateral elephantiasis, the redundant tissue being of fibrous and fatty nature. The increase in size usually ceases at the end of the growth period, but the disproportion constitutes a permanent deformity. In some instances hemihypertrophy occurs in association with congenital dwarfism. Sometimes one extremity only is involved. Treatment is not uniformly effective, but cosmetic operative procedures, such as shortening the longer leg, are indicated in selected cases.

ARTHROGRYPOSIS MULTIPLEX CONGENITA

Arthrogryposis multiplex congenita is an incomplete congenital fibrous ankylosis of many or all of the joints of the limbs. The axial skeleton is frequently spared, but scoliosis may be present. The involvement is usually symmetric (Fig. 2-30) and varies in severity from mild deformities to contractures that render the extremities almost functionless. The etiology is unknown; it is probably not of hereditary nature. Various abnormalities of intrauterine environment have been implicated as

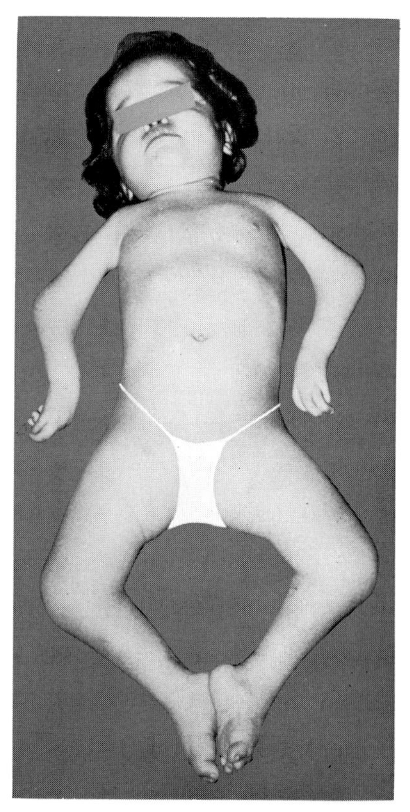

Fig. 2-30. Arthrogryposis multiplex congenita in a girl 2 years of age. Note the internal rotation of the arms and hands, flexion contracture and thickening of the elbows, external rotation of the lower extremities, flexion of the knees, and equinus. Kissing feet and hoop-shaped lower limbs are terms applied to these deformities.

possible causes, but occurrence in several reported instances in only one child of a set of twins seems to refute these hypotheses. Although joint stiffness is the main characteristic of arthrogryposis, the pathologic changes are mostly extra-articular. The joint capsules are thickened and contracted, and the muscles may be decreased in bulk or entirely absent, being replaced by fat. Defective formation or degeneration of the anterior horn cells of the spinal cord has been reported. Just where the primary defect lies, however, remains to be discovered.

Clinical picture. The typical patient shows internal rotation of the arms and flexion contractures of the wrists. The elbows and knees have a fusiform appearance and may be either flexed or extended with only a few degrees of mobility. The hips are often flexed and externally rotated. The patellae are sometimes dislocated or absent. There are usually associated clubfeet and sometimes clubhands. The hands show contractures of the palms and have been likened to walrus flippers. Other congenital anomalies, such as syndactyly and polydactyly, may be present. In some cases absence of the biceps and brachialis muscles makes active flexion of the elbow impossible, and in such instances the child is unable to feed himself.

Treatment. Because of the muscle deficiency and tendency for the deformities to recur after correction, the results of treatment are often unsatisfactory. Nevertheless, although normal function may not be expected, much can be done to improve the child's functional status. The aim of treatment in the lower limb is to obtain stable, well-aligned joints suitable for weight bearing. Arthrogrypotic contractures yield grudgingly to casts and manipulations, and such measures must usually be supplemented by surgical release of soft tissues. The equinovarus deformity is much more resistant to treatment than the usual clubfoot. This is also true of hip dislocation associated with arthrogryposis. In the older child, bone operations such as triple arthrodesis in the foot may be necessary to gain joint alignment. In the upper limb, extension contracture of the elbow is particularly disabling. It may be relieved by surgical release of the joint capsule and lengthening of the triceps tendon. In some instances anterior transfer of the triceps to change its function from elbow extension to flexion is helpful. This is contraindicated in the patient who is totally dependent upon crutches. After these procedures physical therapy should be diligently carried out for a long period of time.

CLEIDOCRANIAL DYSOSTOSIS

Cleidocranial dysostosis (Figs. 2-31 and 2-32) is an uncommon hereditary affection characterized by partial or complete absence of both clavicles, together with changes in the skull and other bones. The sexes are affected with equal frequency.

Pathology. The characteristic changes are a partial or complete absence of both clavicles and an exaggerated development of the transverse diameter of the cranium with delayed ossification of the fontanels. The muscles attached to the clavicle and the chest develop in an anomalous manner. Often other congenital malformations are associated, such as coxa vara and imperfect pubic ossification and spinal segmentation.

Clinical picture. Usually the patient is brought to the physician because something has been found adventitiously to be wrong with one or both shoulders. A defect of the clavicle is demonstrable clinically. Because of extreme relaxation of the shoulder girdle the patient can often bring the tips of the shoulders together below the chin (Fig. 2-31, *B*).

Treatment. As a rule there is little disability, and no treatment is indicated. If the patient complains of pain, however, one or both ends of the clavicles, if present, may be removed.

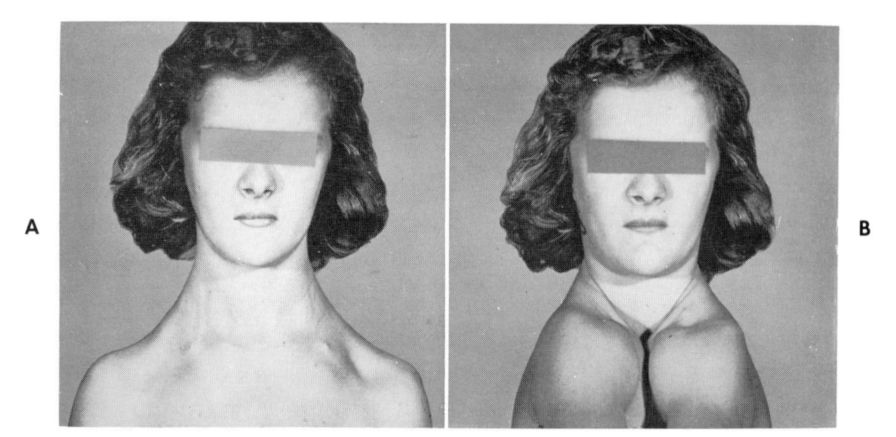

Fig. 2-31. Cleidocranial dysostosis. **A,** Note long neck and low shoulders. A depression in the middle of the forehead and a broadening of the occiput due to delay in fontanel closure are also characteristic. **B,** Absence of the clavicles allows anterior approximation of the shoulders.

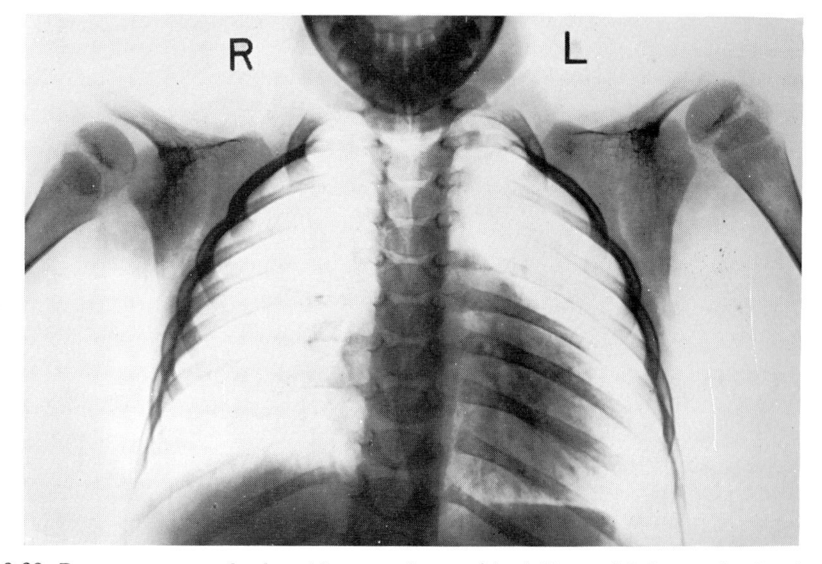

Fig. 2-32. Roentgenogram of a boy 10 years of age with cleidocranial dysostosis, showing complete absence of the clavicles.

CONGENITAL ELEVATION OF THE SCAPULA (CONGENITAL HIGH SCAPULA, SPRENGEL'S DEFORMITY)

Congenital elevation of the scapula, an uncommon anomaly, was first described by Eulenburg in 1863. The scapula is elevated from 1 to 4 inches above its normal position, and usually its inferior angle is rotated medially (Fig. 2-33). The deformity is sometimes bilateral. It is often associated with other congenital anomalies, and particularly with defective development of the cervical vertebrae and upper ribs.

Etiology. Congenital elevation of the scapula is believed to be the result of failure of the scapula to descend to normal position from its high level in the early weeks of gestation.

Pathology. The affected scapula is small. Usually its vertical length is decreased and its width relatively increased. Its upper portion tends to hook forward. The cervical muscles are shortened on the affected side and are changed in direction. In about one fourth of the cases there is union between one of the lower cervical vertebrae and the scapula; such union may consist of bone, cartilage, or fibrous tissue. When a bony connection is present, it extends from the spinous process, lamina, or transverse process of one or more vertebrae to the upper part of the

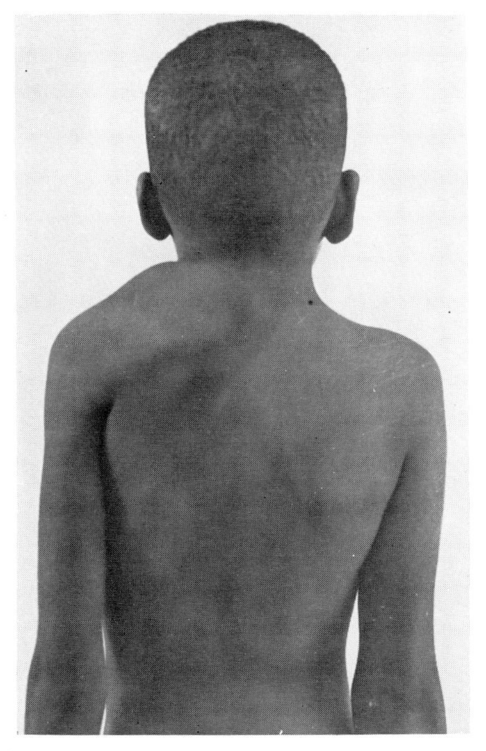

Fig. 2-33. Congenital elevation of the left scapula in a boy 13 years of age. Abduction of the left shoulder was limited at 60 degrees.

vertebral border of the scapula. Such an osseous bridge is called an *omovertebral bone.*

Clinical picture. Asymmetry of the shoulders is often the first evidence of abnormality. In most of the cases abduction of the affected shoulder is greatly restricted. In 10% of the cases torticollis is present, together with lateral curvature of the spine. In bilateral cases the neck appears shortened.

Diagnosis. The diagnosis is made from the physical signs and the roentgenogram. Congenital elevation of the scapula must be differentiated from paralysis of the serratus anterior muscle and from obstetric paralysis.

Prognosis. Without operation no improvement of the deformity is to be anticipated. Surgical treatment is often followed by moderately improved function and appearance.

Treatment. Postural training and exercise of the shoulder may increase the joint function. When motion is greatly limited and the deformity is unsightly, surgical treatment may be indicated. The operation should include exposing the scapula extraperiosteally, excising its deforming supraspinous portion and the omovertebral bone if present, bringing the scapula down to the desired level, and repositioning the muscle attachments to provide anchorage. To be effective the operation must be performed in childhood, preferably before the age of 5 years.

CONGENITAL SYNOSTOSIS OF THE CERVICAL SPINE (KLIPPEL-FEIL SYNDROME)

Congenital synostosis of the cervical spine is a rare malformation resulting from an arrest of development. The outstanding pathologic change is fusion of all or of

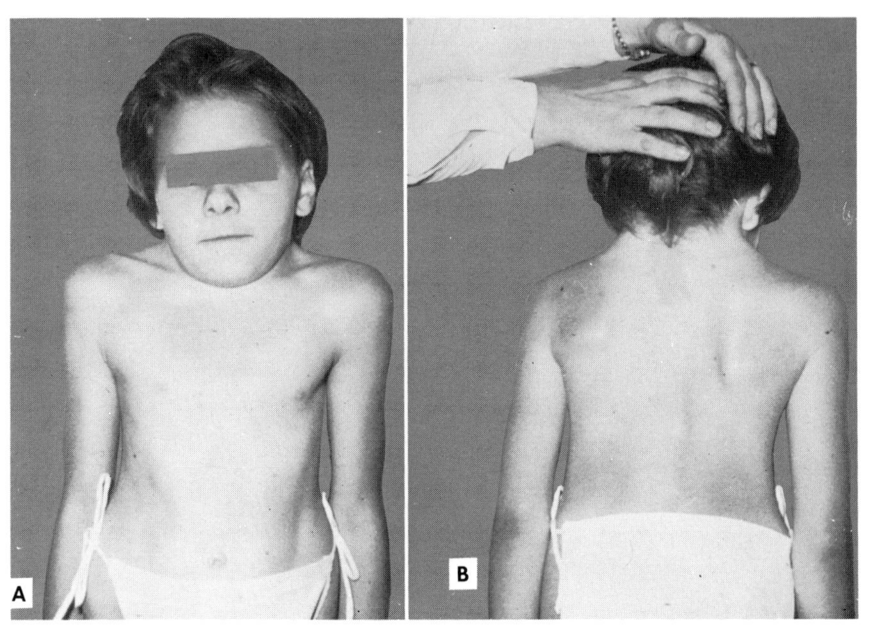

Fig. 2-34. Klippel-Feil syndrome in a girl 7 years of age. **A,** Note short neck with chin resting on chest. **B,** Note short, thick neck and lateral webbing from head to shoulders.

only the lower cervical vertebrae into one homogeneous mass of bone. The posterior portion of some or all of the laminal arches is not developed, resulting in spina bifida, which usually involves the lower cervical vertebrae and one or two of the upper dorsal vertebrae. Cervical ribs, crowding of the upper ribs, and congenital anomalies in other parts of the body are often associated. Roentgenograms show anomalies of the cervical spine, including clefts in its vertebral arches.

Clinical picture. Abnormal shortness of the neck is the most noticeable feature, sometimes causing the head to seem to rest directly on the trunk (Fig. 2-34). There is painless restriction of neck motion in all directions. Occasionally there may be associated neurologic changes due to spinal cord or nerve root compression. Flexion and extension of the cervical spine, which may take place wholly at the joints between occiput and atlas, are better preserved than is lateral motion. The head and neck may be held in an oblique position simulating that of congenital torticollis. Occasionally the trapezius muscles stretch winglike from the mastoid processes to the shoulders, suggesting the term *pterygium colli,* or *web neck.*

Treatment. As a rule no treatment is indicated. However, sometimes a plastic surgical operation to release the webbing of the neck is followed by increased motion. In early childhood a neck brace is occasionally helpful. Attempts to increase mobility by manipulation have proved harmful.

CHAPTER 3 General affections of bone

By its physical properties of strength, resilience, and lightness, bone is admirably adapted to its major function as the supporting framework of the body. Bone is not on this account to be regarded as an inert structural material, however; it is an actively living tissue and like other tissues is constantly undergoing the simultaneous processes of destruction and regeneration.

Functional adaptations of bone. With this continuous breaking down and rebuilding is associated an adaptive remodeling of the bones in response to functional demands that are placed upon them. This far-reaching principle, formulated by Julius Wolff in 1868 and known as *Wolff's law,* holds that: "Every change in the form and the function of bones, or in their function alone, is followed by certain definite changes in their internal architecture, and equally definite changes in their external conformation, in accordance with mathematical laws."

Increased functional demands cause the physicochemical processes of bone formation to outdistance those of resorption, and *bone hypertrophy* results. The thick, heavy bones of athletes and laborers, as contrasted with the lighter bones of sedentary individuals, are illustrative of bone hypertrophy. Of greater clinical concern, however, is the converse condition, *bone atrophy* or *osteoporosis,* which results from a predominance of the catabolic and resorptive processes over those of regeneration. In neither hypertrophy nor atrophy of a bone is its general conformation usually altered.

Bone atrophy, in accordance with Wolff's law, may be caused by disuse (Fig. 3-1). The bone atrophy that constantly develops during therapeutic immobilization of an extremity is a classic example. The cortical bone becomes thinned and the medullary cavity widened, slight narrowing of the shaft occurs, and generalized porosity and loss of bone weight may become severe. These changes appear to represent merely a quantitative variation, however, since the chemical composition of the bone undergoes no demonstrable alteration, the breaking strength in relation to weight is normal, and the power of repair after fracture is retained. Extreme bone atrophy is of particular importance when it involves the growing bones of childhood, since atrophic bones may grow slowly and may never attain full size.

Bone atrophy appears in roentgenograms as a loss of density of the bone shadow

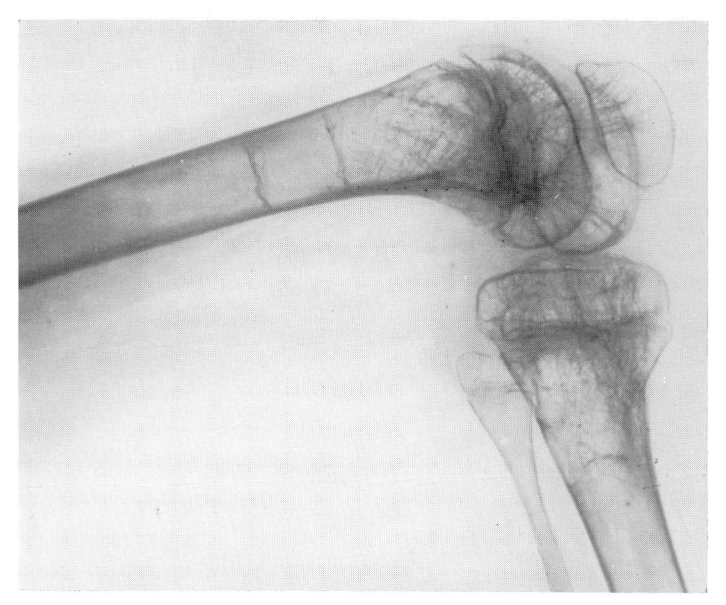

Fig. 3-1. Bone atrophy. Roentgenogram of the knee of a 12-year-old boy following prolonged bed rest after excision of a spinal cord tumor. Note decreased density of all bones, narrowed cortices, and old pathologic fracture near the upper end of the tibia.

as compared with that of the soft tissues. Narrowing of the cortex and attenuation of the bony trabeculae are frequently pronounced. It must be remembered that roentgenographic density is simply an index of the total lime salt content of the bones and gives no direct information regarding the pathologic changes that may be taking place in their organic constituents.

Clear distinction is drawn between the terms *osteoporosis* and *osteomalacia.* Osteoporosis, or bone atrophy, signifies a generalized decrease in bone mass. Osteomalacia, on the other hand, refers to a deficient calcification of bone matrix.

Since the bony skeleton is a living and continually changing tissue, it is, like other tissues, subject to characteristic disorders. These diseases are similar in producing widespread skeletal changes but vary greatly in etiology, pathologic processes, and course. Current biochemical, electron microscopic, and genetic investigations are disclosing much new information on the members of this large, heterogeneous group of bone disorders. With discovery of pertinent etiologic factors should come improvement in our methods of prevention and treatment.

Classification of general skeletal affections. Since the cause of many of the general affections of the skeleton is still unknown or debatable no final etiologic grouping can be constructed. The following classification is subject to change as additional information on the more obscure entities becomes available:

I. Associated with genetic abnormalities
- A. Achondroplasia
- B. Multiple exostoses
- C. Osteosclerosis
  - 1. Osteopetrosis (Albers-Schönberg disease, marble bones)

I. Associated with genetic abnormalities—cont'd
- C. Osteosclerosis—cont'd
  - 2. Osteopathia striata
  - 3. Osteopoikilosis (spotted bones)
- D. Progressive diaphyseal dysplasia (Engelmann's disease)
- E. Progressive myositis ossificans
- F. Osteogenesis imperfecta (fragilitas ossium, brittle bones)
- G. Marfan's syndrome
- H. Hurler's syndrome
- I. Morquio's syndrome
- J. Gaucher's disease

II. Associated with dietary or metabolic abnormalities
- A. Avitaminoses and hypervitaminoses
- B. Rickets (infantile rickets, nutritional rickets)
- C. Osteomalacia (adult rickets)
- D. Osteomalacia associated with renal disorders
  - 1. Vitamin D–resistant rickets (hypophosphatemic rickets)
  - 2. Renal osteodystrophy (renal dwarfism)
- E. Scurvy

III. Associated with endocrine abnormalities
- A. Hypopituitary dwarfism
- B. Hyperpituitarism
  - 1. Gigantism
  - 2. Acromegaly
- C. Cretinism (hypothyroidism)
- D. Hyperparathyroidism (generalized osteitis fibrosa cystica)

IV. Of unknown etiology
- A. Infantile cortical hyperostosis (Caffey's disease)
- B. Histiocytosis (reticuloendotheliosis)
  - 1. Eosinophilic granuloma
  - 2. Hand-Schüller-Christian disease
  - 3. Letterer-Siwe disease
- C. Enchondromatosis (Ollier's disease, dyschondroplasia)
- D. Fibrous dysplasia
  - 1. Polyostotic
  - 2. Monostotic
- E. Melorheostosis
- F. Hypertrophic osteoarthropathy
- G. Osteitis deformans (Paget's disease)
- H. Senile osteoporosis
- I. Posttraumatic painful osteoporosis (Sudeck's atrophy, reflex sympathetic dystrophy)

Affections associated with genetic abnormalities

ACHONDROPLASIA

In achondroplasia, enchondral osteogenesis is retarded and ceases early, membranous osteogenesis is unaffected, and the resulting bodily conformation is that of the typical congenital dwarf of literature and drama. Discovery of a complete achondroplastic skeleton of the early Egyptian period established this as one of the oldest of known diseases.

Etiology and pathology. Achondroplasia is always congenital and often inherited, being transmitted as an autosomal dominant trait. Frequent fresh mutations result in a preponderance of noninherited cases. The exact nature of the inborn defect of localized cartilage growth is unknown. Essential in the pathologic process, which begins early in intrauterine life, is slowed, disorderly development at the epiphyseal cartilage plate that results in retardation of longitudinal growth. The epiphyseal plates are thin; microscopically the zone of provisional calcification fails to show its normal orderly pattern. Membranous bone formation, which accounts for circumferential growth of the shafts of the long bones, proceeds normally. The changes can be recognized in roentgenograms before birth, sometimes even as early as the third month of pregnancy.

Clinical picture. Many achondroplastic infants are stillborn or die in their first year. In children the condition may be first noticed when a disproportion between length of trunk and of limbs becomes evident, the arms and legs appearing short and thickened in contrast to the normally developing torso. The clinical appearance of the achondroplastic dwarf is characteristic and usually permits of ready diagnosis (Fig. 3-2). The adult height is seldom more than 4 feet. The skull is usually brachycephalic, its anteroposterior diameter being less than normal, and the face is small

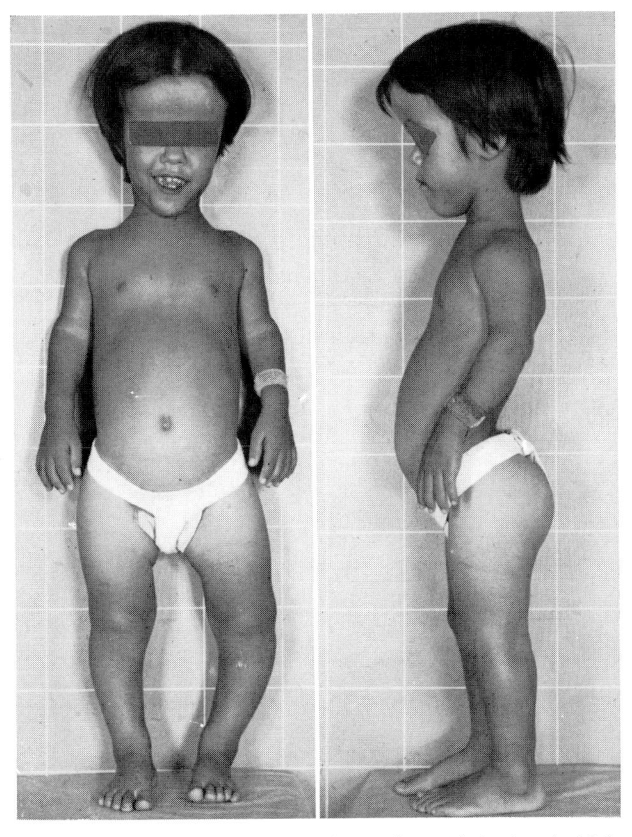

Fig. 3-2. Achondroplastic dwarf 8 years of age. Her height, 37 inches, is 10 inches below average for her age. Note short extremities, long torso, increased lumbar lordosis, protruding abdomen, large head, saddle nose, bowing of left leg, and short, thick fingers.

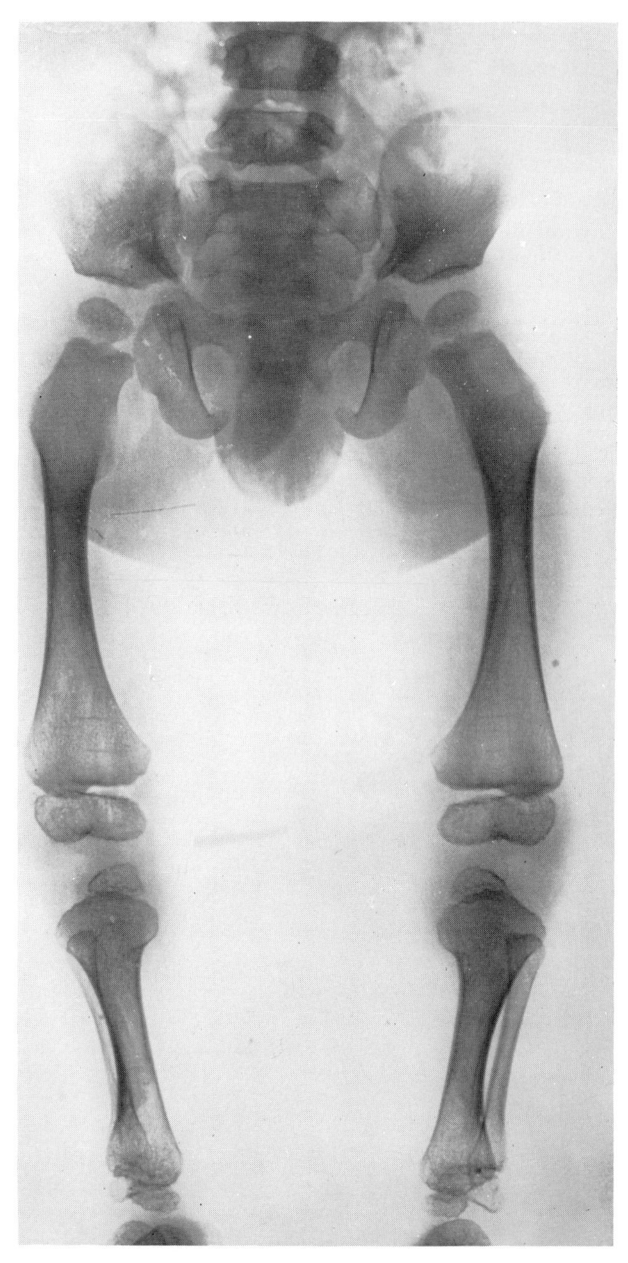

Fig. 3-3. Roentgenogram showing changes due to achondroplasia in a 3-year-old boy. Note the short, thick femurs, tibiae, and fibulae, valgus of the femoral necks, low acetabular angles (Fig. 2-6, *A*), and slight varus of the knees.

in proportion to the cranium. The facial expression resulting from high and broad forehead, flattened nose with depressed bridge, and prominent lower jaw is typical. The hands are short and broad with pudgy fingers of almost equal length, which tend to spread in a radial manner and have occasioned the term *main en trident*. Although growth of the trunk is of nearly normal extent, pathologic anteroposterior and lateral spinal curvatures of secondary nature are common, as is protrusion of the abdomen. Bowing of the femurs and tibiae is a frequent finding. Mental development is usually little impaired, as are the sexual characteristics, although in both respects exceptions occur.

Diagnosis. Differential diagnosis is difficult only in infancy, when rickets and cretinism are to be considered. The frequently normal mentality of the achondroplastic patient forms an obvious contrast with the impaired mental development of the cretin. The roentgenograms, however, are of greatest value in differentiating sharply the short, broad, normally dense shafts and thinned epiphyseal plates of achondroplasia (Fig. 3-3) from the poorly ossified bones and epiphyses of rickets, and from the transverse metaphyseal condensations of cretinism.

Treatment. No effective treatment of achondroplasia is known. Occasionally, especial attention is required for the prevention or correction of specific deformities. In adults, low back pain may require symptomatic measures. Neurologic complications including hydrocephalus, nerve root compression, and paraplegia may necessitate surgical treatment. After infancy the prognosis for length of life is good.

MULTIPLE EXOSTOSES

The affection termed multiple exostoses is characterized by the presence of numerous osteocartilaginous masses that protrude from the metaphyseal portion of the long bones. Associated deformities are broadening of the metaphyses and shortening and bowing of the shafts. The disorder is hereditary in approximately 70% of the cases, being transmitted as an autosomal dominant trait.

Pathology. The exostosis consists of a pedicle of histologically normal bone protruding from the metaphysis and covered by a cartilage cap. During the growing years the exostosis continues to enlarge by enchondral ossification, which takes place beneath the cartilage cap. This aberrant segment of cartilage is believed to have arisen from the epiphyseal plate.

Clinical picture. The exostoses are usually not discovered until several years after birth. They rarely cause symptoms but with gradual enlargement may result in deformity and limitation of motion at the adjacent joints. Shortening of the ulna and curvature of the radius are common findings. The number of exostoses varies from a few to several hundred.

Roentgenographic picture. The usual finding is that of bony outgrowths extending from the metaphyseal region (Fig. 3-4). The marrow cavity of the exostosis is continuous with that of the bone, and its apex is directed away from the adjacent joint. The cartilaginous cap is not visible in the roentgenogram.

Treatment and prognosis. Because the exostoses are so numerous their removal is not feasible. However, lesions that are causing pain or deformity by pressing on adjacent structures should be excised. Usually the exostoses cease to grow when the patient reaches skeletal maturity. Since malignant change occasionally takes place a

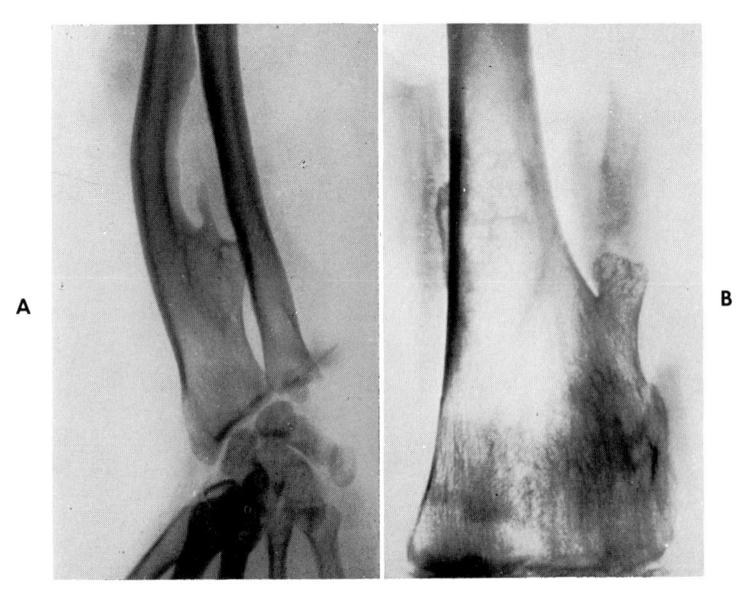

Fig. 3-4. A, Roentgenogram showing multiple exostoses in a woman 28 years of age. Note the exostosis at the lower end of the radius, which encroaches on the interosseous space, and the lateral bowing of the shaft of the radius. **B,** Roentgenogram showing multiple hereditary exostoses in a 9-year-old boy. Note the broadening of the metaphysis and the exostosis above the adductor tubercle of the femur. This patient is the son of the patient shown in **A.**

history of recent enlargement or of pain in a previously symptomless exostosis is an indication for prompt excision.

OSTEOSCLEROSIS

Under the term *osteosclerosis* may be grouped a number of interesting but rare and little-known skeletal diseases characterized by regions of increased bone density. The affected bones have areas of typically opaque roentgenographic appearance; in some of them the cortex is greatly thickened, and the medullary canal may be obliterated. The bone changes do not as a rule cause symptoms and are often discovered as an incidental finding during roentgenographic examinations. The etiology of these diseases is unknown, although the sporadic association of various factors in individual cases has led to a variety of hypotheses. These affections are sometimes hereditary or familial. Symptomatic treatment is occasionally indicated. Osteosclerotic bone changes are sometimes caused by an excessive intake of fluorides.

Osteopetrosis (Albers-Schönberg disease, marble bones)

The areas of greatly increased bone density in this disease are widespread, often including pelvis, vertebrae, skull, and extremities. Failure of osteoclastic resorption of the primary spongiosa seems to be the essential pathologic defect. In the metaphyses, bone is laid down about cores of densely calcified cartilage. As a result, primitive chondro-osteoid persists, not being replaced by normal cancellous bone.

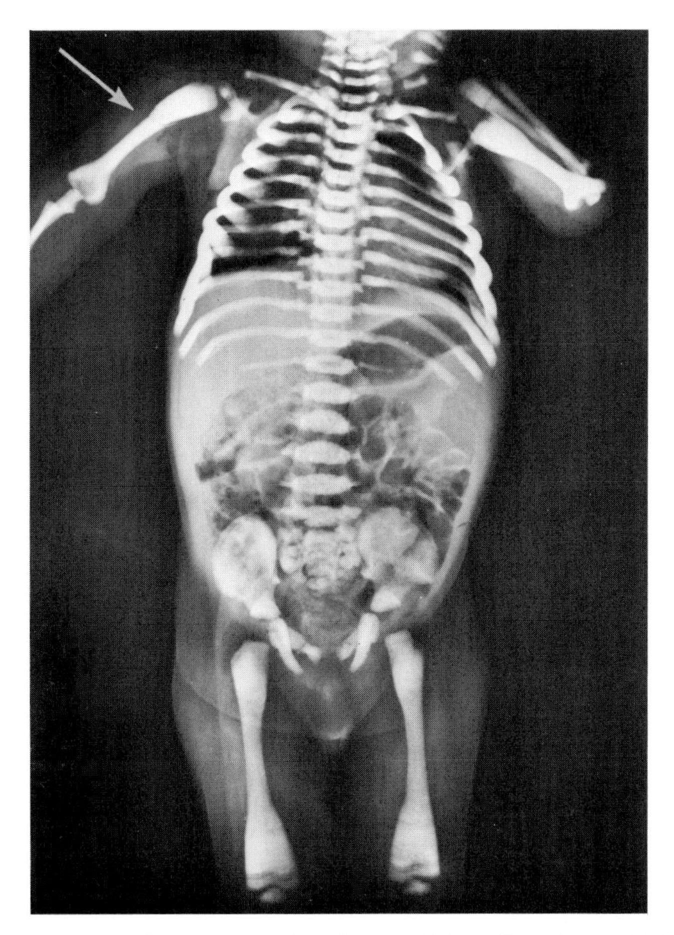

Fig. 3-5. Roentgenogram of osteopetrosis in a 2-year-old boy. Note the greatly increased bone density, obliteration of medullary canals, decreased metaphyseal remodeling, and fracture of a humeral shaft (arrow).

Osteopetrosis may occur in autosomal recessive form with severe, generalized bone changes in the newborn infant that result in early death from osteosclerotic anemia, or in a more benign autosomal dominant form with less extensive bone involvement and relatively good prognosis for length of life. Hydrocephalus, blindness, enlargement of spleen and liver, malnutrition, and progressive osteosclerotic anemia may be present. Transverse pathologic fractures may occur; they ordinarily heal without incident.

In roentgenograms (Fig. 3-5) the affected bones show greatly increased density and loss of normal architecture. The medullary canal is narrowed or obliterated. A bone-within-bone appearance is sometimes observed. Metaphyses of the long bones are splayed and may have a celery-stalk appearance with dense longitudinal striations.

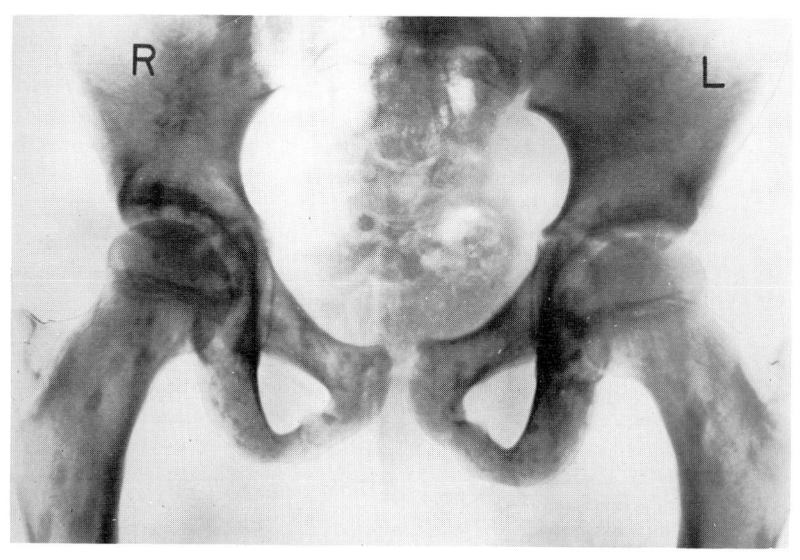

Fig. 3-6. Roentgenogram showing osteopoikilosis in a girl 12 years of age. Note small, dark areas of bone condensation in pelvis and femurs.

Osteopathia striata

Osteopathia striata is a form of osteosclerosis in which coarse longitudinal striations are seen in cancellous bone, especially conspicuous in the metaphyseal areas and extending well into the diaphyses. After epiphyseal closure the striations may extend the entire length of the bone. The skull and clavicles are seldom affected. The bony contour is normal, and no symptoms are referable to the disorder. The cause is unknown, but there are strong hereditary and familial factors.

Osteopoikilosis (spotted bones)

In osteopoikilosis the areas of condensation are small and scattered, giving the affected bones a spotted appearance in roentgenograms (Fig. 3-6). The dense areas consist of thickened trabeculae in the cancellous bone. The metaphyseal and epiphyseal regions are most often involved. Osteopoikilosis is familial and hereditary. Symptoms are slight or absent, and discovery of the lesions may be entirely fortuitous.

PROGRESSIVE DIAPHYSEAL DYSPLASIA (ENGELMANN'S DISEASE)

Progressive diaphyseal dysplasia is a rare affection characterized by symmetric expansion and sclerosis of the diaphyses of the long bones. It is seen most typically in children, occurs with almost equal frequency in males and females, and seems to be inherited as an autosomal dominant trait. The involved bone proliferates by both endosteal and periosteal growth, producing a thickened, fusiform cortex without trabecular pattern and a narrowed medullary cavity (Fig. 3-7). The epiphyses are unaffected. There may be an increase of bone density at the base of the skull.

The patients have leg pain, increased fatigability, muscle weakness, abnormalities

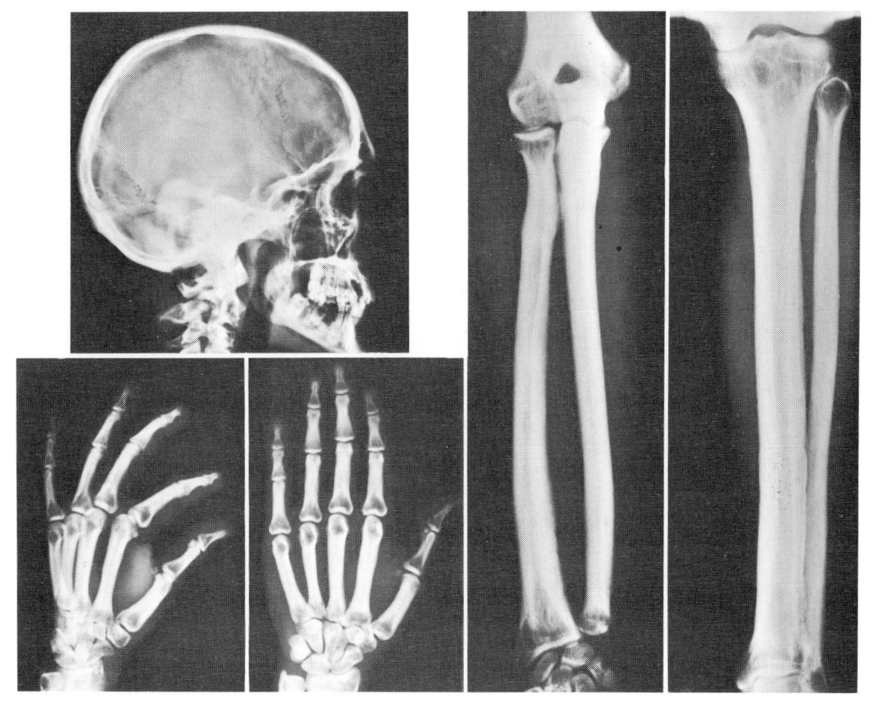

Fig. 3-7. Roentgenograms in progressive diaphyseal dysplasia. Note the dense, thickened cortical bone, narrowed medullary canals, relatively spared epiphyseal areas, and increased bone density at the base of the skull.

of gait, and delayed development. As a rule, all clinical laboratory findings are within normal limits. No specific treatment is known, and the ultimate prognosis has not been determined.

PROGRESSIVE MYOSITIS OSSIFICANS

Although not primarily a disease of the skeleton, progressive myositis ossificans is conveniently considered here because in its advanced stage it is characterized by a transformation of muscles and fasciae into immobile structures of bony consistency (Fig. 3-8). The pathologic changes occur primarily in fasciae, ligaments, tendons, and other connective tissue structures and only secondarily involve muscle. The disease is rare, seen more often in males than females, and probably inherited as a mendelian dominant trait with variable penetrance. The cause and mechanism of the heterotopic calcification are unknown.

Clinical picture. The disease begins in childhood, sometimes as early as the first year. Underdevelopment and malalignment of the great toes are usually present; occasionally the thumbs show similar changes. Onset of the disease may be heralded by the appearance of painful, tender subcutaneous masses that gradually regress, become fixed, and undergo ossification. As a rule, the tissues of the neck or back are first involved by the process of ossification, and although there may be remissions,

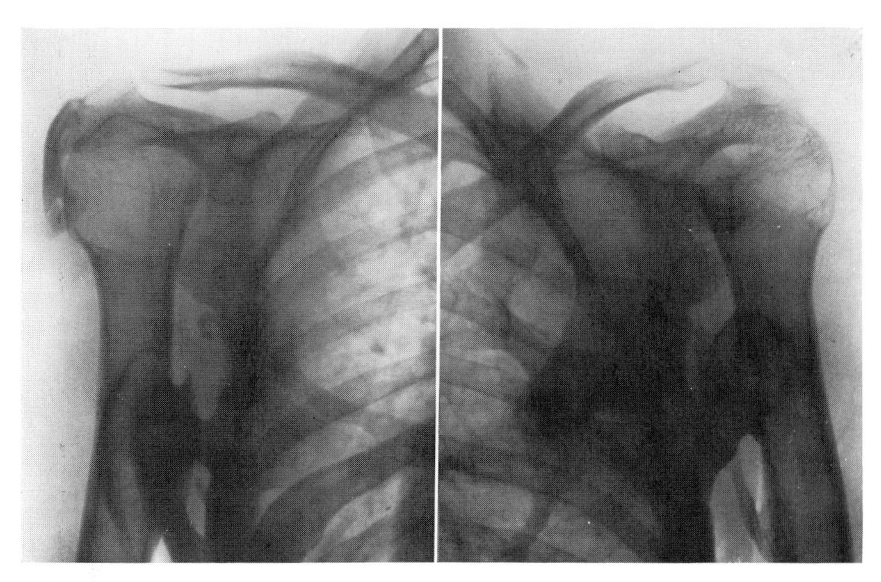

Fig. 3-8. Progressive myositis ossificans. Roentgenograms of the shoulders of a 32-year-old woman who had had gradual progression of joint stiffness and palpable bony masses in various muscle groups since the age of 4 years.

one group of muscles after another gradually becomes stiffened. Ultimately the patient becomes bedridden, and as a rule death ensues from intercurrent infection.

Diagnosis and treatment. The diagnosis presents difficulty only in the early stage of the disease. Muscle biopsy may then be helpful. The localized microdactyly aids in establishing the diagnosis. At the present time no effective treatment is known.

OSTEOGENESIS IMPERFECTA (FRAGILITAS OSSIUM, BRITTLE BONES)

Osteogenesis imperfecta is an uncommon, hereditary, generalized connective tissue disorder in which the occurrence of multiple fractures is rendered inevitable by extreme fragility of the bones. In roentgenograms the bones often show osteoporosis with generalized thinning of the cortex, decreased diameter of the shaft, and coarse trabeculations (Fig. 3-9). The fractures are often undisplaced. In addition to the skeleton, affected tissues include the ligaments, scleras, inner ears, and skin.

Etiology. The cause of the excessive bone fragility and other connective tissue changes is unknown. The basic defect seems to involve bone matrix and to be concerned with an impaired maturation of collagen fibers or the synthesis of an abnormal collagen. Other evidence suggests a qualitative defect in the osteoblasts.

Osteogenesis imperfecta is thought to be inherited as an autosomal dominant trait, with a wide range of expressivity that accounts for the variations in its clinical features. In addition, there seems to be an autosomal recessive form accounting for the small percentage of severe cases with changes obvious at birth.

Pathology. Microscopically, the bone matrix is of abnormal appearance and the

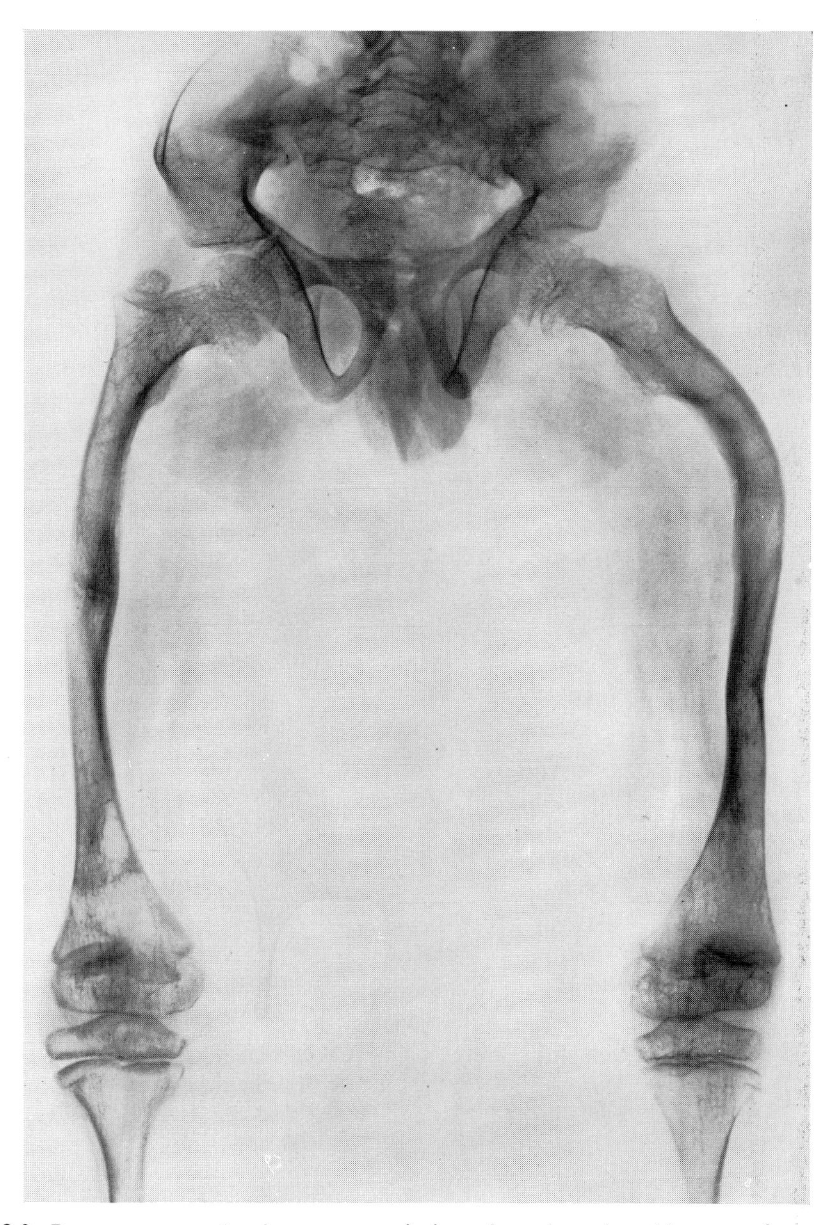

Fig. 3-9. Roentgenogram showing osteogenesis imperfecta in a boy 10 years of age. One femur has been fractured four times and the other twice. Note deformity of femurs, uneven density and thinning of cortices, and heart-shaped pelvis.

compact bone is somewhat like that seen in the fetus or newborn infant. Metaphyses show calcified cartilage rather than true bone or osteoid. Osteoblasts and osteoclasts are usually considered normal in number. Whereas the total content of bone salts is decreased, the calcium-phosphate ratio is normal. In the blood no pathognomonic chemical changes are apparent.

Clinical picture. Broad lines may be drawn clinically between several types of this disease. There is a severe form, to which the term *osteogenesis imperfecta congenita* is applied, in which the disease is obvious at birth from the occurrence of multiple fractures and definite shortening of the extremities as compared with the trunk. In such infants the prognosis for survival is poor. A less severe form of the disease, *osteogenesis imperfecta tarda,* is characterized by the occurrence of fractures first in early or late childhood.

In many of the patients with osteogenesis imperfecta a very definite Wedgewood-blue coloration of the scleras will be noted. This color has been attributed to a decreased opacity of the scleras, which permits the pigmentation of the choroid to

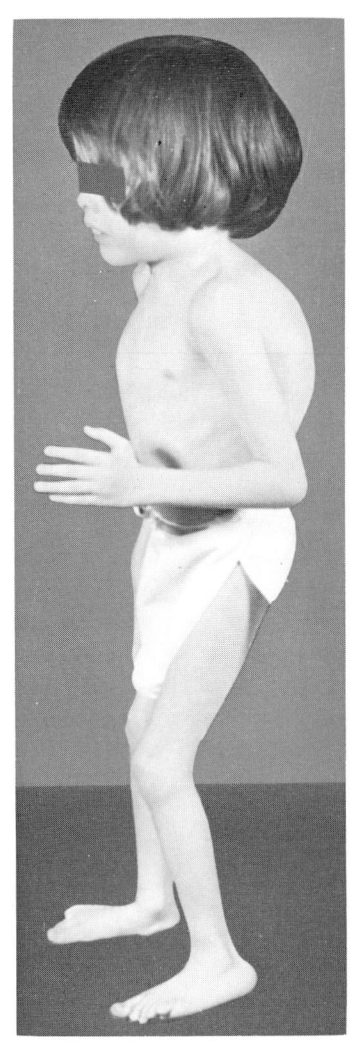

Fig. 3-10. Marfan's syndrome in a girl 6 years of age. Note the long, thin fingers and toes, the severely pronated feet, the slender limbs, and the spinal deformity and barrel-shaped thorax.

show through. Many of these patients also show hypermobility of the joints, thinning of the skin, and in adult years a progressive deafness resembling that of otosclerosis. All such patients are of short stature.

Treatment and prognosis. Since the underlying etiology is unknown, treatment is empiric and, in general, unsatisfactory. Reduction and immobilization after fracture are carried out as usual. An effort should be made to keep the child as active as possible in order to stimulate bone formation.

The prognosis for union of these fractures is excellent. It is a matter of clinical experience that the fragility of the bone does not entail a loss of its capacity to unite after fracture. The fracture callus, however, is of the same poor quality as the bone that makes up the remainder of the skeleton. Occasionally union may be accomplished by exuberant callus formation of tumorlike appearance. With repeated fractures of various bones and abnormal bone growth, it is not surprising that extensive deformities and severe curvatures may ultimately develop. Corrective osteotomies and intramedullary fixation are occasionally indicated. After the age of puberty has passed, the bone fragility and tendency to fracture may become less severe.

MARFAN'S SYNDROME (ARACHNODACTYLY)

Marfan's syndrome, an uncommon affection inherited as an autosomal dominant, is characterized by changes in the bones and joints, the eyes, and the cardiovascular system. Affected persons are tall and slender; their thin fingers and toes, with elongated phalanges, metacarpals, and metatarsals (Fig. 3-10), are sometimes spoken of as spider fingers and toes. In approximately 80% of the cases there is an asymmetry of the skull. About half of the patients have eye abnormalities with dislocation of the lens, and at least a third have a cardiac or aortic lesion. One of the most striking histologic changes is found in the media of the aorta, where fragmentation of the elastic fibers frequently leads to a fatal dissecting aneurysm. Striated muscles are underdeveloped, ligaments are lax, and joints are hypermobile. There may be coxa vara and lateral curvature of the spine that increase as the child grows older. Arachnodactyly is to be differentiated from the Ehlers-Danlos syndrome, which also is characterized by hypermobility of the joints but includes increased elasticity of the skin and fragility of the walls of the blood vessels.

Treatment. There is no specific orthopaedic therapy. Individual deformities or disabilities such as scoliosis should be treated as indicated.

HURLER'S SYNDROME (LIPOCHONDRODYSTROPHY)

Hurler's syndrome is the commonest of the *mucopolysaccharidoses,* a group of related heritable disorders of connective tissue with skeletal manifestations of varying severity. In Hurler's syndrome the mode of inheritance is autosomal recessive. A presumptive defect in connective tissue metabolism results in widespread deposition of dermatan and heparan sulfate in the tissues, and excessive excretion of these mucopolysaccharides in the urine. The characteristic finding on microscopic examination of a number of tissues, including heart, meninges, and periosteum, is the presence of large cells, called Hurler cells, which are distended by material consisting largely of mucopolysaccharides.

Clinical picture. Hurler's syndrome begins in early childhood, causing charac-

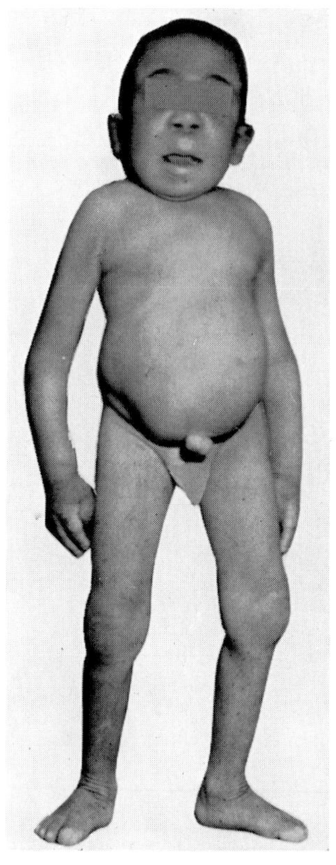

Fig. 3-11. Hurler's syndrome in a boy 12 years of age. Note shortness of neck and trunk, protruding abdomen, and characteristic facies. This patient also had an umbilical hernia. He died of congestive heart failure 3 months after this photograph was taken.

teristic deformities, and is usually fatal in the second decade of life because of heart failure from mucopolysaccharide infiltration. It is characterized by marked shortening of the trunk and neck (Fig. 3-11), a dorsal kyphosis, protruding abdomen, and enlargement of the liver and spleen. The facial features are coarse and grotesque. There are a misshapen, enlarged skull, prominent forehead, widely spaced eyes, prominent supraorbital ridges, saddle nose, thick tongue, and prognathism. The corneas are clouded by multiple deep opacities. There are usually short arms, impaired joint mobility, and flexion deformities of the joints, especially of the fingers, which are thickened. Mucopolysaccharide infiltration of the brain results in mental deficiency.

Diagnosis and treatment. Characteristic roentgenographic changes include abnormalities in development of the cranial bones, abnormal vertebral contours, widening of the medial ends of the clavicles and underdevelopment of their lateral ends, and tapering of the distal phalanges. The diagnosis is substantiated by demonstrating abnormal amounts of dermatan and heparan sulfate in the urine. No effective treatment is known.

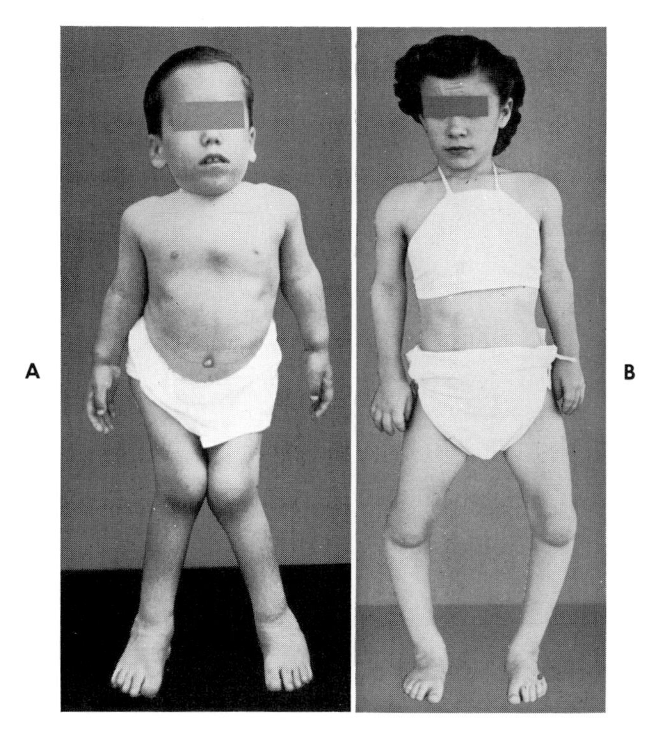

Fig. 3-12. A, Morquio's syndrome in a boy 8 years old whose height, 36 inches, is 13 inches below average for his age. Note large head, flexion of the hips and knees, bilateral genu valgum, and short limbs. **B,** Morquio's syndrome in a girl 13½ years old whose height, 47½ inches, is 14 inches below average for her age. Note large head, bilateral genu varum, and short, stubby fingers and toes.

MORQUIO'S SYNDROME

Morquio's syndrome, like Hurler's syndrome, is a heritable mucopolysaccharidosis that causes characteristic skeletal deformities in childhood and leads to dwarfism (Fig. 3-12). Of autosomal recessive inheritance, the metabolic defect is associated with infiltration of keratan sulfate into the tissues and its excretion in excessive amounts in the urine. Morquio's syndrome is rare; it affects both sexes with about equal frequency and as a rule causes little or no mental impairment.

Clinical picture. Morquio's syndrome is usually unrecognized until the child begins to walk, at which time a waddling gait is noted. There are characteristic changes in the spine and centers of ossification, especially in the capital femoral epiphyses. The femoral heads become enlarged and misshapen, with resulting limitation of external rotation of the hips. The spine is stiff and usually presents a sharp kyphosis at the dorsolumbar junction, which may be confused with the gibbus of a tuberculous spine. There may be a compensatory lumbar lordosis. There is gross deformity of the vertebral bodies, particularly at the dorsolumbar junction, where in lateral roentgenograms the front half of the vertebrae may appear to be missing or may show a slender, midanterior projection. The neck and trunk are short in contrast to the limbs. The phalanges and metacarpal bones may be stubby and the

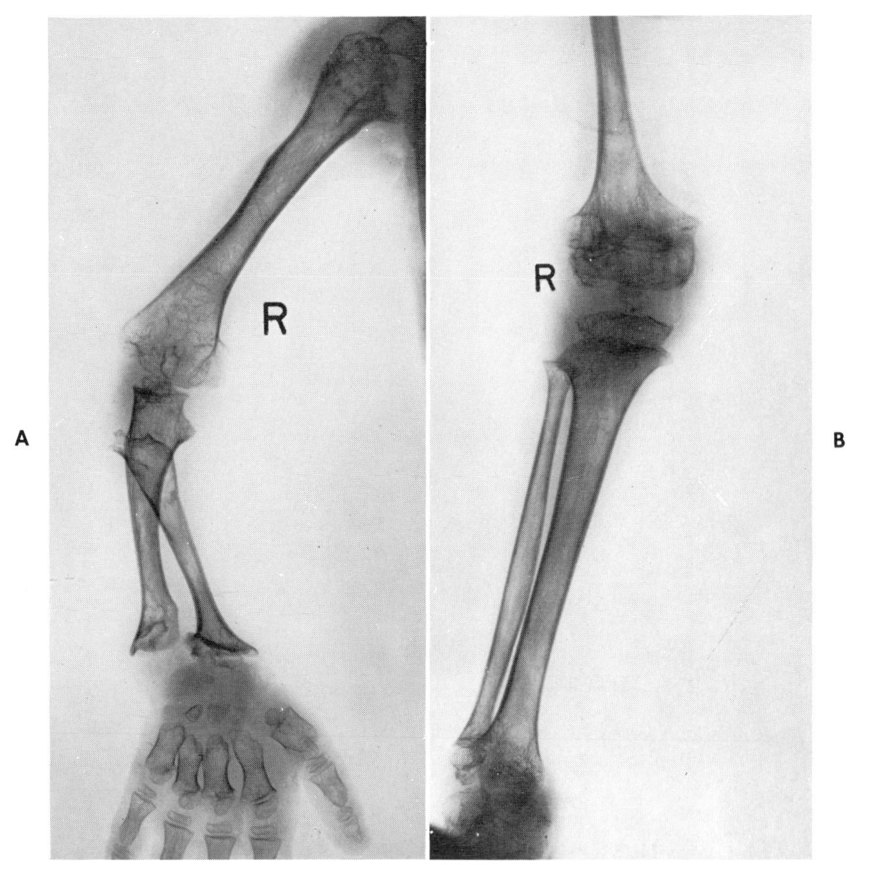

Fig. 3-13. Morquio's syndrome. **A,** Roentgenogram of right upper limb of the patient shown in Fig. 3-12, *A*. Note deformities of epiphyses and long bones. **B,** Roentgenogram of right lower limb of the patient shown in Fig. 3-12, *A*. Note genu valgum, deformity of epiphyses, and thinning of femoral shaft.

ulna and fibula snort (Fig. 3-13). Deformity of the epiphyses and laxity of supporting ligaments tend to cause early disability from degenerative joint disease.

Treatment. The only treatment is that which may be indicated to relieve symptoms or to correct deformities.

GAUCHER'S DISEASE

Gaucher's disease, an uncommon disorder of lipid metabolism and storage, is thought to be caused by an inborn enzymatic defect. In Gaucher's disease, histiocytes laden with a cerebroside known as *kerasin* accumulate in large numbers in the bone marrow, spleen, liver, and other sites. The characteristic cells, large and bizarre under the light microscope, are called *Gaucher cells.* Their presence in the spleen usually leads to extreme splenomegaly.

A hereditary influence is found in about one third of the cases. Gaucher's disease occurs in an acute form in infants and a chronic form in children and adults. The acute, infantile cases are characterized by splenomegaly, hepatomegaly, and retarded physical development. Neurologic manifestations including hypertonia and opisthotonos develop, as well as severe pulmonary symptoms. The infant becomes increasingly cachectic and usually does not survive his first year. Niemann-Pick disease is a similar fatal lipid storage disorder of infancy.

The chronic form is the more common variety of Gaucher's disease. Abdominal discomfort from enlargement of the spleen is usually the first symptom, but anemia or joint pain secondary to bone involvement may occur initially. Hepatomegaly may develop early or late. Brownish discolorations of the skin, yellow deposits in the conjunctivae, and hemorrhagic manifestations also occur.

Roentgenographic picture. Affected bones show reduced density, mottled areas, and radiolucent defects. The lower ends of the femurs appear flared, with thin cortex and wide medullary canal. Similar changes may be seen in other bones, such as the vertebrae, pelvis, tibiae, and humeri. Pathologic fractures may result, and avascular necrosis of the femoral head is a common complication.

Diagnosis and treatment. Diagnosis is made by demonstration of the characteristic cells in the bone marrow. Treatment is palliative; it includes splenectomy to improve the anemia and relieve the abdominal discomfort, local roentgentherapy to relieve skeletal pain, and the use of steroids. Transfusions are indicated for the anemia. The disease often progresses slowly to a fatal termination, but the patients may survive for many years.

Affections associated with dietary or metabolic abnormalities

SKELETAL CHANGES CAUSED BY EXCESSIVE OR DEFICIENT VITAMIN INTAKE

Bone is most affected by the presence or absence of vitamins during its stage of formation. Important effects of the better known vitamins on the skeletal system are summarized here.

Vitamin A. Vitamin A has a specific effect on the osteoblasts, osteoclasts, and epiphyseal chondroblasts of growing bone. It affects the pattern of bone growth. Thick, short bones may result from either a lack or an excess of this vitamin.

Chronic hypervitaminosis A, caused by excessive administration of the vitamin, produces elevation of the periosteum followed by subperiosteal calcification. These changes, together with pain in the extremities and irritability, usually appear in children between the ages of 1 and 3 years. The serum vitamin A level is greatly elevated. The prognosis is excellent after the excessive vitamin A intake has been recognized and discontinued. Hypervitaminosis A may be confused not only with scurvy but also with *infantile cortical hyperostosis (Caffey's disease)*, which is usually accompanied by swelling of the jaw and occurs in infants 1 to 4 months of age (p. 88), and with *progressive diaphyseal dysplasia (Engelmann's disease)*, which is rare, occurs in the older child and young adult, and is associated with a waddling gait and muscular weakness (p. 64).

Vitamin B complex. It has been shown by Warkany that in rats a lack of riboflavin in early pregnancy causes a high incidence of congenital deformities. This has not been proved in human beings. Experimentally it has been shown also that lack of the B complex may cause thinning of epiphyseal cartilage plates and cessation of growth.

Vitamin C (ascorbic acid). Vitamin C is an important factor in regulating the formation of intercellular substances such as osteoid, collagen, and ground substance. A deficiency interferes with osteoblastic activity, resulting in diminished formation of bone matrix, or osteoid. An important clinical manifestation of vitamin C deficiency is hemorrhage, which results from an obscure defect of the capillary walls. Lack of ascorbic acid in the diet causes *scurvy* (discussed on p. 82).

Vitamin D. The principal physiologic action of vitamin D is regulation of the calcium and phosphorus in the blood. This it does primarily by increasing absorption of calcium from the intestine. It promotes the deposition of calcium and phosphorus in the osteoid matrix or protein framework of bone that has been laid down by the osteoblasts. A deficiency of vitamin D produces *rickets* in the young (discussed in next section) and *osteomalacia* in adults (p. 77).

Hypervitaminosis D, an affection resulting from the administration of large doses of the vitamin, is characterized by hypercalcemia, widespread deposition of calcium in the soft tissues, and slowly progressive deossification of the skeleton. In periarticular tissues, bursae, and tendon sheaths, the calcium deposits may reach several centimeters in diameter; in the kidneys they may lead to insufficiency and uremia. Hypervitaminosis D in several respects simulates hyperparathyroidism, excessive doses of the vitamin stimulating bone resorption rather than bone formation. The normal or slightly elevated serum phosphorus in hypervitaminosis D is helpful in differential diagnosis.

RICKETS (INFANTILE RICKETS, NUTRITIONAL RICKETS)

Rickets is a constitutional disease of infancy and childhood caused by a lack of vitamin D and evidenced by bony deformities, which may be striking in degree and widespread in distribution. Its primary skeletal change is *osteomalacia,* a deficient calcification of bone matrix. With improved nutrition and pediatric care the severe cases of infantile rickets that were formerly so common have become rare in highly developed parts of the world.

Etiology. Nutritional rickets may be caused by a vitamin D deficiency either in the diet or resulting from malabsorption defects, such as celiac disease, in the gastrointestinal tract. In pancreatic or biliary disorders, absence of bile salts may impair the absorption of vitamin D, which is fat soluble. Underlying the osteomalacia of rickets is a disorder of calcium metabolism, the exact mechanism of which remains obscure. Phosphate deficiency from hyperparathyroidism secondary to hypocalcemia may play a part. Three factors appear to be important in the prevention of rickets: (1) the antirachitic vitamin D, (2) adequate amounts of calcium and phosphorus in the diet, and (3) sunlight or artificial ultraviolet irradiation to aid endogenous production of vitamin D.

Pathology. The chief characteristic of the pathologic changes is subnormal calcification and a relative increase of osteoid tissue throughout the skeleton. The result is an increased plasticity of the bones that allows them to undergo changes in shape and to become deformed by gravitational and physiologic stresses that would be

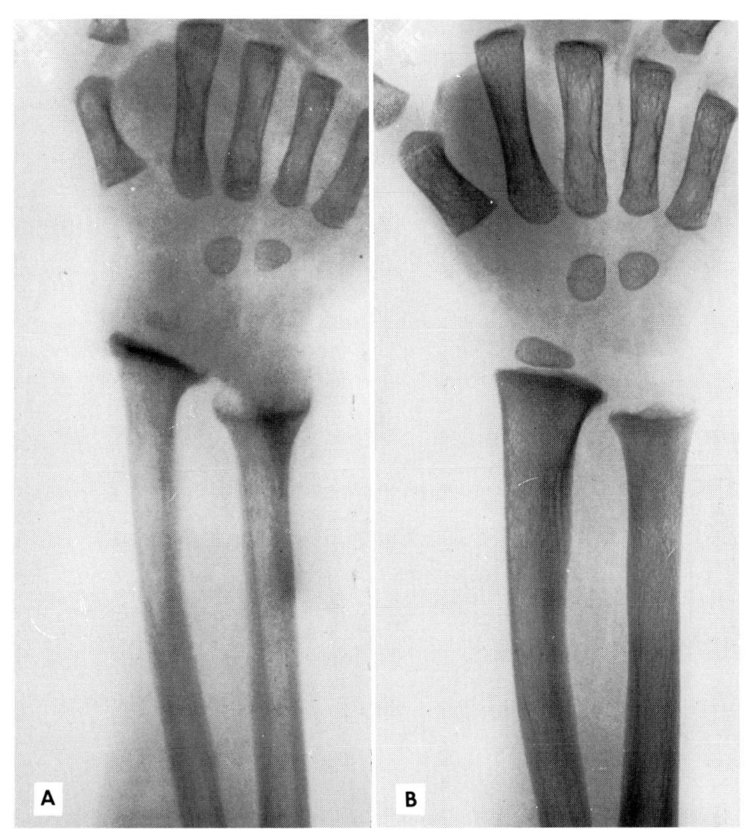

Fig. 3-14. Roentgenograms of rickets. **A,** In a 10-month-old boy with active rickets, the obliquity of the epiphyses and the cupping of the ulnar metaphysis are characteristic. **B,** After 4 months of treatment, inactivity of the rickets is shown by more uniform bone density and improvement of the obliquity and cupping.

withstood by bony structures of normal strength. At the epiphyseal plates, ossification is delayed and disordered. Epiphyseal cartilage cells proliferate and hypertrophy, but there is failure of mineralization at the zone of provisional calcification. The hypertrophic cartilage cells are not destroyed, and vascular invasion is delayed. This failure of the normal process of endochondral ossification results in considerable widening and overgrowth of the epiphyseal cartilage plate. New osteoid tissue produced by osteoblasts is not mineralized.

In active rickets, characteristic blood chemistry findings include a normal or low calcium content, decreased phosphorus, and increased alkaline phosphatase.

Roentgenographic picture. The pathologic changes cause the roentgenographic appearance of a rachitic bone to undergo a cycle of typical alterations as the disease pursues its course.

In the early stages the metaphysis is widened and concave and presents an irregular and indistinct margin (Fig. 3-14, *A*). The epiphyseal line is broad. The epiphysis is obscurely outlined and contains one or more indistinct areas of ossification; as growth proceeds, the epiphysis assumes a vaguely mottled appearance. The shaft may show periosteal thickening and occasionally a fracture line.

In later stages (Fig. 3-14, *B*) the end of the diaphysis loses most of its concavity but remains widened. The epiphysis becomes more distinct in outline and more homogeneous in density. Abnormal curvature of the shaft, with thickening of the cortex on the side of the concavity, is a characteristic finding (Fig. 3-15).

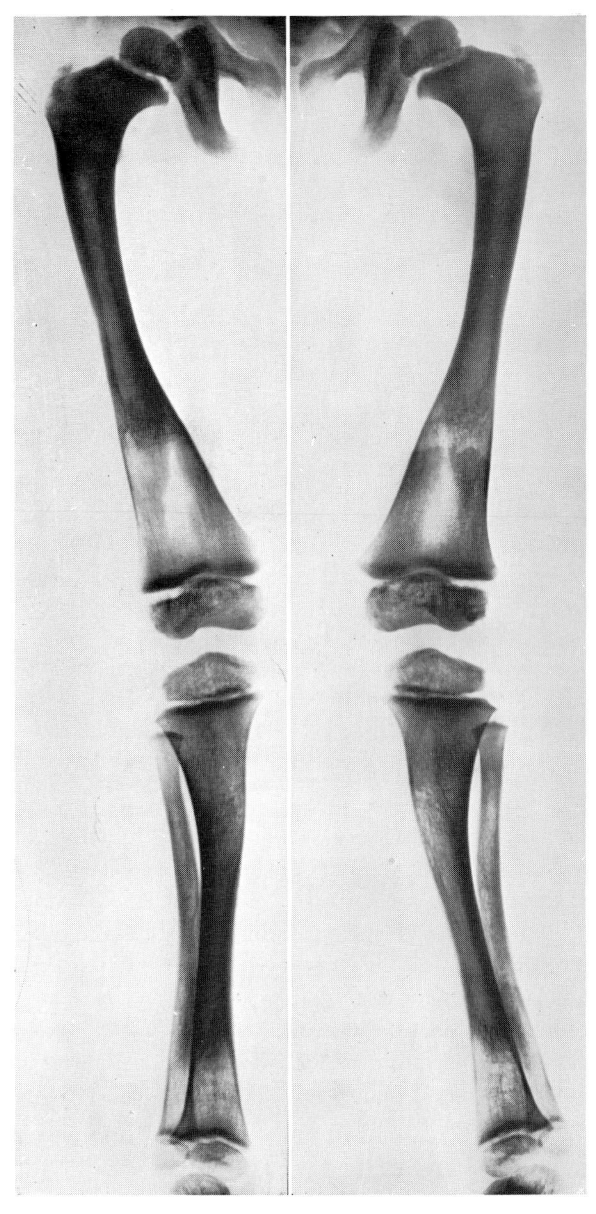

Fig. 3-15. Healed rickets in a patient 6 years of age. Note widening of the epiphyseal lines, distortion of the long bones, and bilateral genu valgum and coxa vara.

Clinical picture. Rickets is usually first recognizable near the end of the first year of life. The early symptoms are of pediatric rather than orthopaedic interest and are not pathognomonic. The patient with severe rickets has a pale skin, flabby subcutaneous tissue, and poorly developed musculature. Because of overgrowth of the epiphyseal cartilage plates, the ends of the long bones and of the ribs are enlarged. This is especially noticeable at the more rapidly growing epiphyses, such as those at the wrist and knee. The liver and spleen are likely to be increased in size, and the abdomen is prominent. The thorax may be of grossly abnormal shape and may show the *rachitic rosary,* due to enlargement of the costochondral junctions, or *Harrison's groove,* a transverse sulcus across the lower portion of the chest, thought to be due to the pull of the diaphragm. The skull is large and, because of its imperfect calcification, may be soft and exhibit a delicate crepitation on palpation, a condition known as *craniotabes.* There is noticeable delay in closure of the fontanels, in development of the teeth, and in acquisition of the ability to stand. Bowleg, knock-knee, coxa vara, scoliosis, and other deformities of the bones occur with great frequency in rickets.

Treatment. Of chief importance in the prevention of rickets and in the treatment of its active stage are adequate dietary calcium, phosphorus, and vitamin D, together with an abundance of sunshine and general hygienic care. An ample intake of vitamin D must be assured by supplementary medication with calciferol or a concentrate of cod-liver oil.

When the disease is established and still in its active stage, orthopaedic measures should be added. For the prevention of deformity, weight bearing and sitting with the legs crossed should be avoided. It is occasionally advisable to keep the patient recumbent. When rickets is controlled by medical treatment, most rachitic deformities in infants will be corrected by growth alone.

Osteoclasis, or fracture without operative exposure, was used frequently in the past for correcting deformity of the long bones in infants and young children but is now seldom employed.

Osteotomy, or operative section of a bone, is the procedure of choice for persistent, severe bowing of the long bones after inadequately treated rickets. Osteotomy is a quick, simple, and accurate procedure requiring a relatively short incision and obviating contusion of the soft tissues. The exact level and degree of the desired correction should be determined preoperatively from analysis of the roentgenograms. After closure a well-padded plaster cast is applied with the extremity in a slightly overcorrected position. It is sometimes advantageous to do a first-stage, incomplete osteotomy and to follow it after 3 weeks by a second-stage osteoclasis, as described by J. R. Moore. Immobilization is continued for from 6 to 12 weeks or until there is satisfactory union, as indicated by adequate callus shown in the roentgenogram and by absence of mobility at the site of the osteotomy. Gradual return to weight bearing is then carried out under observation.

OSTEOMALACIA (ADULT RICKETS)

The term *osteomalacia,* which ordinarily refers to an excess of unmineralized bone matrix, is used also to designate a specific group of nutritional diseases of adult bone, corresponding closely to infantile rickets in pathogenesis. In patients with such osteomalacia, vitamin D is not available in sufficient quantity to provide for the normal mineralization of osteoid that produces bone.

Etiology. The commonest contemporary cause of osteomalacia from vitamin D deficiency is gastrointestinal malabsorption. Since the vitamin is fat soluble, chronic diseases that impair the absorption of fat also reduce absorption of vitamin D. Such disorders, all characterized by steatorrhea, include a number of gastric, intestinal, hepatic, and pancreatic diseases.

Osteomalacia of dietary origin is seen chiefly in northern China, Japan, and northern India. Social customs in these areas force women to remain indoors, and the combination of little sunlight, pregnancy, and diets deficient in vitamin D and calcium leads to severe osteomalacia.

Biochemical picture. The lack of vitamin D impairs absorption of calcium by the small bowel, with the result that the serum calcium tends to fall. This stimulates production of parathyroid hormone by the parathyroid glands, which tends to maintain serum calcium by removing more from the bones, and to decrease the renal tubular reabsorption of phosphorus. Accordingly, in nutritional osteomalacia the serum calcium is usually normal or low and the serum phosphorus is low; the serum alkaline phosphatase is usually elevated.

Pathology. Lack of vitamin D also delays the mineralization of osteoid. In addition, mineralization may be impaired by a low calcium-phosphorus product in the serum. As a result, there is great excess of osteoid on histologic examination, and there may also be evidence of increased bone resorption brought about by secondary hyperparathyroidism. Osteoid seams are greatly widened, cortices are thinned, and grossly the bones may become so soft that they can be cut easily with a knife. Characteristic deformities of the skeleton are the heart-shaped pelvis from pressure at the acetabula, the shortened or telescoped spine, and marked bowing of the tibiae and femurs.

Clinical picture. The patient complains of shooting pains referred to the pelvis, back, or hips and on examination shows tenderness in these regions. Severe, progressive muscular weakness may be present. The softened spine, pelvis, and long bones develop deformities from muscular and gravitational stresses, and pathologic fractures are common.

Roentgenographic picture. Even in early or mild cases, osteomalacic bones may show transverse lines of rarefaction called *Looser's zones* (Fig. 3-16). They occur especially in the large long bones and the pelvis, are sometimes bilaterally symmetric, and probably represent pathologic fatigue fractures with union delayed by lack of mineralization. In later cases of osteomalacia the roentgenograms show a striking loss of calcium and the presence of bony deformities.

Diagnosis. When the diagnosis is not evident from clinical, roentgenographic, and laboratory findings, a bone biopsy is indicated. Hyperparathyroidism and occasionally osteogenesis imperfecta, metastatic carcinoma, multiple myeloma, and senile osteoporosis must be considered in the differential diagnosis.

Treatment. If the osteomalacia is secondary to a malabsorption syndrome, treatment must be directed toward correcting the underlying disease process. If the osteomalacia is dietary in origin, restoration of adequate amounts of vitamin D to the diet may be curative. Calcium and phosphorus supplements may also be given.

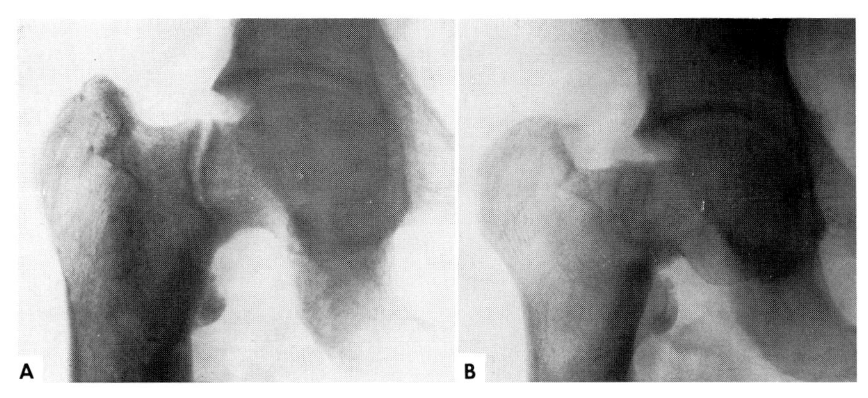

Fig. 3-16. Looser's zone in the femoral neck of a 40-year-old man with osteomalacia. **A,** Note transcervical pseudofracture line and coxa vara. Hip symptoms had been present for 3 years. **B,** Note healing of pseudofracture after a year of treatment consisting of the daily administration of 300,000 units of vitamin D. At this time the patient was bearing full weight on the affected limb without hip symptoms.

Protective measures to forestall deformity should be combined with careful exercise, adequate to counteract the atrophy of disuse. Severe bony deformities may require corrective osteotomies.

OSTEOMALACIA ASSOCIATED WITH RENAL DISORDERS

Defective mineralization of osteoid similar to that occurring in nutritional rickets is seen in a number of chronic affections of the kidneys. A group of related disorders characterized by hypophosphatemia and refractory to ordinary dosage of vitamin D is associated with defective function of the renal tubules; in this group the most common entity, still of debatable pathogenesis, is *vitamin D–resistant rickets* or *hypophosphatemic rickets.* The term *renal osteodystrophy* has been used to designate osteomalacic syndromes associated chiefly with glomerular dysfunction. In any of these renal syndromes, when disturbance of calcium-phosphorus metabolism results in hypocalcemia, the osteomalacia may be complicated by a secondary hyperparathyroidism.

Vitamin D–resistant rickets (hypophosphatemic rickets)

In the vitamin-resistant group of rachitic syndromes, defective reabsorption of phosphate by the renal tubules leads to hyperphosphaturia and hypophosphatemia. The clinical and roentgenographic changes are similar to those of nutritional rickets (p. 74) but respond only to enormous doses of vitamin D. In developed countries vitamin-resistant rickets is seen much more frequently than the true deficiency or nutritional form.

The commonest type of vitamin-resistant rickets is inherited as a sex-linked dominant trait of variable penetrance. It affects boys more often and more severely than girls. The hypophosphatemia is usually ascribed to a defect in renal tubular absorption but may be partly the result of hyperparathyroidism secondary to disturbed calcium-phosphorus metabolism.

A number of similar, less common osteomalacic syndromes are associated with other types of renal tubule defects. In one of these, sometimes called *Fanconi's syndrome,* there is a failure to resorb phosphates, glucose, and amino acids normally, and acidosis is frequently present. These affections occur in both inherited and acquired forms, depending upon the cause of the tubule defect. The serum calcium is normal or low, phosphorus low, and alkaline phosphate increased. Dwarfism and deformity may be severe. The renal changes, initially tubular, may progress to involve the glomeruli and lead finally to death in uremia.

The bony changes of vitamin-resistant rickets may be healed by high doses of vitamin D, but the margin of safety between therapeutic and toxic levels of vitamin D is narrow, and treatment is difficult to regulate. Overdosage of vitamin D, manifested by nausea, weight loss, and hematuria due to hypercalcemia, must be avoided. The typical deformities of rickets may develop (Fig. 3-17). If they persist to adolescence despite medical treatment, orthopaedic treatment like that described for nutritional rickets (p. 77) may be indicated if not prohibited by impaired renal function.

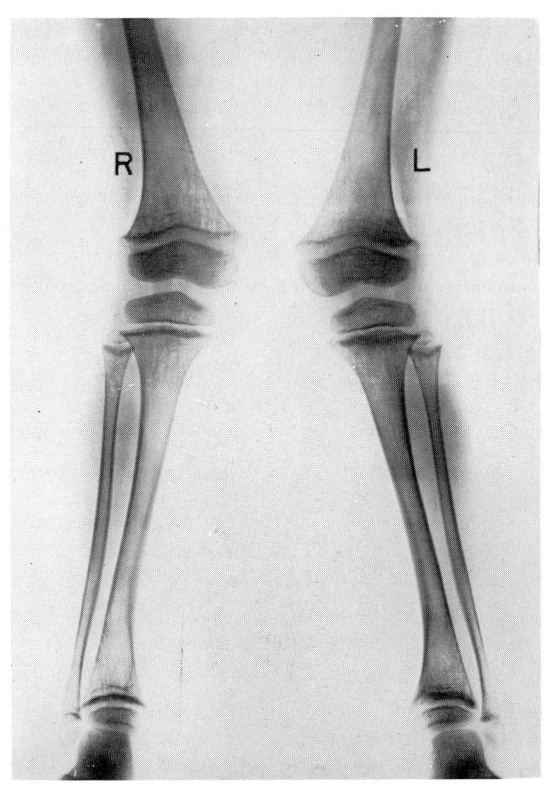

Fig. 3-17. Roentgenogram of vitamin D–resistant rickets in a child 5 years of age. Note widening of epiphyseal lines, increased calcification on the metaphyseal side of the epiphyseal lines, and bilateral genu valgum.

Renal osteodystrophy (renal dwarfism)

Renal osteodystrophy is an uncommon disorder of childhood or adult years characterized by osteomalacic changes similar to those of nutritional rickets and caused by chronic renal insufficiency resulting chiefly from impaired glomerular filtration. In addition, the calcium-phosphorus metabolism is disturbed and a secondary hyperparathyroidism develops; the exact pathogenesis of these changes remains obscure.

Pathology. In children the primary lesion may be a developmental anomaly of the kidney that is followed by a chronic pyelonephritis resulting in uremia; epiphyseal growth is defective and retarded, leading to use of the terms *renal rickets* and *renal dwarfism.* In adults the primary renal lesion is chronic glomerulonephritis. In either children or adults the nephritis is evidenced by a low urinary specific gravity, albuminuria, diminished urea clearance, and an increase of the blood nonprotein nitrogen, chlorides, and uric acid. The serum calcium is usually low or normal, and the phosphorus and alkaline phosphatase are usually high. Postmortem examinations show an advanced bilateral interstitial nephritis. The bones may be dwarfed, softened, and deformed. Microscopically, the metaphyses show excessive osteoid and the shafts a fibrous replacement of bone and marrow. In late cases areas of cystic change, osteoporosis, and osteosclerosis may be present.

Clinical picture. In children the clinical course is marked by gradually developing dwarfism, which is usually not evident until about the fifth year, by thirst and polyuria beginning insidiously and later accompanied by drowsiness, vomiting, and headaches, and by bone deformities that appear near the age of puberty. The roentgenographic changes include widening of the epiphyseal lines and broadening of the diaphyses, as in rickets, but usually to a lesser extent and without the charac-

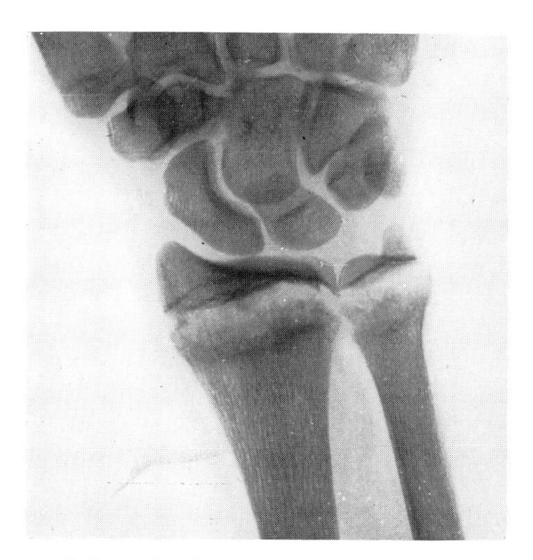

Fig. 3-18. Roentgenogram of the wrist in renal osteodystrophy. Note wide epiphyseal lines of the radius and ulna, normal appearance of epiphyses, and areas of irregular calcification at the metaphyseal margins.

teristic cupping (Fig. 3-18). The epiphyses are not greatly altered but may appear fragmented. Epiphyseal separation is common.

In adults with renal osteodystrophy the symptoms and signs are those of osteomalacia superimposed upon those of uremia.

Prognosis and treatment. The disease is usually progressive and ultimately fatal. In early stages a high-calcium, low-phosphorus diet with adequate doses of vitamin D may be helpful. The recently developed therapeutic measures of long-term hemodialysis and renal transplantation may prolong the patient's life but seem to have no beneficial effect on the skeletal changes. From an orthopaedic point of view it is essential that renal osteodystrophy be recognized in order that the risk of attempting surgical correction of the deformities be appreciated. Braces to prevent the increase of deformity are sometimes indicated.

SCURVY

Scurvy is an acquired constitutional disease caused by an abnormality of nutrition and manifested chiefly by signs related to the bones.

Etiology and pathology. The cause of scurvy is a dietary lack of the antiscorbutic vitamin C. This vitamin occurs in largest amount in the citrus fruits, unboiled milk, and fresh vegetables.

The pathologic changes involve an alteration in certain intercellular materials throughout the tissues, resulting particularly in subperiosteal and submucous hemorrhages.

Clinical picture. Like infantile rickets, scurvy usually makes its clinical appearance in infants between 6 and 18 months of age. Adult cases caused by the same dietary deficiency, however, have in the past been common among sailors and others whose diets have been abnormally restricted, and sporadic cases of scurvy in adults are still seen.

The scorbutic infant, who often has rickets also, is poorly nourished and irritable and experiences extreme pain on motion of the joints and on slightest pressure over the affected bones. In well-marked cases, subperiosteal hemorrhage (Fig. 3-19) causes a palpable thickening of the bone. This thickening may be visible in the roentgenograms but often does not appear until calcification has taken place after dietetic treatment. It must be differentiated from similar roentgenographic changes seen in hypervitaminosis A, Caffey's disease, and syphilis. One of the primary defects in scurvy is in the production of normal bone matrix. At the epiphyseal plate, cartilage development proceeds normally and calcification of cartilage matrix takes place. This zone of provisional calcification is broadened and appears in the roentgenogram as a distinct white line. On the metaphyseal side of this zone, failure of ossification results in an area of diminished bone density through which separation of the epiphysis may occur. Examination of the blood may show the absence of ascorbic acid. In adult cases, and in infantile scurvy when the teeth have already appeared, the gums are likely to be swollen, spongy, and hemorrhagic.

Treatment and prognosis. The treatment of scurvy is primarily of pediatric nature. It consists chiefly of the administration of ascorbic acid and foods such as orange juice that contain relatively large quantities of vitamin C. Clinical improvement is usually rapid, and the prognosis for complete cure is excellent. The ortho-

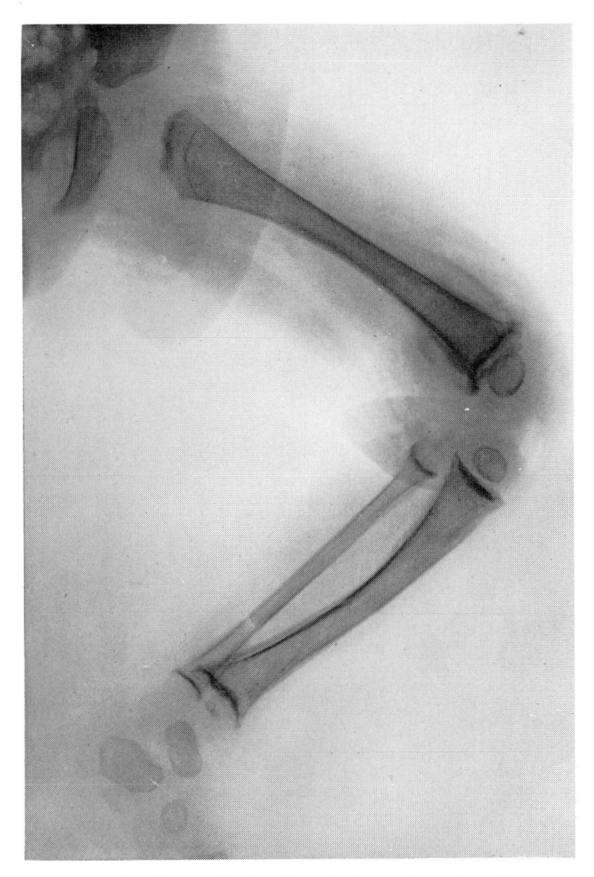

Fig. 3-19. Roentgenogram of a lower limb of an infant, showing changes due to scurvy. Note generalized osteoporosis, calcification in subperiosteal hemorrhages about the femur and tibia, and increased calcification on the metaphyseal side of the epiphyseal lines.

paedic treatment of scurvy is largely of symptomatic type. During the acute stage, recumbency is indicated for protection of the affected bones, and occasionally simple splints are useful to relieve pain and minimize further subperiosteal hemorrhage.

Affections associated with endocrine abnormalities

HYPOPITUITARY DWARFISM

Certain rare congenital anomalies of the anterior part of the pituitary gland or its destruction in early childhood by tumor, histiocytosis, or chronic infection may reduce the production of growth hormone and result in pituitary dwarfism.

Clinical picture. Growth retardation in the congenital case is not manifest at birth but becomes obvious between the ages of 2 and 4 years. Irrespective of age, the proportions between head, trunk, and extremities differ little from those of a normal child (Fig. 3-20). Most pituitary dwarfs do not undergo normal pubertal development. Their intelligence is usually normal.

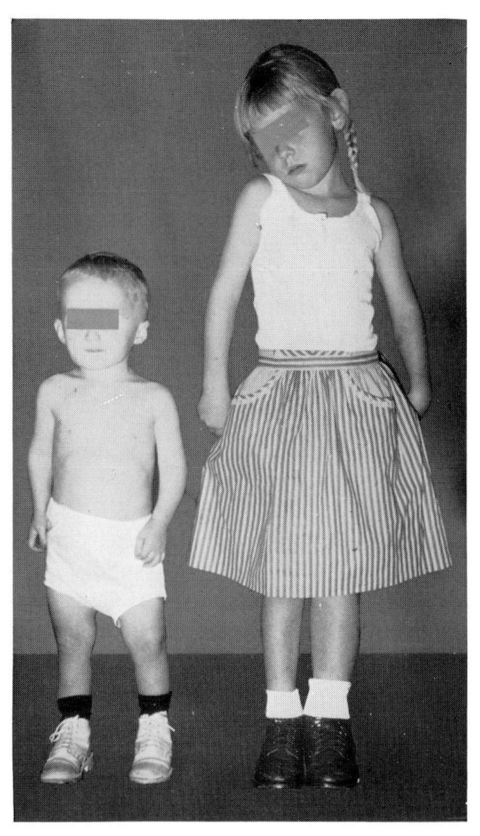

Fig. 3-20. Hypopituitary dwarfism in a 4-year-old boy, whose height is 25 inches. The girl is also 4 years old and has the normal height of 39 inches. The dwarf has a normal facies, as well as head, trunk, and limbs of approximately normal proportions.

Roentgenographic examination shows the fontanels open beyond the second year, osteoporosis of the skull and long bones, and delay in the appearance of ossification centers and in the closure of epiphyses.

Prognosis. The prognosis depends upon the etiologic factor.

Treatment. The advice of an endocrinologist should be sought. Striking improvement has been observed after daily administration of pituitary growth hormone of human origin. Unfortunately the availability of the hormone is quite limited. If there is associated hypogonadism, treatment with anabolic steroids may be instituted at the age when puberty would be expected. Earlier treatment with these hormones may result in maturation and premature closure of the epiphyses.

HYPERPITUITARISM

Gigantism

In pituitary gigantism skeletal overgrowth results from a pathologically increased activity of the anterior portion of the pituitary gland during youth. The overactivity is associated with acidophil cell hyperplasia or adenoma, and it results in an excessive

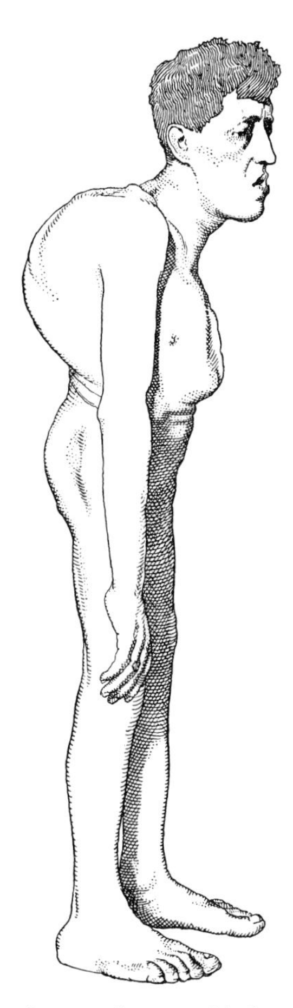

Fig. 3-21. Acromegaly. Note prominence of supraorbital areas, protuberance of jaw, kyphosis, and enlargement of hands and feet. (From clinical material of Dr. W. E. Dandy.)

production of growth hormone. It is thus of orthopaedic interest in demonstrating clearly a close relationship between skeletal development and endocrine function. Subjective symptoms are slight or absent. Treatment of these cases belongs to the endocrinologist and the neurologic surgeon rather than to the orthopaedist.

Acromegaly

The bone changes of acromegaly represent the response of the adult skeleton to excessive production of growth hormone by the anterior lobe of the pituitary gland. Acromegaly in the adult thus corresponds in etiology to *gigantism* in the youth. The usual cause of the pituitary changes in acromegaly is an acidophil adenoma. Gradual bony thickening results from the stimulation of periosteal intramembranous growth.

Clinical picture. The characteristic changes of acromegaly usually appear first in early or middle adult life and are slowly progressive. They include a gradual enlargement of the bones of the skull, jaw, spine, hands, and feet, and associated growth of cartilage and fibrous tissue. The lower jaw protrudes. The orbital and zygomatic arches are prominent, and the nose, ears, lower lip, and tongue are large. Kyphosis of the thoracic spine is commonly present, the hands and feet become massive, and the fingers and toes may be strikingly thickened (Fig. 3-21). The costochondral junctions become enlarged, forming the acromegalic rosary, and with advancing ossification the ribs lengthen and the chest thickens. Most laboratory data are within normal limits. The basal metabolic rate is often increased.

Diagnosis and treatment. The typical deformities are often associated with other evidences of endocrine disorder and with progressive symptoms, such as disturbances of vision, which are secondary local results of the intracranial tumor. These associated symptoms are of diagnostic value, as are roentgenograms showing enlargement of the sella turcica. Myxedema and osteitis deformans are to be excluded. Treatment by surgery, irradiation, or a combination of the two is the prerogative of the neurologic surgeon. The administration of thyroid and sex hormones may be followed by improvement.

CRETINISM (HYPOTHYROIDISM)

Cretinism is a form of dwarfism caused by hypofunction of the thyroid gland. The condition first manifests itself usually in the second six months of life. It may be caused by a congenital lack of thyroid tissue or by inability of the gland to synthesize thyroid hormone. Endemic cases occur frequently in goitrous regions and may be ascribed to a lack of iodine in the mother. In such instances the fetus may develop as a cretin in utero.

The cretin must be carefully differentiated from the rachitic, achondroplastic (Fig. 3-2), and pituitary (Fig. 3-20) dwarfs. Hurler's syndrome (Fig. 3-11) and Down's syndrome should also be excluded. The cretin has a large tongue that often protrudes between his thickened lips, a flattened nose with sunken bridge, puffy eyelids, and a dry skin. The facial expression is characteristically stupid, and usually the mental development is obviously impaired. Ossification of the epiphyses is irregular and delayed. Endochondral bone formation appear to be at fault. Roentgenographic changes of epiphyseal dysgenesis are particularly striking in the hip joints; multiple irregularly shaped areas of calcification give the capital femoral epiphyses a stippled appearance somewhat like that of coxa plana. Although the epiphyseal cartilages may remain ununited for 20 or 30 years, growth of the long bones ceases, and the arms and legs remain short and stubby. The roentgenograms may show transverse bony shadows at the ends of the diaphyses. Of prime diagnostic importance is the low concentration of protein-bound iodine in the serum.

Most of the ill effects of cretinism can be prevented by early recognition and vigorous, continued treatment with thyroid preparations.

HYPERPARATHYROIDISM (GENERALIZED OSTEITIS FIBROSA CYSTICA)

By secreting excessive amounts of parathyroid hormone, adenomatous or hyperplastic parathyroid glands may lead to demineralization of the skeleton, localized areas of bone destruction, characteristic changes in serum calcium and phosphorus levels, nephrolithiasis, and renal insufficiency. The two chief actions of the hormone are raising the level of serum calcium and increasing the renal excretion of phosphate. The mechanism of parathyroid hormone action on bone and kidney is still not completely understood.

Blood chemistry. The serum calcium is consistently increased, to a level frequently exceeding 12 mg.%, and the serum phosphorus is usually decreased. In the presence of renal damage, however, the serum phosphorus may be elevated. The alkaline phosphatase level is high and always rises further as the osseous changes become more severe; this is thought to be the result of increased osteoblastic activity. The excretion of calcium in the urine is also increased.

Pathology. The essential features are a gradual resorption of bone trabeculae and a replacement of bone and marrow by fibrous tissue. Macroscopically this tissue presents a picture of disorderly arrangement with numerous areas of hemorrhage. Hemorrhagic cysts of irregular size and distribution often occur as the result of local degenerative changes; such cysts may have fibrous walls and may contain a grumous, serosanguineous fluid. Bone destruction is manifested microscopically by thin and disappearing lamellae, about which may be grouped numerous osteoclasts; in some places the appearance may suggest giant cell tumor. Other areas may show new bone formation with abundant osteoid and osteoblasts. The gradual loss of cortical bone results in decreased strength, which may lead to curvature, compression, or spontaneous fracture.

Secondary changes in the kidneys are sometimes great. The increased urinary excretion of calcium and phosphorus not infrequently leads to the formation of renal calculi. Chronic nephritis and renal insufficiency may result from precipitation of calcium phosphate within the parenchyma of the kidney.

Clinical picture. Primary hyperparathyroidism is uncommon, may occur at any age, and is seen mostly in the middle years of life. The symptoms are extremely diverse, depending upon the site of major involvement in the individual case, and a number of more or less well-defined clinical types have been described. General lassitude and muscular hypotonia directly due to the hypercalcemia may be the presenting symptoms. Gastrointestinal complaints and urinary disturbances, such as polyuria, are common. Bone disease severe enough to produce symptoms is reported to occur in only 10% to 25% of patients. Skeletal involvement may be evidenced first by deep-seated pain and tenderness, the local swelling of an area of bone expansion, or the disability of a spontaneous fracture. Cases developing in late middle life and ushered in by backache, progressive kyphosis, and decrease in stature form a major group. Deformity of bones other than the vertebrae is usually a late manifestation. In some cases of hyperparathyroidism, particularly those in which renal calculi cause the presenting complaint, skeletal symptoms and bone changes in the roentgenograms may be entirely absent.

Roentgenographic picture. The loss of bone causes a generalized decrease in roentgenographic density, or osteoporosis, which may be complicated by the secondary changes of deformity, local areas of rarefaction, and fractures. In addition to the generalized radiolucency, roentgenograms of the spine may show cupping, wedging, and crushing of the vertebral bodies, as well as various degrees of kyphosis with a pigeon-breast deformation of the chest. The skull may have a moth-eaten appearance. Loss of the lamina dura surrounding the roots of the teeth and resorption of bone at the tufts of the distal phalanges are helpful in early roentgen diagnosis.

Differential diagnosis. The varied clinical picture leads to simulation of numerous other diseases. Skeletal types of hyperparathyroidism must be differentiated from senile osteoporosis, osteomalacia, osteitis deformans, solitary bone cyst, giant cell tumor, fragilitas ossium, fibrous dysplasia, cystic angioma of bone, myasthenia gravis, multiple myeloma, and metastatic malignancy. The biochemical and roentgenographic examinations are usually of chief differential value. Primary hyperparathyroidism must be differentiated from hyperparathyroidism secondary to chronic renal disease.

Treatment. The treatment is parathyroidectomy. After excision of the parathyroid adenoma, great improvement, both in symptoms and in general nutrition, is to be expected; usually regression of the bone changes also occurs. Orthopaedic treatment consists of appropriate measures, such as braces and splints, for the prevention of deformity. Support is especially indicated when the spine is involved. In some cases, after parathyroidectomy has been performed and bony strength has been restored, corrective operations, such as osteotomy, may be indicated.

Affections of unknown etiology

INFANTILE CORTICAL HYPEROSTOSIS (CAFFEY'S DISEASE)

Caffey's disease is a self-limited affection characterized mainly by subperiosteal new bone formation on the shafts of the long bones and mandible in infants under 6 months of age.

Etiology. The cause is unknown. Because of associated fever, leukocytosis, and increased sedimentation rate many observers believe it to be an infection, possibly a virus infection. All cultures, however, have been negative.

Clinical picture. Boys are affected more often than girls. The earliest signs are irritability and swelling of the face, over a clavicle, or in a limb, in whole or in part. The swellings are sudden in onset, tender, and of wooden induration; they appear to be located beneath the subcutaneous fat in the muscular planes. They begin to subside by the time roentgenographic changes appear. The first roentgenographic signs are subperiosteal shadows over the whole or part of the bone; they increase in size and may give the appearance of an enlarged and thickened overall bone structure (Fig. 3-22). The epiphyses, however, are not involved. With healing the proliferative bone disappears, and normal bone structure is usually restored in from a few months to 2 years, the average period being 8 months.

Diagnosis. Infantile cortical hyperostosis must be differentiated from traumatic lesions incurred accidentally or inflicted intentionally (the *battered child syndrome*), from scurvy, hypervitaminosis A, osteomyelitis, syphilis, and neoplasm. Hypervitaminosis A, with which it is most often confused, never occurs in the infant under 1 year of age.

Treatment. No specific treatment is indicated. The prophylactic use of antibiotics is justifiable until the possibility of an infection has been ruled out. Surgical treatment is contraindicated.

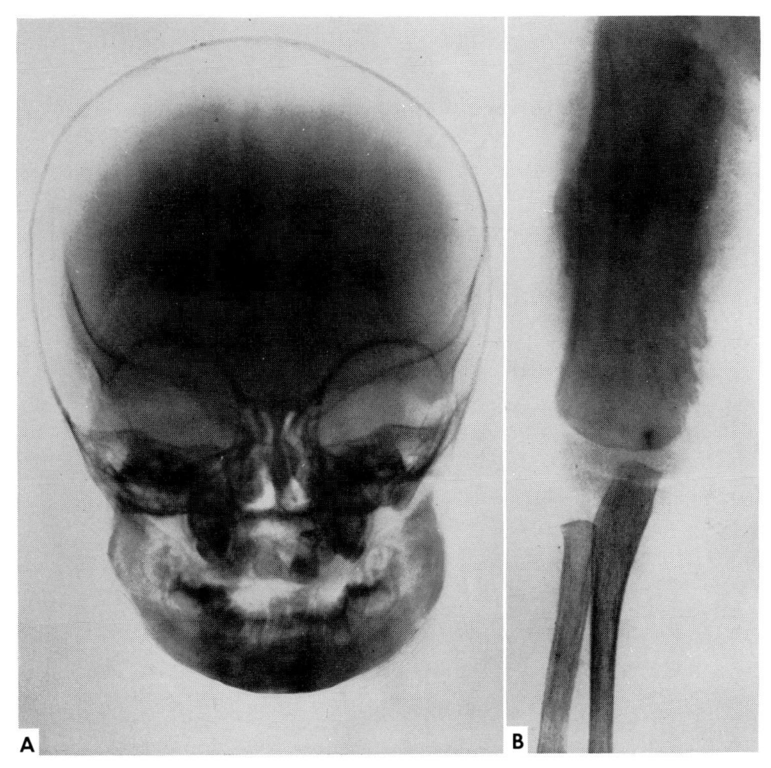

Fig. 3-22. Infantile cortical hyperostosis. **A,** Roentgenogram of the skull of a 7-month-old infant, showing massive subperiosteal overgrowth of the mandible. **B,** Roentgenogram of the upper limb of a 3-month-old infant, showing massive new bone about the humeral shaft. In this infant the mandible, clavicles, and several ribs showed similar but less severe involvement.

HISTIOCYTOSIS (RETICULOENDOTHELIOSIS)

Histiocytic proliferation of unknown cause is common to the three affections to be described in this group: eosinophilic granuloma, Hand-Schüller-Christian disease, and Letterer-Siwe disease. These diseases are believed to be closely related to one another and possibly variants or phases of the same disorder.

Eosinophilic granuloma

A benign inflammatory lesion of bone, eosinophilic granuloma occurs almost exclusively in children and young adults and is much more common in males than in females. It involves both long and flat bones and is usually solitary. The lesion is often discovered accidentally by roentgenographic examination or after pathologic fracture. However, its presence may be heralded by local swelling, tenderness, redness, heat, and pain, with some limitation of function.

The usual roentgenographic finding is a circumscribed area of bone destruction from 1 to 4 cm. in diameter with a sharply punched-out appearance (Fig. 3-23). An involved vertebral body may lose height, becoming greatly thinned; eosinophilic

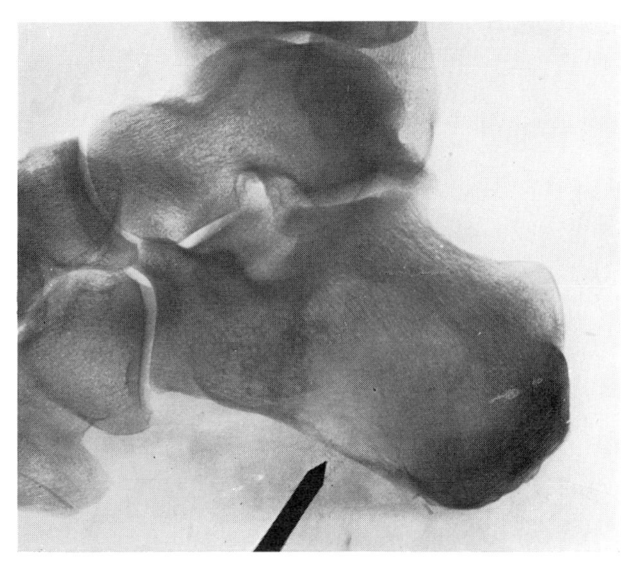

Fig. 3-23. Roentgenogram of eosinophilic granuloma of calcaneus. Note large circumscribed area of radiolucency.

granuloma is one of the causes of vertebra plana (p. 305). On gross pathologic examination the early lesion may have the appearance of soft hemorrhagic granulation tissue; on microscopic examination there are found numbers of large mononuclear histiocytes, which are phagocytic and contain droplets of neutral fat or disintegrated cellular material, and collections of eosinophils from which the disease derives its name. Multinuclear giant cells are often present. Eosinophilia in the peripheral blood is unusual. In the later stages the eosinophils disappear and the histiocytes become foamy. The pathologic tissue is eventually replaced by bone.

Differential diagnosis. Eosinophilic granuloma may be confused with osteoid osteoma, osteomyelitis, Ewing's sarcoma, osteogenic sarcoma, multiple myeloma, bone cyst, and giant cell tumor.

Treatment and prognosis. Without treatment the lesions may undergo spontaneous healing in from a few months to a year. Excision or curettage and packing of the cavity with small bone grafts, when indicated, usually result in complete cure. Roentgentherapy also may be curative but should be considered only for situations in which surgical treatment is not feasible. Occasionally the solitary lesion of eosinophilic granuloma may progress to multiple lesions and a more serious form of histiocytosis.

Hand-Schüller-Christian disease

Hand-Schüller-Christian disease is a rare disorder usually attended by multiple bone defects that render its differential diagnosis from other bone diseases important. The bone lesions contain large numbers of histiocytes of foamlike appearance, which are filled with lipids, predominantly cholesterol and its esters. These lesions are most common in the skull but occasionally are found in other bones, giving rise to characteristic defects visible on roentgenographic examination (Fig. 3-24). They

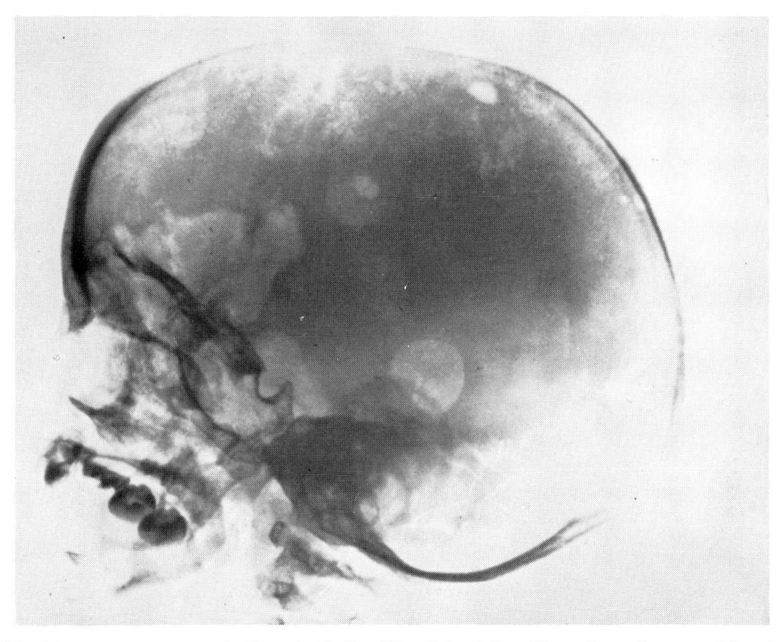

Fig. 3-24. Roentgenogram of the skull in Hand-Schüller-Christian disease. The irregular, punched-out areas contain histiocytes infiltrated with lipids.

may occur also in the skin, tendons or tendon sheaths, and viscera. In well-marked cases these changes may be quite extensive.

The disease usually makes its clinical appearance in childhood. Commonly associated factors, dependent upon the localization of the lesions, are diabetes insipidus, gingivitis, and exophthalmos. Dwarfism and infantilism are sometimes present, and the patient may be jaundiced. Hypercholesterolemia, hyperlipemia, and eosinophilia are not characteristic findings.

Treatment and prognosis. Symptomatic improvement may be induced by roentgentherapy. In several recent cases vinblastine sulfate has seemed to be effective. In spite of treatment, cases with extensive involvement of the bones are often fatal.

Letterer-Siwe disease

Letterer-Siwe disease is a rare disorder of infancy with pathologic changes similar to those of the Hand-Schüller-Christian syndrome, of which it may be an acute form as suggested by its rapid, progressive, and usually fatal course.

The spleen, liver, and lymph nodes are enlarged by proliferation of the reticulum and its derivatives, and there may be anemia, leukopenia, and thrombocytopenia. Skin manifestations occur early in the disease and include petechial hemorrhages that may coalesce and ulcerate.

Roentgenographic findings, especially in the skull, consist of multiple rarefied areas like those of Hand-Schüller-Christian disease. Diagnosis is made by biopsy of bone marrow, skin lesion, or lymph node. Treatment is usually ineffective.

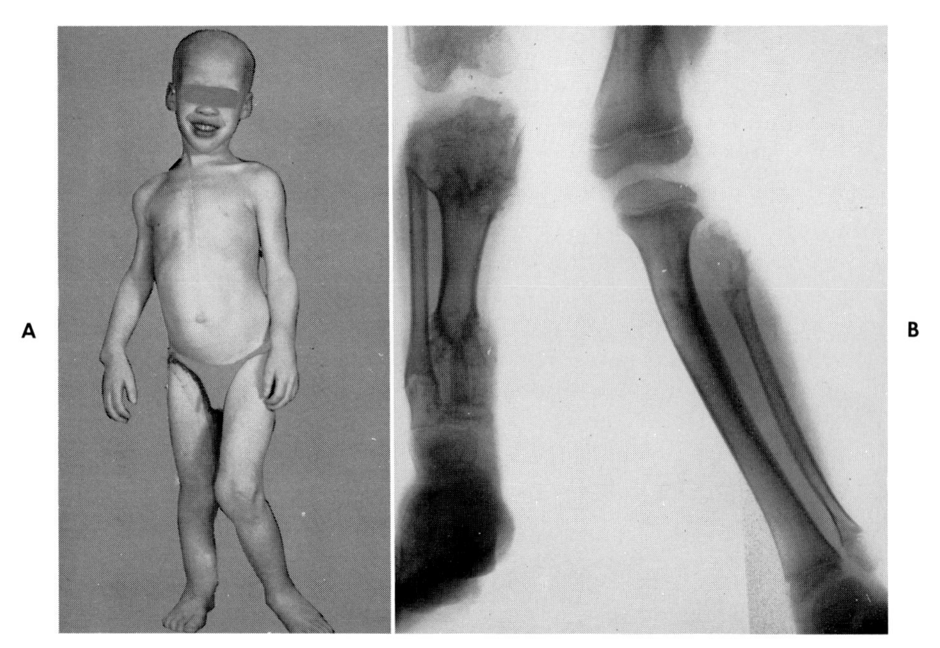

Fig. 3-25. Enchondromatosis (Ollier's disease) in a boy 6 years of age. Involvement is more severe on the right side of the body and is confined mainly to the metaphyseal areas. **A,** Photograph showing multiple deformities including shortening of right lower limb and valgus of left knee. **B,** Roentgenogram of legs showing changes in the long bones.

ENCHONDROMATOSIS (OLLIER'S DISEASE, DYSCHONDROPLASIA)

Enchondromatosis (Fig. 3-25) is an uncommon affection of the skeleton characterized by disorderly and excessive proliferation of cartilage cells at many of the epiphyseal plates. At birth it may be present but not apparent on clinical examination. No genetic basis for enchondromatosis has been demonstrated. The disease has been confused in the literature with multiple exostoses but differs from it in many respects. An obvious difference is that in enchondromatosis the abnormal cartilage masses are within the metaphysis, whereas the lesions of multiple exostoses project from the surface of the metaphysis.

Pathology. Irregularly shaped masses of cartilage, whose cells vary from young forms to mature chondrocytes, are found in the metaphyseal region, presumably remaining in this location because of failure of resorption during the normal process of enchondral ossification. Areas of myxomatous degeneration are frequent, and calcification of portions of the cartilaginous masses may occur.

Clinical picture. The lesions, appearing in early childhood, tend to involve predominantly the bones of one side of the body and leave the opposite side only mildly affected. Thus a common presenting feature is shortening of one leg and an associated limp. Angular deformities at the knees and elbows are frequent. Involvement of the small long bones of the hands and feet may result in enlargement and severe distortion.

Roentgenographic picture. Elongated radiolucent areas extending from the epiphyseal line into the metaphysis and directed in the long axis of the bone are commonly found in the femur, tibia, humerus, and radius. In the short bones of the hands and feet, the radiolucent areas of cartilage may involve the entire bone, and the cortex may be greatly expanded.

Treatment and prognosis. There is no specific treatment for enchondromatosis. The degree of involvement varies from mild to severe. In most extensively involved bones, shortening may be considerable and may be treated surgically or compensated for by an appropriate orthopaedic appliance. On rare occasions angulation at the knee and elbow may be severe enough to require osteotomy. Malignant degeneration of the abnormal cartilage may occur.

FIBROUS DYSPLASIA

Fibrous dysplasia is a chronic affection characterized by the presence of one or many skeletal lesions in which bone is replaced by an abnormal proliferation of fibro-osseous tissue. Its etiology is unknown; it is ordinarily considered to be a noninherited dysplasia of developmental origin.

Clinically fibrous dysplasia is seen in *polyostotic* form, wherein more than one bone is affected, and in *monostotic* form, in which the lesion involves only one bone. The monostotic type may be simply a mild form of the disease, but it has not been shown to progress to the polyostotic type, and the exact relationship between the two remains obscure.

Polyostotic fibrous dysplasia

Polyostotic fibrous dysplasia is somewhat more common in girls than in boys. It is seldom seen in infants but develops usually during the later years of the growth period. The course of the lesions is variable; whereas severely involved cases may progress rapidly in childhood, in many instances the lesions enlarge slowly or not at all after growth has ceased. Unilateral distribution is a common but by no means constant finding. Most frequently involved are the long bones, especially the femur, and the pelvis, ribs, and skull.

The presenting complaint is usually increasing deformity, a limp, pain in the affected area, or disability associated with a pathologic fracture.

Albright's syndrome, seen almost exclusively in girls, is the relatively rare combination of polyostotic fibrous dysplasia, irregularly edged café au lait pigmentation of the skin, and an endocrine disorder evidenced by sexual precocity.

Pathology. The lesions, expanding slowly in cancellous bone, may thin and bulge the overlying cortex but ordinarily do not perforate it because periosteal bone formation continues. Grossly the tissue of the early lesions is gray and of soft, gritty consistency; when more mature, it is yellow and firm. Microscopic sections show a fibroblastic matrix embedding scattered spicules of immature fiber bone. Osteoid may be present, as well as occasional areas of cartilage, and degenerative changes may be superimposed.

The blood chemistry findings are normal except after a major fracture or in the presence of extensive skeletal involvement, when the serum alkaline phosphatase is often elevated.

Roentgenographic picture. The affected bones show radiolucent areas of ground-glass appearance, thin cortices, deformed contours, and fractures. The fractures usually heal well with the exception of those of the upper third of the femur, which cause some of the most crippling features of the disease. After multiple fractures in the upper third, the femoral deformation may become quite severe (Fig. 3-26) and

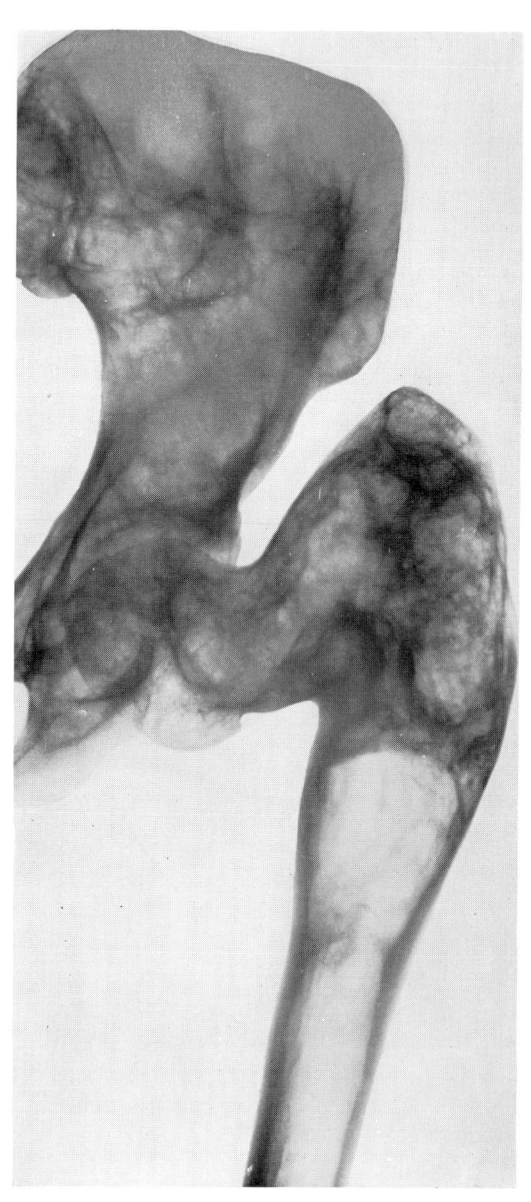

Fig. 3-26. Roentgenogram of polyostotic fibrous dysplasia in a woman 33 years of age. Note the extensive radiolucent areas in femur and pelvis and shepherd's-crook deformity of the upper end of the femur. Almost all bones of the body were involved.

has been called a shepherd's crook deformity. The unilateral bone lesions tend to be segmental, often leaving normal bone adjacent to discrete radiolucent areas.

Treatment. The fractures and deformities often need orthopaedic treatment, especially the bowing and pseudarthrosis of the upper femur. The latter may require osteotomy and grafting. For the underlying disease no specific treatment is available.

Monostotic fibrous dysplasia

In about half of the reported cases of monostotic fibrous dysplasia the lesion has involved a rib, possibly because of the frequency of routine chest roentgenograms. The lesion has been found in many other bones, long and flat, membranous and enchondral, particularly the upper femur, tibia, and mandible. Notably, the bones of the hands and feet seem to escape.

Monostotic fibrous dysplasia may be asymptomatic or may present itself by local swelling, pain, tenderness, or pathologic fracture. The lesion may remain static or slowly enlarge. Roentgenograms show a localized area of rarefaction (Fig. 3-27), which is not pathognomonic. The gross and microscopic appearance of the pathologic tissue is similar to that of the lesions in polyostotic fibrous dysplasia.

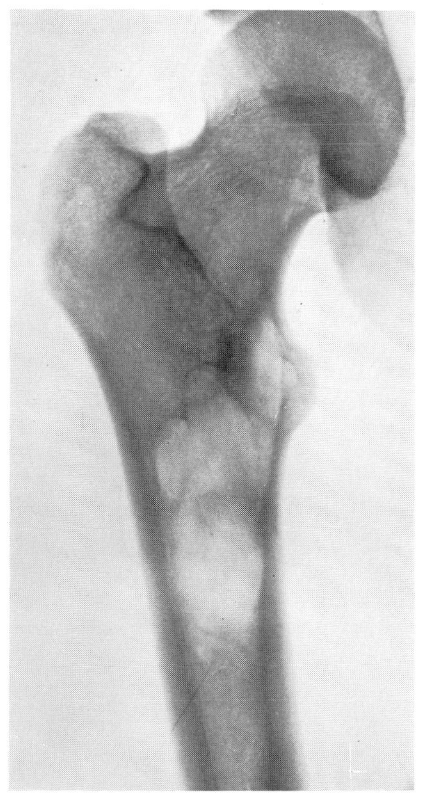

Fig. 3-27. Roentgenogram of monostotic fibrous dysplasia involving the femur. Note multilocular radiolucency extending from above the lesser trochanter down into the shaft.

Treatment. Biopsy is indicated when the diagnosis is in doubt. At the time of biopsy pathologic tissue may be removed by thorough curettement, and the resulting cavity may be packed with small grafts of autogenous bone.

MELORHEOSTOSIS

As the name melorheostosis implies, this rare disease is characterized by a flowing or linear, longitudinal hyperostosis of limb bones. Typical roentgenograms show osteosclerotic streaks and ridges along one side of a long bone; such changes have been likened to melted wax flowing down the side of a burning candle (Fig. 3-28). Although any bones may be involved, the changes are commonly limited to a single limb. Histologically the affected areas show greatly increased bone density. They tend to progress—rapidly in children and slowly in adults. Usually joints also are affected, their mobility being restricted by intra-articular bony changes, small periarticular masses of ectopic bone, or fibrotic contractures.

Melorheostosis is thought to be a congenital abnormality of mesodermal derivatives arising in early embryonic life. No hereditary influence has been found.

In children, contractures may antedate the roentgenographic changes in bone. In later cases, inconstant aching pain and stiffness of affected joints are the major

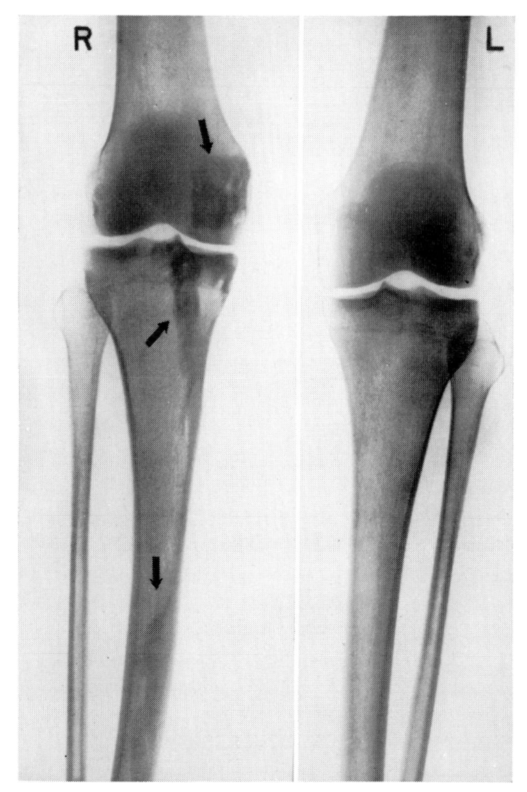

Fig. 3-28. Roentgenogram of melorheostosis in an adult. Note well-demarcated, irregularly shaped areas of increased density in the right femur and tibia.

complaints. Clinical laboratory findings are within normal limits. The treatment includes surgical measures to minimize and correct deformity.

HYPERTROPHIC OSTEOARTHROPATHY

Hypertrophic osteoarthropathy, called also *pulmonary osteoarthropathy of Marie,* is a syndrome that includes clubbing of the fingers, painful swelling and periosteal new bone deposition in the extremities, and in advanced cases arthritic symptoms and signs in peripheral joints. Hypertrophic osteoarthropathy is almost always secondary to chronic visceral disease, usually in the thorax. The most common causes are lung tumors or infections, bronchiectasis, and congenital heart disorders. The mechanism by which these diseases produce the bone and joint changes is not understood but is probably of either circulatory or toxic nature. The characteristic pathologic process is a thickening of both soft tissues and bone; the bony proliferation is brought about by a very chronic ossifying periostitis.

Clinical picture. The affection is more common in men than in women. The first clinical sign is a generalized, symmetric, and painless enlargement of the distal portion of the fingers, due to thickening of the soft tissues (Fig. 3-29). The nails become slightly cyanotic. They may show thickening, ridges, and an increased convexity and may bend down over the ends of the phalanges. They have been called *hippocratic fingers.* Analogous changes may appear in the toes and toenails simultaneously. The affected digits may be painful and tender. Wrists and ankles may become swollen, and the joints may contain increased fluid. The roentgenograms show proliferative bony changes, sometimes with spurring, and in severe cases the shafts of the bones of forearm and leg may be obviously thickened from subperiosteal formation of new bone.

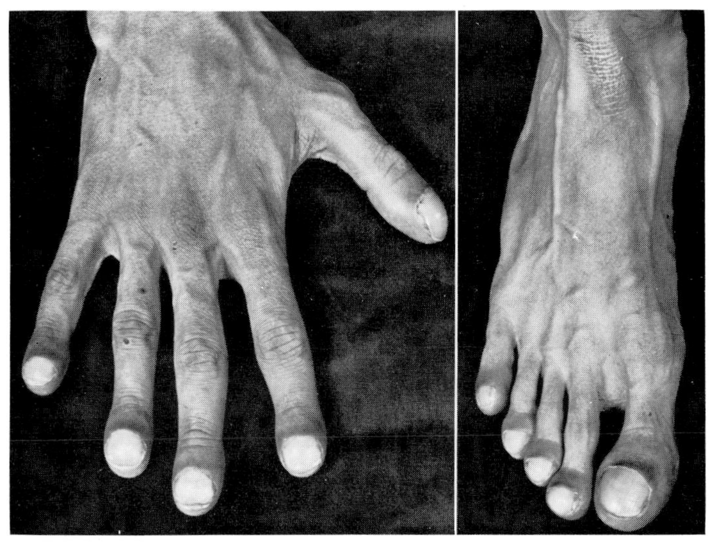

Fig. 3-29. Hypertrophic osteoarthropathy. Note the clubbed appearance of the ends of fingers and toes and the broadening of the nails.

Treatment. The primary visceral disease must be investigated and treated. If it is cured promptly, the symptoms may subside quickly and the hypertrophic changes regress.

OSTEITIS DEFORMANS (PAGET'S DISEASE)

Osteitis deformans was first described by Sir James Paget in 1877. It is a chronic skeletal disease of middle and late life, beginning insidiously and characterized by progressive structural changes and typical deformities occurring in the long bones, spine, pelvis, and cranium. The bone changes consist of accelerated resorption and excessive, abnormal regeneration; among individual cases they vary widely in distribution and severity.

Etiology. The causative agent is unknown. Attempts to establish heredity, infection, or an endocrine defect as the etiologic factor have failed through lack of proof.

Pathology. The initial bone lesion is considered to be resorptive (osteolytic) and to be followed by bone proliferation (osteosclerotic). In active stages the blood flow to bone is greatly increased; when the disease is widespread, this may even lead to high-output cardiac failure. The outstanding gross skeletal changes are a gradually developing thickening and bowing of the shafts of the long bones, particularly the tibiae and the femurs, and a generalized thickening of the cranium. Frequently, osteitis deformans affects only a single bone. In the early stages the bones lose much of their ability to withstand normal stresses, and an increase in their degree of curvature results. In the spine, which is very frequently involved, there is often collapse of one or more of the vertebral bodies, which results in a kyphosis. In later stages more bone matrix is laid down, which calcifies, often producing finally a thick, hard osseous structure. The surface of the involved long bones is of characteristic unevenness and is furrowed by the periosteal vessels. Section discloses a thickened cortical layer that has lost its dense character and sharp outline and encroaches upon the marrow cavity. The microscopic picture of active Paget's disease is characterized by intense osteoblastic and osteoclastic activity, often occurring in the same area at the same time. Bone is constantly being removed and replaced in a disorderly manner. The irregular, wavy cement lines that separate areas of bone formed at different periods result in the characteristic mosaic pattern of Paget's bone. Vascular fibrous tissue often fills the intervals between bone trabeculae and the marrow spaces of affected bones.

In osteitis deformans the serum alkaline phosphatase may reach levels higher than those in any other disease. It is especially increased in active stages of the disease and in patients with multiple bone involvement, and it usually continues elevated as long as the disease remains active. The serum calcium and phosphorus are usually within normal limits.

Clinical picture. In many instances bones affected by Paget's disease remain asymptomatic throughout the individual's life. Estimates including such cases place the overall incidence of the disease at approximately 3% of the adult population. According to several statistical studies, osteitis deformans affects men more frequently than women, and occasionally it occurs in families.

Most patients are between 40 and 60 years of age when the symptoms begin.

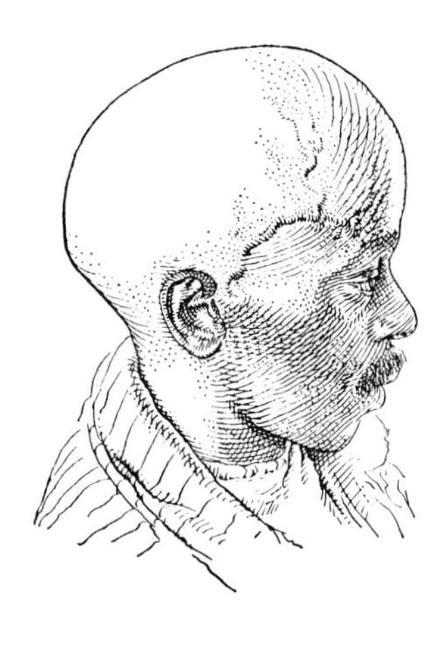
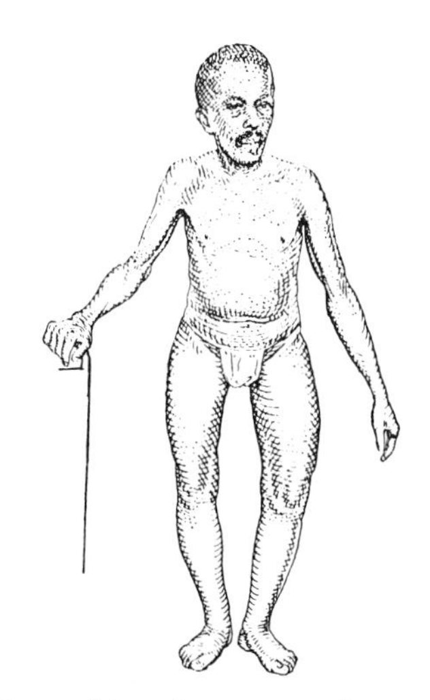

Fig. 3-30. Osteitis deformans. Note massive cranium with enlargement of superficial veins, bowing of the lower limbs, and decreased height. (From clinical material of Dr. G. W. Wagoner.)

The patient may develop severe, intractable pain. He may complain of fatigue or aching in the legs simulating the discomfort of chronic arthritis, may suffer chiefly from headache and backache, may be affected by stiffness or clumsiness; or may simply have noticed that he is becoming bowlegged, that his back is stooped, or that his head is becoming larger, as evidenced by the tight fit of his hat.

The patient suffering from osteitis deformans in its moderately advanced stage presents a characteristic clinical appearance (Fig. 3-30). The head appears massive and too large for the body, whereas the face is relatively small. The head may droop forward, and a long dorsal kyphosis may be present. The chest is likely to be barrel shaped, the lumbar spine flexed, and the legs bowed outward and forward. This posture gives to the arms an exaggerated length, which has been compared to that of an anthropoid ape. There are no mental changes other than those of advancing age. As the disease progresses, there may develop a deafness and tinnitus and an impairment of vision from bony impingement on the cranial nerves as they pass through their foramina. Clinical diagnosis in well-established cases presents little difficulty.

Roentgenographic picture. Roentgenograms are of diagnostic value, as the changes of osteitis deformans are characteristic. The contour of an affected long bone is often altered by an increase of its normal curvature. The cortex is enlarged sometimes to as much as four or five times its normal thickness and presents a blurred, fluffy appearance; the medulla is narrowed, and there is a loss of the normal definitive line between cortex and medulla. The pelvic bones often show an abnormally prominent and coarse trabecular pattern (Fig. 3-31). In cases of suspected osteitis deformans, roentgenograms of the skull should be made since its appearance is often pathognomonic (Fig. 3-32). The bones of the cranium may be greatly thickened, with obliteration of the sutures and vascular channels, indistinct outlines, and uneven density.

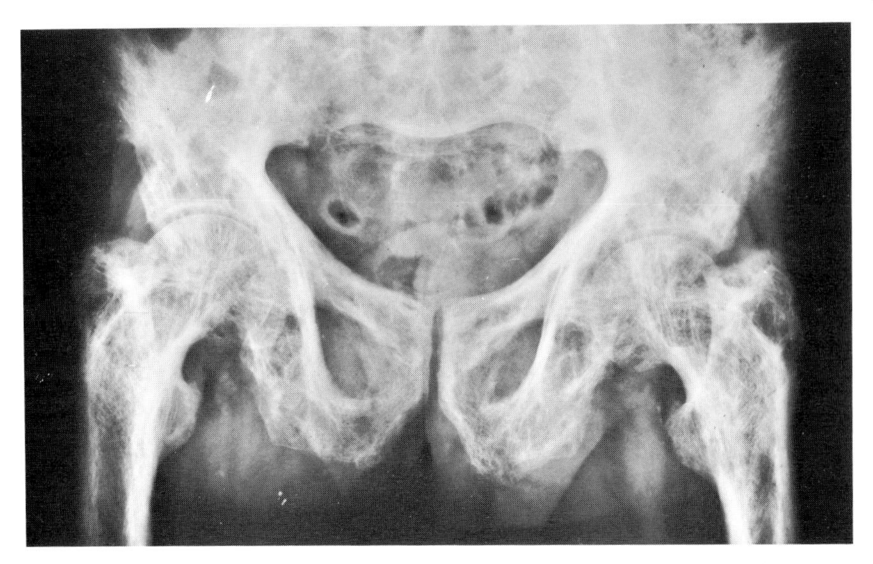

Fig. 3-31. Roentgenogram of the pelvis in advanced osteitis deformans. Note the coarse trabeculation and widespread changes in bone density. The patient, a 70-year-old man, had Paget's disease of many years' duration.

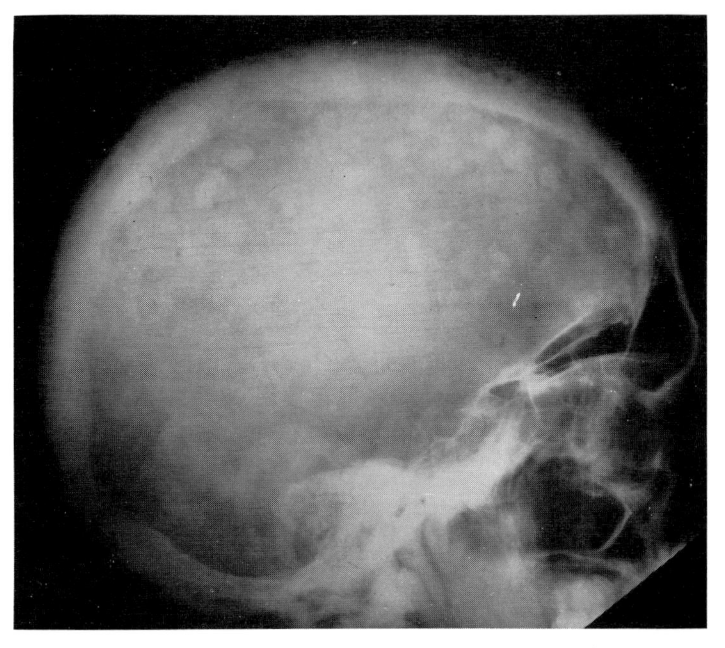

Fig. 3-32. Lateral roentgenogram of skull in osteitis deformans, showing great thickening of calvarium, uneven bone density, and indistinct peripheral margin.

Diagnosis. The characteristic deformities and roentgenographic changes of osteitis deformans are usually adequate to ensure its recognition. A high serum alkaline phosphatase level supports the diagnosis. In the differential diagnosis, hyperparathyroidism, metastatic carcinoma, syphilis, and osteomalacia must be considered. Diagnosis of the localized form of osteitis deformans, in which only a single bone is involved, sometimes presents difficulty. In such cases biopsy is occasionally desirable in order to exclude primary or metastatic tumor or chronic osteomyelitis.

Treatment and prognosis. At present the treatment of osteitis deformans is of symptomatic type. Salicylates are useful for the relief of pain. Early reports suggest that thyrocalcitonin may prove helpful by reducing bone resorption, as may mithramycin, an antibiotic thought to exert a toxic effect on osteoclasts; currently, however, the efficacy and safety of these medications are unproved. In the more active cases braces may be indicated for the prevention of deformity, and rarely a deformity that has become established may be corrected by osteotomy.

The prognosis for length of life is good. In advanced, generalized cases cardiovascular complications are the commonest cause of death. Pathologic fracture is the commonest complication. It occurs most often in the femur and is likely to be slow in uniting. Another serious complication, occurring in from 1% to 15% of the patients in reported series of cases, is the development of malignancy. The commonest type is osteogenic sarcoma. Most of the patients who develop this die within 2 years of diagnosis. Constriction of the spinal cord by the thickened vertebrae is a rare complication.

SENILE OSTEOPOROSIS

When osteoporosis, or generalized decrease in bone mass, occurs in older patients, it is termed *senile osteoporosis*. This is the commonest metabolic affection of bone. It is seen more than twice as often in women as in men and is frequently manifested as an impairment of the bony strength of the spine. It develops when bone that is being lost by normal catabolism is not replaced in equal measure by new bone formation, and it has been thought to result from a decrease in osteoblastic activity. Mild prolonged negative calcium balance resulting from inadequate dietary intake of calcium, a situation common in older people, may be an important factor in the cause of osteoporosis. The constantly negative calcium balance over a period of several years removes calcium from the serum, and this must be continually replaced from bone. The result is a slightly greater degree of bone resorption than of bone formation. The existing bone is chemically normal and adequately mineralized, but the total amount of bone is decreased, trabeculae are thin, and marrow spaces are enlarged. Factors other than the diet may also be involved in causing senile osteoporosis, such as changes in the gonads and adrenal glands and the normal inactivity of the aged. Accordingly senile osteoporosis is possibly related to postmenopausal osteoporosis, in which there is faulty protein metabolism due to deficiency of estrogen, and to the osteoporosis of disuse, or disuse atrophy (Fig. 3-1), in which there is an excessive excretion of urinary calcium.

Clinical picture. Senile osteoporosis is most often discovered when an elderly person, after lifting or bending, develops sudden pain in the lower dorsal or lumbar region of the back, often accompanied by a snap or crack. Roentgenograms will usually show compression of a vertebral body, with increased radiolucency of the whole spine (Fig. 3-33). The vertebrae most often showing collapse are about the lumbodorsal junction. The vertebral bodies often appear biconcave. Deformation of multiple vertebrae may lead to kyphosis and loss of height. The remainder of the skeleton also shows rarefaction but not to the degree present in the spine. A rib or a

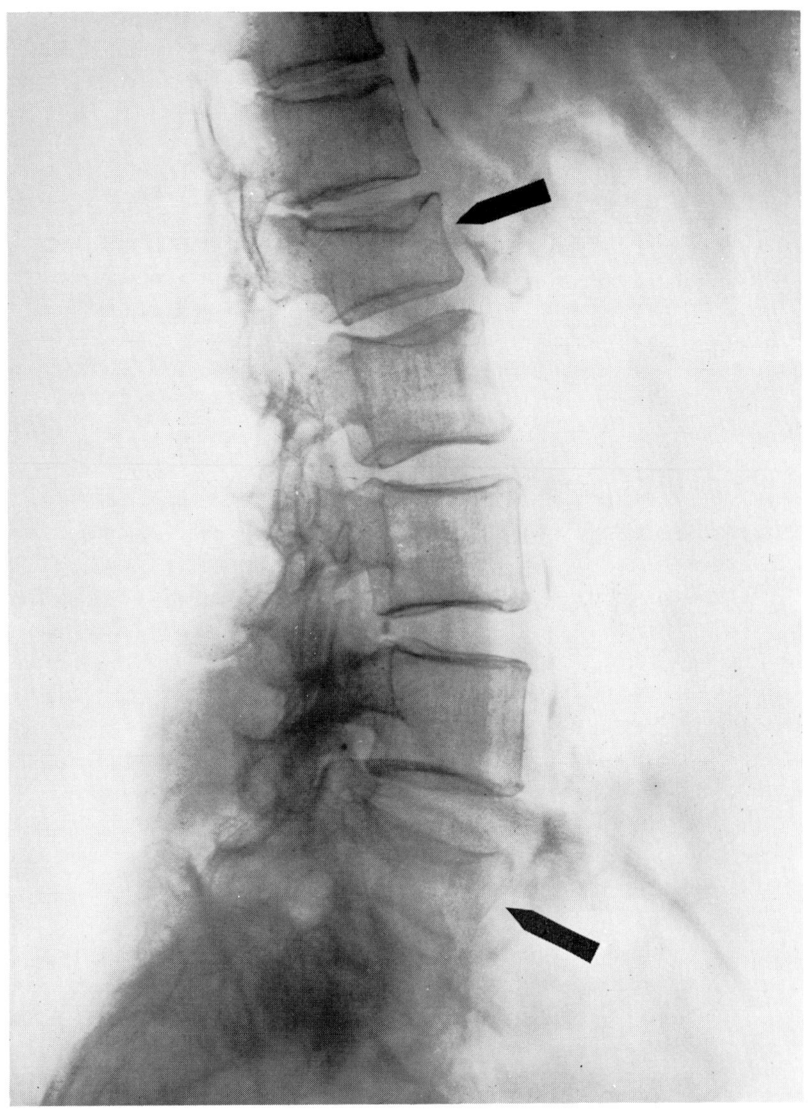

Fig. 3-33. Roentgenogram of senile osteoporosis in a patient 65 years of age. Note greatly decreased density of vertebral bodies, whereas their narrow margins remain distinct. The bodies of the first and fifth lumbar vertebrae, as well as the bodies of three thoracic vertebrae, had sustained compression fractures from minor trauma.

hip may fracture on minimal trauma. The serum levels of calcium, phosphorus, and alkaline phosphatase are within normal limits.

Treatment. A proved, generally accepted drug therapy for senile osteoporosis is not available. Increase in the dietary calcium to ensure a positive calcium balance is indicated. Supplementary vitamin D therapy may be given. Restoration of bone structure is a slow process, requiring years, but often may be hastened a bit by increasing the patient's activity to the limit compatible with his general health. The administration of estrogens and androgens, alone or in combination, relieves the symptoms but has not been proved to result in bone production. Currently medication with fluoride is under trial. Fractures of the vertebrae and other bones should be treated as indicated in the individual case.

POSTTRAUMATIC PAINFUL OSTEOPOROSIS (SUDECK'S ATROPHY, REFLEX SYMPATHETIC DYSTROPHY)

Occasionally an injury of an extremity or of a peripheral nerve is followed by prolonged local pain, vasomotor instability, trophic changes in the soft tissues, and diffuse, patchy rarefaction and atrophy of bone. The trauma may be of quite minor

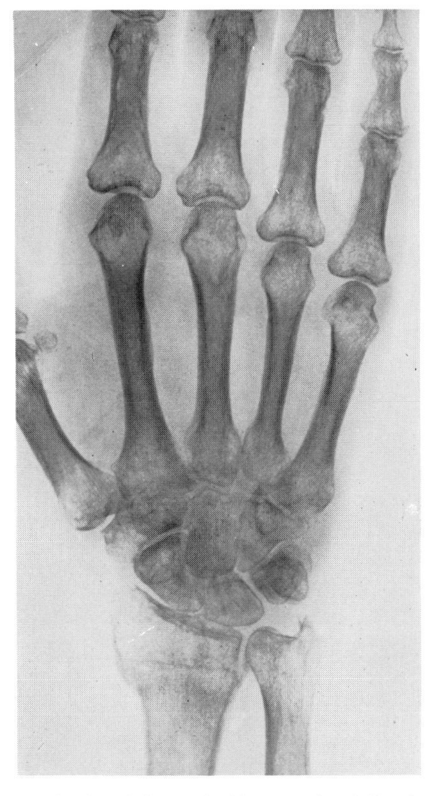

Fig. 3-34. Roentgenogram of the hand in Sudeck's atrophy following a Colles' fracture. Note healed fracture of the distal end of the radius and severe bone atrophy. The patient's disabling pain gradually subsided with treatment, which included a dorsal sympathectomy.

nature, and the disturbance of function is always greater than that expected from the injury alone. This condition, termed posttraumatic painful osteoporosis and first described as an entity by Sudeck in 1900, must be distinguished carefully from simple disuse atrophy of bone, with or without malingering, and from causalgia due to a lesion of a peripheral nerve. The bone atrophy may have an irregular and spotted appearance (Fig. 3-34); it may be band shaped, subchondral, metaphyseal, or diffuse.

Etiology. The pathogenesis is poorly understood. Much of the evidence suggests that disturbed function of the sympathetic nerve supply, due to injury or compression, is the major factor. Accordingly the changes may represent the terminal phase of a neurodystrophic process.

Treatment and prognosis. Although the period of disability is always long, many patients recover spontaneously or with simple treatment, such as brief splinting followed by prolonged physical therapy and active exercise of the extremity. In the more intractable cases, sympatholytic drugs, repeated procaine block of the sympathetic ganglia, ganglionectomy, or periarterial sympathectomy may be indicated. Often some permanent deformity and disability result despite treatment.

CHAPTER 4 Infections of bones and joints

Bone infection or *osteomyelitis* and joint infection or *septic arthritis* may be acute or chronic. To a large extent the nature and degree of the inflammatory reaction, whether in bone or in joint, are determined by the characteristics of the invading organism. Types of bone and joint infection warranting description here include those caused (1) by common pathogenic bacteria *(Staphylococcus, Streptococcus, gonococcus, Salmonella, Brucella,* and others); (2) by *Mycobacterium tuberculosis;* (3) by *Treponema pallidum;* and (4) by pathogenic fungi (*Actinomyces, Blastomyces, Coccidioides, Sporotrichum,* and others).

PYOGENIC OSTEOMYELITIS

Incidence. Osteomyelitis is a common acute illness and a common cause of chronic disability. The incidence of hematogenous osteomyelitis is much higher in children than in adults, and in boys than in girls. Any bone may be involved. Those most commonly affected are the tibia, femur, humerus, and radius.

Routes of infection. In hematogenous osteomyelitis the invading organisms reach the bone by transportation in the bloodstream from a distant focus, such as a furuncle, pustule, infected laceration, or the nasopharynx. In other forms of osteomyelitis the bone is infected by spread from a contiguous focus, such as a pyarthrosis, or by direct introduction through a wound, as in open fracture (Fig. 4-1).

Etiology. In at least 80% of the cases of acute hematogenous osteomyelitis, the infecting organism is the *Staphylococcus aureus.* Infection with hemolytic streptococci is next most common. Trauma, causing minute hemorrhage in bone, may play a part in the etiology by providing a locus in which the organisms of a transient bacteremia may lodge, survive, and multiply. Any debilitating disease may be a predisposing factor.

Pathology. In children, hematogenous osteomyelitis usually starts in the metaphysis. The circulation in this region is relatively sluggish, and here are found end arteries and venous sinusoids in which the bacteria are thought to lodge, causing a septic infarction. Hyperemia, edema, and infiltration of polymorphonuclear leukocytes ensue. Leukocytes destroyed by bacterial toxins liberate proteolytic enzymes. Further necrosis of ischemic tissue culminates in abscess formation. With increasing

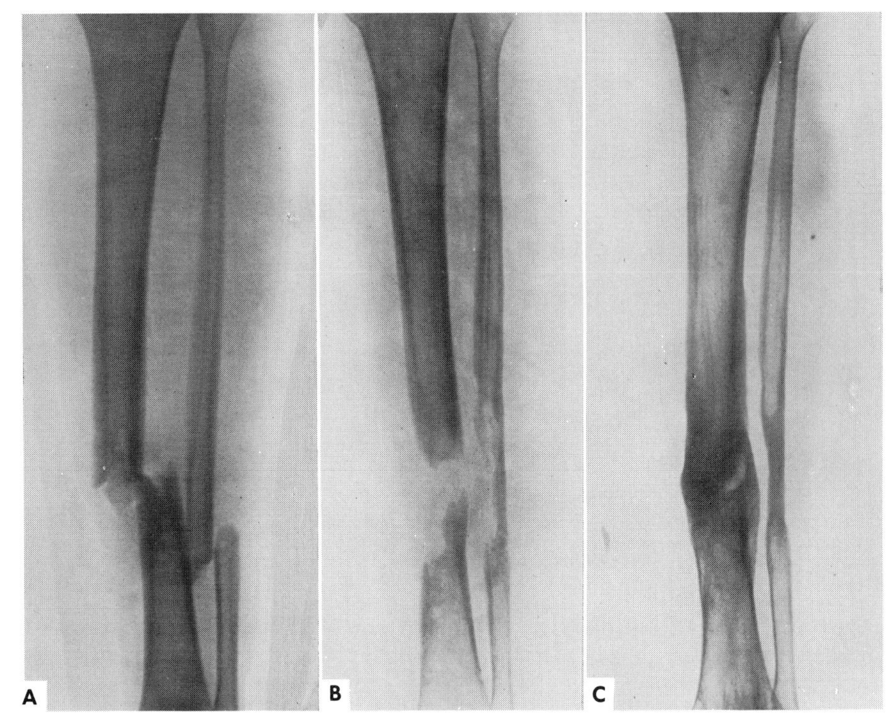

Fig. 4-1. Osteomyelitis secondary to open (compound) fractures of tibia and fibula in a 19-year-old boy. **A,** Fractures on day of injury: before debridement, reduction, and immobilization. **B,** Ten months after fracture: osteomyelitis has involved all fragments, necessitating sequestrectomy, which has produced a wide tibial defect. The limb has remained completely disabled, and there is extensive bone atrophy. **C,** Four years after fracture: bone grafting, performed after all signs of infection had disappeared, has been followed finally by restoration of tibial continuity.

tension the septic process spreads in the directions of least resistance (Fig. 4-2) centrally toward the medullary cavity, and outwardly through vascular channels or erosions of the thin metaphyseal cortex to form a subperiosteal abscess. The subperiosteal pus may spread for some distance along and around the shaft; it may then reenter the medullary cavity or may perforate the periosteum and extend into the overlying soft tissues.

In children the epiphyseal cartilage plate acts as a barrier to the infection, preventing or delaying extension into the epiphyseal end of the bone and thence into the joint. When the metaphyseal area is intracapsular, however, infection may spread quickly into the joint. An important example is the hip joint. Osteomyelitis of the neck of the femur may lead very quickly to the serious complication of septic arthritis of the hip. Often the joints develop a secondary synovitis before they become actually infected.

With accumulation of pus beneath the periosteum, new bone is laid down on its inner surface, forming an *involucrum.* Through openings, or *cloacae,* in this new bone, pus may be discharged. Certain areas of bone may become ischemic from the stripping up of periosteum and the thrombosis of cortical capillaries. Such areas be-

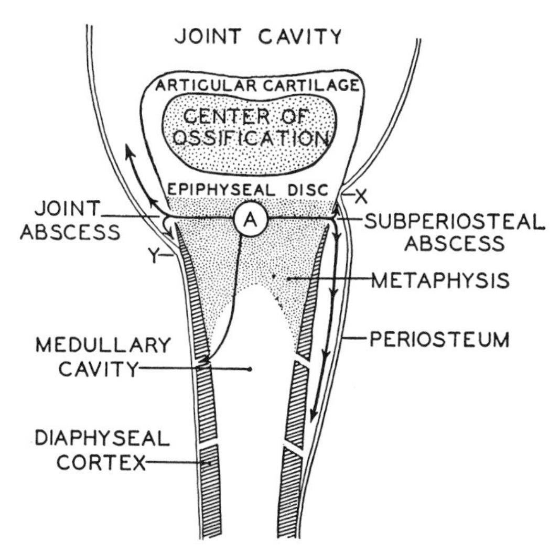

Fig. 4-2. Diagram showing directions of possible spread of infection from focus **A** in metaphysis. **X** and **Y** mark the junction of synovial membrane and periosteum in subperiosteal abscess and joint abscess, respectively. (After Hart.)

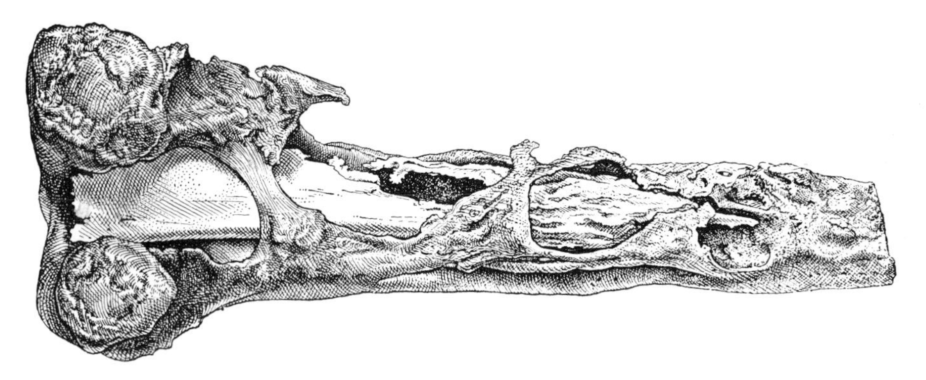

Fig. 4-3. Chronic osteomyelitis of the femur before the days of antibiotics, showing large sequestrum and surrounding involucrum. (Drawing from museum specimen.)

come necrotic and separate, forming *sequestra* (Fig. 4-3). Small sequestra may be extruded through the involucral cloacae into sinuses draining upon the skin surface. It is impossible, however, for the larger sequestra to be extruded, and they may have to be removed surgically. Sequestra may become surrounded by granulation tissue, gradually disintegrated by proteolytic enzymes in the purulent exudate, and absorbed.

Accordingly the microscopic picture of acute osteomyelitis is characterized by masses of polymorphonuclear leukocytes, areas of necrosis, and cell detritus about bone trabeculae. In later stages of the disease microscopic sections show fibrotic bone marrow, many round cells, the dead bone and empty lacunae of sequestra, osteoclasts and granulation tissue about sequestra, and the new trabeculae and osteoblasts of the involucrum.

In extensive osteomyelitic involvement of the shaft of a long bone, pathologic fracture occasionally occurs. This is particularly likely to happen if stress is placed upon the bone before sufficiently mature involucrum has restored its strength.

Often pyogenic osteomyelitis is not accompanied by the formation of sequestra. Atypical infections from relatively avirulent organisms are not uncommon. The defenses of the body and early, adequate antibiotic therapy frequently overcome the infection before extensive bone necrosis occurs. Osteomyelitis in infancy rarely results in gross sequestration.

Clinical picture. Acute osteomyelitis usually starts with malaise, general weakness, and aching, followed quickly by an elevation of body temperature and intense pain in the affected region. In the early stages, when the infectious process is within a deeply situated bone, no sharply localized tenderness may be found. Ordinarily there is protective muscle spasm, the joint nearest the disease focus being held in flexion. When a lower limb is involved, the child often refuses to bear weight on it.

Acute osteomyelitis is usually attended by septicemia; the blood culture becomes positive and the patient extremely ill. A leukocytosis of 15,000 to 40,000 cells, depending on the severity of the infection, may be present, and the erythrocyte sedimentation rate is greatly increased. As the infection advances, tenderness over the bony focus usually becomes localized and acute. Swelling, increased heat, and redness are unlikely to be present if the soft tissues overlying the infected bone are thick. The clinical diagnosis of acute osteomyelitis must often be made, and essential early antibiotic treatment must be started, solely on the triad of sepsis, local pain, and local tenderness. Confirmatory roentgenographic changes do not appear until a week or two later; irreparable damage may be done if therapy is delayed until they become obvious.

When the defenses of the body, aided by antibiotics, are sufficient to overcome the infection, the acute inflammatory symptoms will subside; this is the usual course. Otherwise the symptoms may persist until pus is released by a surgical procedure or liberates itself spontaneously. Afterward the fever decreases gradually and may continue for several weeks.

In infants, acute osteomyelitis may be attended by little or no fever. The chief signs are irritability, loss of function of the affected extremity, tenderness, and swelling. The clinical course is relatively rapid.

If the initial therapy has been ineffective, perhaps because it was started too late, subsidence of the acute symptoms may be followed by the draining sinus or sinuses of *chronic osteomyelitis*. The drainage will persist as long as infected granulation tissue and dead bone, acting as a foreign body, remain. Accumulation of pus from inadequate drainage may cause periodic exacerbations of symptoms. Prolonged disuse of the infected limb leads to muscle atrophy, and if the joints are not properly splinted, contractures often develop. There may be no general symptoms at first, but when purulent drainage has continued for a period of years, nephritis or amyloid disease may develop. The constant inflammation of epithelial cells at the mouth of such a sinus may lead to a squamous cell carcinoma. Recurrence of infection may take place many years after apparently complete healing of an osteomyelitis. Increased circulation associated with osteomyelitis of a long bone in childhood may stimulate growth to produce slightly increased length; if the epiphyseal plate is infected, however, growth may cease or occur in an asymmetric manner leading to deformity.

Roentgenographic picture. Bone changes are seldom visible in the roentgenograms in less than 10 days from the time of onset; they appear later in adults than in children, and several weeks sometimes elapse before definite bony abnormality can be demonstrated. Often the first change in the affected bone is an area of haziness or of mottling in the metaphysis, caused by the destruction of trabeculae. This is soon followed by new bone formation under the periosteum. Later there may be destructive areas in the shaft (Fig. 4-4) and evidences of sequestration. Roentgenographic changes may be atypical or absent when the course of the infection has been modified by early antibiotic treatment.

In roentgenograms of chronic osteomyelitis, sequestra appear as irregularly shaped islands more dense than the surrounding, viable bone.

Differential diagnosis. Acute osteomyelitis must be distinguished from acute rheumatic fever with its migratory polyarthritis. Acute cellulitis must be differentiated. Pyogenic arthritis causes more swelling and spasm than does aseptic synovitis secondary to osteomyelitis and can be identified by aspiration of the joint, smear, and culture. In diagnosing chronic osteomyelitis one should rule out Ewing's sarcoma, osteogenic sarcoma, osteoid-osteoma, fungus infections, and tuberculosis. In adults, vertebral osteomyelitis is sometimes a difficult diagnostic problem.

Prognosis. With antibiotic treatment the mortality of hematogenous osteomyelitis has been reduced from more than 20% to perhaps 1% or less. The most serious cases are those in infants and young children with septicemia. The course of the disease depends largely upon the promptness with which effective antibiotic therapy

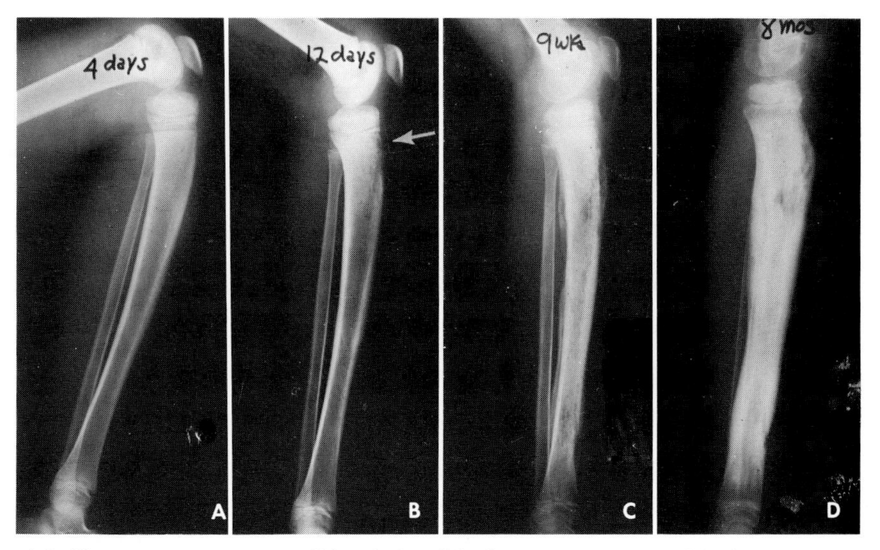

Fig. 4-4. Hematogenous osteomyelitis of the tibia in an 8-year-old child. **A,** Normal roentgenographic findings at 4 days after acute clinical onset. **B,** Early localized destruction in metaphysis (arrow) at 12 days. **C,** Extensive diaphyseal destruction and involucra at 9 weeks. **D,** Massive new bone associated with progress of healing at 8 months.

can be started. Frequently the bacteremia disappears quickly, and the local infection is controlled before extensive necrosis of bone can take place. Less favorable cases progress to chronic osteomyelitis with its periodic exacerbations. In chronic cases of long standing, nephritis or amyloid disease may prove fatal.

Treatment. The type of treatment depends upon the stage to which the septic process has progressed.

Acute stage. In the acute stage the patient is suffering not only from a local bone lesion but also from a bloodstream infection with severe general reaction. Early treatment is essential; however, every effort should be made to recover the infecting organism prior to giving antibiotics. Such efforts should include one or more blood cultures and aspiration of the abscess if one is present. When this has been done, antibiotic therapy should be started immediately without awaiting the results of culture. Since the offending organism is a staphylococcus in the vast majority of cases and a streptococcus in most others, penicillin is the drug of choice. In view of the constant increase in infections caused by penicillin G–resistant staphylococci, however, it is advisable to include an additional drug that is effective against these organisms. At present, methicillin or one of the other semisynthetic penicillinase-resistant antibiotics seems to be most useful for this purpose. In general, bactericidal drugs are more effective and preferable to those that are only bacteriostatic. Penicillin should be given in large doses. Intramuscular injection may be used, but in the septic child, higher blood levels can be maintained by continuous intravenous infusion. If these drugs prove ineffective, as shown by lack of clinical response in 24 to 48 hours, a different antibiotic should be started. By this time the results of antibiotic sensitivity testing of the organism cultured from the blood or from a subperiosteal abscess may be available. Because of the frequency of toxic reactions, allergic phenomena, and bacterial resistance, selection of the proper antibiotic from the large number of available preparations is often difficult. As a rule, antibiotic therapy should be continued for at least 3 weeks.

The acutely ill patient requires constant bed rest and skilled nursing care. His course should be followed by blood cell counts and blood cultures. Attention to hydration and electrolyte replacement is important. Anemia may require transfusions.

The affected limb should be put at rest in good functional position by means of traction or the support of a splint or plaster shell. Hot moist dressings may be applied.

With early intensive treatment of the patient's general condition, including antibiotic therapy, it is possible in most cases to avoid surgical operation upon the infected bone. If, however, delay of several days in starting treatment with antibiotics has so reduced their effectiveness that satisfactory clinical improvement does not occur after 24 to 48 hours of intensive therapy, surgical drainage of the infected bone may be advisable.

After the local signs, of which the most useful is fluctuation, indicate that a subperiosteal abscess has formed, it should be aspirated of as much pus as can be obtained. The pus should be examined by Gram stain and cultured. An antibiotic may be instilled directly into the abscess cavity. If the abscess recurs, it may be best to effect surgical drainage. As the acute condition subsides, a cast must usually be applied, especially if the destruction of bone has been extensive. The patient must be followed carefully during his convalescence.

Chronic stage. In the chronic stage the patient is no longer obviously ill but has a local infection of low grade that may be manifested by persistent drainage from one or more cutaneous sinuses and by recurrent episodes of local pain, increased heat, and swelling. Chronic osteomyelitis is still seen frequently despite improved treatment of early hematogenous infections and of compound fractures. The drainage from sinuses should be cultured, the sensitivities of the causative organism and any secondary invaders determined, and appropriate antibiotic treatment started. Antibiotics may be relatively ineffective, however, because of the poor circulation in scarred soft tissues and sclerotic bone.

When surgical treatment is contemplated, it is often helpful to visualize sinus tracts, abscess cavities, and sequestra by roentgenograms made after injection of the sinus with a contrast medium or by laminagrams of the diseased bone. Surgical removal of dead bone, or *sequestrectomy,* is indicated when the patient is in good general condition, the necrotic bone is roentgenographically well separated, and an adequate involucrum has been formed. In the absence of persistent diffuse infectious changes in the affected bone, individual sequestra are best removed through sinuses or small incisions. If roentgenograms show most of the bone to be still involved by chronic infection, however, the much more extensive surgical procedure of *saucerization* may be indicated. It includes removal of all scar tissue, infected granulation tissue, sequestra, sclerotic bone, and overhanging bone edges; it leaves a long flat depression in the bone with a grossly clean and freely bleeding surface. Continuous irrigation of the infected area with antibiotic solutions through catheters inserted at the time of surgery is often helpful. In some cases closure of the soft tissues is practicable; in others it may be preferable to pack the wound with petrolatum gauze, allowing it to heal from within outward (the *Orr method*). In either case an ample cast is applied to rest the affected structures and protect the weakened bone. Later, skin grafting for coverage or even bone grafting for strength may be indicated.

In the chronic stage it is often difficult to obliterate large infected cavities, especially those in the thick lower end of the femur or upper end of the tibia. Skin grafting may be required before bone surgery can be done. Sometimes it is possible to close such cavities by removing completely the superficial layer of bone lining the cavity, together with all infected soft tissue, and allowing the remaining soft tissue to collapse into the cavity. It may be necessary to carry out a bone-filling operation of transplanting muscle and fascia or fat into the cavity. Sometimes cancellous bone chips are packed into the cavity after complete saucerization.

Amputation is occasionally indicated in chronic osteomyelitis. There are a few adult patients in whom chronic infection over a period of many years has responded so poorly to the usual treatment that their general health and function can be improved greatly by amputation.

BRODIE'S ABSCESS

Brodie's abscess is a localized form of chronic osteomyelitis dating back to a recognized or unrecognized bacteremia, which may have preceded the clinical appearance of the abscess by many years. The abscess has a thin wall of fibrous tissue and sclerotic bone; it contains pus that on culture may be sterile or may contain

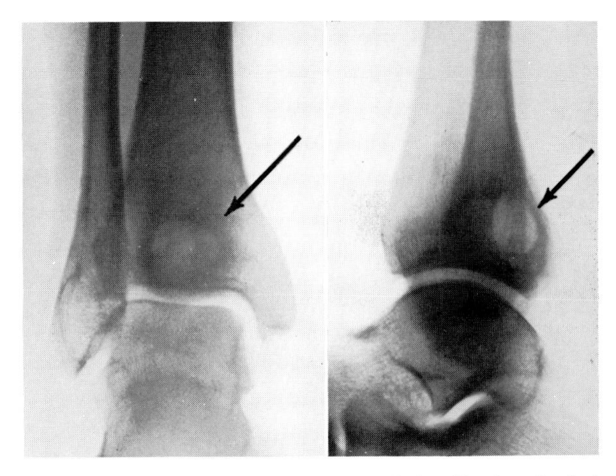

Fig. 4-5. Roentgenograms showing a Brodie's abscess of the distal end of the tibia in a man 29 years of age. Note the circular, well-demarcated area of decreased density.

staphylococci of low virulence. The commonest site is the lower end of the tibia. These abscesses occur most often in older children and young adults. Clinically the onset is gradual. The chief symptom is local pain, often worse at night, and usually there are slightly increased heat and tenderness over the site of the lesion.

Diagnosis. The roentgenographic appearance of an area of decreased density surrounded by sclerotic bone (Fig. 4-5) confirms the clinical diagnosis. Brodie's abscess is to be differentiated from osteogenic sarcoma, osteoid-osteoma, eosinophilic granuloma, bone cyst, and fibrous dysplasia.

Treatment. The treatment consists of operation and the local and systemic administration of an antibiotic. The abscess should be thoroughly cleaned out or excised. Usually the wound may be closed, after which a cast is applied.

OSTEITIS PUBIS

Osteitis pubis is a painful, usually nonsuppurative affection of the pubic symphysis, occurring most often after prostatic operations in older men. A few cases have been reported in young adults after local trauma and in women during or after pregnancy. Urinary tract organisms have been cultured from the diseased area in some cases. However, it is questionable whether osteitis pubis is a low-grade infectious process or a noninfectious sequel of trauma.

Clinical picture. Characteristically, 2 to 8 weeks after prostatic surgery there is sudden onset of pain over one side of the symphysis, with pubic tenderness. The pain rapidly becomes more severe, especially with activity, and is worse on coughing, defecation, and urination. The patient tends to lie in one position, with flexed hips and severe adductor muscle spasm. Recovery is often spontaneous but slow.

Roentgenographic picture. Several weeks after the clinical onset the margins of the symphysis become irregular and the pubic bodies and rami osteoporotic. Decalcification may extend laterally to the ischium and even to the acetabulum. Early the symphyseal interval may appear widened; much later it may be narrowed or obliterated.

Differential diagnosis. Osteitis pubis must be distinguished from strain of the adductor muscles, from frank osteomyelitis with its more severe local and systemic reactions, and from metastatic malignancy and tuberculosis.

Treatment. The pain is treated by bed rest, appropriate medication, and local application of heat or cold. A short bilateral hip spica cast may afford much relief. An antibiotic should be given if signs of infection are present. Accumulation of pus may require aspiration or incision and drainage.

PYOGENIC ARTHRITIS

Pyogenic, suppurative, or septic arthritis results from the activity of pus-forming bacteria in a synovial joint. Since it may rapidly and permanently destroy articular cartilage, pyogenic arthritis demands prompt recognition and vigorous treatment.

Incidence. Pyogenic arthritis is more common in infants and children than in adults but is seen at all ages. Males are affected more commonly than females. The hip and knee are the most frequent sites of infection.

Etiology. Trauma probably influences localization of the infection. As in osteomyelitis, *Staphylococcus aureus* and hemolytic streptococci are the common infecting agents. Less commonly, joints are infected by any of a variety of microorganisms including pneumococcus, gonococcus, meningococcus, *Salmonella, Brucella,* colon bacilli, and many others.

Routes of infection. Organisms reach the joint by one of three routes: (1) via the bloodstream in septicemia or by transfer from a distant focus, such as an infected abrasion or furuncle; (2) by direct or lymphatic extension from a neighboring infected area, such as an osteomyelitis; or (3) by direct inoculation of infected material through a penetrating wound. Of the last type is infection from a contaminated aspiration or injection, a grave complication than can almost always be prevented by the use of strictest aseptic technic when a needle must be introduced into a joint.

Pathology. In the early stages of joint infection the synovial membrane becomes hyperemic, edematous, and infiltrated with inflammatory cells. The synovial fluid, greatly increased in volume, may at first be thin and cloudy. It contains large numbers of polymorphonuclear leukocytes, the count frequently being higher than 50,000 per cubic millimeter. Its protein content is increased, and its sugar content is decreased.

If the infection is unchecked, the synovial fluid becomes thicker and frank pus may be found within the joint cavity. Rapid destruction of articular cartilage takes place. A fibrinous exudate forms on the synovial surface, to be replaced later by inflammatory granulation tissue. Cartilage erosion, especially at points of pressure, leads to inflammatory changes in the underlying bone and later to frank osteomyelitis. The amount of destruction depends upon the virulence of the organism and the length of time during which the infection has been present. Pathologic dislocation sometimes occurs. Septic joints of this type often progress to a fibrous or bony ankylosis (Figs. 5-14 and 5-15).

In addition to the local inflammatory process the widespread pathologic changes associated with septicemia may be present.

Clinical picture. The symptoms vary with the degree of the joint inflammation.

In mild cases there may be only moderate swelling and slight pain. There is an increase of local heat, and the joint usually becomes flexed because of protective muscle spasm. Palpation of the distended joint capsule yields the sensation of fluctuation, and often the outline of the capsule can be defined. Attempts to move the joint are accompanied by pain. Fever and leukocytosis are present but not of great degree.

In more virulent infections the symptoms and physical findings are more striking. There is usually a severe systemic reaction with a considerable elevation of temperature (104° to 105° F., 40° to 41° C.) and leukocytosis. The joint is extremely painful on examination. The infection may spread to neighboring structures, giving rise to brawny induration and thickening of the periarticular tissues.

Differential diagnosis. To make a definite diagnosis of pyogenic arthritis it is necessary to aspirate the joint fluid, examine a smear microscopically, and culture the fluid. Acute osteomyelitis, periarticular cellulitis, and purulent bursitis are often confused with pyogenic arthritis. Gonococcal arthritis must be remembered. Frequently pyogenic arthritis must be differentiated from acute rheumatic fever. Acute rheumatoid arthritis, gout, tuberculosis, scurvy, and hemophilia must also be excluded.

Acute appendicitis may cause psoas spasm, resulting in flexion of the right hip, which may suggest pyogenic arthritis of the hip. Acute pelvic inflammatory disease may cause symptoms referable to either hip joint.

Treatment. Therapy should be prompt and vigorous. If fluid aspirated from the joint is cloudy, several hundred thousand units of penicillin in saline solution may be instilled immediately into the joint and systemic penicillin treatment started. Giving high doses of penicillin by continuous intravenous infusion may be preferable to intra-articular injection. A different antibiotic should be substituted later if culture and sensitivity tests so indicate.

The inflamed joint structures should be put at rest by means of traction, splint, or bivalved cast.

Aspiration, culture, gentle irrigation, and instillation of antibiotic may be done daily and later at longer intervals as the local and systemic signs of infection subside. However, if such improvement does not begin within 24 hours and continue, and in particular, if the aspirate remains purulent and the patient febrile, immediate surgical drainage is indicated. Irrigation with an antibiotic may be done at operation and repeated, with the use of small tubes implanted at surgery, several times a day or continuously for several days after operation. A windowed cast may be used to support the joint in its best functional position. Because the hip joint is relatively inaccessible for aspiration and irrigation and because distention of the hip capsule by pus may lead to necrosis of the femoral head, early surgical drainage is usually the best initial treatment for pyogenic infections of the hip. After the joint inflammation has subsided and the postoperative cast has been removed, mobility and strength are restored by gradually increasing exercise. When severe damage of intra-articular structures has made ankylosis inevitable, care should be taken that it occur in the position of maximum usefulness (p. 164).

Prognosis. Joint effusions are rapidly absorbed and recovery takes place quickly and completely if early, effective antibiotic therapy has been carried out. When a purulent exudate has formed, however, some limitation of motion usually results and occasionally the final outcome is bony ankylosis. Pyarthrosis associated with septicemia, which is seen more frequently in children than in adults, may lead to a fatal outcome unless the infection is quickly controlled by therapy.

PYOGENIC ARTHRITIS OF THE HIP IN INFANTS

Pyogenic arthritis of the hip in infants warrants individual emphasis because of its difficulty of diagnosis, need of prompt treatment, and serious consequences if inadequately handled. Although the infection is usually blood borne, it may be introduced by inadvertent penetration of the joint capsule associated with procedures requiring puncture of the femoral artery or vein.

Pathology. In all probability the infecting organisms usually enter the hip joint from an expanding focus in the intracapsular metaphysis of the femoral neck. In some instances they may lodge first in synovial membrane or bony portion of the capital epiphysis. The epiphysis, which in infancy is wholly or largely cartilaginous, may be severely damaged or destroyed by proteolytic enzymes in the exudate and by impairment of its blood supply resulting from increased intra-articular pressure. Pathologic dislocation is a common sequel.

Diagnosis. The systemic reaction, especially in premature infants, may be deceptively mild, without fever or leukocytosis. Permanent, irreparable joint damage may occur unless the diagnosis is made early. It should be suspected from the local, acute inflammatory signs of swelling, tenderness, and pain on motion and should be confirmed by prompt aspiration.

Treatment. Unless the diagnosis has been made and local and systemic antibiotic treatment has been started as early as a day or two after onset, prompt surgical drainage is usually indicated, followed by rest in a spica cast with the hip in extension, slight internal rotation, and moderate abduction to forestall subluxation.

GONOCOCCAL ARTHRITIS

Gonococcal arthritis is usually a metastatic sequel of inadequately treated acute gonorrheal urethritis. Since urethritis is less commonly recognized and treated in women than in men, gonococcal arthritis is more common in women.

Clinical picture. The joint involvement usually appears 2 to 3 weeks after the onset of a urethral or vaginal discharge, but it may occur much later. The arthritis may be polyarticular or monarticular. Fleeting pains are sometimes present in multiple joints for several days before obvious infection becomes localized in a single joint. The knee is affected most commonly, and next the ankle. Periarticular inflammation may involve tendon sheaths and bursae.

In acute cases the joint is extremely painful and tender, with redness, increased heat, swelling, glossy overlying skin, and severe muscle spasm; the systemic reaction includes fever and severe leukocytosis. Cases of more chronic type are often characterized chiefly by pain in several joints, swelling, restricted mobility, and little systemic reaction.

Diagnosis. In acute cases the diagnosis is suggested by the clinical picture and by identification of the organisms in the urethral or vaginal discharge. It is important to exclude *Staphylococcus* and *Streptococcus* infection and acute rheumatic fever. Early diagnostic aspiration of the affected joint and bacteriologic study of the aspirate should be carried out. Within the first 10 days of the infection gonococci will often be found in the joint fluid. In chronic or subacute cases there may be some difficulty in finding the organisms and in differentiating the condition from rheumatoid arthritis, Reiter's syndrome, and tuberculosis. Immunofluorescent studies of gonococcal antibodies may establish the diagnosis.

Treatment. Therapy should be directed toward cure of the primary focus, as well as the inflamed joint. Penicillin should be given systemically in large doses. The affected joint may require elevation, heat, support in good functional position, aspiration, gentle lavage, and instillation of penicillin. After the infection has subsided the joint should be gradually mobilized. Joints that do not receive early treatment may develop fibrous or even bony ankylosis.

SALMONELLA OSTEOMYELITIS AND ARTHRITIS

In many countries typhoid osteomyelitis, which is caused by *Salmonella typhosa,* is now rare. Infections of bone or joint by other members of the *Salmonella* group are seen occasionally. Their incidence is relatively high in children with sickle-cell anemia, perhaps because small infarcts in the gut may allow intestinal organisms to enter the bloodstream. Several bones may be involved, with local pain, tenderness, and roentgenographic changes, in the later stages of an enteric fever; a chronic localized osteomyelitis resembling Brodie's abscess may make its clinical appearance years after a *Salmonella* septicemia; or a low-grade pyogenic osteomyelitis or arthritis may occur without recognized primary infection. The diagnosis is made on culture of the organism. The treatment is similar to that of other forms of pyogenic osteomyelitis and arthritis. Chloramphenicol has proved effective in most *Salmonella* infections.

BRUCELLA OSTEOMYELITIS AND ARTHRITIS

Human brucellosis, an uncommon disease in the United States, is seen chiefly as a result of direct contact with cattle or swine or of drinking unpasteurized milk. Its orthopaedic manifestations are of two types: (1) excessive fatigue, aches, and pains without local infectious foci and (2) localized osteomyelitis, arthritis, and bursitis. Of the bursae, the prepatellar is most commonly affected and, of the joints, the hip or knee. Involvement of the spine, usually at the lumbar level, tends to destroy contiguous vertebral surfaces and disk, ending in bony ankylosis of the bodies. The diagnosis may be suspected from the clinical picture, strengthened by serum agglutination testing, and proved by culture. *Brucella* is cultured more readily from tissue removed at operation than from aspirates. The treatment includes use of the tetracycline drugs and streptomycin.

TUBERCULOSIS OF BONES AND JOINTS

Tuberculosis of bones and joints is a localized, progressively destructive disease resulting from the activity of *Mycobacterium tuberculosis* in bone or articular structures. It is secondary to tuberculous infection in another part of the body, such as lungs or lymph nodes.

Incidence. In recent decades better methods of prevention and treatment have reduced the incidence of active tuberculous infections and their complications. In many areas osteoarticular tuberculosis, which constitutes no more than 2% or 3% of tuberculous infections in general, is no longer a common disease.

Tuberculosis of bones and joints may occur at any age. In the United States in recent years the ratio of childhood to adult cases has greatly decreased. It is commonest in groups whose living conditions are substandard. Although any bone or joint may be involved, the most frequent sites are, in order, the spine, the hip, and the knee.

Routes of infection. Tuberculous joints are seen in patients with recognized pulmonary tuberculosis, in whom the mycobacteria have traveled from lung to bone or joint via the bloodstream. In some instances they may spread via lymphatic channels or by direct extension. Bone and joint tuberculosis is seen also in individuals who have no obvious primary lesion. In such cases the initial focus may be a bronchial lymph node that the mycobacteria reach after being inhaled or a mesenteric node they reach after being ingested. In many countries infected cattle and unpasteurized milk are still important sources of tuberculous infection.

Pathology. An outstanding characteristic of tuberculous infection of bone is destruction with little tendency toward the formation of new bone. The process usually begins in subchondral or metaphyseal bone and then by gradual extension along the line of least resistance enters the joint. In some instances it starts in the synovial membrane and involves secondarily the bone beneath the joint surface.

Localization of the mycobacteria in bone leads to the formation of tubercles. Microscopically the typical tubercle contains epithelioid cells and one or more Langhans' giant cells with their peripherally placed nuclei. In the earliest stages polymorphonuclear leukocytes are present. Later, numerous lymphocytes infiltrate the marginal areas. Adjacent tubercles may coalesce to form larger ones, and the expanding lesion gradually destroys and replaces surrounding tissues. Necrosis of ischemic and toxic origin may take place near the center of such a granuloma, progressing slowly to liquefaction, whereas about the periphery fibroblasts lay down collagen that tends ultimately to wall off the lesion.

As the destructive process near the end of a bone advances, it soon penetrates into the nearest joint along an edge of its articular cartilage. Rapid dissemination of the tuberculous material then takes place over synovial membrane and cartilage. Although relatively resistant to tuberculous destruction, the articular cartilage may be undermined at its synovial margins, gradually separated from the underlying bone, and destroyed. These changes are accompanied by the formation of a purulent, cheesy exudate containing remnants of necrotic tissue. The bone becomes disintegrated and destroyed on its exposed joint surfaces. If the destructive process is unchecked by treatment, the increasing exudate may dissect the surrounding joint structures along the planes of least resistance, enter the soft tissues to form an abscess, and penetrate the overlying skin to produce a chronically draining sinus that soon becomes infected secondarily with pyogenic bacteria.

If the infection is overcome, the exudate is slowly absorbed and replaced by fibrous tissue. Bone about the edges of the lesion is stimulated to proliferate, and gradually the evidences of active infection subside. Tuberculous organisms in the tissues may remain viable for long periods.

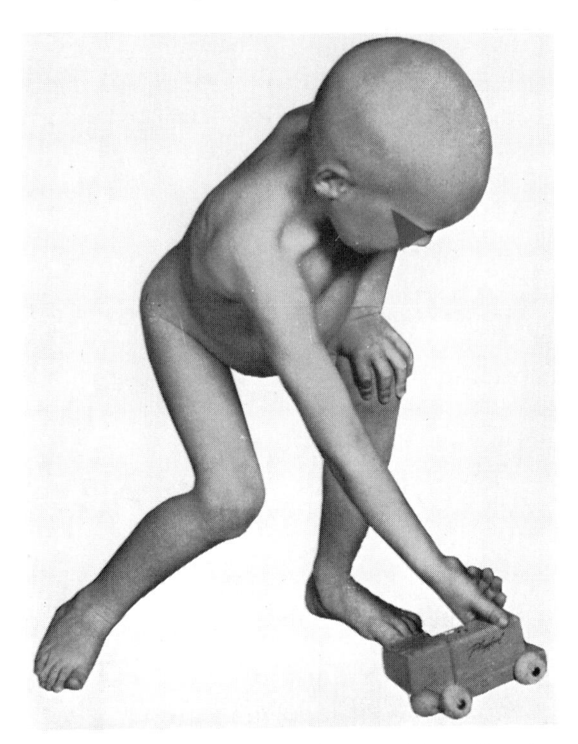

Fig. 4-6. Tuberculosis of the lumbar spine in a 4-year-old boy. Note how, in picking up an object from the floor, the patient protects his spine in extension by flexing hips and knees and propping with his left hand.

Tuberculous abscesses are called cold abscesses because of the absence of an acute inflammatory reaction. As the focus feeding a deep tuberculous abscess subsides with treatment, the pus is slowly absorbed. The abscess wall becomes shrunken and fibrotic; it may become partially calcified.

Except in very early infections, tuberculosis completely within bone and without extension into a joint is unusual. It is seen occasionally in phalanges, metacarpals, metatarsals, or ribs, and rarely in other bones.

Clinical picture. The onset of local symptoms is insidious. As a rule the involvement is monarticular. Spontaneous pain, pain on motion of the affected joint, and restriction of its mobility by spasm (Fig. 4-6) may be the first evidences of tuberculosis; they are followed quickly by muscle atrophy. In other cases pain may be slight or absent. A painless kyphosis, or even the appearance of a cold abscess, may be the first manifestation of spinal tuberculosis, and a painless limp may be the first sign of involvement in the lower limb. In superficial joints, swelling is an early sign; they usually show no redness, little heat, and only slight tenderness. There is usually a slight afternoon elevation of temperature. A mild leukocytosis may be present. In late stages severe deformity may result from contractures and bone destruction.

Roentgenographic picture. Early changes are slight and often not diagnostic. Later the bones show decalcification, faint joint outlines, and irregular notching

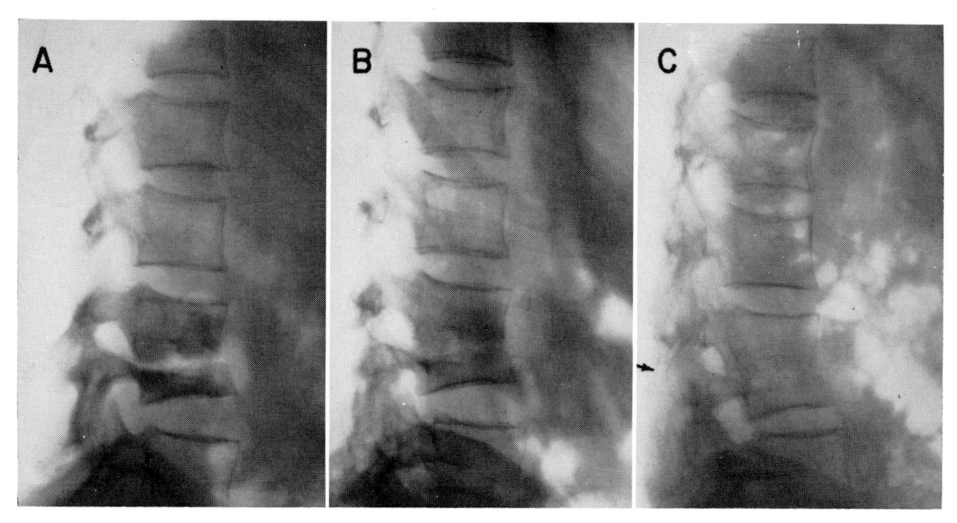

Fig. 4-7. Lateral roentgenograms of tuberculosis of the spine in a woman 54 years of age. **A,** Note the destruction of bodies of the third and fourth lumbar vertebrae and the narrowing and erosion at the disk space. **B,** Nine months later: the patient has taken antituberculous drugs and has had a posterior spine arthrodesis. **C,** Sixteen months after arthrodesis: note bony mass (arrow) that unites the posterior elements. Spontaneous fusion of the involved vertebral bodies has taken place.

of the joint surfaces. There is sometimes a circumscribed area of decreased density without surrounding sclerosis. In tuberculosis of the spine (Fig. 4-7), loss of disk space and erosion and collapse of the vertebral bodies are seen. The most characteristic feature of the roentgenograms, however, is the almost complete lack of bone regeneration in the early cases. In later, healing stages of the infection new bone may be seen, and the joint outlines again become sharply defined.

Diagnosis. Since early, accurate diagnosis is of the greatest importance, free use should be made of aspiration and biopsy technics. Positive diagnosis depends upon the identification of *Mycobacterium tuberculosis* in smear and culture or the demonstration of tubercles in microscopic sections.

The intracutaneous tuberculin test often provides useful information. Especially in children, a positive test is suggestive but not conclusive evidence of tuberculosis somewhere in the body. Rarely is tuberculosis present when the tuberculin skin test is negative and remains so even when repeated with increased concentrations of tuberculin.

Treatment. The treatment of bone and joint tuberculosis includes general measures, which apply to all cases, and local measures that vary with the extent and site of the lesion.

General measures. Of first importance is antituberculous drug therapy. To be most effective it must be started early, before its access to bacilli in the lesion has been blocked by avascular necrotic tissue, pus, or fibrosis. The most useful antituberculous drugs are isoniazid (isonicotinic acid hydrazide), streptomycin, and para-aminosalicylic acid. Since the effectiveness of an individual drug may decrease through the development of resistant strains, it is customary to prescribe two antibiotics concurrently at the start of treatment, whereas the third is reserved, possibly to be added at the time of surgical operation. As a rule, drug therapy should be continued for two years or more.

Isoniazid is the most effective of current antituberculous drugs. Except when a specific allergy exists, it should be used in all cases. Isoniazid may be given by mouth or by intramuscular injection. The usual dosage is at least 300 mg. daily for adults, and 10 mg. per kilogram of body weight for children. The toxic complications of isoniazid, which include mild skin and neurologic changes, are uncommon. *Streptomycin* is given intramuscularly in doses that vary from 1 Gm. twice a week to 2 Gm. daily, depending upon the age of the patient, the nature of the infection, and the imminence of toxic reaction to the drug. Such reactions include vertigo and deafness that is sometimes permanent. *Para-aminosalicylic acid* is given by mouth in doses of 10 to 20 Gm. a day.

Other general measures, including nutritious diet and prolonged bed rest, are especially beneficial for debilitated patients, patients with concomitant pulmonary infections, and patients with multiple bone and joint foci.

Local measures. The object of the local treatment of bone and joint tuberculosis is to make conditions at the site of the lesion optimal for prompt and permanent healing. Usually this is accomplished by (1) initial immobilization to provide local rest from tension and compression stresses; (2) extirpation of the lesion, when feasible and essential to allow approximation of healthy tissues; and (3) arthrodesis if destruction of the joint has made permanent support necessary.

Immobilization of some degree is usually indicated from the time the diagnosis is first suspected to the end of the treatment period. It may be secured simply by rest in bed on fracture board and foam rubber mattress, or by traction, brace, or plaster cast. Rarely a case is seen so early that its small focus of tuberculous infection can be reached readily by antibiotics in the blood and healed spontaneously, requiring only prolonged external immobilization.

For most patients, however, surgical treatment is essential. The optimal time for operation is when the patient's resistance is strong (as indicated by apyrexia, a low sedimentation rate, and a high lymphocyte-monocyte ratio) and when the lesion is no longer advancing. Virtually complete excision of early lesions, such as a tuberculous synovitis or a small bone focus, is occasionally possible, leaving articular surfaces adequate for preserving joint mobility. Most patients are seen much later, however, when the joint structures have been extensively destroyed, when painless movement and stability cannot be restored, and when arthrodesis and bone grafting are indicated as the only practicable means of curing the infection and minimizing the permanent disability. With the arthrodesis, excision of the pathologic tissue can often be combined. Postoperatively the antituberculous chemotherapy should be continued for two years or longer, until mycobacterial activity can be presumed to be no longer present. The affected joint requires appropriate protection by means of a cast or brace until healing is sound and mature. The period of postoperative immobilization may last several months to a year or more. The patient's activity is increased gradually. To prevent possible recurrence he should be reexamined at increasing intervals for at least several years. Tuberculous sinuses and abscesses that persist despite medical treatment can often be successfully excised after the bone or joint lesion has regressed with immobilization and antibiotics. Abscesses that are causing symptoms but cannot be resected may be evacuated. Amputation is occasionally indicated, most often for extensive, secondarily infected tuberculosis of the ankle and foot of an adult.

Prognosis. The prognosis in bone and joint tuberculosis depends much upon early diagnosis and treatment. Most of the patients recover. Retention of joint mobility, however, is unusual; especially is this true in adults.

The outlook is relatively grave when the patient also has active pulmonary tuberculosis or when he has involvement of more than one major joint. Prolonged suppuration may lead to amyloidosis. Tuberculosis of the spine or hip is more serious than that of other joints.

TUBERCULOSIS OF THE SPINE

The spine is the commonest site of bone and joint tuberculosis, being involved in more than half the cases. Any level of the spine may be affected; the greatest incidence is in the lower thoracic region. Spinal tuberculosis is sometimes called *Pott's disease* because in 1779 Sir Percivall Pott described a painful deformity of the spine, accompanied by paraplegia, which is thought to have been tuberculous.

Pathology. Tuberculosis of the spine is thought to begin in the cancellous bone of a vertebral body (Fig. 4-8). Uncommonly it may start in posterior arch, process, or contiguous joint structures. The infection may extend beneath the anterior longitudinal ligament or directly across affected, narrowed disks to involve several adjacent vertebrae. In some instances multiple foci are separated by uninfected vertebrae. With increasing destruction the strength of the involved vertebral body becomes so impaired that it collapses under continued stress, producing a posterior protrusion of the spine, or *kyphosis* (Fig. 4-9). Healing takes place by gradual fibrosis and by new bone formation with resulting bony ankylosis of the collapsed vertebrae.

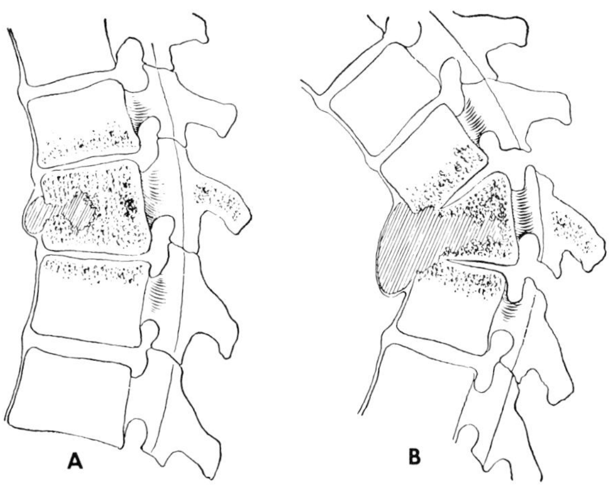

Fig. 4-8. Tuberculosis of the spine. Stages in the process of bone destruction and abscess formation: **A,** early; **B,** late. (After Calot.)

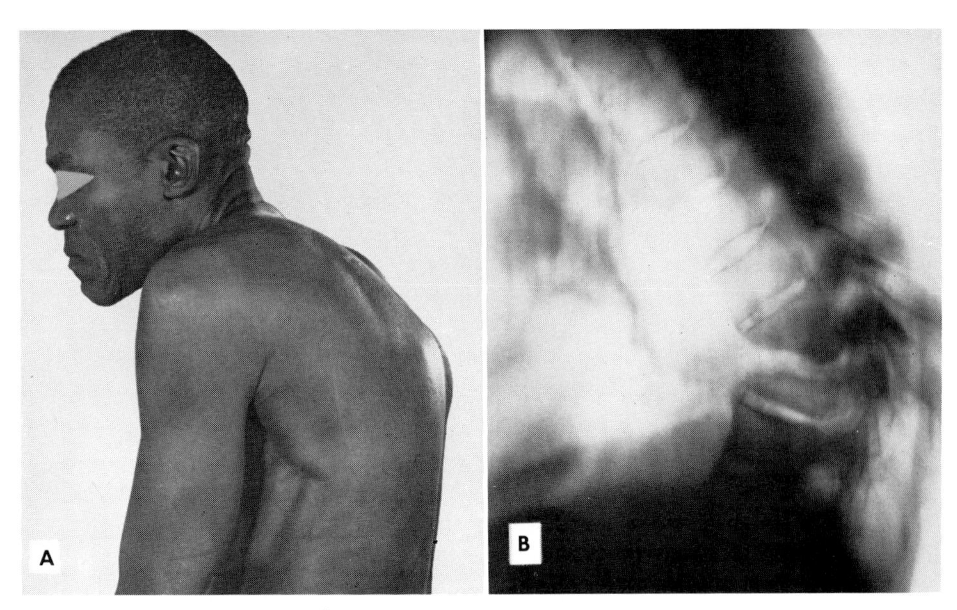

Fig. 4-9. Tuberculosis of the dorsal spine in a man 51 years of age. **A,** Photograph showing angular kyphosis. **B,** Lateral laminagram showing destruction and collapse of the ninth and tenth dorsal vertebrae.

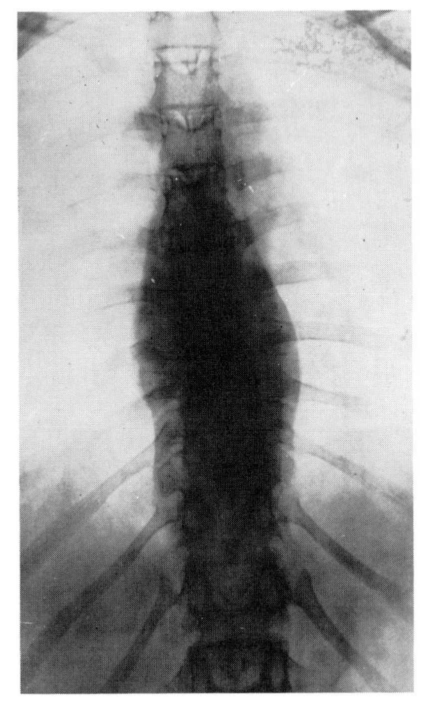

Fig. 4-10. Anteroposterior roentgenogram showing large spinal abscess in a 33-year-old man who had tuberculosis of the midthoracic vertebrae.

Abscesses. Some degree of abscess formation doubtless occurs in every case of vertebral tuberculosis. When treatment of the bony focus is delayed or ineffective, the abscess may become quite large. It may appear in anteroposterior roentgenograms as a fusiform or flask-shaped shadow encircling the spine at and just below the level of the involved vertebrae (Fig. 4-10). These abscesses tend to gravitate along fascial planes and to present themselves in characteristic locations. The typical lumbar abscess descends along the psoas fascial sheath; it may point posteriorly, laterally, or in the groin or upper, inner thigh as a *psoas abscess.* Untreated, such abscesses may rupture through the skin, producing sinuses that soon become secondarily infected.

Paraplegia. Paralysis of the lower limbs is seen most frequently in tuberculosis of the upper or middle thoracic vertebrae, where the spinal cord is relatively large in diameter, the spinal canal narrow, and the spine slightly kyphotic. The spinal cord may be compressed by an abscess, a caseating or granulating mass, or an edge of bone or disk that protrudes posteriorly as the kyphosis increases. Edema of the cord and thrombosis of local vessels may play a part. In some cases tuberculous granulation tissue has been shown to extend through the meninges to the spinal cord.

The first symptom may be a gait disturbance from inability to control the legs and feet. The paraplegia is usually of spastic type with hyperactive deep reflexes. The sphincters are sometimes involved; sensory changes are usually slight.

Treatment. The general antituberculous treatment, outlined previously (p. 119), is essential. The local treatment often consists of three phases: preoperative recumbency, surgical treatment, and postoperative immobilization.

Preoperative recumbency. Rest on a foam rubber mattress over a fracture board may be adequate to relieve stresses on the diseased vertebrae. More efficient support of the spine may be obtained by having the patient lie on a Bradford frame, which is a pipe frame across which canvas is tightly stretched. Body casts provide excellent immobilization. In some orthopaedic institutions full-length posterior plaster shells are used. Lesions of the cervical spine are often treated by recumbency with head traction or with a brace incorporating the head, neck, and trunk. Nonoperative treatment is ordinarily continued for at least several weeks and often much longer. If sufficiently prolonged, it may result in spontaneous fusion and healing of the affected vertebrae, in which case surgical treatment may be unnecessary.

Surgical treatment. Arthrodesis, or surgical fusion, of the spine is usually indicated to provide relatively prompt internal splinting and permanent rest of the affected vertebrae. The surgical technic that has been used most frequently is the posterior arthrodesis devised in 1912 by Hibbs. It consists of splitting the spinous processes into small fragments, chipping up the outer surfaces of the laminal arches, removing the cartilage from the articular facets, and placing the bone chips as bridges across the denuded bone of the posterior surfaces of the vertebral arches (Fig. 4-11). It is often advisable, especially in adults, to reinforce this bridge of bone chips with additional autogenous bone from the iliac crest or the tibia.

The more aggressive surgical treatment of approaching the spine anteriorly is advocated by Hodgson and others. It makes possible direct evacuation of pus, diseased tissue, and sequestra. Stability of the spine is then restored by the insertion of strut

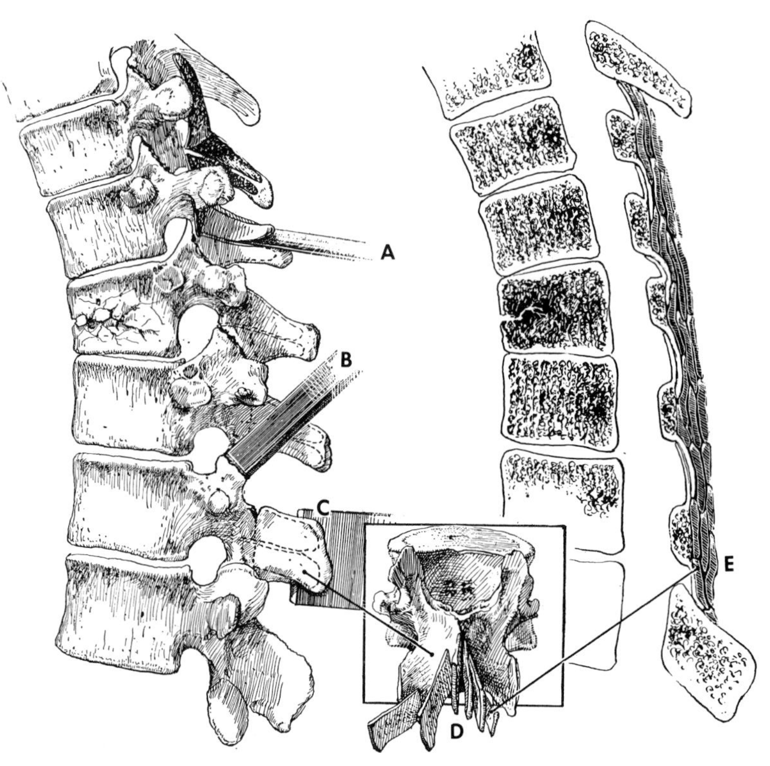

Fig. 4-11. Hibbs method of stabilization of the spine. **A,** Gouge splitting off and turning down chips from spinous processes and laminae. **B,** Chisel removing cartilage from articular facets. **C,** Splitting of spinous processes. **D,** Posterior view showing splitting of spinous processes. **E,** Sagittal section of spine showing continuous layer of small grafts posterior to the denuded laminae.

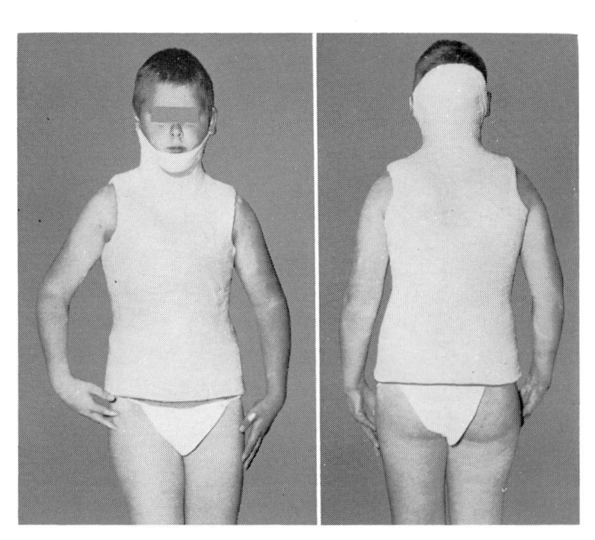

Fig. 4-12. Modified Calot jacket (sometimes called Minerva jacket) for immobilization of upper dorsal or lower cervical spine.

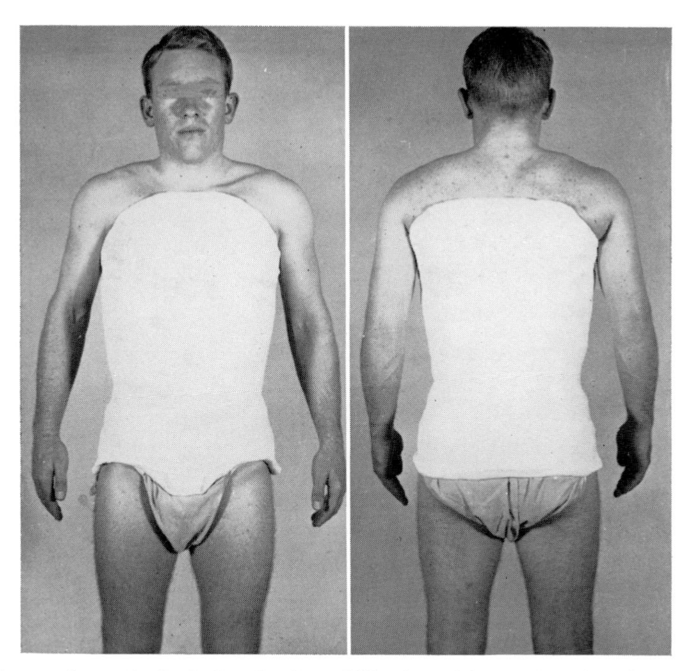

Fig. 4-13. Sayre plaster body jacket for immobilization of lower dorsal or lumbar spine.

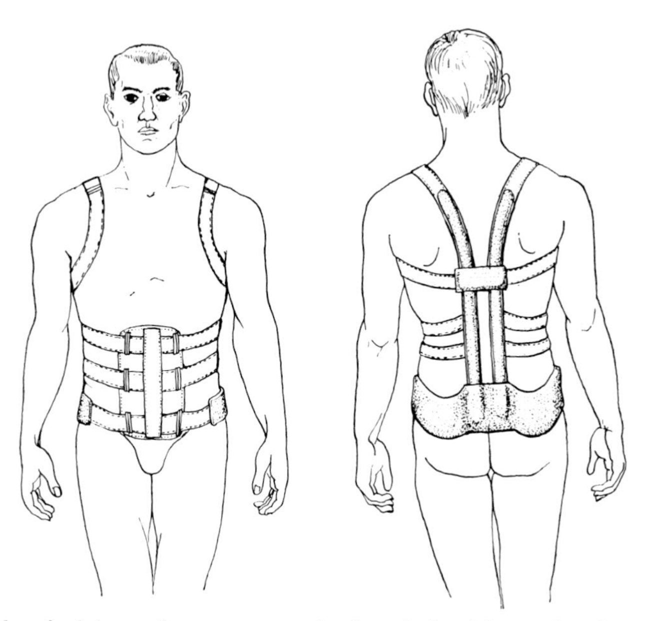

Fig. 4-14. High spinal brace for treatment of tuberculosis of lower dorsal or upper lumbar spine (modification of Taylor brace).

bone grafts between the vertebral bodies above and below the diseased area. This technic may be particularly helpful in patients with paraplegia. Evacuation of the paravertebral abscess, curettage of diseased bone, and relief of pressure on the spinal cord may also be accomplished by means of a costotransversectomy. Laminectomy is indicated rarely in patients with paraplegia.

Postoperative immobilization. Recumbency and antibiotic therapy are continued as healing of the spine takes place. A period of 3 to 6 months in bed is often advisable, after which the patient is allowed to be up with a plaster body jacket (Figs. 4-12 and 4-13) or a brace (Fig. 4-14) for perhaps another 6 months, the support being discarded when roentgenograms show mature bony fusion.

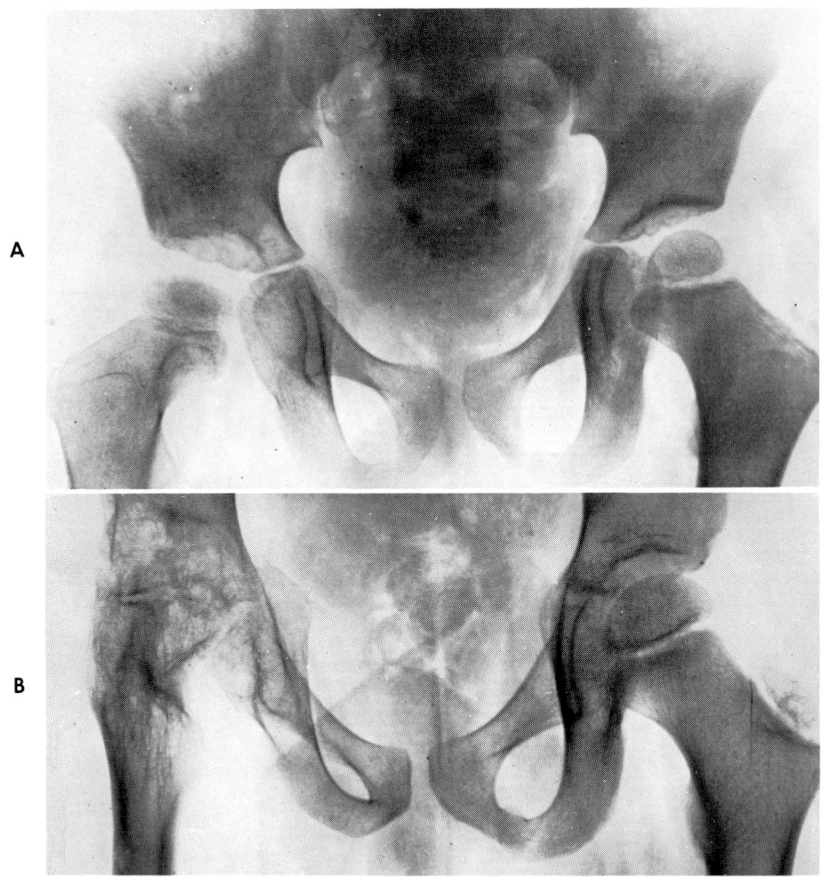

Fig. 4-15. Roentgenograms of tuberculosis of the right hip. **A,** Note changes in the inferior portion of the femoral neck, osteoporosis of the upper end of the femur and acetabulum, flattening of the capital epiphysis, and lateral displacement of the femoral head with relation to the acetabulum. **B,** Four years later, and 2 years after arthrodesis of the right hip: note continuous bony structure from the pelvis to the upper end of the femur, upward displacement of the femur, and marked atrophy of the femoral shaft.

TUBERCULOSIS OF THE JOINTS OF THE LIMBS

Of the limb joints, the hip is most commonly involved by tuberculosis, and next the knee. Tuberculosis of the ankle or foot sometimes responds poorly to treatment, especially in adults, and may even end in amputation. Although tuberculosis of the upper limb may involve any of the joints, it is much less common than that of the lower limb.

Tuberculosis of the hip (Fig. 4-15) may be considered typical of the infected limb joint. As in the case of spinal involvement, the clinical picture and the treatment vary greatly with the promptness with which the lesion has been recognized.

Clinical picture. In the *early case,* with a small tuberculous focus that is intraosseous or intrasynovial, there may be no symptoms except a limp and slight discomfort in the hip or referred pain in the knee. The only physical signs may be restriction of passive mobility by spasm, slight muscle atrophy, and possibly some tenderness over the hip. The tuberculin test is positive and the erythrocyte sedimentation rate in all probability increased. When a case is diagnosed and treated at this early stage, the patient has a good prognosis for retaining hip function.

In the *late case,* with tuberculous destruction of the joint surfaces, there may be severe muscle spasm, deformity, swelling of the joint, muscle atrophy, and some

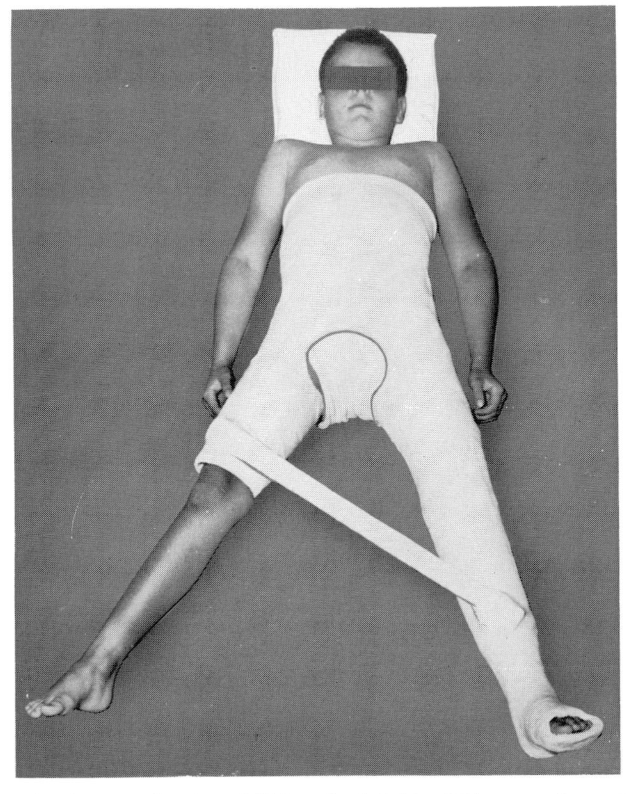

Fig. 4-16. Bilateral spica cast for immobilizing the left hip. This type of cast is useful in acute and chronic infections, injuries, and postoperative treatment.

shortening of the extremity. An abscess may appear, and the disability is severe.

Treatment. In the *early case,* in which intra-articular changes are minimal, healing without loss of joint mobility is the goal of treatment. Antituberculous drugs are given for many months, and the joint is put at rest by means of traction or cast (Fig. 4-16). Early operation may be indicated to eradicate necrotic tissue and pus about or in the joint. After the lesion has healed, weight bearing is cautiously resumed.

In the *late case* and in cases that have responded poorly to early treatment, extensive intra-articular destruction has usually occurred, painless mobility cannot be regained, and a fused hip becomes the therapeutic goal. A hip solidly ankylosed in good functional position is permanently strong, painless, and unlikely to undergo reactivation of the infection. As a rule, the optimal position (p. 165) consists of about 5 degrees of abduction, 5 degrees of external rotation, and 10 to 25 degrees of flexion. When intra-articular structures have been destroyed by tuberculosis, arthrodesis is usually advised as soon as the patient's general condition will permit. Tuberculous tissue is cleaned out of the joint, the cancellous bone of the femoral head is placed in contact with that of the acetabulum, and an iliac graft is fixed from ilium to femur. Subtrochanteric osteotomy may be done to relieve stress at the hip joint line. The operation must be followed by an ample and closely fitting bilateral spica cast for at least 6 to 9 months, or until roentgenograms show strong bony union, after which a gradual return to weight bearing under supervision is allowed.

SYPHILIS OF BONES AND JOINTS

Syphilis is still a common infection and a serious public health problem. However, with early penicillin treatment, premarital serologic testing, and supervised prenatal care, syphilitic bone and joint infections have become rare in many parts of the world. They occur in both congenital and acquired forms of syphilis. They may mimic many other diseases and are of chief interest from the standpoint of differential diagnosis. Several types warranting mention here affect different age groups.

In early infancy, *osteochondritis* is a characteristic manifestation of congenital syphilis. The lesions are often symmetric, involving the ends of the large long bones. Metaphyseal tissues may be replaced by syphilitic granulations, and the process of endochondral ossification is disturbed. Radiographically an irregular deposit of lime salts appears along the epiphyseal line. If the disease is unchecked, it may cause suppuration and the development of frank syphilitic osteomyelitis. Occasionally the epiphysis separates from the shaft and extreme distortion results. Clinically the joints become swollen and tender, but no local or general febrile reaction is present. Often the infant appears to be partially paralyzed; this condition has been called *pseudoparalysis.* If proper antisyphilitic therapy is instituted, the prognosis for recovery is good.

In childhood, *periostitis* of localized nature from congenital syphilis may produce a hard, dense enlargement of the convex side of the shaft of a long bone. When the tibia is so involved, the condition has been termed *saber shin.* Also in childhood a painless, bilateral, serous synovitis of the knee, called *Clutton's joints,* may develop gradually, persist for years without causing disability, and subside at about the twentieth year without residuals. Its recognition is important in obviating exploratory operation and in leading to antisyphilitic medication.

In adults, *periostitis, osteomyelitis,* and *arthritis* are seen in association with the *gumma,* or necrotizing, ischemic, proliferative lesion characteristic of late syphilis. There is little or no pain, and acute inflammatory signs are absent. The skull as well as the long bones may be involved. There may be diffuse thickening of the bony cortex with formation of many small hyperostoses, or the process may lead to suppuration and simulate chronic pyogenic arthritis. Pathologic fracture sometimes occurs. Joint involvement may produce a severe destructive arthritis. If adequate antisyphilitic treatment is begun early in the course of the osteomyelitis or arthritis, however, the prognosis for recovery is good.

Syphilitic disease of bone must be differentiated from sarcoma, tuberculosis, and metastatic carcinoma. Useful differential diagnostic data are obtained by serologic testing, roentgenographic findings, biopsy, and the response to antisyphilitic therapy.

FUNGUS INFECTIONS OF BONES AND JOINTS

Although the fungi that may infect bones and joints are of many varieties, mycotic osteomyelitis and arthritis are uncommon or rare entities. In some instances the disease is systemic, and the organisms reach a bone or joint focus through the bloodstream; in others, involvement takes place by direct extension from infected overlying tissues. The lesions, in general, are infectious granulomas; as a rule they are osteolytic, and differentiation from tuberculosis is sometimes difficult. Whereas serologic and skin tests may be helpful, final diagnosis depends upon identification of the organism in smear, culture, or histologic section. Some of these infections respond to specific antibiotic therapy; for some, surgical drainage, extirpation, or amputation is indicated.

Actinomycosis occurs about the jaw in more than half of the cases, involving soft tissue primarily and spreading later to the mandible. Indurated areas and deep abscesses are characteristic. A purulent exudate containing mycotic colonies, the so-called sulfur granules, is often extruded through multiple sinus tracts.

Blastomycosis may be disseminated widely from the lungs via the bloodstream, causing a chronic destructive osteomyelitis of vertebrae, ribs, skull, and other bones. In the spine, blastomycotic lesions simulate those of tuberculosis.

Coccidioidomycosis is endemic in the southwestern part of the United States. Primary pulmonary infection may be followed by a systemic phase characterized by multiple bone, joint, and visceral lesions and a grave prognosis.

Sporotrichosis is a disease of gardeners and farmers since the fungus is a saprophyte of plants and trees. A common type of infection begins with a trivial skin wound such as a thorn prick, which thereafter leads to a granulomatous lesion of skin and subcutaneous tissue. Dissemination via the bloodstream may produce a chronic osteomyelitis or arthritis similar in many respects to that caused by tuberculosis. The treatment often comprises a combination of chemotherapy and excision of infected tissue.

CHAPTER 5

Chronic arthritis

The most important cause of pains and aches in the joints is *chronic arthritis,* which in this context includes *rheumatoid arthritis, osteoarthritis,* and several related, less common forms of chronic joint disease.

In individual cases the disability caused by chronic arthritis varies from trivial and temporary to complete and permanent. In the aggregate, disability from chronic arthritis is enormous because of its high incidence in adult years and its chronicity. Statistical studies have indicated that in the United States more than thirteen million persons have chronic arthritis, and of these 10% are severely disabled.

The causes of chronic arthritis have not been established. In recent years investigation of this and related problems has been greatly accelerated. In the United States this research has largely been guided and financed by a number of scientific, lay, and governmental organizations dedicated to clearer understanding of arthritis and to better treatment of arthritic patients.

Most general practitioners and many specialists are concerned with evaluating and treating the patient who has chronic arthritis. To his therapy, orthopaedics can usually make a number of helpful contributions. In many early cases pain can be relieved and deformity prevented by the use of appropriate physical therapy, braces, casts, and orthopaedic surgical procedures of prophylactic nature, such as synovectomy. In later cases major improvement of deformity and disability can often be accomplished by carefully chosen orthopaedic operations and postoperative rehabilitation.

Until more accurate knowledge of its causes and the interrelationships of its subvarieties becomes available, no completely satisfactory classification of chronic arthritis can be devised. Two major types that can be clearly differentiated, however, and that warrant detailed consideration are *rheumatoid arthritis* and *osteoarthritis (degenerative joint disease).*

RHEUMATOID ARTHRITIS

Rheumatoid arthritis is a connective tissue disease characterized by chronic inflammatory changes in the synovial membranes and other structures, by migratory swelling and stiffness of the joints in its early stage, and by more or less deformity, ankylosis, and invalidism in its late stage.

Incidence. Rheumatoid arthritis is estimated to afflict 3.6 million persons in the United States. It occurs throughout the world, and its prevalence varies little with differences in climate. Women are affected almost three times as often as men; 80% of the cases begin in persons between 25 and 50 years of age; the highest incidence is found between 35 and 40 years.

Etiology. The cause of rheumatoid arthritis has not been determined. A slight familial tendency has been proved. Hypotheses of the etiologic factor have included infection, abnormality of the peripheral circulation, endocrine imbalance, metabolic disturbance, allergic phenomenon, faulty adaptation to physical or psychic stress, and many other concepts. Recent investigations suggest that autoimmune mechanisms may be the underlying cause and that proteolytic enzymes released from disrupted lysosomes within the joint may play a part in the chronic synovial inflammation and the destruction of articular cartilage. Despite the tremendous amount of data provided by recent laboratory and clinical investigation, no hypothesis is yet supported by enough evidence to be accepted as the true or major explanation of the disease.

Rheumatoid arthritis is currently regarded as belonging to a group of connective tissue diseases that exhibit somewhat similar clinical and pathologic changes. Other members of the group include systemic lupus erythematosus, polyarteritis nodosa, dermatomyositis, progressive systemic sclerosis, and rheumatic fever. Fundamental relationship between these diseases, however, is questionable.

Pathology. Of the pathologic changes that may be found in many parts of the body in rheumatoid arthritis, the most constant and most characteristic affect the joints.

The earliest change is inflammation of the synovial membrane. It becomes congested and edematous, is infiltrated first by polymorphonuclear leukocytes and later by lymphocytes and plasma cells, and is further thickened by the proliferation of synovial cells and villous hypertrophy. Peripheral portions of the articular cartilage are involved in the inflammatory reaction, undergo lysis, and are replaced by granulation tissue. Similar inflammatory changes take place in the subchondral marrow spaces, and there is resorption of bony trabeculae. A thin layer of granulation tissue, called a *pannus,* spreads from the synovial lining over parts of the articular cartilage, covering and eroding its surface. Enzymes released from the pannus are thought to cause the erosion. Adhesions of granulation tissue may undergo fibrosis and restrict joint mobility. The synovial fluid tends to be cloudy; its viscosity is low and its white cell content increased. In later stages the discolored and thinned articular cartilage becomes eroded, exposing areas of cancellous bone. Maturing granulation tissue between the bone ends may lead to fibrous ankylosis or, with ossification, to solid bony ankylosis. Unless preventive measures have been taken, the position of ankylosis may be one of deformity and severe disability. More frequently, especially in the joints of the fingers, the losses of cartilage and bone, together with damage to articular capsules and ligaments, produce instability and pathologic subluxation.

In the tendon sheaths, inflammatory changes similar to those in the joints lead to effusion and synovial proliferation. Smaller tendons become frayed; they may rupture, especially near a joint, where they may encounter friction over a bony prominence.

Many patients with rheumatoid arthritis show more or less typical pathologic changes in extra-articular tissues. The most characteristic histologic lesion is the granuloma-like *rheumatoid nodule* of chronically inflamed connective tissue, which has been found in the subcutaneous tissues in 15% to 20% of the cases. In the skeletal muscles, focal infiltrations of lymphocytes are a common finding, together with atrophy and fibrosis. Cardiac changes including pericarditis, myocarditis, coronary arteritis, and granulomatous inflammation of the valves, rarely of sufficient magnitude to cause symptoms, have been demonstrated frequently at necropsy.

Symptoms. The type of onset and the progression of rheumatoid arthritis vary widely. In 75% of the cases the onset is insidious. The joint symptoms may be preceded by periods of fatigue and lassitude. The joints may feel stiff, especially in the mornings. There may be fleeting pains in one or more joints. These are followed by joint symptoms of more persistent type, consisting first of constant pain, swelling, and stiffness. Additional joints may become involved. Later severe muscle atrophy develops. The joints may become deformed and ankylosed. The proximal interphalangeal and metacarpophalangeal joints of the hands are often affected first, and next the wrists and the knees. The pain is usually not limited to the joint but may be felt in adjacent muscles, tendons, and ligaments. The symptoms tend to be worse with bad weather, overexercise, and fatigue.

About one fourth of the cases begin more acutely, with an abrupt onset of pain, swelling, and tenderness in one or several joints. These changes may progress rapidly to produce severe deformities within a few weeks, or they may subside to a less acute phase.

The constitutional symptoms frequently include weakness and fatigue. Malaise and fever are common during periods of acute joint inflammation. The disease often seems to be influenced by emotional and psychologic factors.

Physical signs. Periarticular swelling is usually the earliest local sign. It is most characteristically found in the smaller joints of the hands and feet, especially in the

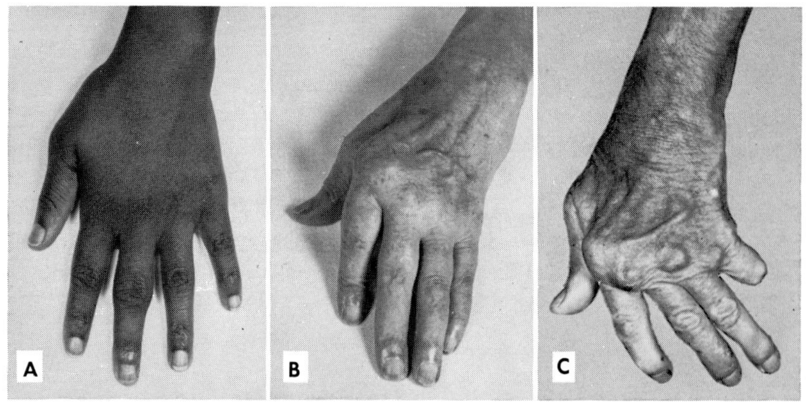

Fig. 5-1. Rheumatoid arthritis of the hand. **A,** Early stage. Note fusiform swelling of the proximal interphalangeal joints, especially that of the middle finger. **B,** Moderate involvement. Note swelling from chronic synovitis of the metacarpophalangeal joints, and early ulnar drift. **C,** Advanced stage. Note marked ulnar drift and subluxation of the metacarpophalangeal joints with extension of the proximal interphalangeal joints and flexion of the distal joints. Note also the deformed position of the thumb.

middle joints of the fingers, where it produces a spindle-shaped appearance (Fig. 5-1). Any joint of the body may be involved. In early stages the joints may be acutely tender; later they become less sensitive. The tenderness may extend into the adjoining muscles and tendon sheaths. There is slight increase of surface temperature over the joints, but redness is unusual. Muscular weakness and atrophy are quite characteristic. Flexion deformity may result from positions assumed in an effort to relieve pain.

In the extremities the skin may become thin and glossy. The fingers are often cool and moist. In 15% to 20% of the chronic cases, rheumatoid nodules occur in the subcutaneous tissue, especially about the elbows and the phalangeal joints; these nodules are usually less than a centimeter in diameter, firm, and nontender. Generalized lymphadenopathy of moderate degree is a common finding.

There are many phases and *variants* of rheumatoid arthritis, as well as possibly unrelated diseases that closely simulate it. Among them are the following:

1. *Juvenile rheumatoid arthritis,* or *Still's disease* (Fig. 5-2), an uncommon, crippling disease of childhood, with which is associated enlargement of the lymph nodes and spleen.
2. *Felty's syndrome,* a severe arthritis associated with leukopenia and splenomegaly.
3. *Reiter's syndrome,* an ill-defined disease occurring chiefly in adult males and characterized by the triad of polyarthritis, nongonorrheal urethritis, and conjunctivitis.
4. *Psoriatic arthritis,* the intimate association of rheumatoid arthritis and psoriasis. In these patients exacerbations and remissions of the two diseases coincide, the distal interphalangeal as well as other joints are involved, and resorption of bone may occur at the ends of the phalanges.
5. *Palindromic rheumatism,* characterized by repeated, brief episodes of acute arthritis without residual joint damage. This may be an atypical or prodromal form of rheumatoid arthritis.
6. *Intermittent hydrarthrosis,* a recurring joint effusion usually in the knee and

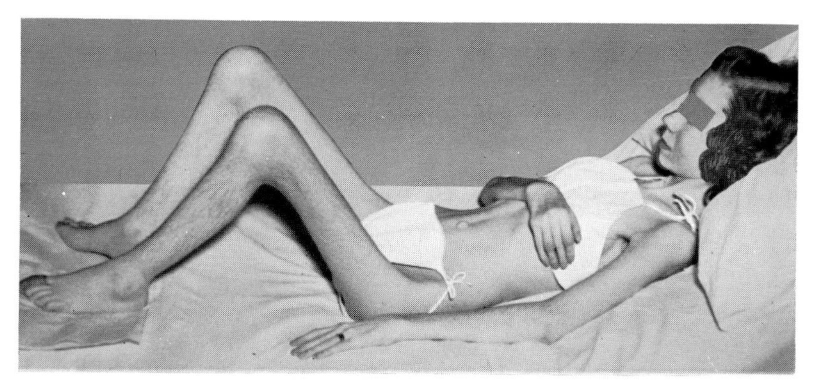

Fig. 5-2. Juvenile rheumatoid arthritis (Still's disease) in a 16-year-old girl. Note enlargement of the knees and ankles, flexion contractures of the hips and knees, equinus contractures, and severe muscle atrophy in the right hand, thighs, and legs.

possibly representing a mild phase of rheumatoid arthritis. It is accompanied by characteristic local symptoms and signs (p. 370).

Clinical course. Many patients improve spontaneously, but their recovery may be hastened by treatment. When the disease is of short duration, there may be no objective joint changes. A long course, marked by exacerbations and remissions, is frequently seen. In about half of the cases the disease is progressive, with severe joint involvement and permanent changes in the joint tissues. Deformity may develop because the joint is held in the position of greatest comfort—the knees, for example, are likely to develop a flexion deformity. Maintenance of the flexed position may be followed by capsular contractures, adhesions, and destructive changes within the joint. Subluxation and dislocation are common, especially in the joints of fingers and toes, and partial or complete ankylosis may occur. However, at any stage the disease may become quiescent, in which event gradual improvement of the general health follows.

Roentgenographic picture. Swelling of the tissues about the affected joint is usually visible in the roentgenogram. In early stages the joint cavity may be distended by increased synovial fluid. Bone atrophy, evidenced by rarefaction and greatest in the epiphyseal areas, gradually becomes conspicuous. The contour of the ar-

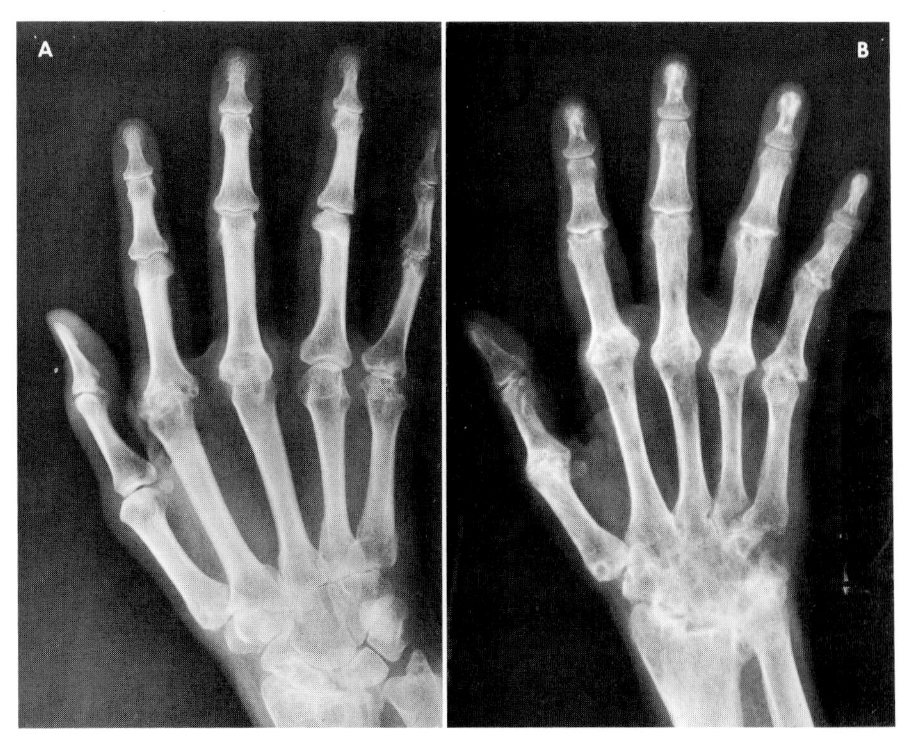

Fig. 5-3. Roentgenograms of rheumatoid arthritis in the hand and wrist. **A,** Moderate changes ranging from atrophic bone areas and narrowed cartilage spaces to subluxation of second and third metacarpophalangeal joints. **B,** More advanced case with severe destructive changes including multiple subluxations in the digits and ankyloses in the carpus.

ticular surfaces is not necessarily altered. The cartilage space later becomes narrowed, however, and small areas of bone absorption, so-called punched-out areas, may be seen (Fig. 5-3); the edges still remain distinct. Subluxation or dislocation may be evident. In advanced stages the articular margins may show bony spurs, ridges, and prominences. These hypertrophic features of the late case resemble the changes of osteoarthritis and tend to obscure the true atrophic character of the bone lesions. In late stages the articular surfaces may become ankylosed.

Laboratory findings. During active phases of the disease the erythrocyte sedimentation rate is usually elevated and a slight leukocytosis is occasionally present. In these patients there is usually a normocytic, hypochromic anemia refractory to iron.

The serum globulin is increased in relation to the albumin. In most cases the serum, apparently because of the presence of abnormal macroglobulins called *rheumatoid factors,* will agglutinate or flocculate suspended particles such as hemolytic streptococci, sheep erythrocytes sensitized with amboceptor, and latex or bentonite particles sensitized with human gamma globulin. Several tests based on this property of the serum in rheumatoid arthritis have a diagnostic specificity as high as 85%. Their results may not be positive, however, in the first weeks or months of the disease. They may be positive in certain connective tissue diseases and other disease states besides rheumatoid arthritis, but as a rule they are negative in juvenile rheumatoid arthritis, ankylosing spondylitis, and psoriatic arthritis.

The synovial fluid may appear cloudy. When acetic acid is added, its mucin clot is friable. Its viscosity is decreased. Its white cell count is high, varying with the severity of the arthritis. During acute attacks, polymorphonuclear leukocytes are most numerous; in chronic stages most of the cells are lymphocytes.

Differential diagnosis. As a rule, late rheumatoid arthritis is easily recognized; in early cases the diagnosis may be obscure. Affections requiring differentiation include acute rheumatic fever, osteoarthritis, pyogenic arthritis, gonococcal arthritis, tuberculosis, gout, and systemic lupus erythematosus.

Rheumatic fever in its acute inflammatory stage may present a migratory polyarthritis simulating that of early rheumatoid arthritis. The patient with acute rheumatic fever is more likely to have had a streptococcic throat infection and to have electrocardiographic changes and a high fever. Serologic tests may aid in the differentiation. After the acute phase of rheumatic fever the joints recover completely, contrasting with their usual course in rheumatoid arthritis.

Osteoarthritis, or degenerative joint disease, may be confused with rheumatoid arthritis, especially in the aged. The following are the most significant points in differentiation: (1) rheumatoid arthritis is a systemic disease and the patients are sick, whereas in osteoarthritis they are not sick; (2) acute inflammatory signs and cutaneous changes are more commonly associated with rheumatoid arthritis; (3) weight-bearing joints are more often involved in osteoarthritis; (4) in the fingers the proximal interphalangeal joints are more often involved in rheumatoid arthritis and the distal interphalangeal joints in osteoarthritis (Heberden's nodes, Fig. 5-4); (5) subcutaneous nodules are not seen in osteoarthritis; and (6) osteoporosis and bony ankylosis are uncommon in osteoarthritis. Erythrocyte sedimentation and serologic tests aid in differentiation.

In pyogenic arthritis, as a rule, a single large joint is involved. High fever and leukocytosis may distinguish the early case from rheumatoid arthritis, and destructive changes in the roentgenograms, the late case. Aspiration of the joint may yield pus, and a positive culture confirms the diagnosis.

Gonococcal arthritis may simulate rheumatoid arthritis. Careful questioning and thorough examination of the patient are important. Frequently in acute gonorrheal arthritis the gonococcus can be demonstrated in the joint fluid.

Joint tuberculosis in its early stages may be confused with rheumatoid arthritis; however, tuberculosis is more often monarticular, is more insidious in its onset, and is likely to show more bone destruction roentgenographically. Culture of the joint exudate or biopsy of the synovial membrane may be necessary to establish the diagnosis.

Gout must frequently be considered in the diagnosis. Its high blood uric acid is characteristic, especially in the absence of an increase of the nonprotein or urea nitrogen. It may be possible to demonstrate urate crystals in the joint aspirate by means of polarized light microscopy. In gout the joint quickly loses its tenderness between attacks, and the great toe is often the first part of the body to be affected (Fig. 5-13). Chronic gout is more often confused with rheumatoid arthritis than is acute gout. Roentgenographically, the punched-out areas in bone characteristic of gout may be found also in rheumatoid arthritis. Biopsy of synovial membrane or suspected tophi may be indicated.

Systemic lupus erythematosus may be difficult to distinguish from rheumatoid arthritis. The exact relationship between these two diseases of connective tissue is unknown. In lupus erythematosus the patient usually shows minimal joint changes, severe systemic symptoms, and the characteristic lupus erythematosus cell phenomenon.

Treatment. The treatment of rheumatoid arthritis must be suited to the individual patient after a careful survey of the various factors in his particular case. The most important elements in the treatment are general measures, drug therapy, local treatment of joints, prevention of deformity, and correction of deformity.

General measures. During acute stages of rheumatoid arthritis, the patient should be at complete rest in bed. A firm mattress and fracture board should be used. Attention must be paid to the patient's posture in bed. The ankles should not be held in equinus by tight sheets, the knees supported in flexion on pillows, nor the neck, low back, or hips kept in a flexed position for long periods while the patient sits up. A program of exercises for the recumbent patient should be prescribed and carried out regularly, with emphasis on major muscle groups in the trunk and lower limbs. Such exercises counteract muscle weakness from disuse while in bed and stimulate cardiac and respiratory function; they will facilitate the return to standing and walking after the acute illness has subsided. It should be explained early to the patient that the period of bed rest is only the necessary first step in a program of treatment planned for restoring him as quickly as possible to maximum functional capacity. His activity is increased progressively as his joints and his general health improve.

Other general measures, including attention to the patient's diet and correction of any anemia or infection, are indicated in rheumatoid arthritis as in other chronic debilitating diseases. Demonstration of the physician's sympathetic, reassuring interest in his patient's problems is an important part of the treatment.

Drug therapy. As yet no specific, curative drug for rheumatoid arthritis has been found. Among preparations in current use are salicylates, phenylbutazone, antimalarials, gold salts, steroids, and indomethacin.

The salicylates are most helpful in alleviating pain and suppressing inflammation. Plain, coated, or buffered aspirin is the most commonly used drug. It may be given in dosage of one to three 325 mg. (5-grain) tablets two to six times a day, depending upon the severity of the arthritis and the patient's tolerance.

Phenylbutazone will often relieve arthritic symptoms, but it is in no way curative. It may be given in an initial dosage of 600 mg. a day and gradually reduced to 100 to 200 mg. daily. If the toxic effects of a drug rash, thrombocytopenia, leukopenia, or symptoms suggesting a peptic ulcer appear, use of the drug should be promptly discontinued. Many rheumatologists consider phenylbutazone too dangerous to be given over a long period of time.

Antimalarial compounds, such as chloroquine phosphate, often seem to exert a slow but long-lasting suppressive effect in rheumatoid arthritis. They may produce toxic reactions, however, which may include retinal damage and which necessitate prompt discontinuance of the drug.

Treatment with soluble gold salts, or chrysotherapy, may suppress or even arrest the active rheumatoid process and so induce lengthy remission. Chrysotherapy is especially effective in early cases. Its mode of action is poorly understood. Gold salts are given by intramuscular injection in small doses over a long period of time. The usefulness of gold therapy is distinctly limited by its occasionally dangerous toxic effects on kidneys, intestines, and skin; toxic effects appear in from 10% to 20% of the treated patients.

The steroid drugs, cortisone and the new synthetic hormones, have a very limited place in the systemic treatment of rheumatoid arthritis. They are potent anti-inflammatory agents that may produce dramatic alleviation of symptoms and signs, reduction of fever, reversal of abnormal laboratory findings, and psychologic improvement. However, they do not alter the fundamental pathologic process of rheumatoid arthritis. Their suppressive action is temporary, whereas their prolonged administration may lead to adverse changes like those of adrenocortical hyperfunction in Cushing's syndrome. The long list of these undesirable effects includes such formidable complications as skeletal demineralization, pathologic fracture, peptic ulcer, edema from sodium and water retention, hypopotassemia and muscle weakness, depressive psychosis, reactivation of infections such as tuberculosis, and atrophy of the adrenal cortex. Continued investigation may discover a steroid derivative less likely to cause these complications. Currently, steroids should be reserved for temporary use in small doses as supplementary treatment for patients in whose joints rheumatoid inflammation is not adequately suppressed by full doses of salicylates.

Indomethacin is an anti-inflammatory drug sometimes helpful in rheumatoid arthritis. The recommended dosage is 25 to 50 mg. three times a day. Prolonged treatment with indomethacin may lead to gastrointestinal bleeding or ulceration. Indomethacin is contraindicated in children.

Local treatment of joints. Physical therapy is helpful in all stages of rheumatoid arthritis. Local applications of heat by means of warm, moist compresses or packs, infrared radiation, whirlpool baths, or diathermy may alleviate the acute pain. Paraffin baths are most helpful for arthritic hands and feet. Heat should be followed by gentle massage of the muscles above and below the affected joints. To minimize joint stiffness and muscle atrophy, active exercise of the affected joints, without weight bearing, should be carried out several times a day. In these exercises the joints should be put through their full range of voluntary motion. Pain and fatigue must be avoided. The amount of exercise may be increased gradually if there is no unfavorable reaction. After the acute stage of the disease has passed, most patients can be helped by occupational therapy. Posture exercises are often indicated.

In acute arthritis the pain is sometimes so severe that it is advisable to provide rest for the joint by immobilizing it in a light splint or plaster cast. Continuous immobilization periods of a few days to 2 weeks may be associated with significant decrease in the signs of inflammation. When the acute inflammatory reaction has subsided, a program of gentle, active exercises should be carried out, the casts or splints being removed for increasing periods as pain and tenderness decrease. Aspiration of synovial fluid will sometimes relieve the pain of an acutely distended joint capsule. Intra-articular injection of an anti-inflammatory steroid such as hydrocortisone may be helpful. Despite all technical precautions, multiple or repeated injections entail risk of infection. The repeated injection of corticosteroids into a weight-bearing joint may so relieve the pain that the patient damages the atrophic joint through overuse, producing articular changes similar to those of neuropathic arthropathy.

Excision of the lining membrane of a joint, or *synovectomy,* may be indicated as a local prophylactic measure in rheumatoid arthritis when synovial inflammation persists for months despite thorough nonsurgical treatment and seems likely to lead to serious damage of the articular cartilage. In such cases, synovectomy removes tissue that theoretically is producing immunologic abnormalities as well as cartilage-destroying enzymes. The exact indications and results of prophylactic synovectomy, however, have yet to be established. Synovectomy may be indicated for any of the limb joints; it is used most commonly for the knee, wrist, and metacarpophalangeal joints. *Tenosynovectomy,* the excision of tendon sheath lining, which in rheumatoid arthritis may be chronically inflamed and greatly thickened, is of most value at the wrist and metacarpophalangeal joints.

Prevention of deformity. The prevention of deformity in rheumatoid arthritis is of the utmost importance because of the frequency of subluxation and of fibrous and bony ankylosis. A most important preventive measure is to alternate periods of splinting in good functional position with periods of active range-of-motion exercises. When an ankylosis cannot be averted, however, care should be taken that the joint become stiff in the best position for function (p. 164). This can be accomplished by fixing the joint in its optimal position by means of traction, a plaster splint, or a brace.

Correction of deformity. Most deformities from rheumatoid arthritis can be either greatly improved or corrected by orthopaedic means without resort to surgery. A common example is flexion contracture of the knee, a disfiguring and disabling deformity making it impossible for the patient to stand or walk in anything approaching normal fashion. Often the knee flexion can be gradually overcome by skin traction without too much discomfort; care must be taken to prevent posterior subluxation of the upper end of the tibia. An alternative treatment is the use of a long leg cast, which is cut transversely back of the knee and wedged at intervals of a few days into increasing extension or is straightened gradually with turnbuckles. Such casts must be well padded and used with caution to avoid causing pressure sores. Forceful manipulation to correct the deformity, with or without anesthesia, is usually best avoided because of the likelihood of causing additional injury to already damaged and weakened joint structures. To prevent recurrence of the flexion deformity the goal of treatment must be complete extension, and it should usually be maintained for at least several months by means of a long leg brace.

Arthritic deformities that fail to respond to closed methods can often be corrected by surgical means. Such operations must be chosen with great regard for the qualifications and needs of the individual patient. A resistant knee flexion contracture may be treated by *capsulotomy,* or surgical release of the constricted or adherent joint capsule. Secondary contracture of an overlying tendon may require *tenotomy* and lengthening. A knee joint chronically enlarged by arthritic thickening of its lining membrane may be improved by *synovectomy.* The crippling deformity of bony ankylosis in flexion is correctable by *osteotomy.*

In some joints the stiffness that follows rheumatoid arthritis can be treated successfully by *arthroplasty* (p. 166). As a rule the hip, elbow, metacarpophalangeal joints, and, to a less extent, the knee are the joints most favorable for arthroplasty. To mobilize the stiffened hip the orthopaedist must decide whether to advise a cup or interposition arthroplasty, a partial or complete replacement arthroplasty, an osteotomy, or other type of reconstructive operation. Mobility can be restored to an ankylosed jaw by resecting the condyle of the mandible. Surgical fusion, or *arthrodesis* (p. 164), may be indicated for an arthritic joint when maximum stability and freedom from pain are desired.

A spine partially ankylosed in a position of kyphosis can sometimes be improved by mobilization with exercises and recumbency in hyperextension; a strong back brace is then applied to help maintain the corrected position. Bony ankylosis in severe kyphosis may be improved by vertebral osteotomy; the operation must be done with great care to prevent injury of the spinal cord. For the wrists and fingers, intermittent support in functional position and elastic or spring tension splinting to overcome contractures are of the greatest value. More advanced arthritic disabilities of the hand can often be helped by carefully selected surgical procedures, followed by splinting and therapeutic exercises.

The patient who has been severely disabled by arthritis is usually most grateful for any improvement of joint position or motion, although it be only the ability to put one hand behind the head to comb the hair. When a deformity is permanent, even its partial correction may be extremely helpful.

Prognosis. When rheumatoid arthritis is recognized early as a systemic disease and is treated with vigor and persistence, reasonably good results can usually be obtained. Patients with slight changes in the joints may recover spontaneously; in late cases, long remissions may occur. With proper orthopaedic therapy the serious consequences of the disease can usually be prevented or greatly ameliorated.

OSTEOARTHRITIS (DEGENERATIVE JOINT DISEASE)

Osteoarthritis is a form of chronic arthritis found commonly in middle-aged and elderly people, affecting especially the weight-bearing joints, and characterized by degenerative changes in articular cartilage and bony overgrowth at the joint margins.

Etiology. The cause of osteoarthritis is obscure; it seems to include mechanical, dystrophic, and genetic factors. Degenerative changes in articular cartilage are more common and more severe with advancing age, in the weight-bearing joints, and in joints that have become incongruent or have been used abnormally. The fact that osteoarthritic changes are often localized to only part of a single joint suggests that there are causative factors other than age and attrition. Mechanical injury, which may consist of a single major trauma or of repeated minor traumas, may cause intra-articular changes that act as a predisposing or aggravating cause. In people over 45 years of age whose parents have suffered from arthritis late in life, there is a greater susceptibility than in those who have no family history of arthritis. Although no clear-cut relation between endocrine abnormalities and osteoarthritis has been established, degenerative joint changes occurring in women at the menopause have sometimes been spoken of as *climacteric* or *menopausal arthritis.*

Pathology. In osteoarthritis, in contradistinction to rheumatoid arthritis, characteristic early changes appear in articular cartilage rather than in the synovial membrane. The cartilage seems to lose some of its ability to withstand mechanical stresses. A number of microscopic degenerative changes have been described, such as swelling of the chondrocytes and chondromucoid softening of the matrix. Grossly the cartilage undergoes splitting, fibrillation, gradual thinning, and widespread degeneration. In late stages the underlying bone may become denuded. About the cartilaginous edges, reactive chondro-osseous spicules, or *osteophytes,* form quite early. These appear in roentgenograms as spurs and lipping. Occasionally the osteophytes become detached and form loose bodies within the joint. As degeneration of the cartilage progresses, the underlying bone usually becomes eburnated and deformed. There are often significant changes of shape in the end of the bone, increased bony deposits in and about the areas of cartilaginous degeneration, and erosion of bone beneath the destroyed cartilage. Subluxation may occur. The synovial membrane and periarticular tissues often show little change but may become considerably thickened. The amount of synovial fluid may remain normal but may increase considerably if the joint is subjected to excessive use.

Symptoms and signs. In the early stages of osteoarthritis the patient usually complains of stiffness of one or more joints, associated with an aching pain in or about the affected joint. The involvement is more often monarticular than polyarticular. This stiffness tends to become less noticeable after moderate use of the joint. Continued use is followed, however, by great discomfort, which may be relieved by rest, support, and heat. The patient tires easily on exertion. His symptoms are worse in cold, wet weather. In this stage there is slight enlargement of the affected joints, which may be slightly tender about their margins; these changes are usually most noticeable in the fingers and knees. Bony enlargement of the distal interphalangeal joints *(Heberden's nodes)* is one of the commonest signs (Fig. 5-4).

Later in the course of the disease there are considerable limitation of joint mo-

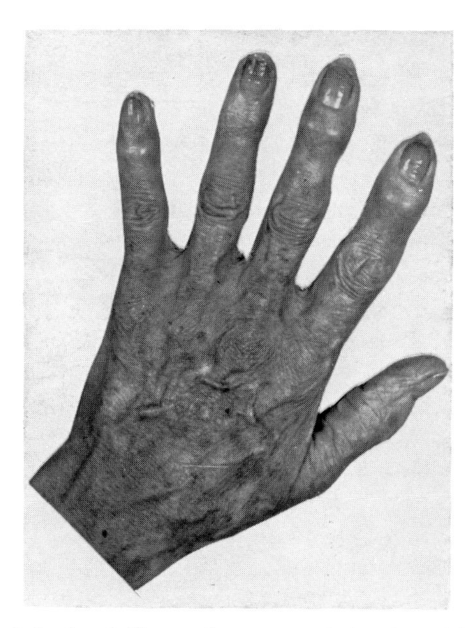

Fig. 5-4. Osteoarthritis of the hand. Note enlargement of the distal joints of the index, middle, and little fingers (Heberden's nodes).

tion and disability, especially in the larger weight-bearing joints. Pain may then be present while the joint is resting, as well as when it is in motion. Malalignment of the joint is a frequent result of the irregular degeneration and loss of articular cartilage. Crepitation may be noticed frequently, and loose intra-articular fragments, especially in the knee, may produce transient locking. Examination at this stage reveals moderate swelling and puffiness and a loss of the normal joint contour.

A tendency to early fatigue is characteristic of osteoarthritis. Many of the patients are obese. There may be a disturbance of body mechanics with a protuberant abdomen and abnormal weight-bearing lines in the feet, knees, and hips, resulting in chronic strain of these joints. Generalized arteriosclerosis is often associated.

Clinical course. The symptoms of osteoarthritis usually exhibit a slowly progressive and often intermittent course. Periods of improvement are frequent, especially under proper treatment. Trauma may intensify and prolong the symptoms. More disability follows involvement of the spine, hips, and knees than that of the upper limbs. As the disease progresses in the spine, there may be restriction of the normal range of motion, localized pain from impingement of bony overgrowths, and radiating pain from pressure of bony overgrowths on nerve roots. Osteoarthritis may progress to the point of causing an extreme disability of one or more joints but, unlike rheumatoid arthritis, seldom results in bony ankylosis except in the spine.

Roentgenographic picture. Roentgenographic signs of osteoarthritis may antedate the symptoms, or the reverse may occur. Thinning of the degenerated cartilage appears as narrowing of the cartilage space. Proliferation of bone produces sharpening of intra-articular prominences, spurs, lipping, bridging between adjacent vertebral bodies, and condensation of articular margins. At the same time cystlike spaces frequently appear in the subjacent cancellous bone. Bone atrophy may be noticeable.

Laboratory findings. There are no diagnostic laboratory findings. The blood picture, including the sedimentation rate, is usually normal. The synovial fluid may contain cartilage fibrils and a somewhat increased number of white cells.

Diagnosis. The diagnosis can usually be made after careful analysis of the history and of the joint changes as shown by physical and roentgenographic examination. Osteoarthritis is sometimes difficult to distinguish from rheumatoid arthritis, and the two may coexist. Osteoarthritis may be simulated also by gouty, tuberculous, or neuropathic joint disease. Osteitis deformans should be excluded. Occasionally neoplasms must be considered in the diagnosis. In doubtful cases biopsy may be indicated.

Treatment. Treatment includes general measures, drug therapy, local treatment of joints, prevention of deformity, and surgical measures.

General measures. It is important to relieve the patient's apprehension and secure his constructive attitude toward treatment by explanation and reassurance. Usually he may be told that although his joint structures cannot be restored to normal, the osteoarthritic process is essentially benign and will not shorten his life or cause complete joint stiffness, and that as a rule, the symptoms can be satisfactorily controlled by appropriate treatment.

Adequate rest is important. Afternoon rest periods may be helpful. The patient must be taught to adjust his physical activity to a level that his affected joint or joints will tolerate. He should steer a middle course, not allowing himself to become either fatigued from overexercise or weak from inactivity. In some instances a change of occupation is required.

The patient should also be instructed in a simple program of brief range-of-motion, nonweight-bearing exercises. Such exercises, done once or twice daily, will do much to maintain joint mobility, ward off symptoms, and improve body mechanics.

Since obesity is often an aggravating factor when the weight-bearing joints are affected, gradual reduction of weight by dieting is frequently indicated.

Drug therapy. There is no medication specific for osteoarthritis. Salicylates, as in other joint diseases, relieve pain. Aspirin may be given in doses of one to three 325 mg. (5-grain) tablets three to five times daily. Phenylbutazone may be used but because of its potential toxicity must be carefully controlled. Steroid hormones have proved to be of no permanent value and should not be given systemically in osteoarthritis.

Local treatment of joints. Some degree of rest of the involved joint is nearly always beneficial. For weight-bearing joints, either a restriction of standing and walking or the temporary use of crutches or cane may be indicated. Local rest may be facilitated by means of a splint or brace, but no joint should be completely fixed for long periods. Usually physical therapy proves helpful in relieving the symptoms. Heat in the form of baking, hot packs, whirlpool or paraffin baths, or diathermy is indicated. Massage of the muscles above and below the affected joints may make the patient more comfortable. Active, nonweight-bearing exercises are helpful in preventing muscle atrophy but should not be continued to the point of fatigue.

Injections of a steroid such as hydrocortisone into painful joints sometimes give temporary relief; such injections may be repeated several times. Frequent injections, however, greatly increase the risk of infection and can lead to destructive changes similar to those of neuropathic joint disease.

Prevention of deformity. The likelihood of developing fixed deformity is much less in osteoarthritis than in rheumatoid arthritis. Most of the deformity is caused by changes in the shape of the articular ends of the bones. At times, however, the hips may become flexed and adducted, the knees may develop flexion, varus, or valgus deformity, and the feet may become pronated. Selected exercises and splinting will do much to prevent such deformities.

Surgical measures. In osteoarthritis, surgical treatment is often indicated if the joint changes have become so advanced that nonoperative measures no longer control the symptoms effectively. Operations are used primarily to relieve pain on motion and secondarily to correct deformity, restore mobility, or provide stability.

Loose bodies, which occasionally appear in osteoarthritic joints and cause pain and locking, should be removed. Synovectomy may be indicated if the lining membrane is chronically thickened and joint movement is painful. Osteotomy is often helpful in improving alignment and relieving pain. Arthrodesis is ordinarily followed by permanent freedom from pain. Arthroplasty may improve the pain without producing stiffness; it is especially useful for one major joint when the involvement is bilateral. Careful selection must be made among these and other surgical procedures in the treatment of the osteoarthritic hip (p. 152) and the osteoarthritic knee (p. 154).

Prognosis. The prognosis for relief of the symptoms and control of the arthritic process is better in osteoarthritis than in rheumatoid arthritis. The damaged joints can be made less disabling and less painful by well-advised treatment. Despite roentgenographic evidence of advanced changes, the patient will experience remarkably little pain when a suitable therapeutic regimen is prescribed and diligently followed.

CHRONIC ARTHRITIS OF INDIVIDUAL JOINTS

Chronic arthritis of the spine, hip, knee, and hand presents individual problems that warrant separate consideration.

Chronic arthritis of the spine

Rheumatoid arthritis

Incidence. Rheumatoid arthritis of the spine often accompanies rheumatoid arthritis of the joints of the extremities but may occur independently. It is more common in women than in men and is seen most frequently in the early decades of adult life, between the ages of 25 and 45 years.

Pathology. The pathologic findings are essentially the same as those described for rheumatoid arthritis elsewhere in the body. The initial change is thought to be a synovitis of the facet joints. The vertebrae and cartilages become atrophic. In the cervical spine, subluxations of slight degree are common. In the later stages of rheumatoid arthritis the spine is more likely to develop fibrous ankylosis than bony ankylosis. The subcutaneous nodules of rheumatoid arthritis may be present.

Clinical picture. Back pain and stiffness usually begin insidiously and increase slowly. The discomfort is worse on bending over and lifting and is relieved but does not subside completely with recumbency. The symptoms often begin in the cervical or upper dorsal region; in other patients the lower dorsal or upper lumbar level may be affected first. At times the pain may radiate to the front of the chest and abdomen or down the legs. Coughing and sneezing often cause sharp pain in the back. It is not uncommon for the symptoms to appear in recurrent attacks that may follow minor infections, excessive exercise, or slight trauma. As the condition progresses, muscle spasm and contractures may lead to dorsal kyphosis and secondary postural changes in the cervical and lumbar regions. The stiffness gradually becomes greater but usually does not progress to bony ankylosis. As a rule, rheumatoid arthritis of the spine is accompanied by rheumatoid arthritis in several joints of the extremities; often the smaller joints are so involved.

Diagnosis. The diagnosis is based on the character of the onset, the findings on clinical examination, positive serologic tests for the rheumatoid factor, the roentgenographic evidence, and the frequent association with rheumatoid involvement of other joints.

Differential diagnosis. Rheumatoid arthritis of the spine must be differentiated from other causes of pain and stiffness in the back. Most important of these are ankylosing spondylitis, osteoarthritis, tuberculosis, malignancy, intervertebral disk lesions, vertebral epiphysitis, fracture, sprain, and visceral disease with referred pain.

Treatment. Indicated general measures and drug therapy are discussed on p. 136. Local measures used for rheumatoid arthritis of the spine are similar to those for ankylosing spondylitis (p. 147).

Ankylosing spondylitis (Strümpell-Marie arthritis)

Ankylosing spondylitis, although sometimes described under rheumatoid arthritis, has a number of characteristics that distinguish it as a separate entity. It was first described clinically in 1884 by Strümpell in Germany and in 1898 by Marie in France. It is characterized by ossification of the ligaments of the spine and by involvement also of the hips and shoulders. The sacroiliac joints are affected early in its course. Its cause is unknown. Factors related to sex and age may play a part since ankylosing spondylitis is ten times as common in men as in women and usually begins between the ages of 20 and 30 years.

Pathology. The pathologic changes begin like those of rheumatoid arthritis and proceed gradually to an extensive bony ankylosis. Progressive ossification occurs in the capsular and other intervertebral ligaments, fusing the lower and often in later stages the entire spine into a single mass. The costovertebral joints may become ankylosed. In the intervertebral disks, peripheral ossification takes place much earlier than does central ossification. The vertebral bodies tend to become osteoporotic; their contour changes little except in the dorsal spine, where they often become wedge shaped.

There may be irregular spur formation and lipping of the vertebrae. The typical gross specimen has an appearance suggesting that liquid bone has been poured upon the anterior surface of the vertebral column and has congealed as it flowed down (Fig. 5-5).

Clinical picture. The onset may be associated with acute pain, but on the contrary, stiffness without pain is sometimes the first symptom. The pain first appears

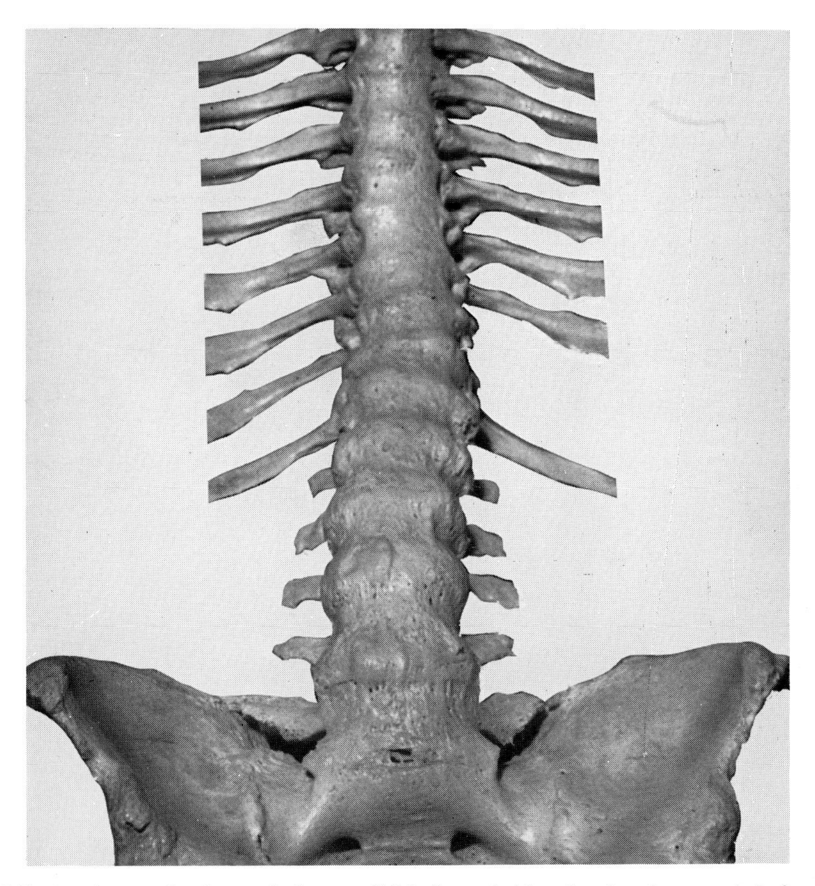

Fig. 5-5. Specimen of advanced Strümpell-Marie arthritis, showing bony ankylosis of the vertebrae and sacroiliac joints.

in the hips, buttocks, or lumbosacral region. Low back pain that comes on while the patient is recumbent in early morning hours suggests the diagnosis. Decreased spinal mobility and diminished chest expansion are characteristic signs. As the disease progresses, kyphosis and flattening of the chest frequently develop unless preventive measures are taken. The pain, followed by stiffening, may progress very slowly from the low back farther and farther upward until finally the entire spine becomes ankylosed (Fig. 5-6). The pain tends to lessen as the stiffness increases. The arthritic process may extend to involve the hips and shoulders. If the hips become stiff, a characteristic gait develops in which the pelvis is rotated from side to side to compensate for the lack of hip motion. In typical cases there is no involvement of the smaller joints.

Roentgenographic picture. Roentgenographic changes may not appear until many months after the onset of symptoms. As a rule, the earliest findings are changes in density along the margins of the sacroiliac joints. As the disease progresses, small erosions and narrowing appear in the apophyseal joints. Late changes include obliteration of the sacroiliac and apophyseal joint spaces, ossification of the periphery

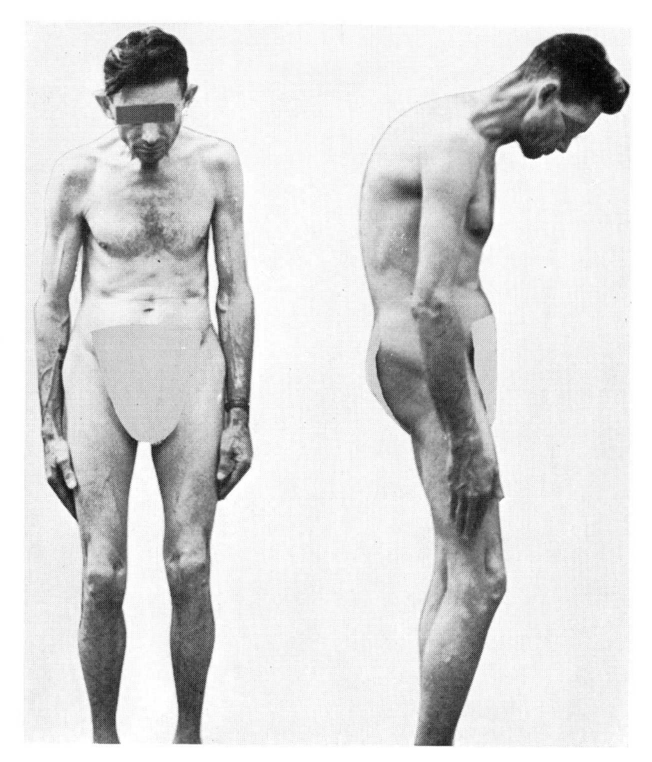

Fig. 5-6. Strümpell-Marie arthritis in a 46-year-old man with ankylosis of the entire spine in faulty position.

of the annulus fibrosus (giving the appearance sometimes called bamboo spine), and bony bridging of the laminae and spinous processes (Fig. 5-7).

Diagnosis. The diagnostic criteria are similar to those of rheumatoid arthritis of the spine. Characteristic features, however, are the high incidence in young men, the onset of symptoms and signs in the low back and their slow proximal progression, the roentgenographic changes in the sacroiliac joints, the absence of arthritis in the smaller joints of the extremities, and the absence of subcutaneous rheumatoid nodules.

The sedimentation rate is elevated in 80% of the cases. There is often a mild hypochromic anemia. Tests for rheumatoid factor are negative.

Differential diagnosis. The differential diagnosis is similar to that of rheumatoid arthritis.

Treatment. The treatment of ankylosing spondylitis includes general measures, local nonsurgical measures, and surgical treatment.

General measures. General measures include ample rest and a well-balanced diet. Control of pain may be provided by regular doses of aspirin. If the pain is intense and salicylates prove inadequate, phenylbutazone may be used; it is particularly effective in this disease, but proper regard must be paid to its toxicity. Satisfactory relief of the symptoms and signs by administration of indomethacin has been reported. Only rarely is steroid therapy indicated for ankylosing spondylitis.

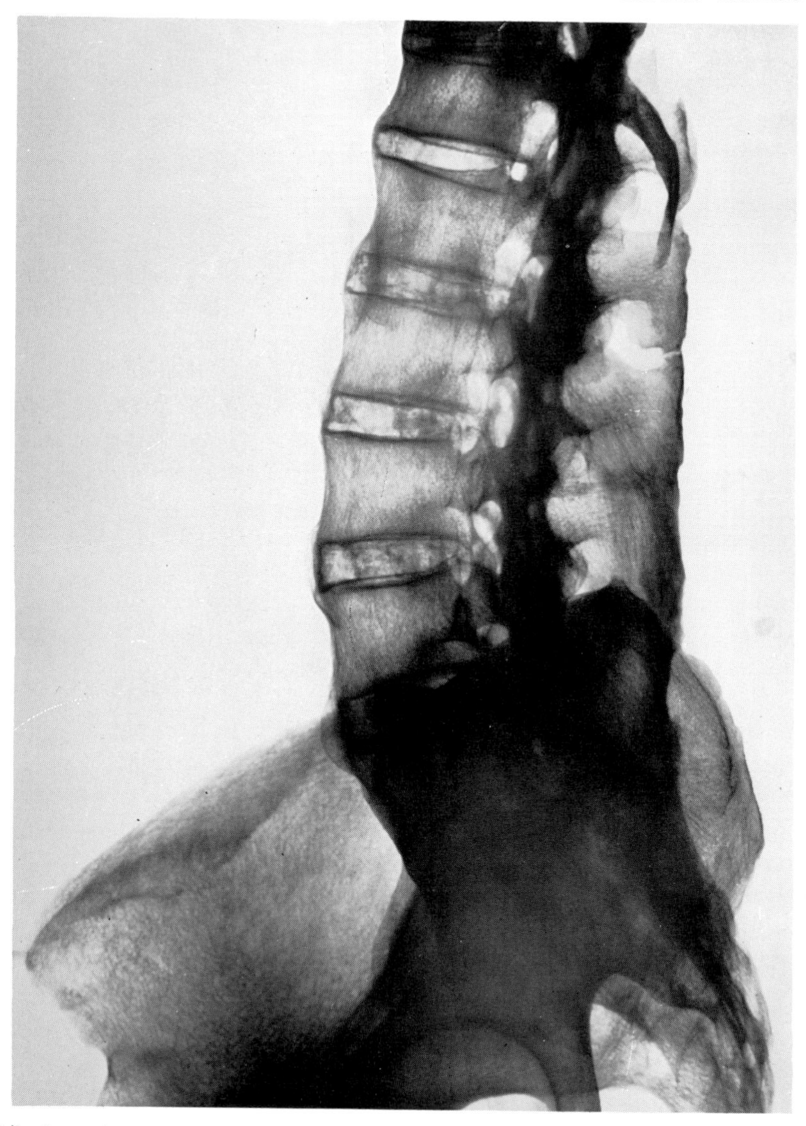

Fig. 5-7. Lateral roentgenogram of the specimen in Fig. 5-5, showing ossification between the vertebrae anteriorly and posteriorly as well as in the intervertebral disks.

Local nonsurgical measures. Local treatment is used for the relief of pain and the prevention and correction of deformity. During an acutely painful stage of the disease, bed rest is indicated. Because of the great tendency to kyphosis a firm mattress or bed board should be used, and the patient should rest without a pillow. Daily massage of the back, preceded by hot packs or other form of heat, may be helpful. Hyperextension exercises should be given, and the position of extension should be maintained. When the patient begins to stand, the use of a three-point brace helps to prevent the development of kyphosis. Breathing exercises are useful in maintaining vital capacity and in lessening chest deformity.

Surgical treatment. In late cases, osteotomy of the spin may be indicated to improve a severe, incapacitating flexion deformity. The procedure entails risk of damaging the spinal cord. Moderate improvement may be obtained more safely by bilateral pelvic osteotomy.

Osteoarthritis

Incidence and etiology. Osteoarthritis of the spine is seen most often in stocky and obese persons over 40 years of age and is more frequent in males than in females. It is much more common than either rheumatoid arthritis of the spine or ankylosing spondylitis. It is most often observed in the lumbar spine. The repeated minor traumas of constant use of the back probably constitute an important causative factor. Faulty body mechanics probably play a part by putting added strain upon localized areas of the spine.

Pathology. The pathologic changes of osteoarthritis described previously (p. 140) take place in the spinal diarthroses, or intervertebral joints. Although the articular cartilage becomes frayed and thinned, these joints do not undergo bony ankylosis. Marked eburnation of bone may occur about the articular facets.

Thinning of the intervertebral disks and lipping and spur formation at the anterolateral margins of the vertebral bodies result from disk degeneration and reactive bone production; although these changes, often termed *spondylosis* or *spondylophytosis,* usually accompany osteoarthritis of the spinal diarthroses, they probably constitute a related but distinct degenerative process (p. 373). In later stages the vertebral bodies become flattened and much new bone may develop about their margins, producing so-called bridging and leaf formation. Such changes, demonstrable in roentgenograms, are present to some degree in a high percentage of old people who have never had back pain; this is especially true of heavy individuals.

Clinical picture. After slight trauma or strain there may be complaint of pain in the lower part of the back, with considerable stiffness and lameness. Stiffness may be most noticeable on first getting up in the morning. Pain that radiates around toward the chest or abdomen or down the legs or arms is often present; such pain may be a result of compression of nerve roots, caused by small outgrowths of bone near the intervertebral foramina. Radiating pain is particularly common in involvement of the lumbosacral joints and of the intervertebral foramina of the cervical spine. Examination of the patient's back is likely to show muscle spasm, restricted mobility, pain at the extremes of motion, and very little tenderness. Deformity of the lumbar spine, often consisting of lateral curvature and a decrease of the normal lordosis, may develop. Ankylosis of some areas of the spine may follow. The lumbar and lower dorsal spine is more often involved than higher levels of the back and frequently is the first to become stiffened.

Diagnosis. The diagnosis is made from the history, the physical findings, and the associated roentgenographic changes of lipping, spurs, and irregular bone formation about the vertebral margins (Fig. 5-8).

Differential diagnosis. The differential diagnosis is similar to that of rheumatoid arthritis of the spine. Osteoarthritis must also be differentiated from neurogenic arthropathy and from osteitis deformans. It should also be differentiated from the localized vertebral lipping associated with collapse or fibrosis of an intervertebral disk

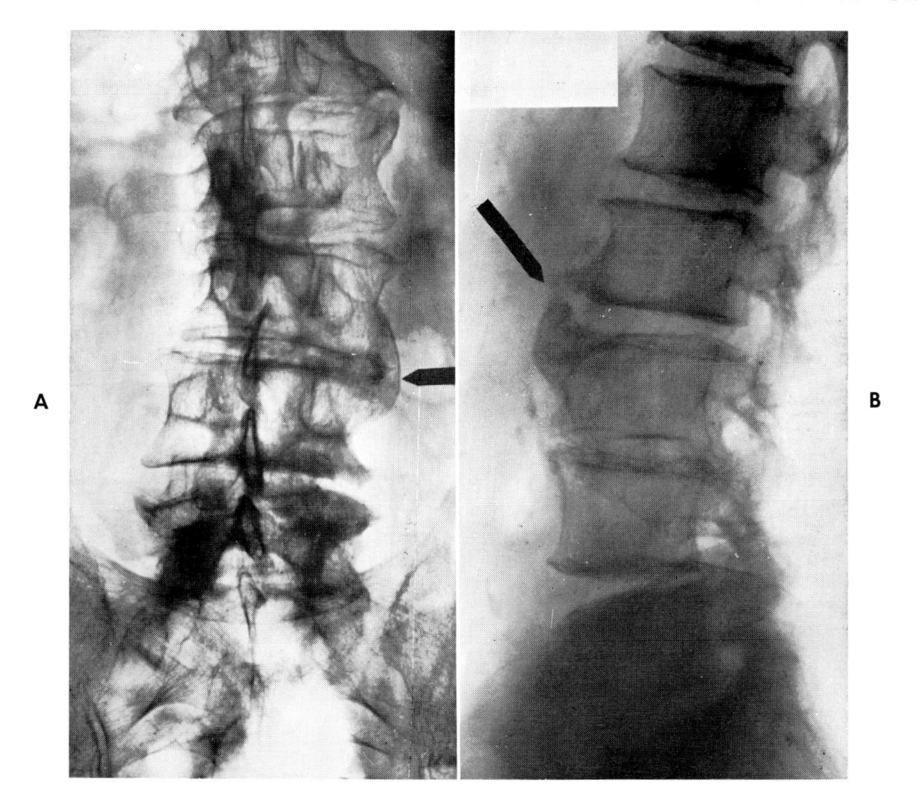

Fig. 5-8. Osteoarthritis of the lumbar spine in a man 70 years of age. **A,** Anteroposterior roentgenogram showing deformity of the spine with spur formation and a large bridge of new bone. **B,** Lateral roentgenogram showing spur formation, almost complete anterior bridging between the second and third lumbar vertebrae, and narrowing of the third lumbar interspace.

called spondylosis, and from the traction spurs that may follow localized injury of an intervertebral articulation.

Treatment. Many of the measures used in treating ligamentous and muscular strains of the low back (p. 314) are applicable to osteoarthritis. During painful periods, rest and restriction of activity are necessary. Salicylates are useful in relieving discomfort, as are the local application of hot packs and the use of massage. Obesity should be corrected by dietary measures. A well-fitting back support is helpful in controlling pain and in preventing recurrent attacks.

Spinal operations are seldom necessary. When there is localized pain with roentgenographic evidence of bone changes over a small segment of the spine and when nonsurgical treatment has failed to produce a satisfactory result, an arthrodesis may be indicated for permanent relief of the discomfort.

Chronic arthritis of the hip

The hip joint may be involved primarily in either rheumatoid arthritis or osteoarthritis. In either type the hip affection is often a part of a widespread arthritic process. Osteoarthritis virtually limited to the hip, however, is a common, painful, and disabling affection of degenerative nature in middle and late adult years.

Osteoarthritis

Etiology. Although the cause of osteoarthritis of the hip is not thoroughly understood, trauma associated with incongruity of the joint surfaces is thought to be the most important factor in many cases. Osteoarthritis may not follow immediately after a major injury or repeated minor injuries of the hip but may develop insidiously years later. It is often a late sequel of congenital dysplasia of the hip, reduced congenital dislocation, or congenital coxa vara. It often appears many years after an epiphyseal disturbance such as coxa plana or slipping of the capital femoral epiphysis. Avascular necrosis of the femoral head, of whatever origin, is likely to result in distortion of the articular surface, leading sooner or later to osteoarthritis of the hip.

Pathology. The initial changes, which have been shown to begin in areas of the femoral head not subject to weight-bearing pressure, are fibrillation, erosion, and thinning of the articular cartilage. The cartilage becomes irregularly worn away over the head and in the acetabulum and is replaced by hard, eburnated bone. Proliferation of new bone around the head may take place, in the form of a collar. New bone is formed also around the margin of the acetabulum. The acetabulum grows larger as the head of the femur flattens. The femoral neck may become short and broad and often shows increased anteversion. In addition to spurring and gross changes

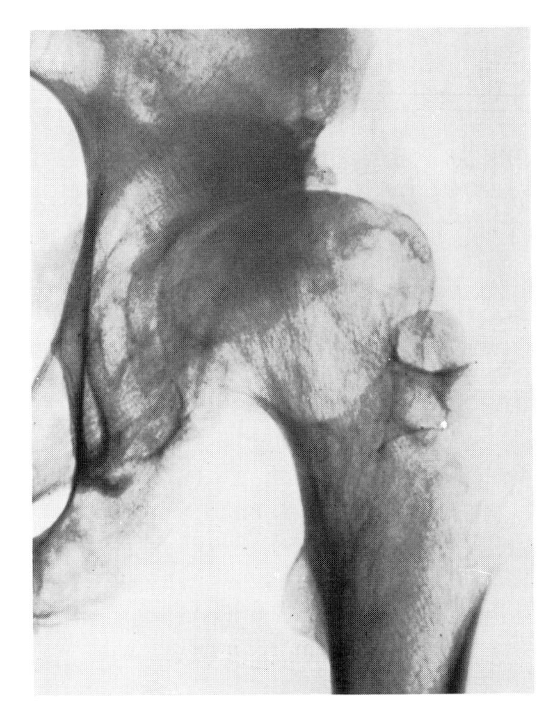

Fig. 5-9. Anteroposterior roentgenogram showing osteoarthritis of the hip in a 57-year-old man. Note narrowing of the articular cartilage space, flattening and widening of the femoral head, bony proliferation about the acetabular margin, and areas of rarefaction and of sclerosis in the head and about the acetabulum. The oblique acetabular roof and shallow acetabulum suggest that dysplasia preceded the osteoarthritis.

of contour, roentgenograms show narrowing of the cartilage space, irregular areas of increased bone density, mottling of the head and neck, and numerous small cyst-like areas near the articular surfaces (Fig. 5-9). The synovial membrane usually becomes thick and fibrous. It may be completely replaced by fibrous connective tissue.

Clinical picture. Often both hips are involved; when this is the case, the changes are usually more advanced on one side than on the other.

The onset is insidious. Although stiffness without pain may be the first symptom, most patients complain initially of aching in the hip region, together with slight stiffness or pain down the thigh to the knee after excessive exercise; the pain is completely relieved with rest. After standing or sitting there may be a slight catching sensation and pain in the hip on starting to walk. As the condition progresses, these symptoms appear most often in the morning and wear off during the day. After activity, however, the symptoms recur with increasing regularity. Gradually muscle spasm develops, causing flexion and adduction of the hip. Associated with the limitation of motion, there is usually discomfort on pressure about the joint. Later, slight shortening of the leg and a limp become noticeable. Creaking and grating may develop in the joint with motion, and the muscles atrophy. Later there is considerable pain on weight bearing. Radiating pain often extends down the front of the thigh to the knee but is sometimes greater on the posterior aspect. The limb gradually becomes externally rotated, and the prominence of the greater trochanter increases. Abduction is impaired early, and internal rotation, as well as all other motions, becomes decreased in range. As the deformity of the hip increases, the added stress on the low back may cause troublesome symptoms. Arthritic changes may develop in the lumbar spine as a result of strain. Recurrent, protracted periods of severe pain and disability are characteristic of the later stages.

Diagnosis. As a rule, the diagnosis is not difficult. The roentgenographic findings, as well as the history and clinical signs, are characteristic.

Differential diagnosis. Ischemic necrosis of the femoral head, tuberculosis, neuropathic joint disease, rheumatoid arthritis, and subacute or late forms of pyogenic arthritis may simulate osteoarthritis of the hip.

Prognosis. There is little possibility of complete restoration of function. If treatment is started early and is continued faithfully, however, the pain may be relieved. The mobility of the hip usually becomes quite restricted, and the limitation may progress to complete stiffness. When a flexion and adduction deformity develops, the disability is severe because of postural shortening of the leg.

Treatment. The treatment includes general measures, local nonsurgical measures, and surgical treatment.

General measures. General measures include weight reduction by low-calorie diet, when the patient is obese, and regulation of physical activity to ensure adequate rest from prolonged weight bearing. For limited periods analgesic medication may be indicated.

Local nonsurgical measures. In early cases the relatively mild hip pain can often be relieved for long periods by conservative forms of local treatment. Heat and massage are often helpful. Active, nonweight-bearing exercises of the hip should be carried out daily in an effort to preserve its range of motion. If the extremity has become shortened, a heel lift may be tried. Weight-bearing stresses on the affected hip can be lessened by using a cane in the opposite hand; for short periods crutches are sometimes indicated. As a rule, ischial-weight-bearing braces are difficult for the middle-aged patient to wear effectively and comfortably. For temporary relief during an acutely painful episode, intra-articular injection of a steroid such as hydrocortisone may be helpful; repeated injections are contraindicated by the danger of introducing infection. For hip pain that persists despite these measures, a period of several weeks of recumbency with traction on the affected lower extremity and daily physical therapy may be most helpful. On returning to activity the patient can continue to use the traction at night.

Surgical treatment. In the later stages of osteoarthritis of the hip the only effective treatment is surgical. Operation is indicated if the patient's pain is intolerable despite thorough trial of other forms of treatment and if his general condition is good enough to withstand major surgery. Of the many operations devised for the osteoarthritic hip, none completely abolishes the symptoms. Selection of the procedure best suited to the needs of the individual patient requires experience and judgment.

Since the primary object of treatment is to relieve hip pain, *neurectomy,* or section of the sensory nerves to the hip joint, is sometimes done when more strenuous surgery or long convalescence is contraindicated. The results of sensory neurectomy, however, have been undependable and often disappointing.

The most useful types of surgical treatment are arthroplasty, arthrodesis, and osteotomy.

By *arthroplasty* both pain and stiffness in the osteoarthritic hip can often be satisfactorily reduced. Arthroplasty is used more freely in older than in younger individuals because its early good results sometimes deteriorate in later years. The *cup* or *mold interposition arthroplasty* (Fig. 5-19), pioneered by Smith-Petersen, includes placing a metal cup between the reshaped acetabulum and remodeled head. In *femoral replacement arthroplasty* (Fig. 5-20), the head of the femur is excised and replaced by a metal prosthesis with a long stem that traverses the medullary canal of lower neck and upper shaft. Operations to replace both acetabular and femoral surfaces of the joint with an articulated metal prosthesis have been developed and are undergoing clinical trial; their early results are encouraging.

Since *arthrodesis* provides complete and permanent relief of hip pain it is usually the operation of choice for young, active individuals with monarticular disease, despite the drawbacks of long postoperative convalescence, hip stiffness, and possible later discomfort in the low back. The procedure usually includes complete removal of articular cartilage from the acetabulum and femoral head, bridging from ilium to femur with massive bone grafts, metallic internal fixation, and the application of a hip spica cast.

The pain of osteoarthritis of the hip can often be relieved by a subtrochanteric *osteotomy,* which presumably redistributes to some extent the weight-bearing stresses in and about the joint. McMurray's operation of osteotomy at the level of the lesser trochanter and medial displacement of the femoral shaft, followed by internal fixation or a hip spica cast or both, is frequently used. Other procedures include varus or valgus osteotomy, in which an alteration is made in the neck-shaft angle of the femur. This shifts the area of maximum pressure on the femoral head and often improves hip function.

Chronic arthritis of the knee

The knee joint may be affected in both major types of chronic arthritis. Usually both knees are involved. The etiologic factors and pathologic changes are essentially the same as those previously discussed (pp. 131 and 140).

In osteoarthritis the knee is the most frequent of the large joints to be involved. Most of the affected individuals are heavy, and the increased weight is probably an important predisposing factor. Old ligamentous injuries, resulting in instability of the knee, and old meniscus derangements may also lead to osteoarthritis.

Clinical picture. Chronic arthritis of the knee often starts with slight stiffness in the joint after sitting, or a tightness in the back of the knee associated with creaking. It becomes difficult to walk up stairs. Tenderness may be accompanied by swelling and increase of joint fluid, which together with muscle atrophy give the knee a spindle-shaped contour. Attempts to straighten the knee may cause pain, and a flexion contracture may develop. Occasionally in chronic cases the tibia may become subluxated backward; this occurs more often in rheumatoid arthritis than in osteoarthritis. In osteoarthritis the roentgenographic appearance of the knee (Fig. 5-10) is typical, with spurs and lipping about the margins of the femur and tibia as well as around the articular surface of the patella. Within the joint are often found loose bodies that have broken off from the small marginal projections of bone. Advanced destructive changes may lead to instability of the joint and a varus or valgus deformity that increases when the patient bears weight on the knee.

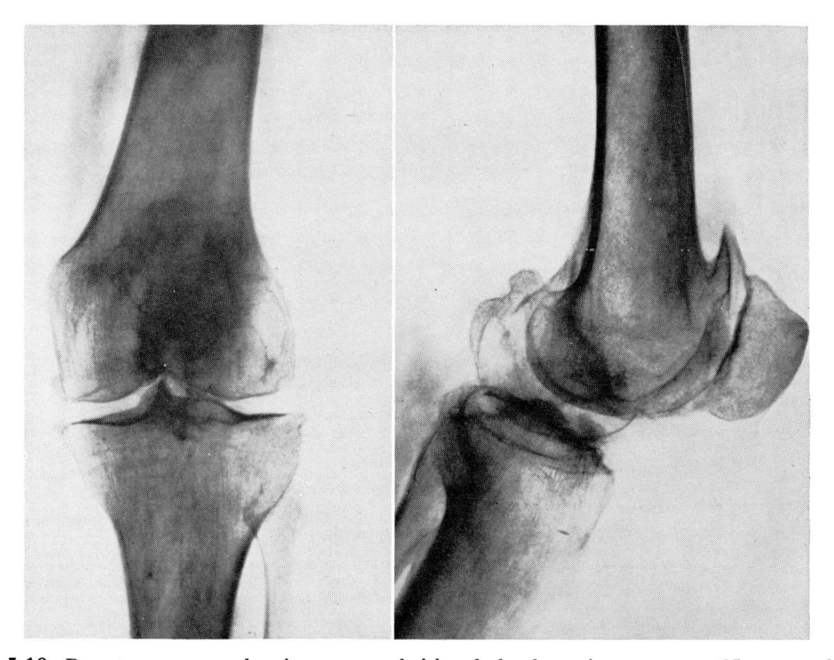

Fig. 5-10. Roentgenograms showing osteoarthritis of the knee in a woman 65 years of age. Note spurs about the margins of the femoral and tibial condyles, the tibial spine, and the upper and lower poles of the patella, as well as a possible loose body in the intercondylar fossa.

Diagnosis. The diagnosis is made from the clinical findings and the roentgenograms. In differential diagnosis, synovitis of other types, tuberculosis, and neuropathic joint disease must be considered.

Prognosis. In advanced rheumatoid arthritis of the knee the outlook for recovery of function is poor. In early osteoarthritis the symptoms can be relieved, although slow progression of the degenerative changes usually continues.

Treatment. The general treatment has been outlined (pp. 136 and 142). Locally, rest is important; the strain of weight bearing should be relieved. The knee should be protected with an elastic support, brace, or cast, and crutches may be advisable. Hot applications may relieve the pain. Regulated exercise of quadriceps and hamstring muscles will maintain muscle tone and assist in preventing contracture.

In rheumatoid arthritis, knee flexion contracture may be very difficult to correct. Traction or wedged casts may be used, followed by a brace to maintain extension. Posterior *capsulotomy* may be necessary. If the knee has retained mobility in flexion, supracondylar *osteotomy* of the femur will allow correction of the unsightly flexion deformity, preserve the range of motion, and improve stance and gait. At the same operation any valgus or varus deformity may be corrected. Bony ankylosis in flexion is also correctable by osteotomy.

In rheumatoid arthritis, *synovectomy* may be indicated when the inflammatory process has persisted despite a thorough trial of nonsurgical treatment, or when it has subsided but left persistent pain, synovial thickening, and restriction of knee motion. For the knee with moderate or localized destructive changes, arthroplastic technics of inserting metallic interposition devices to replace one or both of the tibial or femoral condyles are being developed and show considerable promise in carefully selected cases. Advanced arthritic changes with pain and instability may be treated best by *arthrodesis* of the knee when the homolateral hip and ankle and the opposite knee have retained a useful range of painless motion. When all of the articular structures have undergone severe destruction, when the pain of weight bearing is severe, and when no great amount of walking will be required, excision and replacement of the joint with a hinged metallic prosthesis (Fig. 5-21) may be indicated as a salvage procedure. This operation is especially useful in late rheumatoid arthritis of the knees.

In osteoarthritis with pain and crepitation limited to the anterior aspect of the knee and unrelieved by nonoperative treatment, *patellectomy* is often indicated. In selected cases it may be advisable to combine patellectomy, synovectomy, and excision of degenerated cartilages and bony spurs; after so extensive an operation several weeks of traction and physical therapy are essential. Varus or valgus deformity may indicate a high tibial *osteotomy* to correct the tilt and so relieve instability and pain on weight bearing. Similar benefit can sometimes be achieved by an operation to elevate an osteoarthritic tibial condyle or to replace it with a metallic prosthetic device. The late results of these recently devised arthroplastic technics are yet to be thoroughly evaluated.

Chronic arthritis of the hand

Rheumatoid arthritis, which often begins in the proximal interphalangeal joints of the fingers, leads frequently to severe deformity and disability of the hands.

Osteoarthritis, affecting chiefly the distal joints, may cause disfigurement but rarely results in severe loss of function.

Rheumatoid arthritis and its surgical treatment

In the hand and wrist the lesions of rheumatoid arthritis are characteristic and progressively disabling. The thumb is often drawn into adduction, the fingers tend to deviate toward the ulnar side, and individual digits may develop grotesque deformities and severe restriction of function. Arthritic destruction at the wrist may result in dorsal subluxation of the distal end of the ulna, medial subluxation of the carpus on the radius, and radial deviation of the hand. The inflammatory synovial reaction not only damages the joints but also invades tendon sheaths and tendons to produce a variety of deformities, many of which can be helped by surgical means.

Metacarpophalangeal arthritis. The metacarpophalangeal joints and their surrounding structures are particularly vulnerable to destruction by rheumatoid arthritis. Inflamed synovium distends the joint capsule and the overlying dorsal hood. This permits the extensor tendons to shift to the ulnar side of the metacarpophalangeal joints. As a result, progressive ulnar drift takes place. Continued synovitis with stretching of the capsule and collateral ligaments results in volar subluxation of the proximal phalanx on the metacarpal head. Early synovectomy may prevent or postpone the onset of severe deformities. If ulnar deviation of the fingers is present, synovectomy may be combined with realignment of the extensor tendons by imbrication of the radial side of the dorsal hoods. When subluxation and instability of the metacarpophalangeal joint have developed, arthroplasty or arthrodesis may be needed.

Intrinsic contracture. The severe periarticular inflammatory reaction of rheumatoid arthritis often leads to contracture of the small intrinsic muscles. Such contracture results in flexion of the metacarpophalangeal joints and hyperextension of the proximal interphalangeal joints, followed by a secondary flexion deformity of the distal joints. This is referred to as a *swan-neck* or *intrinsic-plus deformity* (Fig. 5-11). The surgical release of intrinsic contractures may improve both the function and the appearance of the hand.

Involvement of tendons and nerves. The extensor tendons may be invaded by inflamed synovium and so weakened that tendon rupture occurs. An arthritic nodule may form within a tendon and impair its strength. Other contributing causes of tendon ruptures are abrasion and attrition over a prominent bony spicule, and ischemia from compression by a hypertrophied paratenon within an inelastic fascial tunnel. Ruptures are seen most frequently in the extensors of the little, ring, and middle fingers and the long extensor of the thumb. They occur frequently at the wrist, beneath the dorsal carpal ligament. Function can often be restored by tendon transfer and, occasionally, by tendon suture after removal of an arthritic spur; as a rule, tenosynovectomy and synovectomy are carried out at the same operation.

Arthritic destruction of the middle slip of an extensor tendon, and volar subluxation of the lateral bands, may lead to a boutonnière deformity, in which the proximal interphalangeal joint becomes flexed and the distal joint hyperextended. Surgical repair of the defective extensor mechanism back of the proximal interphalangeal joint may be helpful.

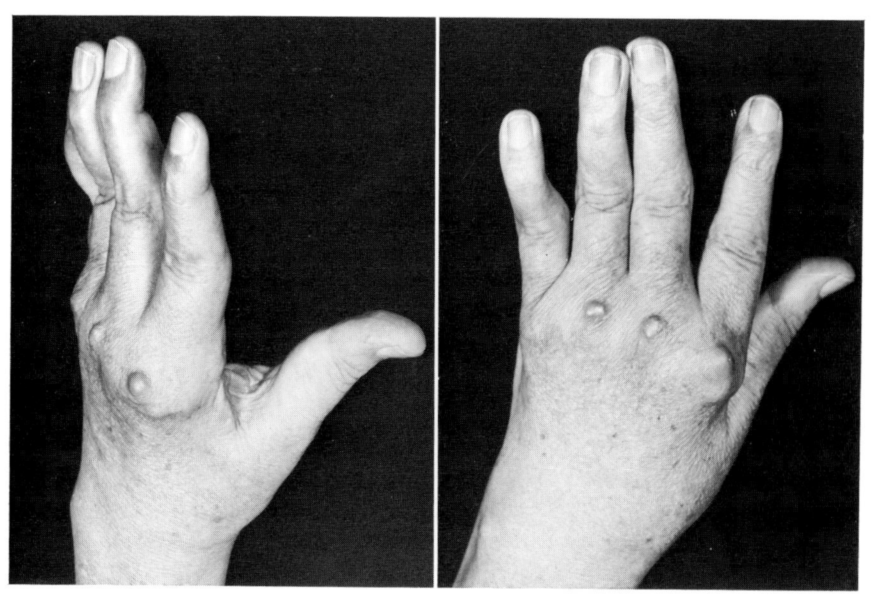

Fig. 5-11. Swan-neck deformities of the fingers in rheumatoid arthritis. Note the flexed proximal, hyperextended middle, and flexed distal joints associated with contracture of the intrinsic muscles. Note also several subcutaneous rheumatoid nodules and atrophy of the first dorsal interosseous muscle.

Chronic synovitis in the flexor tendon sheaths may produce the symptoms of a trigger finger (p. 435). The finger may lock in either flexed or extended position. To restore function, incision of the fibrous sheath is required. The synovial lining of the flexor tendons within the carpal tunnel may become so inflamed that swelling in this area compresses the median nerve, producing the carpal tunnel syndrome (p. 205). To prevent or relieve median nerve paralysis, surgical division of the transverse carpal ligament may be necessary.

The many hand deformities caused by rheumatoid arthritis must be evaluated on an individual basis and in the context of the patient's overall arthritic problems. Surgical procedures, chosen with these considerations in mind and followed by selected exercises and dynamic splinting as needed to prevent recurrent stiffness or deformity, often afford the disabled patient significant improvement in the use of his hands.

HEMOPHILIC ARTHRITIS

Prolonged and repeated hemorrhage into joints is a common manifestation of hemophilia, a hereditary disease of the blood that occurs in males and is carried in females by a sex-linked recessive gene. Repeated hemarthroses lead eventually to articular changes similar in many respects to those of other types of chronic arthritis.

In hemophilia the excessive bleeding results from a defect in the clotting mechanism caused by lack of an essential serum protein. In the classic and commonest type of hemophilia, hemophilia A, the deficient protein is antihemophilic globulin, or clot-

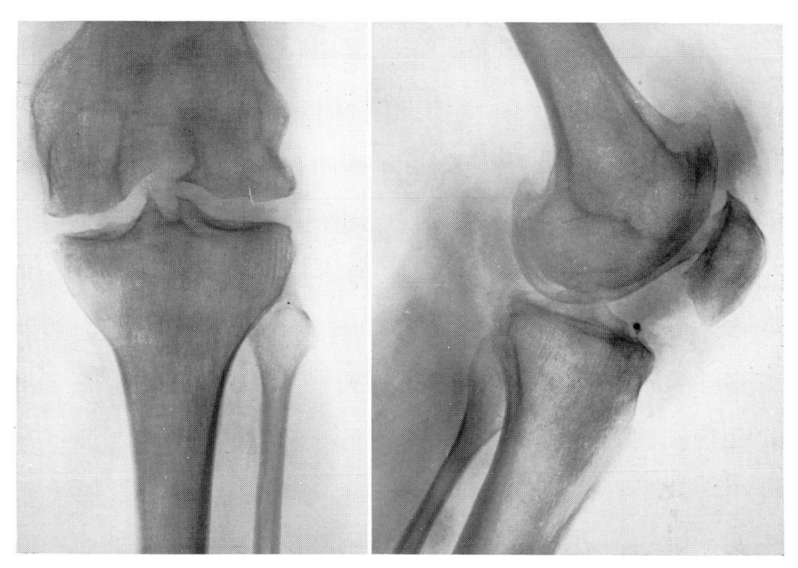

Fig. 5-12. Roentgenograms of the knee of a 29-year-old man, showing characteristic changes in hemophilic arthritis. In the anteroposterior view note the wide intercondylar fossa and spurs about the tibial spine; the lateral view shows spurs on the articular surface of the patella and about the anterior and posterior margins of the articular surface of the femur.

ting factor VIII. One fifth as common is hemophilia B, or *Christmas disease,* in which the deficient clotting factor is plasma thromboplastin component, or factor IX. The hemophilic patients requiring most orthopaedic attention have moderate or severe clotting defects with serum antihemophilic globulin levels of less than 5% of the normal value.

Although the knee is most commonly affected, the ankle, elbow, shoulder, hip, and joints of the fingers are also subject to frequent intra-articular hemorrhages.

Pathology. Repeated hemorrhage into a joint, precipitated by the trauma of motion, weight bearing, or minor external violence, may be followed by incomplete absorption of the blood and later by organization. Changes in the synovial membrane include increased vascularity, hyperplasia, and thickening. Iron pigment granules are found in the highly vascular subsynovial layer. The articular cartilage shows extensive degeneration with fibrillation, thinning, and fragmentation of collagen fibers. The subchondral bone becomes atrophic with broken trabeculae and blood-filled cysts. Accelerated growth in ossification centers may lead to deformity. In late cases the pathologic changes may resemble those of rheumatoid arthritis, and the roentgenograms (Fig. 5-12) may show areas of spur formation and of cavitation in the atrophic bone. Widening of the intercondylar fossa of the femur is a common roentgenographic finding.

Massive subperiosteal hemorrhage, resembling that of scurvy, may occur near the end of a long bone. Slow, prolonged subperiosteal bleeding may lead to an expanding process of local bone destruction and new bone formation, called a *hemophilic pseudotumor,* that simulates a malignant neoplasm. Pseudotumors occur in no more than 1% or 2% of hemophiliacs, however, and only in severe degrees of clotting deficiency.

Clinical picture. The patient is a male whose first episode of hemarthrosis has occurred, as a rule, during early childhood. At first the involvement may be monarticular. In cases of recent hemorrhage the joint capsule is distended, tender, and fluctuant, and motion of the joint causes discomfort. Increased local heat and discoloration of the skin are sometimes observed. Extensive hemorrhage is often accompanied by fever. Repeated hemorrhage is eventually followed by chronic swelling, discomfort on motion, muscle atrophy, and contracture.

Hemorrhages may occur in any part of the body. Repeated bleeding into the calf muscles may lead to fibrosis and severe equinus deformity. Hemorrhage into the iliacus muscle sometimes causes hip flexion contracture and femoral nerve palsy. The urine often is discolored and contains red blood cells from hemorrhage in the urogenital system.

Diagnosis. Diagnosis of the acute hemarthrosis is made on the physical findings, family history, and prolonged clotting time. Late cases are distinguishable from other types of chronic arthropathy by the typical history and hematologic findings.

Prognosis. The initial hemorrhage is usually absorbed, and the prognosis for future joint function depends largely on the severity of the clotting defect and on the patient's cooperation and good fortune in avoiding further episodes of trauma. Successive hemorrhages are likely to be followed by increasing disability, and the affected joint may finally present a clinical and pathologic picture similar to that of late rheumatoid arthritis. Improved methods of treatment, utilizing recently developed preparations to supply the missing coagulation factor, should greatly reduce the incidence of severe hemophilic crippling.

Treatment. In cases of acute hemarthrosis the extremity should be immobilized at once; to accomplish this a large cotton pressure dressing and splint may be applied. Elevation and an ice pack may be helpful. Many authorities regard either aspiration of blood from the joint or instillation of preparations such as hyaluronidase or hydrocortisone as inadvisable in the great majority of cases. If the bleeding is extensive, concentrated human antihemophilic factor should be administered intravenously. Transfusion with fresh whole blood or freshly frozen plasma is also helpful in controlling active hemorrhage.

Subsequent hematologic treatment is based on the type and degree of the coagulation defect. After most of the intra-articular blood has been absorbed, muscle-setting exercises should be started with extreme caution. A cast to protect the affected joint from friction and weight-bearing strain for several weeks is often indicated; usually this should be followed by a brace. Whether joint motion and weight bearing should be allowed early, whether they should be permitted constantly, and for how long a brace may be needed will vary with the individual case. To minimize muscle atrophy, nonweight-bearing exercises should be carefully increased.

Late cases of hemophilic arthropathy usually present disabling and unsightly deformity. Flexion contracture of the knee is most frequent; with it may be associated valgus or varus, torsion, and posterior subluxation of the tibia. Traction is ineffective if the deformity is of long duration, and casts should be applied for gradual correction by cautious posterior wedging or the use of a turnbuckle or similar attachment. Such correction should be followed by effective bracing, supervised exercises, and gradual return to normal levels of activity.

Until recently the surgical correction of hemophilic deformities has been extremely hazardous because of the danger of fatal postoperative hemorrhage. With potent antihemophilic concentrates now available, reconstructive surgery is relatively safe. Problems currently unsolved include the danger of serum hepatitis from derivatives of pooled plasma, and in occasional cases the inhibition of antihemophilic factor by circulating antibodies.

GOUTY ARTHRITIS

Gout is a familial disorder of purine metabolism in which uric acid, the normal end product, is involved. Gout is characterized by hyperuricemia and the deposition of sodium urate in the tissues. Uratic deposits in and about joints lead to severe forms of acute and chronic arthritis. The cause of gout remains obscure. Approximately 95% of the patients are males; they are usually past the age of 30 years. Rarely is gout observed in Negroes.

Clinical picture. Early stages of gout are marked by recurring attacks of acute monarticular pain and inflammation, and the later stages by chronic deforming articular changes. The attacks are precipitated by excessive intake of meats and

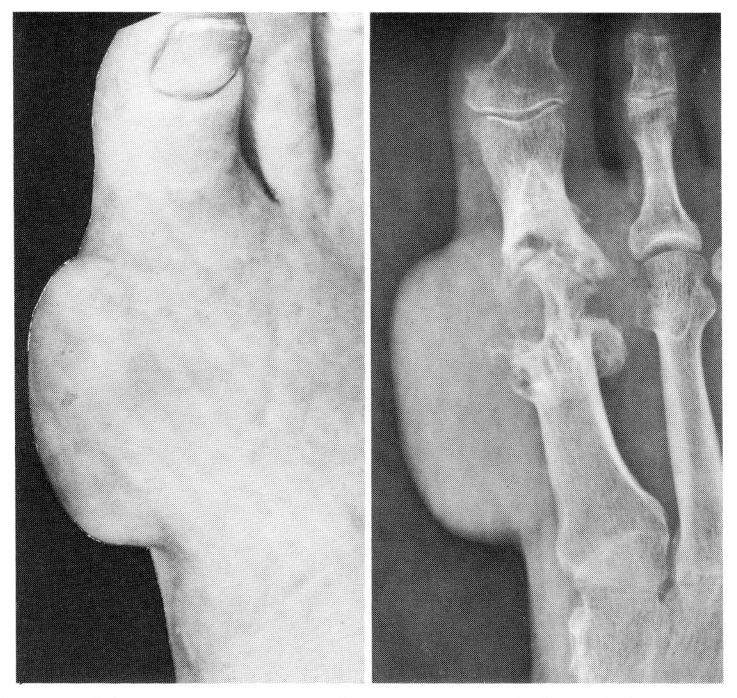

Fig. 5-13. Gout of long duration. The tophaceous mass at the base of the great toe, as well as the destructive bone and joint changes shown in the roentgenogram, are associated with extensive urate deposits.

other foods high in protein, fatigue, overindulgence in alcohol, and occasionally by trauma or surgery of the involved joint. Between attacks the patient is symptom free unless permanent joint damage has taken place.

The joints most often involved are those of the foot (classically the first metatarsophalangeal joint, Fig. 5-13), hand, wrist, knee, and elbow. Seldom is the hip, shoulder, or spine affected. In acute attacks the involved joint is usually red, swollen, warm, tender, and extremely painful on motion; veins in the overlying skin may be distended.

Pathology. The joint lesions consist of creamy or chalky deposits of sodium urate surrounded by foreign body inflammatory reaction; they occur in synovium, ligaments, articular cartilage, and periarticular bone. Such urate deposits, known as *tophi,* are sometimes found in the cartilage of the ears. They may occur also in subcutaneous tissue, fascia, kidneys, heart, and other viscera.

Roentgenographic picture. In early gout, roentgenograms usually show no bone changes. In later cases typical small, punched-out areas, which contain uratic deposits, often appear near the ends of the bones at the affected joint.

Laboratory findings. The blood uric acid is usually elevated (6 to 15 mg. per 100 ml.). The hyperuricemia is essentially unchanged before, during, and after an acute attack of gouty arthritis. The diagnosis of gout can be made by finding the needle-shaped microcrystals characteristic of monosodium urate, best seen by polarized light microscopy, in synovial fluid or in synovial tissue that has been fixed in alcohol.

Treatment. Although no cure for the metabolic defect of gout is known, certain drugs provide dramatic relief of the pain and swelling of acute attacks. Colchicine is most useful; it may be given hourly in doses of 0.5 mg. until the pain subsides or until toxic gastrointestinal upsets supervene. Phenylbutazone or indomethacin often relieves the symptoms within 48 hours or less.

In chronic gout, long-term medical supervision and treatment are indicated. A low-purine diet and avoidance of excessive alcohol and trauma may help to prevent acute attacks. Tophaceous deposits can be prevented or reversed by prolonged use of the xanthine oxidase inhibitor, allopurinol, which reduces the formation of uric acid, or by the use of a uricosuric agent such as probenecid. In late cases, large tophi may require surgical extirpation. As a rule gout, if properly treated, does not alter life expectancy.

CHONDROCALCINOSIS (PSEUDOGOUT)

Chondrocalcinosis, or pseudogout, is an uncommon form of arthritis, of unknown cause, associated with the deposition of crystals of calcium pyrophosphate within articular cartilage. It is characterized by acutely painful attacks similar to those of gout and by the gradual development of chronic articular changes like those of osteoarthritis. The knee is most frequently affected, and next the hip. Serum chemistry levels, including the uric acid content, are within normal limits. The diagnosis is based on the roentgenographic finding of calcification within the substance of articular cartilage and, in the case of the knee, within the fibrocartilaginous menisci. Microscopic examination of synovial fluid aspirated during an acute attack shows the rod-shaped and rhombic crystals of calcium pyrophosphate;

they may be distinguished from the urate crystals of gout by their shape and by the fact that, unlike urate crystals, they are not digested by uricase. No curative treatment is available, but the acute attacks may be relieved by phenylbutazone or indomethacin.

OCHRONOTIC ARTHRITIS

Ochronosis, a rare hereditary disease associated with alcaptonuria, is manifested by progressive degenerative arthritic changes that are most severe in the spine and proximal joints. Because of an enzyme deficiency, homogentisic acid, an intermediate in tyrosine metabolism, accumulates in the blood and leads to the deposition of a dark pigment in articular cartilage and other tissues. Roentgenographic findings include the changes of severe degenerative arthritis, particularly in the spine, where calcification of intervertebral disks is common. Clinically, the symptoms of arthritis usually appear in the fourth decade of life, frequently beginning in the spine. Acute radicular symptoms sometimes simulate those due to rupture of an intervertebral

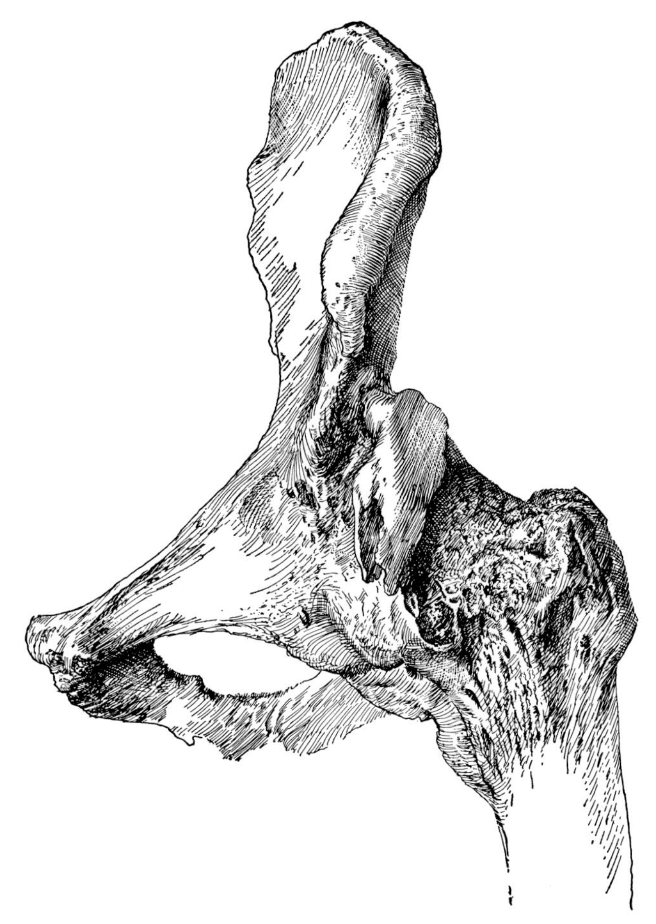

Fig. 5-14. Bony ankylosis of the hip, the end result of a severe pyogenic arthritis. (Drawing from museum specimen.)

disk. No therapy specific for the enzymatic defect is available, and treatment of the associated chronic arthritis is largely symptomatic.

ANKYLOSIS

Ankylosis is a restriction of the normal range of motion of a joint by tissue changes within or without the joint cavity. Ankylosis may occur with the joint in a position relatively favorable for function or in an attitude of deformity.

Etiology. Ankylosis is often a result of the incomplete healing or restoration of joint structures damaged by chronic arthritis, infection, or trauma. A severe burn about a joint sometimes results in ankylosis. Ankylosis is of two kinds: (1) fibrous, partial, or false (when connective tissue adhesions between the articular surfaces or extra-articular tissues are not accompanied by actual bony union); and (2) bony, complete, or true (when solid new bone has been formed between the articular surfaces). Fibrous ankylosis may follow joint inflammations, intra-articular fractures, repeated intra-articular hemorrhages, or extra-articular changes resulting from infection, trauma, or prolonged immobilization. Bony ankylosis is often the end result of a more severe infection or a more advanced rheumatoid arthritis, which has destroyed the cartilaginous surfaces and allowed the apposition of denuded bones.

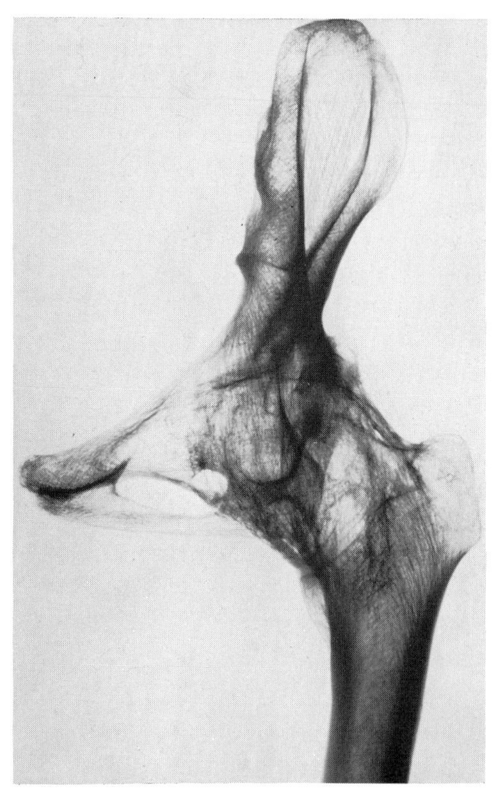

Fig. 5-15. Bony ankylosis of the hip. Roentgenogram of the specimen illustrated in Fig. 5-14, showing bony trabeculae extending from the pelvis to the femur.

Clinical picture. In fibrous ankylosis a small amount of motion may be present, and pain may be experienced when the joint is manipulated. In bony ankylosis (Fig. 5-14) no joint motion is possible. Often it is difficult to determine whether the ankylosis is true or false. Roentgenograms of the joint should always be examined, but at times they are deceptive. The films may appear to show a solid bony fusion when definite joint motion can be elicited clinically or can be demonstrated later at operation. The most reliable roentgenographic criterion of bony ankylosis is the presence of fine bony trabeculae that can be traced directly across the region of the former joint space (Fig. 5-15).

Treatment. In the early stages of joint involvement the best assurance against the development of ankylosis is proper treatment of the primary disease or injury. In pyogenic arthritis it is often necessary to place the joint absolutely at rest in order to minimize injury of its articular surfaces. In rheumatoid arthritis the joint may be best treated by alternating periods of splinting and of regulated exercise. When the knee or the hip is involved, traction applied below the joint and designed to lessen the pressure of the apposed articular surfaces is often an effective aid in preventing joint destruction and ankylosis. In order to minimize the danger of ankylosis, casts used in the treatment of fractures should immobilize no joint without good reason, should be discarded as soon as early union will permit, and should be followed by active exercise of all affected joints.

If fibrous ankylosis has taken place, it may be possible by means of physical therapy to preserve or increase a useful range of motion in the joint. Vigorous active exercise is essential; preliminary heat and massage may be of some help. Passive motion should never be forced if the movement is accompanied by pain, since additional damage and stiffening may result and since the discomfort may serve to discourage further endeavors. Electrical stimulation of muscles about the joint and insistence on active contraction of these muscles will often help. A joint that by gentle physical therapy can be restored even partially to motion will usually preserve its regained function.

If forcible mobilization of a partially ankylosed joint becomes necessary, the manipulation should be done gently under anesthesia. In an occasional case it may be desirable to perform repeated manipulations. If the adhesions and contractures can be overcome with a minimum of effort, the prognosis for improved joint function is good provided that muscle tone and strength are adequate, but if considerable force is used the outcome is at best doubtful. The bones about these partially ankylosed and little-used joints are very atrophic and weak. Care should therefore be taken not to exert too much force, as the bones fracture easily. The smaller joints do not respond well to manipulation. Of great importance is the patient's cooperation in a program of vigorous active exercise after the manipulation.

If ankylosis has occurred with the joint in good position, arthroplasty, or other operation for mobilization, may be indicated. When the ankylosis has occurred in a faulty position, however, operations for improvement of the deformity, as well as those for restoration of joint motion, must be kept under consideration. In a partial extension ankylosis of the knee such as sometimes develops after a fracture of the lower half of the femur, with contracture of the quadriceps muscle, a lengthening of the quadriceps tendon (Bennett operation) or a resection of the scar tissue about the quadriceps tendon and muscle fibers (Thompson operation) may be followed by a useful increase of the range of flexion. In young children, quadriceps contracture that sometimes follows intramuscular injections may require surgical release.

OPTIMUM POSITIONS OF JOINT FIXATION

In many patients with joint disease it becomes impossible to prevent ankylosis. Should ankylosis become inevitable, the position most suitable for future function of the extremity must be determined early, and efforts must be made to prevent ankylosis from occurring in any less favorable position. This is a vital factor in preserving function, as well as in preventing disfigurement. Many orthopaedic operations become necessary only because ankylosis in poor functional position has been allowed to take place.

The therapeutic procedure of producing a bony ankylosis by surgery is called *arthrodesis.* Arthrodesis may be indicated (1) to eradicate an intra-articular disease, as in advanced tuberculosis; (2) to abolish intractable pain on motion, as in chronic arthritis with severe intra-articular damage; or (3) to provide permanent stability for a joint rendered flail by paralysis.

In arthrodesis, care must be taken that the ankylosis occur with the joint in the position most favorable for function. With variations depending upon the circumstances of the individual patient, an optimal position for each joint has been recognized.

Spine

The vertebral column should be so supported that its anteroposterior curves remain relatively normal. Ankylosis should never be allowed to take place with the spine in a flexed or laterally deviated position.

Shoulder

In children the glenohumeral joint should be allowed to ankylose in from 50 to 75 degrees of abduction; in adults abduction of 50 degrees is preferable. The arm

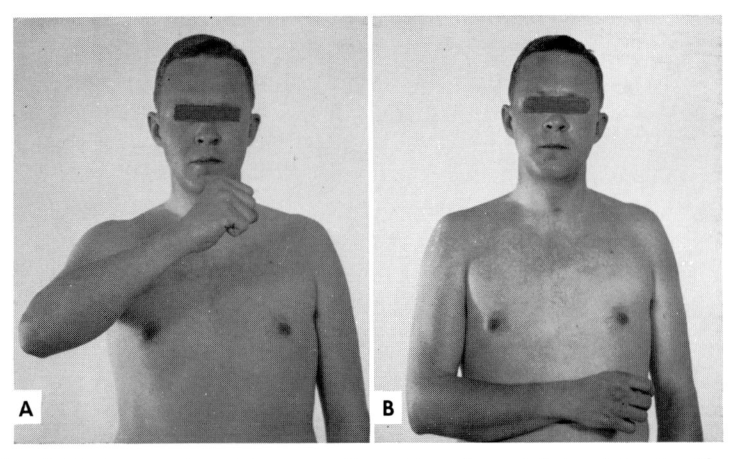

Fig. 5-16. Correct, **A,** and incorrect, **B,** positions for ankylosis of the shoulder.

should also be brought forward 45 degrees from its resting position at the side of the body and should be so rotated that the hand approaches the face when the elbow is flexed (Fig. 5-16).

Elbow

The elbow should be allowed to ankylose at a right angle with the forearm in a position midway between supination and pronation. When both elbows are affected, one should be allowed to ankylose at about 100 degrees of flexion and the other at 80 degrees of flexion.

Wrist

The wrist should be allowed to ankylose in from 15 to 25 degrees of dorsiflexion (Fig. 5-17). This is most important, as ankylosis in flexion severely limits the usefulness of the hand.

Hip

The hip is ankylosed preferably in about 5 degrees of abduction, 5 degrees of external rotation, and 10 to 25 degrees of flexion. The most suitable degree of flexion depends largely upon the habits of the individual. If the patient's occupation is sedentary, ankylosis of the hip in full extension would obviously be undesirable.

Knee

The knee may be ankylosed in full extension, but in women and in men of sedentary occupation flexion of from 20 to 30 degrees forms a less awkward position. In children full extension is essential because with growth an increase of flexion is likely to occur.

Ankle

The ankle should be allowed to ankylose in neutral zero position with the foot at a right angle to the leg, or in slight equinus to allow for the height of the shoe heel (Fig. 5-18). If the affected leg is shortened, slight equinus provides a more serviceable extremity. Care should be taken to avoid rotation deformity, by aligning the foot with the ankle. When the ankle, knee, or hip undergoes ankylosis, the position should be such that a line drawn from the anterior superior iliac spine to the middle of the patella and projected to the foot passes either through or close to the second toe.

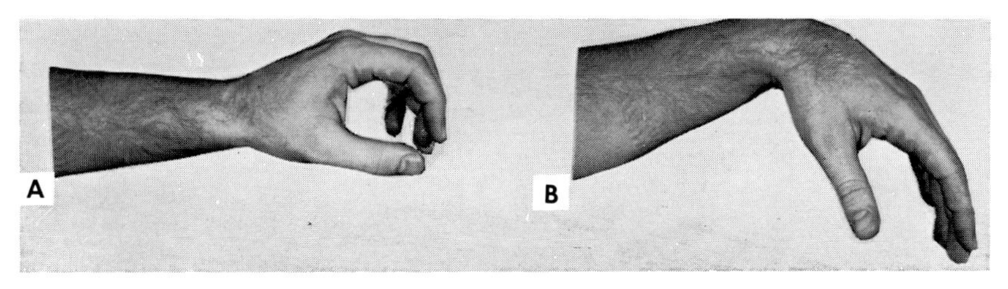

Fig. 5-17. Correct, **A,** and incorrect, **B,** positions for ankylosis of the wrist.

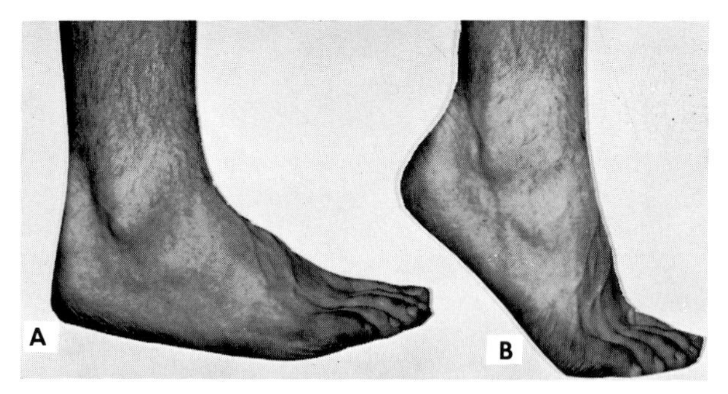

Fig. 5-18. Correct, **A,** and incorrect, **B,** positions for ankylosis of the ankle. Slight equinus is usually desirable to allow for the height of the shoe heel and to facilitate push-off in walking.

ARTHROPLASTY

Arthroplasty is a surgical procedure for restoring motion to a stiffened joint. At the same time the necessary stability of the joint and its freedom from pain must be preserved. In properly selected cases the results of arthroplasty may be quite satisfactory. Best results are obtained in the hip, elbow, metacarpophalangeal joints, jaw, and less reliably, the knee. Of the many arthroplastic technics that have been described, few attempt to restore the detailed anatomy of the joint.

Arthroplasty is most likely to be successful when the ankylosis has resulted from trauma or osteoarthritis. In joints ankylosed from rheumatoid arthritis there is often a tendency for some degree of stiffness to recur gradually after arthroplasty. In many rheumatoid patients, however, the ankylosis is bilateral and even a partial restoration of mobility is most helpful. Ankylosis after pyogenic or tuberculous arthritis is relatively unfavorable for arthroplasty because of the danger of reactivating the infection. In selected cases arthroplasty may be done after the infection has remained quiescent for at least a year and with care to provide appropriate antibiotic coverage.

The most favorable patient for arthroplasty is a well-motivated adult with ankylosis of only one joint and good musculature above and below it. Arthroplasty is most likely to be successful when the joint has become ankylosed in a relatively normal position and when little bony overgrowth has developed about it. Arthroplasty is contraindicated in children before the epiphyses have united and is usually inadvisable for any weight-bearing joint of a person whose occupation requires heavy manual labor.

In arthroplasty sufficient bone must be removed to permit free movement in the desired directions. The bone ends should often be separated ½ inch or more; they are shaped, smoothed, and usually covered with some type of interposing material *(interposition arthroplasty).* For many years fascia lata was the most popular interposing material. In arthroplasty of the hip, Smith-Petersen's technic of inserting a cup of inert metal over the head of the femur (Fig. 5-19) has given good results. Postoperatively, traction is helpful. Early mobilization is essential and should be encouraged by an extended program of physical therapy with emphasis upon active exercise. Improvement of mobility and strength after arthroplasty may continue for several years, during which the patient should be kept under periodic observation.

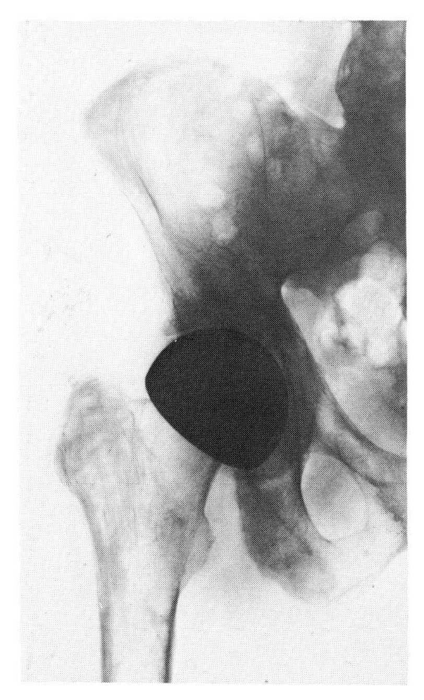

Fig. 5-19

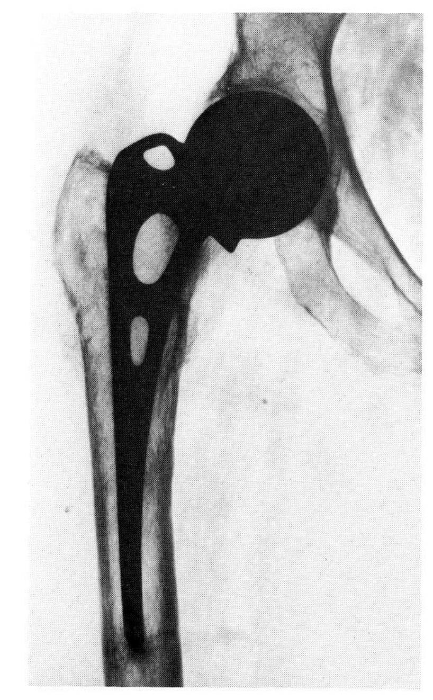

Fig. 5-20

Fig. 5-19. Roentgenogram of the hip following interposition arthroplasty with Smith-Petersen metal cup in a 41-year-old man who had had pain and partial ankylosis of the hip from Strümpell-Marie arthritis.

Fig. 5-20. Roentgenogram of the hip following femoral replacement arthroplasty with Austin Moore metal prosthesis in a 75-year-old woman who had had nonunion of a femoral neck fracture and avascular necrosis of the femoral head.

Femoral *replacement arthroplasty* of the hip is much used in the treatment of chronic arthritis with disabling pain, avascular necrosis of the femoral head, and high, vertical fractures of the femoral neck in elderly persons. In this technic the head and neck of the femur are replaced by a metal prosthesis with a stem extending distally through the base of the neck and down into the medullary canal of the upper part of the shaft (Fig. 5-20). When done on proper indication by a well-trained surgeon, this type of partial replacement arthroplasty often produces quite satisfactory early relief of disabling hip symptoms. Some of these hips become painful several years after operation, however, when repeated weight-bearing stresses may lead to loosening of the prosthetic stem in the femur and to erosive changes in the acetabulum.

When a joint has been irrevocably damaged by destructive changes in both of its component bones, it may be possible to excise the bone ends and replace them by an articulated internal prosthesis. This procedure, sometimes called *total replacement arthroplasty,* is being widely investigated at this time. At the hip both acetabular cup and femoral head can be replaced by metal or metal-and-plastic prostheses, with or without an attempt to bind foreign body to host tissue with a bone cement.

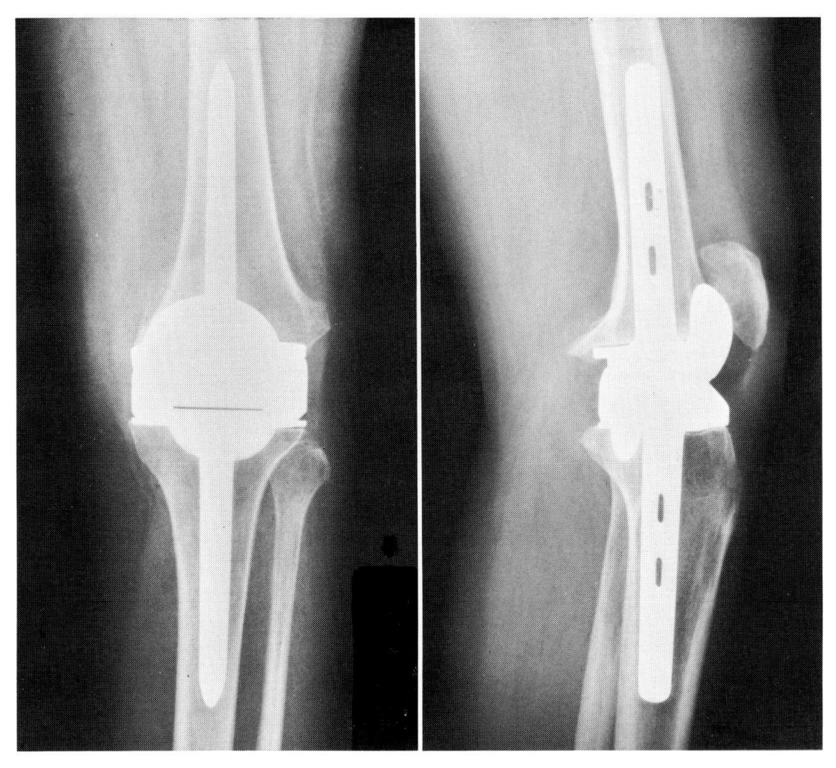

Fig. 5-21. Roentgenograms of total replacement arthroplasty of the knee with Walldius prosthesis, a salvage operation sometimes indicated when the articular structures have suffered extreme destructive changes.

At the knee it is possible to replace the distal end of the femur and the tibial plateau by a metallic hinged prosthesis (Fig. 5-21). Similarly, flexible plastic intramedullary inserts can be used to replace destroyed metacarpophalangeal or interphalangeal joints. Procedures of this type must be regarded currently as promising rather than of proved, permanent value. They seem likely, with further development, to be most useful as salvage procedures.

CHAPTER 6 Neuromuscular affections

Affections of the nervous system frequently result in muscle imbalance, weakness, and impaired function of the trunk and extremities. Many of these disorders are relentlessly progressive, but some are static or progress so slowly that much can be done to help the patient and to reduce his disability. Deformities that develop as the result of muscle imbalance can often be corrected. Muscle weakness can frequently be compensated for by braces or by tendon transplantation. Unstable and flail joints can be made solid by arthrodesis. Appropriate training in the use of crutches, appliances, or prostheses can sometimes permit a severely paralyzed patient to become self-sufficient.

In this chapter a number of neurologic and muscular disorders amenable to orthopaedic treatment will be discussed according to the following outline:

A. Involvement of the brain
   1. Cerebral palsy
   2. Acquired spastic paralysis in adults (cerebral vascular accidents)
   3. Neuromuscular disabilities of psychiatric origin (hysterical paralysis)
B. Involvement of the spinal cord
   1. Traumatic paraplegia and quadriplegia
   2. Poliomyelitis
   3. Progressive muscular atrophy (Aran-Duchenne type)
   4. Infantile spinal muscular atrophy (amyotonia congenita)
   5. Friedreich's ataxia
   6. Subacute combined sclerosis
   7. Neuropathic disease of bones and joints
   8. Spina bifida
C. Involvement of peripheral nerves
   1. Peripheral nerve injuries
   2. Neuritis (peripheral neuropathy)
   3. Guillain-Barré syndrome (postinfectious polyneuritis)
   4. Hereditary muscular atrophy of peroneal type (Charcot-Marie-Tooth disease)
D. Involvement of muscles
   1. Progressive muscular dystrophy

Involvement of the brain

CEREBRAL PALSY

Cerebral palsy is a common and extremely disabling affection seen most frequently in infants and children and characterized clinically by disturbance of volun-

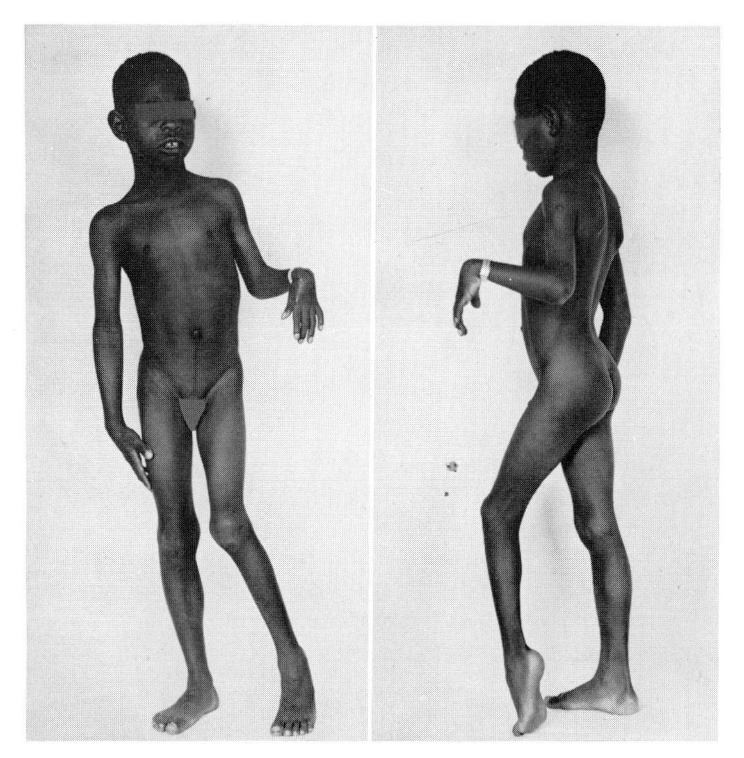

Fig. 6-1. Spastic left hemiplegia in cerebral palsy. Note abnormal positioning and generalized underdevelopment of the left limbs as compared with the right limbs.

tary motor function. It includes a large variety of nonprogressive neurologic entities; the essential pathologic change is destruction or congenital absence of upper motor neurons. Mental impairment may be present. Involvement of a single limb is termed *monoplegia;* involvement of both limbs of the same side of the body is called *hemiplegia* (Fig. 6-1); involvement of both lower limbs, *paraplegia;* and involvement of the four limbs, *quadriplegia* (Fig. 6-2). The term *spastic diplegia* is often used to denote bilateral involvement in which spastic paralysis affects the lower limbs much more than the upper limbs. Statistics from a large series of cases showed quadriplegia in about 60%, hemiplegia in 20%, and paraplegia in 10%. In reported series the incidence of cerebral palsy is given as 1 to 5 in every 1,000 live births; the incidence does not vary with sex.

Etiology. The causes of cerebral palsy may be classified chronologically into three groups:

1. *Antenatal,* consisting of congenital defects resulting from arrested development of the cerebrum and pyramidal tracts in utero, certain inherited metabolic defects, maternal infectious diseases such as rubella, toxoplasmosis, and syphilis, excessive irradiation, and toxemia of pregnancy
2. *Natal,* the most frequent type, consisting of nerve cell injury following cerebral hemorrhage due to trauma at birth, or anoxia or hypoxia following winding of the umbilical cord around the baby's neck, too heavy sedation of the mother, or aspiration of mucus by the child

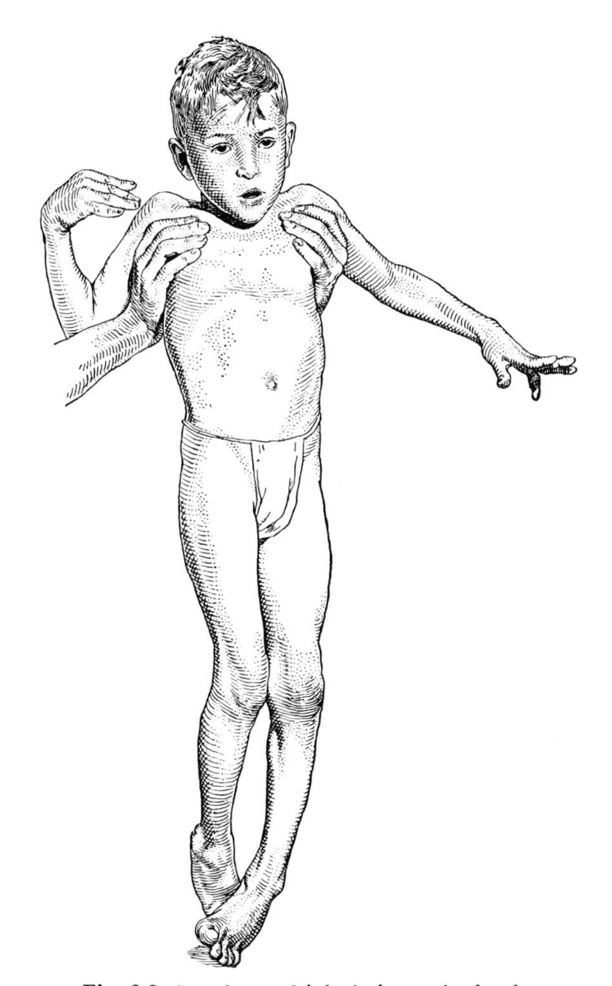

Fig. 6-2. Spastic quadriplegia in cerebral palsy.

3. *Postnatal,* consisting chiefly of infectious, vascular, or traumatic lesions, such as those of encephalitis, meningitis, vascular accidents, and Rh incompatibility associated with jaundice of the newborn, or *kernicterus*

The antenatal type is common in premature babies. Among children with cerebral palsy the number of premature births is definitely above the normal expectancy; in one series it was reported to be three times as great. There are an unusually high percentage of complications of pregnancy and a high percentage of previous abortions. Most children with postnatally acquired cerebral palsy are affected before the age of 3 years; in 30% of these children an infectious disease is the etiologic factor.

Pathology. Although frank contusion and laceration of cortical tissue may be present, the factor of intracranial hemorrhage is believed to play a major role in the pathologic process in most cases. The hemorrhage may occur directly from a vessel ruptured when the dura mater is torn, or it may be of widespread petechial type resulting from increased tension when the venous circulation is temporarily obstructed. Degeneration of the injured nerve cells and fibers ensues and is followed by sclerosis. Grossly the affected areas may show atrophy and softening. The lesions may be more severe either in the cerebral cortex or in the basal ganglia. Degeneration of corresponding tracts in the spinal cord is a constant finding. A severe diffuse injury resulting in multiple hemorrhages through all parts of the brain may produce rigidity with mental retardation. A single hemorrhage or infarct involving the motor cortex may cause spastic hemiplegia or quadriplegia. Kernicterus affects the basal ganglia and produces athetosis. Injury of the cerebellum produces ataxia.

Pathologic changes may occur in the spinal cord as the result of injury below the level of the brain. Traction applied during delivery may cause such damage. The clinical picture is in many respects similar to that of birth injury of the cerebrum or basal ganglia, and it is therefore convenient to consider cases of primary cord involvement together with those resulting from intracranial lesions.

Clinical picture. Intracranial hemorrhage in the newborn infant may be evidenced by severe asphyxia, bulging fontanels, refusal to nurse, spasticity or flaccidity, and convulsions. Lumbar puncture shows the spinal fluid to be bloody and under increased pressure. Many of the infants die shortly after birth. In milder cases the child may be considered normal until unusual delay in holding up the head, sitting, or standing becomes obvious.

Cases of cerebral palsy in older infants and children may be divided into several major groups, according to the location of the lesion within the central nervous system. These groups are characterized by outstanding clinical manifestations, of which the most important are spasticity, athetosis, and ataxia. In many cases the lesions are widespread, however, and the symptoms are of mixed type. Other forms of motor disturbance may also be present. Outstanding among these are the following: (1) *tension,* or widespread muscular hypertonicity or rigidity; (2) *tremor,* or rhythmic involuntary contractions limited to certain muscle groups; and (3) *overflow,* of which an example is involuntary motion of the facial muscles during voluntary movement of the arm.

Spasticity. From 50% to 65% of the cases of cerebral palsy fall into this group. Of the patients with spasticity, nearly one half are hemiplegic. In spasticity the lesion is cortical and the pyramidal tracts are involved. The evidences of spasticity are characteristic. The involved muscles are hyperirritable and contract on the slightest stimulation. The tendon reflexes are hyperactive, the Babinski reflex is positive, and clonus usually is easily elicited. Antagonistic muscle groups, hypersensitive to the stimulus of stretching, contract simultaneously with the protagonists; this results in difficult and inaccurate voluntary movements and in increased muscular resistance to passive manipulation. The involuntary contraction of a spastic muscle when it is suddenly stretched is a useful diagnostic sign; it is called the *stretch reflex.*

An equinus limp is a frequent finding. The equinus deformity may be due not to weakness of the dorsiflexor muscles but to simultaneous contraction of the normally more powerful plantar flexor muscles and to the effect of gravity. In hemiplegia the patient can usually walk, but the affected lower limb may trail and be of little use; the affected upper limb is usually severely involved.

In spastic diplegia, commonly found in children who have been born prematurely, the so-called *scissors gait* or *cross-legged progression* occurs, characterized by disabling adduction of the hips resulting from the greater strength of the adductor over the abductor muscle groups and augmented by the force of gravity. Involvement of the upper limbs is usually mild, and these children often have normal or near-normal intelligence.

Characteristic deformities develop as a result of the continued contraction of antagonistic muscles of unequal power. The hips may become flexed, adducted, and internally rotated. The knees are flexed. The feet usually assume an equinus or equinovarus position. The shoulders tend to become internally rotated and adducted and the elbows flexed. The forearm is pronated; the wrist and fingers become flexed and the thumb adducted. When the patient attempts to sit erect, unsightly bowing of the spine and forward protrusion of the head may take place.

Muscle atrophy may not be conspicuous. Asymmetry of the limbs sometimes becomes obvious, however, from retardation of growth of the affected limb due to trophic disturbances. Sensory defects may also be present.

Athetosis. Athetoid patients make up from 25% to 40% of the cases of cerebral palsy. The lesion is subcortical, the basal ganglia and extrapyramidal tracts being affected. Athetosis is a more or less constant involuntary contraction of successive muscles. These arrhythmic, purposeless contractions become superimposed upon voluntary movements and cause incoordination. The reflexes are normal, and the stretch reflex is absent. In athetosis due to an Rh factor incompatibility, there are a hearing loss and a limitation of vertical eye motion; intelligence may be average, and the child can become a good lip reader. Kernicterus is responsible for about half of the cerebral palsy patients with athetosis.

Ataxia. Cases of ataxia, or primary incoordination, are much less numerous than those of spasticity or of athetosis. The lesion is subcortical, and probably cerebellar in most instances. Characteristic clinical manifestations are incoordinated movements, impaired balance, and nystagmus. The child walks with a broad base and may have an intention tremor. The tendon reflexes are normal or hypoactive.

Mental status. Some degree of mental impairment, which may be either true deficiency or simply retardation, is usually present. Statistics have shown that the mentality of from 30% to 50% of all cerebral palsy patients is either seriously retarded or otherwise defective and that approximately 75% have a mentality below average. Some observers, however, believe these figures for mental impairment are too high. Cerebral palsy with spasticity and mental deficiency, of which the ill-tempered, feebleminded, drooling child with scissors gait is the classic example, was described by Little in 1843 and is sometimes called *Little's disease.* Mental deficiency is greatest in the spastic group and more so in quadriplegic patients than in hemiplegic ones. Some patients present obvious idiocy, with characteristically stupid facies and thick, unintelligible speech. In other cases, however, the appearance of subnormal mental endowment is unquestionably exaggerated by inability to control the facial or speech musculature, or by the deficient education that has resulted from absence of the normal locomotor activities of the growing child. Estimation of the mental capacity in the individual case is essential since it is of primary importance in determining prognosis and treatment.

Diagnosis. Accurate diagnosis is important in order to provide a basis for selecting proper therapeutic goals and procedures. History of difficult, prolonged, or instrumental delivery or of a definite episode of infectious disease during early childhood, although often not obtainable, may be useful auxiliary evidence. The spinal fluid should be examined when congenital syphilis is suspected. Developmental arrest of the central nervous system, hydrocephalus, brain tumor, atypical forms of poliomyelitis, and degenerative diseases of the central nervous system should be considered and excluded. Observation over a period of several months is often useful in ruling out the possibility of a disease process of progressive nature. For proper planning of treatment an accurate differentiation of the type of cerebral palsy also is necessary.

Prognosis. In untreated cases slow spontaneous improvement in use of the extremities often takes place as motor experience and better control are acquired during the years of childhood. Well-ordered treatment always results, however, in more rapid and more extensive improvement. Since the ability to cooperate and to learn is essential to effective treatment, the degree of improvement will depend to a large extent upon the mental capacity of the individual patient. Thus patients with paraplegia due to spinal lesions without cerebral involvement may achieve considerable function under treatment, whereas patients exhibiting severe mental deficiency improve little, form difficult nursing problems, and often succumb early to intercurrent infection. The prognosis of the patient with frequent convulsive seizures is unfavorable unless the seizures can be controlled with drug therapy.

Treatment. The treatment of cerebral palsy has both nonsurgical and surgical aspects. A first step is the formulation of a realistic goal based on appraisal of the potentialities of the individual patient. Affected children with adequate mental capacity may require a comprehensive program designed to achieve physical capability and social competence. The treatment program should develop speech, self-help, locomotion, psychologic adjustment, and appropriate education. The orthopaedic treatment is essentially a process of habilitation and, of necessity, embraces a long program of motor training during which resort to operation is made only as an auxiliary measure. Treatment should begin as soon as the diagnosis has been made and should start and continue at home as long as possible. Subsequent therapy is governed largely by the patient's response. When the necessary discipline, good hygiene, and freedom from excitement and worry cannot be obtained at home, the therapy can be given more successfully in an institution.

Nonsurgical measures. Motor education is an important part of the treatment of most cases of cerebral palsy. The first step is to teach the child how to relax voluntarily; this is particularly important for the athetoid patient. When muscular relaxation can be initiated at will and continued for as long as several minutes, training in maintaining balance and in performing simple movements is begun. In infants and in older patients with severe involvement, attention is concentrated first upon learning to sit or stand. A Thomas collar is sometimes provided to afford helpful support and prevent stretching of the neck muscles. A light corset is sometimes indicated. Braces and plaster shells, worn a part of each day or during the night, are often helpful in the treatment of mild spastic deformities. The most important element of the treatment, however, is a program of exercises selected for the individual patient and carried out daily over a long period of time. Simple reciprocal motions of the extremities are started, at first passive and later active, and finally the patient is taught gradually to combine these simple movements into composite ones, such as those of walking. Many aids to activity have been devised, such as ski shoes that allow the patient to stand and shuffle along without fear of falling. In many cases, daily speech training is important.

Many drugs have been tried in the treatment of cerebral palsy. Anticonvulsant medications have been extremely helpful in patients with seizures. Drugs to control muscle tone in spastic and athetoid patients have been less successful. Curare-like drugs relax muscle but are dangerous because of the narrow margin between therapeutic and toxic dosages. Tranquilizing drugs have had limited usefulness and are effective in a small percentage of cerebral palsy patients.

During the long period of training, the details of treatment must be adjusted to the needs of the individual patient; experience, judgment, and patience on the part of the specially trained physical therapist are of the utmost importance. Occupational therapy forms an important part of the treatment of older patients, especially in motor training of the upper extremity.

Surgical measures. As a rule, orthopaedic operations for cerebral palsy are applicable chiefly to cases characterized by spasticity. In many such cases surgical measures serve as a useful adjunct to motor reeducation. Operation cannot be regarded as adequate treatment unless accompanied by conservative measures, however, and is contraindicated when the patient possesses too little mental capacity to cooperate in necessary postoperative training.

The object of surgical treatment in cerebral palsy is to diminish muscle spasm, equalize the power of opposing muscles, stabilize poorly controlled joints, and correct deformity. Of the great variety of procedures that have been devised and employed, many have proved with further experience to be unsatisfactory. In a series of approximately 2,500 surgically treated cases from many different clinics, there were 50% successful results and 50% failures. In this series almost all of the operations upon athetoid patients were failures; however, other clinics have reported favorable results in carefully selected cases. The procedures now in common use may be classified as operations upon motor nerves, operations upon muscles and tendons, and operations upon bone. Of these three groups, the operations upon bone are followed by the most uniformly successful results.

Operations upon motor nerves (neurectomy, Stöffel operation). Division or partial excision of the motor nerves of spastic muscles is helpful in selected cases. Attempt is made not to paralyze a muscle group and so abolish its entire function but to produce in the stronger muscles a loss of power sufficient to result in improved balance and hence increased capability for muscle training and function. This type of operation has produced better results in treatment of the lower extremity than of the upper extremity. It has proved particularly suitable when applied to the branches of the obturator nerve for adductor muscle spasm and to the branches of the tibial nerve for spasm of the plantar flexors. In selected cases of spastic pronation of the forearm and flexion of the wrist, section of appropriate branches of the median nerve is a helpful procedure. Recurrences due to nerve regeneration are frequent, however; accordingly neurectomy is less popular today than it has been in the past.

Operations upon muscles and tendons. Procedures on muscles and tendons include tendon transplantation, tendon lengthening, tenotomy, and myotomy. They are sometimes combined advantageously with section of motor nerves to the spastic muscle group. In each case adequate preliminary correction of the deformity must be carried out, care being taken in the release of fibrous contractures of all involved periarticular structures. After operation the immobilization in corrected position and the subsequent physical therapy must be meticulously carried out.

The type of operation must be chosen carefully with regard to the nature and location of the deformity in the individual case. Some of the more commonly used procedures warrant brief description.

AT THE HIP. In cases of severe adduction deformity, tenotomy of a portion of the adductor muscles, followed by immobilization of the thighs for a period of about 6 weeks in wide abduction, often results in gratifying improvement of the gait. Release of the gracilis muscle may be particularly important. Division of part of the obturator nerve can be carried out during the same procedure.

AT THE KNEE. A number of operations have been devised to correct incomplete active extension of the knee on standing. Occasionally it is sufficient to lengthen one or more of the hamstring tendons. If the patellar ligament has become elongated, it may be shortened by plication or the tibial tubercle may be transplanted distally *(Chandler operation)*. Good results have also followed section of the patellar retinacula and transplantation of the hamstring insertions into the femoral condyles *(Eggers operation)*.

AT THE FOOT. Transplantation of the anterior tibial tendon to the lateral side of the dorsum of the foot may be used to improve spastic varus deformity. Lengthening of the Achilles tendon for talipes equinus is indicated only when there is unquestionable structural shortening, since spastic equinus would be all too easily converted into the more disabling deformity of calcaneus type. However, when actual shortening is present, Achilles tendon lengthening is one of the more frequently used and successful operations in the treatment of cerebral palsy. If the ankle can be dorsiflexed passively while the knee is flexed, but not while the knee is extended, the tightness can be attributed to the gastrocnemius alone, which, unlike the soleus, spans the knee; in this situation, sectioning the gastrocnemius tendon alone, or releasing the gastrocnemius origins from the femoral condyles, is preferable to lengthening the Achilles tendon. After any of these operations for spastic equinus, the ankle should be immobilized in a plaster cast in 90 degrees of dorsiflexion.

AT THE FOREARM. Pronation contracture of spastic origin may be relieved by suitable transplantation of the tendon of the pronator teres muscle.

AT THE WRIST AND HAND. Tenotomy, tenodesis, and tendon transplantation may be helpful in carefully selected cases. Rarely are complicated tendon transplantations successful in cerebral palsy.

Operations upon bone. Only in selected cases are bone operations indicated. In persistent varus or valgus deformity of the foot, subtalar arthrodesis greatly improves stability and gait. When the wrist is flexed and cannot be held voluntarily in a neutral position, radiocarpal arthrodesis in a position of 10 degrees of extension improves the appearance and in some cases the function of the hand.

ACQUIRED SPASTIC PARALYSIS IN ADULTS (CEREBRAL VASCULAR ACCIDENTS, STROKES)

Acquired lesions of the motor areas of the cerebral cortex often produce in adults a pathologic and clinical picture similar to that of the spastic type of cerebral palsy seen in children.

Etiology. Cerebral vascular accidents, or strokes, which include embolism, thrombosis, and hemorrhage, cause most cases of acquired spastic paralysis in adults. The underlying causes are arteriosclerosis and hypertension. Other agents that may damage the upper motor neurons to produce spasticity include trauma, encephalitis, and brain tumors.

Incidence. As deaths from infectious disease decline and the average length of life increases, the incidence of cardiovascular degenerative disease and cerebral vascular accidents increases. Accurate statistics are not available, but in the United States the number of living hemiplegic victims of stroke has been estimated at more than a million.

Clinical picture. As in cerebral palsy in children, the distribution and degree of the paralysis, as well as associated mental or speech impairment, depend on the location and extent of the brain damage. The common clinical picture is that of spastic hemiplegia, in which spasticity of one side of the body has resulted from injury of upper motor neurons of the cerebral cortex of the opposite side. In such cases the affected upper limb may be awkward and useless, the shoulder being held in adduction and the elbow in extension, whereas the fingers are no longer under voluntary control. Tendon reflexes are hyperactive, clonus may be present, and sensation may be impaired. Frequently enough power remains about the hip and knee to allow walking, but leg weakness and spastic equinus make the gait slow, awkward, and precarious.

Prognosis. Extensive initial paralysis may improve slowly for several months, leaving a permanent motor disability that varies in degree with the individual case. Orthopaedic treatment can often lessen the deformities and improve the motor abilities of these patients to a very worthwhile degree.

Treatment. A careful selection of appropriate physical therapy, bracing, or surgical procedures may be required. Passive and assistive exercises are usually indicated to prevent or overcome contractures. Early supervised exercise and activity tend to minimize physical disability. Positioning and splinting joints in functional attitudes may prevent deforming contractures. In spastic hemiplegia a short double-upright leg brace may greatly improve the gait; the brace may be fitted with an adjustable spring to exert dorsiflexion (Fig. 9-11) if not contraindicated by a brisk stretch reflex. Spasticity can be selectively controlled in some cases by intraneural phenol injections, but the results are seldom lasting. Some patients can be significantly helped by surgical measures such as tendon lengthening or transfer, neurectomy, fasciotomy, and arthrodesis. Heel cord lengthening is the most commonly used procedure.

NEUROMUSCULAR DISABILITIES OF PSYCHIATRIC ORIGIN (HYSTERICAL PARALYSIS)

Cases of orthopaedic disability due partially or entirely to psychosomatic or psychogenic causes are encountered with considerable frequency. They are especially common in the armed forces during wartime. Often a part in the causation is played by the element of financial compensation or the desire to be relieved of some unpleasant duty. Insurance and liability claims are involved in many of these cases. Functional or hysterical paralysis and deformity may simulate a variety of primary osteoarticular or neuromuscular diseases. Hence these disorders, although primarily of psychiatric interest, are of importance in the differential diagnosis of orthopaedic affections.

Clinical picture. Hysterical paralysis may occur in either sex and at any age. The history is likely to include discrepancies that suggest the diagnosis. The joint symptoms are varied, may be out of all proportion to the inciting trauma, and may change unnaturally from day to day. The physical findings also are inconsistent. Spasm and tenderness may be excessive, whereas heat, redness, and swelling are completely absent, and the signs may change as soon as the patient's attention is diverted.

Diagnosis. In diagnosis it must be remembered that organic and functional elements frequently coexist and that in late hysterical cases a secondary structural element of circulatory changes and of contractures is almost always present. Examination under an anesthetic and roentgenographic visualization of the involved structures are of help. The diagnosis of hysteria should never be made until every effort to establish the presence of an organic lesion has been exhausted. The patient should be kept under observation for a considerable period, and repeated diagnostic examinations should be made.

Treatment. The treatment of functional paralysis and deformity should be largely psychiatric. It is of fundamental importance that the confidence and active interest of the patient be secured. After this has been done it is usually advisable to place the patient on a regimen of gradually increasing corrective physical therapy. Emphasis should be placed on active motion and active correction. Occupational therapy will often prove of value. When the element of compensation or disability insurance is present, termination of the payments and closure of the case are often followed quickly by relief of the disability.

Involvement of the spinal cord

TRAUMATIC PARAPLEGIA AND QUADRIPLEGIA

Complete transection of the spinal cord results in paralysis of all muscles supplied by motor neurons situated below the level of the lesion and in loss of skin sensation in all areas supplied by sensory neurons below the lesion. Since neuronal regeneration does not take place in the central nervous system, both motor and sensory paralyses are permanent.

Although muscle power and tactile sensation cannot be restored in these severely crippling conditions, much can be done from an orthopaedic standpoint to minimize deformity, prevent complications, and improve the patients' physical abilities and general welfare.

Incidence. The number of paraplegics in the United States, according to the 1963-1965 National Health Survey of the United States Public Health Service, was 81,000, and the incidence is believed to be increasing each year. Recent medical advances have greatly increased the number of surviving quadriplegics. A distressing factor in traumatic paraplegia and quadriplegia is that most of the patients have been injured in their late teens or early adult years. The economic loss associated with paraplegia and quadriplegia is staggering. The costs of medical care, rehabilitative facilities, wage loss, and tax loss for a young paraplegic have been estimated to amount to a total cost to society of more than $300,000.

Etiology. Most transections of the spinal cord are the result of vertebral fractures and dislocations. In America in recent years, automobile accidents have been the greatest single cause of such injuries. Wider use of seat belts, safer cars, and more stringent laws for traffic safety may reduce the incidence of these severe disabilities. Falls from a height, blows from heavy falling objects, and dives into shallow water are other common causes of paraplegia or quadriplegia. In war and, much less commonly, in civil life the spinal cord may be sectioned in wounds from gunshot and other projectiles.

Clinical picture. Immediately after the injury the areas below the lesion may show flaccid paralysis and complete loss of skin sensation. Called spinal shock, this stage usually lasts from 4 to 6 weeks, after which the affected muscles develop permanent spasticity and the skin sensibility may progress distally to reach a slightly lower permanent level.

The extent of paralysis, and with it the patient's signs, treatment, and orthopaedic rehabilitation potential, depend upon the level at which the spinal cord has been divided. To understand whether a patient with section of the spinal cord at the level of a fractured vertebra will retain enough power in major muscles of the lower limbs

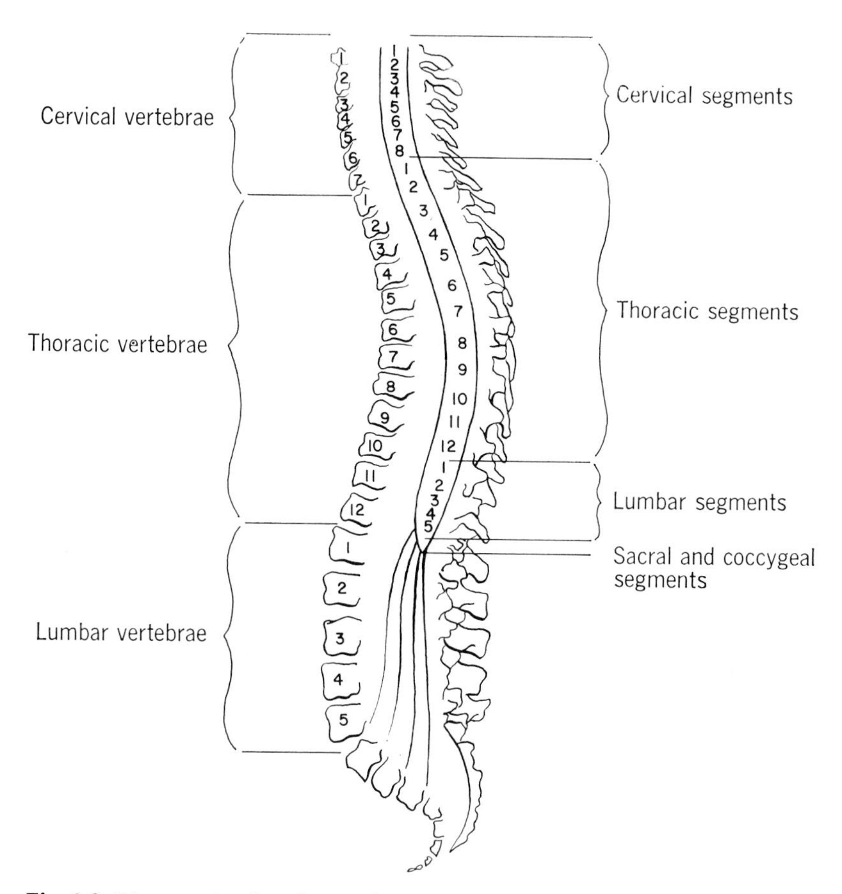

Fig. 6-3. Diagram showing the relationship of vertebrae to spinal cord segments.

to walk without or with braces, the physician will need to recall the relative levels of respective vertebrae and spinal cord segments (Fig. 6-3) as well as the segmental innervation of the limb muscles. Armed with this information, he can evaluate the patient's paralysis and disability, determine whether walking is an attainable and practical goal for him, and prescribe bracing and other measures to enable him to reach such a goal. Similarly, in traumatic quadriplegia from a lower cervical dislocation the cord segment involved and the segmental innervation of upper limb muscles determine the degree of residual function in arms and hands.

Among the serious complications of spinal cord injuries is paralytic incontinence of the bladder and rectum. Infections of the urinary tract are frequent hazards. Another serious complication is decubitus ulcers that result from the combination of confinement to bed or chair, anesthetic skin, and inability to shift position easily.

Treatment. As a rule, laminectomy is inadvisable. Rarely, in early cases with signs of incomplete cord section, surgical exploration may be done in an attempt to relieve pressure on the cord. At first the patient must be turned every 2 hours day and night to prevent pressure sores. The urinary bladder must be emptied by catheter or expression. The paralyzed limbs must be supported in functional position to prevent deforming contractures. Passive exercises tend to preserve mobility in the flail joints; active resistive exercises strengthen remaining muscle groups that must take on increased function.

Bowel and bladder training are an important part of the early treatment. Extensive decubiti may require skin coverage or resection of underlying bony prominences.

As the patient begins to sit or stand, his damaged spine may require the support of brace or arthrodesis. In the limbs, contractures and spasticity may indicate release or neurectomy. With long leg braces, crutches, and gait training, the paraplegic who is strongly motivated and whose cord lesion is not at too high a level will return to walking and may become self-supporting. His restoration to maximum physical and social function is often the culmination of the work of representatives from many disciplines: orthopaedics, neurosurgery, urology, psychiatry, physical therapy, physiatry, orthotics, occupational therapy, and vocational counseling.

POLIOMYELITIS

With the development and widespread use of prophylactic immunization, poliomyelitis has been reduced from a common, terribly crippling disease to a rare cause of paralysis. In America its epidemics have vanished, but sporadic cases still occur.

There remain some victims of poliomyelitis who still require orthopaedic treatment to decrease their deformities and improve their locomotor abilities. Especially useful are the reconstructive operations on tendons and the stabilizing operations on bone. Moreover, the surgical procedures devised initially for salvaging limbs crippled by poliomyelitis remain of greatest value in the surgical treatment of all similar types of paralytic crippling. Such paralyses are of flaccid type and result from lesions that destroy the lower motor neuron.

Poliomyelitis is an acute infectious disease caused by a group of viruses that attack the nervous system and have an especial tendency to affect the anterior horns of the gray matter of the spinal cord. The infection is often accompanied by paralysis, which may be mild, or severe, widespread, and permanent. The disease occurs in infants, children, and young adults. Most cases occur in midsummer and early fall. The alimentary tract, particularly its upper portion, the mouth, pharynx, and esophagus, is considered to be the chief pathway for entry of the virus and spread of the disease.

Pathology. The greatest changes appear in the anterior horn cells of the gray matter, especially those in the lumbar enlargement of the cord. These changes result in a motor paralysis. The cell bodies of these areas show evidence of atrophy and distintegration and may be replaced by scar tissue. If all the neurons innervating a particular muscle are destroyed, the muscle will be completely paralyzed and flaccid. All of the neurons may recover, leading to complete return of motor power, or recovery of only a part of the neurons may take place with the result that the muscle regains some power but remains permanently weakened. In the chronic stage, when the motor cells have been replaced by scar tissue, degeneration of the peripheral nerve fibers occurs and is accompanied by atrophy of muscle, tendon, and bone.

Clinical picture. The disease may be abortive (without involvement of the central

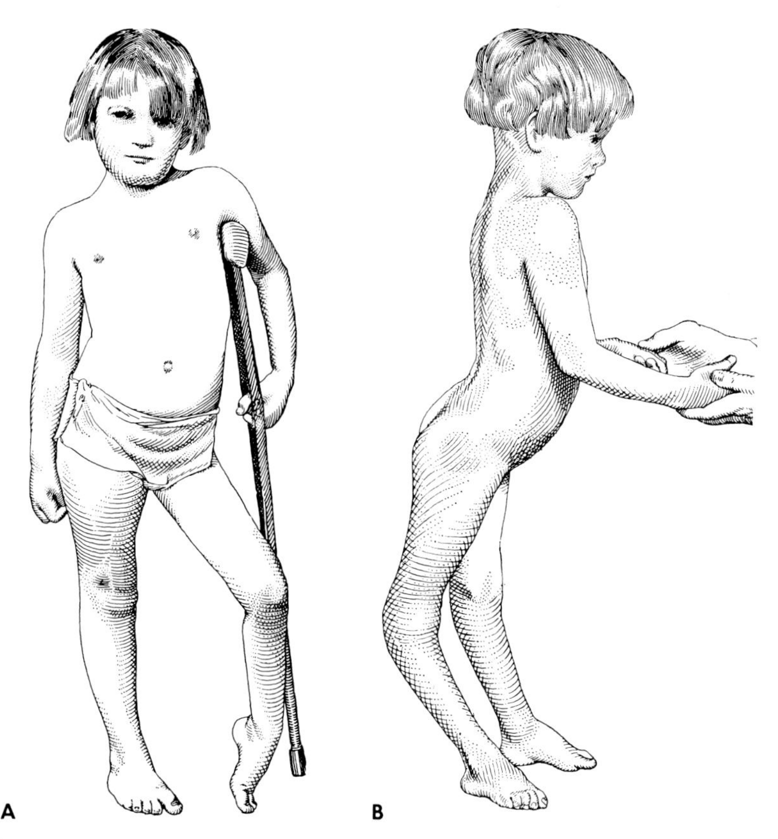

Fig. 6-4. Deformities in residual stage of poliomyelitis. **A,** Note flexion contracture of the left hip and knee and equinovarus deformity of the left foot. **B,** Note increased lumbar lordosis and bilateral hip flexion contracture, genu recurvatum (back-knee), and equinus. (**A** after Ombredanne and Mathieu.)

nervous system), it may be nonparalytic (without gross evidence of flaccid paralysis), or it may be paralytic (with varying degrees of muscle weakness or paralysis). Upper spinal involvement is usually termed *bulbar poliomyelitis*.

The average incubation period is from 7 to 14 days. After the onset of the disease there are three distinct stages: (1) the acute stage, which lasts from 1 to 4 weeks and is characterized early by fever, headache, and gastrointestinal symptoms and a few days later by muscle stiffness, pain, and progressive motor paralysis; (2) the stage of convalescence and recovery, which lasts from 6 months to 2 years; and (3) the residual stage, during which little or no spontaneous improvement of the paralysis occurs (Fig. 6-4). The symptoms and physical findings in each stage are detailed in pediatric and neurology texts.

Diagnosis. The diagnosis is frequently not made until after paralysis appears. It is usually based on the history of slight systemic upset, stiffness of the neck, muscle tenderness, paralysis, and findings on examination of the spinal fluid. It should be verified by isolation of the virus from fecal specimens or pharyngeal swabbings.

Differential diagnosis. In the *acute stage* poliomyelitis is to be differentiated from other febrile affections that may be associated with paralysis, such as other enterovirus infections, diphtheria, the Guillain-Barré syndrome, and other forms of polyneuritis. In the *chronic stage* it is to be differentiated from: (1) primary diseases of the nervous system, such as polyneuritis, encephalitis, meningitis, brain tumor, cerebral palsy, transverse myelitis, and infantile spinal muscular atrophy; (2) diseases accompanied by pseudoparalysis or spasm, such as hypervitaminosis A, infantile cortical hyperostosis, scurvy, and rickets; (3) other conditions that present deformity and loss of function, such as congenital clubfoot and congenital dislocation of the hip; and (4) peripheral nerve injuries, including obstetric paralysis.

Prognosis. In bulbar cases the mortality is high. Death may be caused by hypoxia from partial paralysis of the respiratory muscles, with pooling of secretions in the throat, or from involvement of the respiratory or the circulatory center in the medulla.

The prognosis as to muscle function is uncertain. The extent of recovery of muscle power depends upon survival of the affected nerve cells. Spontaneous improvement, if it occurs, starts within a few weeks after the onset of the paralysis and is most rapid during the first 6 months. Complete recovery after extensive paralysis is unusual, but after moderate paralysis it is quite common.

Treatment. The treatment varies with the stage of the disease.

Acute stage. Constant rest in bed with the joints supported in a neutral position is important. Hot moist packs may be helpful in relieving pain. In some clinics good results in relieving sustained muscle contraction by the use of muscle-relaxant drugs have been reported.

When, in bulbar cases, the respiratory muscles become involved, the patient should be kept quiet with his head down. The administration of oxygen is essential if the patient is hypoxic. Keeping the airway open may require frequent suction. If respiration becomes difficult, from paralysis of the primary respiratory muscles, the patient should be placed in a respirator until recovery of muscle power is sufficient to permit maintenance of normal respiration. Occasionally tracheotomy becomes necessary.

Convalescent stage. The convalescent stage, extending from the subsidence of muscle tenderness to the maximum recovery of muscle power, may last for 2 years or longer. Gentle active exercises should be started early and gradually increased. Resistive exercises may be of great value at this stage. In hydrotherapy the support of the water allows muscles to be moved more effectively and provides a pleasant way of carrying out the exercises. The exercises may be preceded by heat and massage. Selected exercises establish better coordination between partially paralyzed and normal muscle groups. As the patient returns to walking, gait instruction and training should be provided. Great care should be taken to prevent the development of deformities by proper positioning of the joints and early active corrective exercises.

In the convalescent stage, braces and other supportive devices (Fig. 6-5) are often necessary (1) to enable the patient to stand and walk, (2) to prevent deformity and malposition, and (3) to help in stretching certain groups of muscles that have become contracted. A walking caliper brace with or without a lock joint at the knee is indicated in paralysis involving the thigh muscles. An ankle brace with a right-angle stop joint may be used for paralysis of the dorsiflexor muscles of the ankle. In paralysis of the invertors or evertors of the foot, a lateral or medial single-bar type of brace with a T strap attached to the shoe may be used.

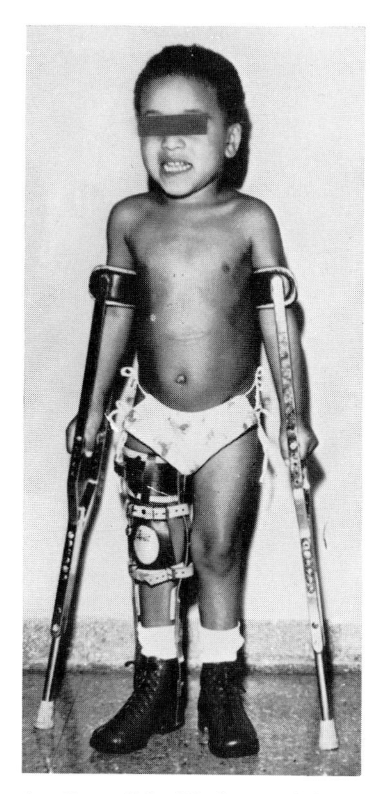

Fig. 6-5. Convalescent stage of poliomyelitis. Weakened right lower limb is protected by use of a long leg brace and crutches of adjustable length.

Residual stage. In the residual stage operations are performed to reduce or eliminate the need for braces by correcting deformities, improving muscle balance, and stabilizing joints. In general, procedures of the last two types should not be done until 18 months to 2 years after the onset of the disease.

Surgical measures. The correction of deformities and disabilities of the joints after poliomyelitis is still a part of the surgical work on crippled children's services. The procedures, when selected on the basis of sound indications, performed correctly, and followed by adequate aftercare, are productive of gratifying results.

1. *Operations for the correction of deformities of long standing:* Because of muscle imbalance major deformities of the trunk and extremities may gradually develop. Common among them are flexion contractures of the hips and knees and equinus deformities of the ankles (Fig. 6-4). Such deformities can at times be overcome with gradual stretching by means of casts and exercises. Frequently, however, they become so resistant that surgical release of fascia, muscle, or tendons must be done. When a deformity has been corrected, steps must be taken to prevent its recurrence. This can be done by bracing, by surgical stabilization of the bony structures, or by restoring muscle balance by means of tendon transplantation.

2. *Muscle and tendon transplantations:* In selected cases, improvement in muscle balance may be obtained by muscle and tendon transplantations, or transferences. In tendon transference it is important: (1) that any deformity of the joint should have been corrected previously; (2) that the tendon used should arise from a muscle with good power, preferably equal to that of the paralyzed muscle; (3) that the tendon should pass to its new insertion in a straight line through subcutaneous fat or through a tendon sheath; (4) that the tendon should be inserted under slight tension directly into bone; and (5) that, after the period of postoperative immobilization, supervised exercises should be carried out for the patient to learn optimum use of the transferred muscle-tendon unit.

An example of a helpful tendon transplantation is one used for paralysis of the dorsiflexors of the foot. The peroneus brevis, if of good or normal strength, may be brought forward and inserted into the dorsum of the foot, its original function of eversion and plantar flexion thus being changed to that of dorsiflexion. Similarly, in a hand with paralytic wrist drop, one of the wrist flexors can be rerouted and inserted posteriorly for conversion to a dorsiflexor.

3. *Operations to increase the stability of joints:* Most useful procedures in the surgical treatment of poliomyelitis and other types of flaccid paralysis are operations that restore stability to flail joints. Arthrodesis is the most important of these surgical procedures.

In the lower limb one of the most effective stabilizing procedures is *triple arthrodesis,* or arthrodesis of the talocalcaneal, talonavicular, and calcaneocuboid joints. This operation eliminates the inversion and eversion that take place at the subtalar joint, thus giving stability to the foot when the muscles are not strong enough to support it and control its position (Fig. 6-6).

4. *Operations to equalize the length of the legs:* After poliomyelitis, considerable shortening of the affected leg may be produced by the retardation of bone growth associated with muscle atrophy and diminished blood flow to the limb. The degree of shortening depends upon the severity of the paralysis, the age at which

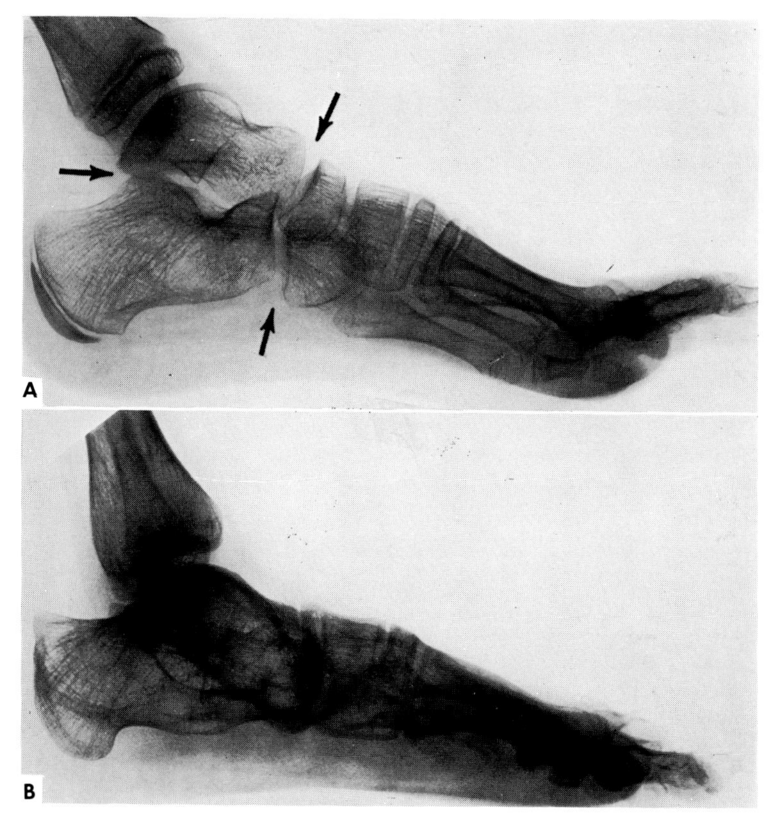

Fig. 6-6. Roentgenograms of the foot before and after triple arthrodesis for weakness and instability from poliomyelitis. **A,** Arrows show the three joints that are to be fused. **B,** Seven years after arthrodesis. The talus, calcaneus, navicular, and cuboid now form a single bony mass providing stability for the foot.

paralysis occurs, and other factors that are poorly understood. The correction of leg length discrepancies is discussed on p. 378.

PROGRESSIVE MUSCULAR ATROPHY (ARAN-DUCHENNE TYPE)

Progressive muscular atrophy is a primary disease of the spinal cord, characterized by a slow degeneration of the anterior horn cells, which produces muscular wasting without sensory losses. It is found most often in adults between 25 and 45 years of age.

Clinical picture. The first evidence of the disorder is usually atrophy of the intrinsic muscles of the hand. The thenar eminence becomes flattened, the interosseous areas deepen, and with progression of the disease a clawhand develops. Gradually the paralysis extends to the muscles of the arms, shoulders, back, hips, and thighs. Other cases may first show atrophy and weakness in the shoulders, and rarely the legs are first involved. As the condition progresses, the deep reflexes may become hypoactive or absent. Fibrillation of the muscles is a frequent finding and is of diagnostic importance. Muscle biopsy and electromyography may aid in establishing the diagnosis.

When the disease has partially destroyed the pyramidal tracts, as evidenced by spasticity of the legs, hyperactive reflexes, and a positive Babinski sign, it is called *amyotrophic lateral sclerosis.*

Treatment. The orthopaedic treatment is usually conservative, consisting of exercises and braces. However, if the disease becomes nonprogressive, stabilization of a joint to improve function may be considered.

INFANTILE SPINAL MUSCULAR ATROPHY (AMYOTONIA CONGENITA)

Infantile spinal muscular atrophy is characterized by extreme muscular weakness and hypotonia secondary to gradual degeneration or a developmental defect of the anterior horn cells of the spinal cord. When fully developed at birth, it has usually been called amyotonia congenita, or *Oppenheim's disease,* whereas it is often called *Werdnig-Hoffmann disease* when the onset follows several weeks or months of normal or nearly normal development. Deep tendon reflexes are absent or gradually lost. These infants will almost never learn to walk. There is inadequate muscle strength to prevent sagging of the head and legs when such an infant is lifted under the back (Fig. 6-7). Postural deformities, including a severe scoliosis, may develop.

Infantile spinal muscular atrophy should be differentiated from *benign congenital hypotonia,* a syndrome in which there are also severe muscle weakness and hypotonia in infancy, with moderate to great retardation of sitting and walking. In these patients deep tendon reflexes are at least weakly present, and by the fourth to sixth year these youngsters have usually developed normal strength and appearance. The cause of benign congenital hypotonia is unknown; it may be a retardation of muscle maturation.

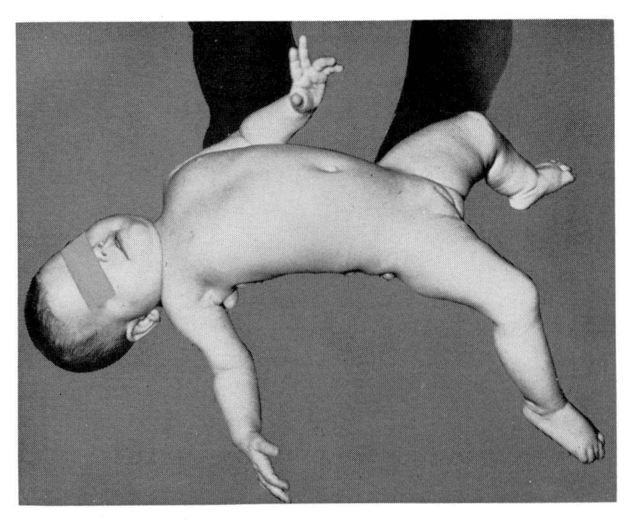

Fig. 6-7. Infantile spinal muscular atrophy in a girl 6 months of age. As the infant is picked up with support under the shoulders and hips, the head and lower limbs fall backward. Because of muscular weakness the infant cannot hold her head and lower limbs straight.

Treatment. Massage and exercises may be of some benefit. Braces are sometimes indicated for support of the spine and legs.

FRIEDREICH'S ATAXIA

Friedreich's ataxia is a progressive disease, usually hereditary or familial, which often affects several children of the same family. It develops in childhood, about 90% of the cases appearing before the fifteenth year. The essential pathologic change is an extensive degeneration or sclerosis of the posterior columns of the spinal cord. There is also involvement of the spinocerebellar tracts and later of the corticospinal tracts.

Clinical picture. The affection is characterized by weakness of the legs, ataxia, and a swaying, irregular gait with the feet placed widely apart. Nystagmus is characteristically present. The speech is usually thick. Equinovarus deformity of the feet is almost always present, and lateral curvature of the spine is often associated. Early in the disorder the deep reflexes are decreased or absent, but the Babinski sign is present. Later there is often a loss of position and vibratory sensations in the lower limbs, and disturbances of other types of sensation are sometimes observed.

Treatment. At present no curative treatment is known. Braces are sometimes indicated. Occasionally, when the disease is no longer progressive, surgical stabilization of the feet is useful to improve the gait.

SUBACUTE COMBINED SCLEROSIS

In pernicious anemia and deficiency diseases such as pellagra, lesions involving the posterior and lateral columns lead to the development of numbness and tingling in the hands and feet, which may be followed by the gradual onset of weakness. Later an obvious flaccid or spastic paralysis of the legs may be present, and secondary deformities are likely to appear.

Treatment. Deformities secondary to the paralysis may respond to physical therapy, and braces may be of benefit. Rarely is lengthening of the Achilles tendons or stabilization of the feet indicated in an effort to improve the gait. Vitamin B_{12} is indicated when the primary disease is pernicious anemia; niacin is indicated when it is pellagra.

NEUROPATHIC DISEASE OF BONES AND JOINTS

Etiology. Chronic disease of the spinal cord may cause extensive trophic changes in the bones and joints. Such osteoarticular lesions are seen typically in association with the spinal cord involvement of *tabes dorsalis* and of *syringomyelia.* Statistics show that joint symptoms occur in from 3% to 4% of patients with tabes dorsalis and in from 10% to 40% of those with syringomyelia. The smaller joints of the foot are frequently involved in patients with *diabetic neuropathy.* Similar joint changes occur less commonly with other affections of the spinal cord, including traumatic conditions, congenital malformations such as spina bifida with myelomeningocele, tumor, acute myelitis, and progressive muscular atrophy. They have been reported after involvement of the peripheral nerves by trauma, toxic neuritis, or leprosy, and in association with the cerebral changes of general paresis or of cerebral hemorrhage with hemiplegia. Although more than one factor is probably involved, repeated minor trauma of the anesthetic joint seems to be most important in the pathogenesis. The patient, unable to appreciate normally painful stimuli arising in the joint, fails to protect it, voluntarily or involuntarily, from harmful forces.

Pathology. The changes in neuropathic osteopathy include thinning of the cortical bone and osteoporosis, which may lead to spontaneous, painless fracture. Healing of such fractures occasionally takes places with the formation of an enormous amount of callus, which may show degenerative changes leading to a second fracture.

The neuropathic arthropathies, often called *Charcot's joints* because they were first described by Jean Martin Charcot in 1866, are most frequently encountered in tabes. The articular cartilages and adjacent bone surfaces become worn away, while at the same time hypertrophic, sclerotic changes take place at the joint edges, and loose bodies of irregular shape and size appear. Severe deformity and instability result from mushrooming of the bone, relaxation of the ligaments, and accumulation of intra-articular fluid. The roentgenographic picture of fully developed cases is striking (Fig. 6-8). Both atrophic and hypertrophic bone changes are present. The joint surfaces appear extensively eroded and deformed. The bone margins are jagged, blurred, and sclerotic, and there may be many irregular pieces of detached bone.

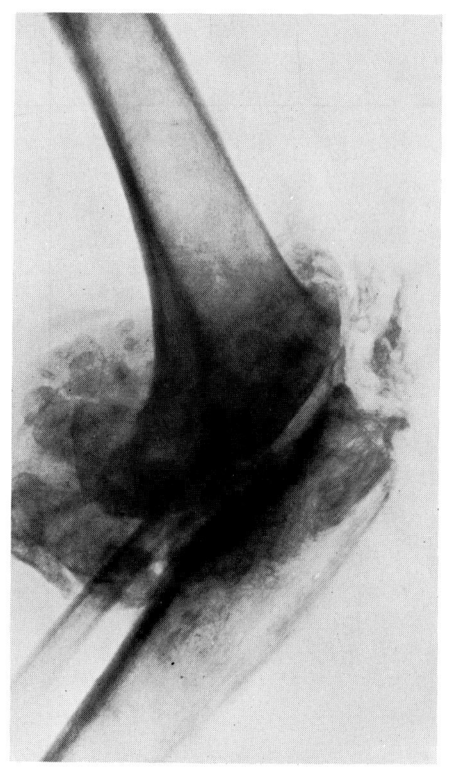

Fig. 6-8. Roentgenogram of advanced neuropathic disease of the knee (Charcot joint). Note erosion of articular ends of the femur and tibia, subluxation, and multiple loose fragments of bone.

Clinical picture. In neuropathic arthropathy of limb joints the early signs are instability, false motion, and swelling. These may increase rapidly or slowly. Pain, although a common complaint, is mild in comparison with the severe destructive changes in the joint. Examination shows a tense or boggy, nontender swelling, not of uniform consistency but containing indurated masses. As a rule the range of joint motion is increased. Although most frequently only a single joint is involved, bilateral symmetric lesions are occasionally seen, and rarely a large number of joints may be affected, especially in syringomyelia. The knee, hip, shoulder, tarsus, elbow, wrist, and ankle are most frequently involved, in this order. Lesions of the spine are also encountered. Seventy-five percent of the tabetic arthropathies occur in the lower limb, whereas 80% of the joints involved in syringomyelia are those of the upper limb.

Diagnosis. In advanced cases the diagnosis is often obvious, but in the early stages it may be difficult. In neuropathic arthropathy the local triad of swelling, instability, and absence of significant pain should always suggest the diagnosis, and the roentgenographic evidence is most helpful. The findings of the general history and physical examination are of confirmatory value. In the case of tabes dorsalis, the Argyll Robertson pupil, absent knee or ankle jerk, and diminished position and vibratory sensations are important findings. Suggestive of syringomyelia are loss of pain and temperature sensations and atrophy of the small muscles of the hand.

Treatment. In tabes dorsalis, antisyphilitic treatment is a major consideration, although it may be too late to have any effect on the disintegrated joint. In neuropathic arthropathy, protection for the affected joint is afforded by means of a brace, such as a walking caliper splint for the knee. Crutches may be necessary. With conservative treatment the progress of the disease may sometimes be retarded to a gratifying degree. Arthrodesis of the affected joint is not always successful but in some cases has given excellent results, particularly at the knee. Rarely, in extreme involvement of the knee or ankle, the impossibility of securing adequate stability either by apparatus or by arthrodesis may make amputation the procedure of choice.

SPINA BIFIDA

Spina bifida is a congenital anomaly consisting of a developmental gap or defect in one or more of the vertebral arches (Fig. 6-9), through which the contents of the spinal canal may protrude and with which partial or complete paralysis of the legs may be associated. The term *myelodysplasia* is used to designate a variety of clinical syndromes that include a developmental defect of the spinal cord and a peripheral neurologic defect. Spina bifida sufficiently severe to produce clinical deformity is present in about 1 in 1,000 births. Of these infants, the overall survival rate is about 50%. The projecting meningeal sac usually appears posteriorly (Fig. 6-10). Occasionally it extends anteriorly into the pelvis, abdomen, or thorax; the condition is then called an *anterior spina bifida.*

Pathology. The essential pathologic characteristic of spina bifida is incomplete development of the roof of the neural arch due to failure of complete fusion of the embryonic neural canal.

Each vertebra has three centers of ossification: one for the body and one for

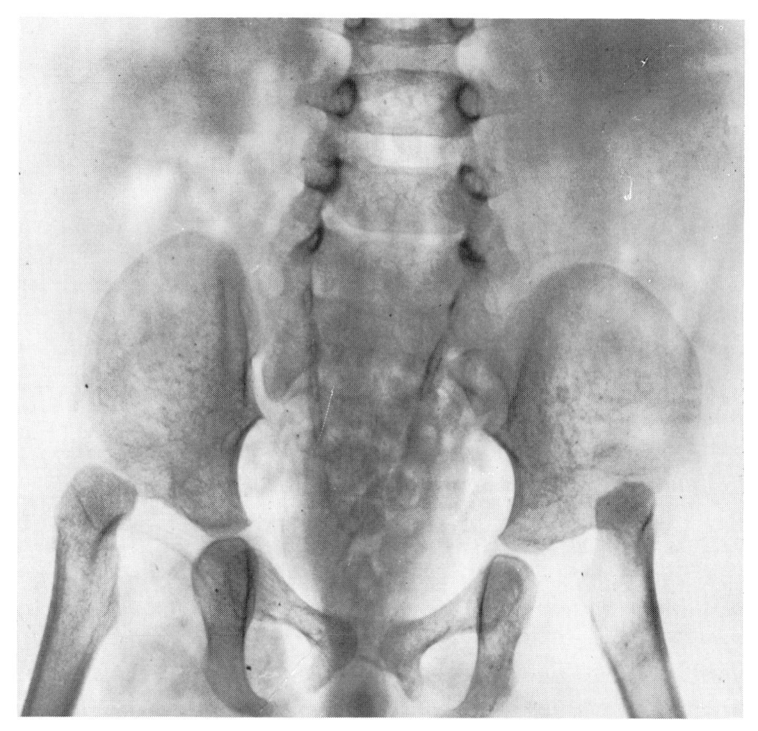

Fig. 6-9. Spina bifida in a boy 3 years of age. Note wide defect in the laminal arches of the lower three lumbar vertebrae and sacrum, and bilateral dislocation of the hips. In early infancy this patient had had surgical treatment of his myelomeningocele.

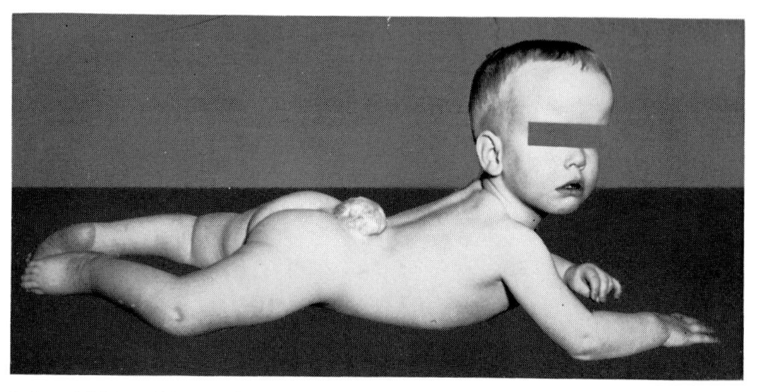

Fig. 6-10. Spina bifida with myelomeningocele and hydrocephalus in a boy 2½ years of age. There were associated contractures of the knees and clubfeet.

each half of the neural arch. The neural arch is formed by the posterior, midline fusion of the two laminae arising from their separate centers of ossification. Union of the laminae begins in the thoracic region and extends in both directions along the length of the developing spine. The lumbosacral and cervical regions are the last to unite, and it is in these two areas, particularly the former, that faulty closure, resulting in spina bifida, is seen most frequently.

In the normal human embryo the entire neural tube is closed at the end of the third week, and by the eleventh week the partially ossified neural arches of the vertebrae are closed from the first cervical to the third or fourth sacral segments. The spinal cord and the vertebral column are of equal length until the twelfth week. With further growth of the fetus, however, the vertebral canal becomes proportionately longer, so that in the adult the conus is at the level of the twelfth thoracic or first lumbar vertebra. When the nerve roots or the spinal cord is involved in a spina bifida, the upward migration of the cord is prevented. This may be associated with the *Arnold-Chiari malformation,* an elongation of the brain stem and portions of the cerebellum downward through the foramen magnum. Spina bifida may also be associated with *diastematomyelia,* in which a segment of the spinal cord is divided longitudinally by a midline bony spur or fibrous band. Diastematomyelia also secures the dura and prevents migration of the spinal cord.

The defect of spina bifida may be large enough to allow the meninges or the spinal cord to form a protruding soft tissue tumor. According to the extent of the pathologic changes, spina bifida has been classified into five types: spina bifida occulta and spina bifida with, respectively, meningocele, myelomeningocele, syringomyelocele, and myelocele.

Spina bifida occulta. A defect is present in the spinous process and laminae of one or more vertebrae (Figs. 11-4 and 11-6), but there is no obvious swelling or protrusion. The skin may remain attached to the membranes, nerve roots, or cord by fibrous tissue called the *membrana reuniens.* These subjacent anatomic changes may be suggested by alterations of the skin such as indentations, pigmentations, telangiectases, or hairy patches (hypertrichoses). There may also be associated tumors inside or outside the vertebral canal; among them are lipomas, angiomas, and dermoid cysts. As the growth of the spinal column later exceeds that of the spinal cord, adherence of the overlying structures to the cord may impede its normal ascent. In this manner, as well as through compression of the nerve roots by soft tissue at the site of the laminal defect, paralysis of gradual onset and incomplete type may result at a later age, particularly at a time of rapid growth. In the vast majority of cases, however, spina bifida occulta is discovered as an incidental finding in routine roentgenograms of the low back. In such situations it is usually of no clinical significance.

Spina bifida with meningocele. Through the defect of the incompletely closed arch of one or more vertebrae, the meninges herniate and are covered by only a thin, parchmentlike layer of skin. The hernial sac contains only cerebrospinal fluid; transillumination will reveal no nerve tissue.

Spina bifida with myelomeningocele. With the bony defect is associated a hernial sac that contains, in addition to cerebrospinal fluid, the spinal cord, nerve roots, or both, either free or attached to the walls of the sac. As compared with meningocele, myelomeningocele is usually much greater in extent, more subject to ulceration if untreated, and more often associated with neurologic deficit in the lower extremities, bowel and bladder incontinence, and hydrocephalus.

Spina bifida with syringomyelocele. Syringomyelocele is a severe type of myelomeningocele in which the herniated spinal cord contains a central canal greatly dilated and distended with cerebrospinal fluid.

Spina bifida with myelocele. Also known as *rachischisis,* this condition includes the most severe forms of spina bifida. As a result of the absence of laminae and pedicles, the wide bone defect forms an open groove, partially lined by imperfectly formed spinal cord tissue through which cerebrospinal fluid drains. As a consequence infection quickly takes place, and the infant succumbs early.

Of these five types, the last four are characterized by the presence of a soft tissue tumor due to herniation of the meninges and have been called *spina bifida manifesta.*

Clinical picture. The clinical picture presents one or more of three features: (1) the tumor, or protruding soft tissue mass, (2) the neurologic manifestations, and (3) the associated deformities.

The hernial protrusion is located in the midline, most commonly in the lumbar or lumbosacral region. It may vary greatly in size and may increase with violent expiration as in crying. It is usually translucent on transillumination. A defect in the underlying bony structure is apparent on roentgenographic examination.

The neurologic manifestations may be absent in spina bifida occulta and absent or minimal in meningocele. In myelomeningocele they may consist of severe motor involvement or complete paraplegia with bowel and bladder incontinence, together with extensive sensory disturbance resulting in trophic ulcerations. The motor involvement is usually a flaccid paralysis and occasionally may be a spastic paralysis.

The most frequently associated deformities are clawfoot, clubfoot, contractures of the knees, malformations of the spine with varying degrees of scoliosis, kyphosis, or lordosis, dislocation of one or both hips, and hydrocephalus (Fig. 6-10).

Diagnosis. The diagnosis is obvious in cases with a protruding hernial mass in the lumbar region. Among conditions to be excluded are lipomas and neurofibromas. Spina bifida of the occult type may be diagnosed only as an incidental finding in roentgenograms made for other purposes; however, palpation of the area will sometimes demonstrate changes as compared with the normal findings at higher levels of the spine.

Treatment. The treatment of severe cases requires the combined efforts of the orthopaedist, urologist, neurosurgeon, and physical therapist. Early neurosurgical closure of the dural sac, fascial reinforcement, and full-thickness skin coverage are usually indicated. Drainage of cerebrospinal fluid, ulceration, paraplegia, absence of the anal reflex, constant dribbling of urine, poor general health, or an already existing hydrocephalus may complicate or delay operation.

Spina bifida occulta may require operation in later childhood if neurologic signs secondary to involvement of the spinal cord by traction or compression develop. Surgery then consists of careful excision of the membrana reuniens and of any tumor that may be present. The prognosis following this procedure is usually good.

The degree of disability and the indications for orthopaedic treatment vary with the level of the lesion in the spinal cord. Paralysis and deformity in the lower limbs may require extensive orthopaedic treatment with casts, braces (Fig. 6-11), or operations. These procedures should be carried out in a manner similar to that described for corresponding deformities and disabilities following poliomyelitis. Paralytic dislocation of the hip, a frequent problem when neurologic function is lost below the third or fourth lumbar segment of the spinal cord, can often be

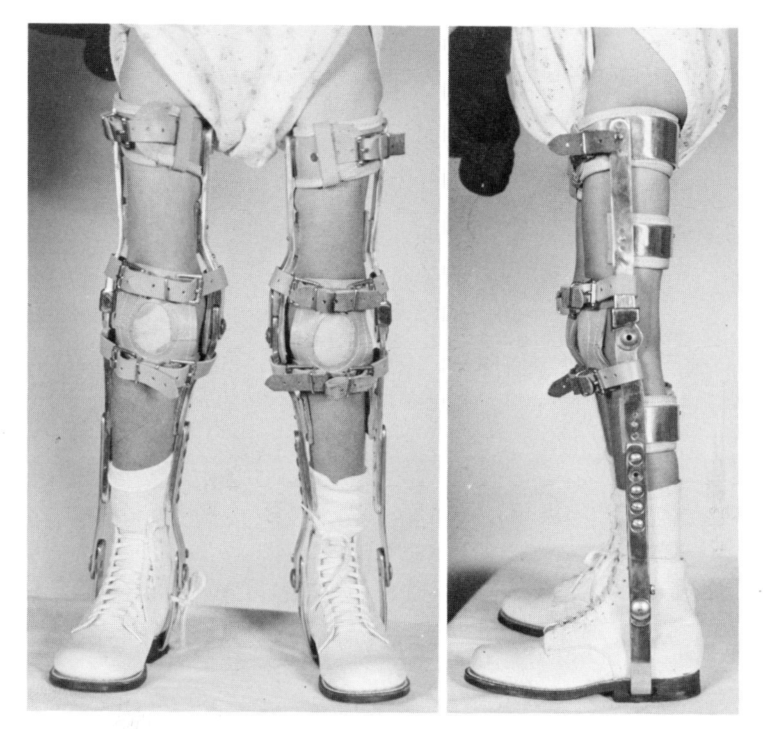

Fig. 6-11. Long leg braces for a child with flaccid paralysis of both lower limbs due to spina bifida. Extension lock at the knee joints may be released for sitting. Ankle stops prevent foot drop.

treated successfully by open reduction, innominate osteotomy, or acetabuloplasty to deepen the hip socket, and transfer of the iliopsoas tendon through the ilium to the greater trochanter to correct the muscle imbalance.

Care must be taken to avoid pressure ulcers. Training the severely involved patient to walk with the aid of crutches and braces and with a three-point or four-point gait in the manner of a traumatic or poliomyelitic paraplegic may greatly facilitate his treatment and decrease his invalidism.

Prognosis. With early closure of the dural sac, control of hydrocephalus, prevention of urinary tract infection, and adequate orthopaedic treatment, the formerly dismal outlook for these severely handicapped children has improved considerably in recent years.

Involvement of peripheral nerves

PERIPHERAL NERVE INJURIES

Injury of the peripheral nerves is a frequent and serious complication of traumatic lesions of the bones and joints. Every injured limb must be examined for the possibility of nerve damage, and in a considerable proportion of cases the neurologic lesion will prove of far greater significance than the osteoarticular lesion.

A short review of the structure of peripheral nerves and their reactions to injury forms a convenient preliminary to consideration of lesions of the individual nerves.

Anatomy. A peripheral nerve is composed of a tremendous number of fibers enclosed in a connective tissue covering. The individual fibers are separated by a loose fibrous tissue termed the *endoneurium.* Fibers are grouped into bundles enclosed by the *perineurium.* The entire peripheral nerve is enclosed by the collagenous *epineurium,* which also sends connective tissue septa between the nerve bundles. The central portion of the individual nerve fiber, or *axon,* is the *axis cylinder;* its protoplasm, which is a prolongation of the cytoplasm of the nerve cell, extends the entire length of the nerve. Its extremely thin membrane is continuous with the nerve cell membrane. Surrounding the axis cylinder is a layer of myelin of variable thickness. The axon and myelin in turn are enveloped by the thin Schwann cell membrane. Nerve fibers vary in size. The largest, heavily myelinated axons are motor fibers. Next in size are fibers concerned with proprioception and cutaneous sensation. Pain fibers and those of the autonomic system are very small. There is a definite relationship between fiber size and extent of myelination on the one hand and electrical properties on the other; the large, heavily myelinated fibers conduct impulses at a much more rapid rate than do the small, thinly myelinated fibers.

Mechanism of injury. Traumatization of a limb may involve the nerves in any of several types of injury. These may be classified according to the time of their occurrence with relation to that of the original injury.

At the time of the original injury. The nerves may be directly bruised, lacerated, or completely severed by the rough edge of a fractured bone or by a sharp object from without, such as the blade of a knife or a piece of glass. A portion of the nerve may be torn away; this happens not infrequently in severe gunshot wounds. An additional type of immediate injury is the severe stretch, tear, or even complete rupture of the nerve from strong traction on the limb; this type of injury occurs, for instance, in *obstetric paralysis.*

Several hours after the original injury. Temporary loss of nerve function may be occasioned by ischemia or by compression of the nerve from edema or hemorrhage of gradual development.

At the time of treatment of the original injury. Nerve damage is sometimes a result of excessive traumatization during the transportation of the patient when a splint has not been applied or during the reduction of a dislocation or fracture.

Several weeks or more after the original injury. The nerve may undergo gradual compression from cicatricial fibrous tissue or rarely from bony callus.

Several months or more after the original injury. Friction or stretching of the nerve is sometimes the result of a deformity caused by the original injury, as in late traumatic ulnar neuritis.

Gross pathology of the injured nerve. After an old injury the damaged nerve is often recognizable only with difficulty because of extensive scarring of its sheath and the adjacent connective tissue. In infected wounds the adhesions are often particularly extensive, and in the proximal segment an ascending neuritis sometimes develops, causing widespread degenerative changes that are followed later by fibrosis. The appearance of the injured nerve varies especially with the degree of completeness of the tear. After complete division the nerve ends retract; a *neuroma,* or small bulb-like growth of nerve fibers and connective tissue, forms at the end of the proximal segment; and the end of the distal segment may become narrowed and atrophic or may enlarge to form a *pseudoneuroma,* which is composed of connective tissue only. After incomplete laceration the injured nerve may show a fusiform enlargement, formed by the proliferation of axons, Schwann cells, and fibrous tissue and termed a nerve spindle, or *neuroma in continuity.* In some cases there will be found a very thin segment that may suggest an incomplete division but that represents essentially a complete one, since all nerve fibers have ruptured and degenerated, only connective tissue sheaths remaining intact. In other cases long segments of the nerve fibers may degenerate from friction or stretching but produce no macroscopic changes of any kind.

Degeneration and regeneration of nerves after section. The distal segment of a peripheral nerve fiber that has been severed undergoes characteristic changes referred to as wallerian degeneration. In the first three days after nerve section no microscopic changes are noted, and the distal segment will still conduct if artificially stimulated. From the third day until the end of the first week there is degeneration of the axoplasm and of the myelin, which gathers into small globules. Schwann cells begin to proliferate, and an invasion of macrophages occurs. The fiber will no longer conduct, and enzymes associated with acetylcholine metabolism disappear. In the following weeks the formation of Schwann cell tubules continues, awaiting regenerating axons from the proximal segment.

In the process of regeneration the axis cylinder of the proximal segment, still under the trophic influence of the nerve cell, gradually grows across the gap, enters the empty neurilemmal sheath, and continues to grow distally toward the end-organ while gradual restoration of the myelin sheath takes place. The rate of growth of the axis cylinder, under favorable circumstances, is said to be from 1 to 2 mm. per day.

Since anatomic and functional recovery of the injured nerve is directly dependent upon reestablishment of the pathway from nerve cell to end-organ by means of the regenerating axis cylinder, it is evident that the degree of recovery after any nerve injury will depend largely upon the ability of the axis cylinders to traverse the gap successfully and to find suitable waiting neurilemmas. Hence recovery will not take place if the defect is too extensive or if too much scar tissue is present.

Classification of nerve injuries. Three types of peripheral nerve injury have been described by Seddon: neurapraxia, axonotmesis, and neurotmesis.

Neurapraxia. In neurapraxia there is nerve contusion or compression resulting in segmental damage but leaving the axon structurally intact. Since large fibers are especially vulnerable to this type of injury, motor paralysis in the milder cases may be more prominent than loss of pain sensation. Nerve conductivity is impaired and there may be localized demyelination, but recovery is rapid and always complete. Examples of neurapraxia include crutch palsy, and radial nerve paralysis from sleeping with an arm hanging over the back of a chair.

Axonotmesis. Axonotmesis is an interruption of the axons without damage of the nerve sheath. This type of lesion results from more severe or prolonged compression than that which causes neurapraxia. Since the axons are damaged, recovery must be preceded by wallerian degeneration of the distal segment. In the electromyogram, fibrillation is noted 2 to 3 weeks after the injury. Since the epineurium and perineurium are intact, however, the regenerating Schwann cell sheaths are properly oriented and allow the new axons to find their way into correct channels. Recovery is usually satisfactory if excessive scar tissue does not develop; the time required for recovery is related to the distance of the lesion from the end-organ.

Neurotmesis. Neurotmesis, in which there is complete interruption of nerve fibers and their connective tissue coverings, is the most severe form of nerve injury in Seddon's classification. There is total paralysis of the involved muscles, and after 2 weeks their response to faradic stimulation is lost. Electromyographic findings and strength-duration curves show changes typical of degeneration. Cutaneous sensation is lost over the *autonomous zone,* which is the area supplied exclusively by the injured nerve. Examples of neurotmesis are nerve transections in knife and bullet wounds. The prognosis is much poorer than in the milder nerve injuries and depends largely upon the accuracy of surgical repair.

Recovery phase. The recovery of larger, more heavily myelinated fibers is poorer than that of smaller ones. Accordingly the return of motor function is often less complete than that of pain sensation, and touch and proprioception occupy an intermediate position. In axonotmesis and especially in neurotmesis faulty reinnervation may occur, the developing fibers entering inappropriate axon sheaths. Regenerating fibers are often of smaller diameter than the original fibers. Accordingly, incomplete or inappropriate motor innervation may result, and sensory function may be disturbed in several ways. In some instances severe pain and burning sensations, occurring spontaneously or provoked by minimal stimuli, result in the disabling condition known as *causalgia.*

As the regenerating axons extend distally, there develops a hypersensitivity that is manifested by a tingling sensation when the regenerating portion of the nerve is tapped. This is called Tinel's sign; usually it appears at the site of the lesion a few weeks after injury and extends distally with nerve regeneration. In favorable cases the return of sensation is followed by motor recovery of variable degree. The prognosis for functional recovery after complete injury is much better in children than in adults.

Treatment of peripheral nerve injuries. The treatment of nerve injuries is either nonsurgical or surgical, and every case involves a decision as to whether spontaneous recovery can reasonably be expected or whether surgical exploration is the wiser course. A detailed history of the injury and a thorough neurologic examination are essential. Of all the factors involved in the therapy of these cases, perhaps the most important are (1) surgical judgment in deciding when operation is indicated and (2) surgical technic in carrying out the procedure with a minimum of trauma and hemorrhage. The value of physical therapy in nonsurgical treatment and in postoperative care is not to be underestimated.

Nonsurgical treatment. Treatment without operation is always indicated when there is evidence of progressive spontaneous improvement of nerve function, and it is to be employed tentatively when such improvement may be expected after a probably incomplete division of a nerve. It should be explained to the patient at the outset that the course of treatment may be prolonged.

The essentials of nonsurgical treatment are two: (1) support of the paralyzed muscles and (2) physical therapy. The affected muscles must be kept in a position of relaxation; they should not be stretched by normal opponents. To this end, light retentive apparatus suited to the particular case is to be worn most of the time. *Cock-up splints* in radial paralysis and *foot-drop braces* in peroneal paralysis are common examples. After a short initial period of rest, physical therapy should be used daily in an attempt to preserve muscle nutrition until regeneration occurs and to prevent the development of adhesions and contractures. It consists at first of heat and massage, passive exercises, and electrical stimulation; later, as recovery begins, muscle reeducation and active exercises are added. The progressive recovery of power in paralyzed muscles usually observed in these cases is a source of gratification that makes the extended course of treatment well worthwhile.

Surgical treatment. Operations for exploration or repair of a damaged nerve may be (1) *primary,* that is, performed immediately after a lacerating injury, or (2) *secondary,* performed after the original wound has healed or when progress in the gradual recovery of nerve function has been unsatisfactory.

It is sometimes possible at the primary operation, as in the repair of a clean laceration, to suture the ends of the sectioned nerve. This is usually inadvisable in the presence of open fractures or bullet wounds. In such cases a secondary nerve suture should be performed after the wound has become perfectly healed and free of infection and after scarring at the severed nerve ends has matured. The interval from nerve injury to nerve suture may vary from several weeks to several months. During this interval the muscles and joints should be kept in as good condition as possible by means of daily physical therapy.

In some cases injury may have resulted in such a long defect in the nerve that it is impossible to bring its two ends together. Nerve grafting has been used in such instances with limited success. Autogenous nerve grafting, with the resected sural nerve, has been more effective than repair with bank or cadaver nerve. It should be noted that nerve grafting requires that instead of one gap the regenerating axons must bridge two, that at the proximal suture line and that at the distal line; this decreases the probability of complete functional return after nerve grafting.

The indications for exploratory or secondary operation vary within certain recognized limits. When complete nerve division is suggested by the history and by extensive and persistent sensory and motor losses, exploration should be performed within the first month. If separation, or neurotmesis, has occurred, satisfactory regeneration will not take place without approximation of the nerve ends by suture. The time of appearance of the nerve signs with relation to the time of injury is sometimes of value in differentiating actual rupture from a temporary interruption of function by edema or hemorrhage. When the signs of incomplete interruption are present, nonsurgical treatment should be adopted tentatively, but exploration should usually be carried out if definite improvement has not been shown after 3 or 4 months.

In operations upon peripheral nerves, scrupulous precautions to preserve asepsis and hemostasis and to avoid traumatization are essential. *Neurolysis,* or freeing of the nerve from cicatricial adhesions and constrictions, is a common and often very beneficial procedure. It is sometimes necessary to compensate for the shortening of a divided nerve by fixing the adjacent joint temporarily in a suitable position or by transplanting the nerve to a shorter course. Postoperatively braces and physical therapy are used just as in the nonsurgical treatment. Postoperative recovery of nerve function is slow and, as a rule, never quite complete, depending upon many factors, such as the duration of the paralysis, the age of the patient, the type of nerve involved, the level at which it is injured, and the degree of secondary change in muscles, tendons, and joints. Often with careful protection and physical therapy, however, improvement can be expected to continue for as long as 2 or 3 years.

INJURIES OF INDIVIDUAL NERVES

Theoretically, every peripheral nerve is susceptible to mechanical injury. Practically, however, and particularly in civil life, the frequent and important lesions are limited to a few of the major trunks and to nerves in the hand. Nerve injury occurs far more commonly in the upper limb than in the lower limb. In the upper limb, lesions of the radial, ulnar, and median nerves are of major significance, and in the lower limb injuries of the sciatic and peroneal nerves are most important. Lesions of less commonly injured nerves are of great interest in the individual case.

Accessory nerve (eleventh cranial nerve)

Etiology. The accessory nerve is sometimes divided by lacerating or perforating wounds of the neck or during neck dissections.

Clinical picture. If the nerve is divided in the posterior triangle after emerging from the sternocleidomastoid muscle, only the trapezius muscle is paralyzed. This produces an unsightly deformity consisting of change in the contour of the neck, drooping of the shoulder, and occasionally slight winging of the scapula. Abduction of the arm is at first impaired but later may be partially restored by the action of other muscles of the shoulder girdle. If the nerve is divided in the anterior triangle of the neck, the sternocleidomastoid muscle also is paralyzed, which results in little loss of power but in a noticeable asymmetry.

Treatment. Unless primary suture can be accomplished, the nerve lesion is as a rule irreparable. If winging of the scapula is conspicuous, the deformity may be lessened by surgical measures to limit its displacement.

Brachial plexus

Etiology. The nerves of the brachial plexus are subject to the following types of injury: traction lesions; friction, contusion, or compression lesions; and penetrating lesions.

Traction lesions. If the head is laterally flexed while the opposite shoulder is fixed, or if the shoulder is violently depressed while the head is fixed, the trunks of the brachial plexus become taut and, if the force increases, may be stretched, torn, or even completely ruptured. Such injury occurring during birth causes the common and important *obstetric paralysis,* or *birth palsy.* Similar lesions in adults may result from severe stretching of the brachial plexus incident to the trauma of falls or heavy blows upon the shoulder.

Friction, contusion, or compression lesions. Nerve injuries of this type are of particular interest in connection with two clinical entities: (1) *cervical rib* or spasm of the scalenus anterior muscle, which may cause chronic friction or compression (p. 406); and (2) *dislocations or fractures at the shoulder joint,* including displaced fractures of the clavicle, in which the nerves may be injured by the original trauma, by secondary edema or hemorrhage following the original trauma, or by manipulation during reduction of the displacement.

Penetrating lesions. Gunshot wounds form the commonest cause of this type of nerve injury. In the early treatment of these cases, hemorrhage from associated injury of the great vessels is often an important factor.

Clinical picture. The symptoms and signs of lesions of the brachial plexus vary widely with the type of trauma and the part of the plexus injured. It is therefore convenient to consider the more important clinical entities separately.

Obstetric paralysis

Obstetric paralysis, or birth palsy, is a paralysis of the muscles of the upper limb resulting from mechanical injury of the nerve roots of the brachial plexus during birth and is most commonly seen in infants born after a prolonged and difficult labor. The nerve injury may vary from slight stretching (neurapraxia or axonotmesis) to complete rupture (neurotmesis) of one or more of the components of the plexus. Edema and hemorrhage follow, and later cicatricial fibrosis occurs. In some instances the plexus injury is accompanied by fracture of the upper end of the humerus or by soft tissue injury of the shoulder. Clinically the newborn infant may present, in addition to paralysis of the arm, transient spasticity of the other arm and the legs as a result of the hematomyelia accompanying avulsion of the nerve roots. Inequality of

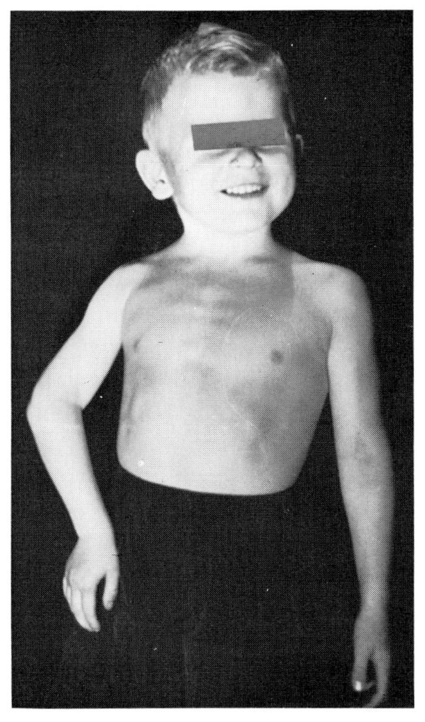

Fig. 6-12. Right obstetric paralysis in a boy 5 years of age. Note atrophy, internal rotation of the shoulder, and pronation of the forearm.

the pupils may be present from stretching or tearing of the cervical sympathetic nerves. Three main types of paralysis are encountered, depending upon the location of the injury: Erb-Duchenne or upper arm paralysis, Klumpke or lower arm paralysis, and paralysis of the entire arm.

Erb-Duchenne or upper arm paralysis. Upper arm paralysis is by far the commonest of these types and is caused by injury of the fifth and sixth cervical roots. Accordingly the deltoid muscle, the external rotator muscles of the shoulder, and the biceps, brachialis, supinator, and brachioradialis muscles are usually involved, and the limb occupies a typical position, with the shoulder internally rotated and the forearm pronated (Fig. 6-12). Movements of the wrist and of the fingers are not affected. There may be slight sensory changes, which in the infant are difficult to evaluate.

Klumpke or lower arm paralysis. Paralysis of the lower arm is much less common; it is due to injury of the eighth cervical and first dorsal roots. The intrinsic muscles of the hand and sometimes the long flexor muscles of the fingers are paralyzed. The upper arm is not affected. A homolateral *Horner's syndrome* is often present because of involvement of the cervical sympathetic fibers in the first dorsal root; it is characterized by slight ptosis of the eyelid, enophthalmos, and miosis.

Paralysis of the entire arm. In this type of obstetric paralysis, which in frequency occupies a position between the other two varieties, the limb is often completely flaccid and powerless. Extensive sensory losses occur, which in the infant are difficult to demonstrate.

Differential diagnosis. It must be remembered that at birth any of several injuries about the shoulder can cause a flail arm. Separation of the upper humeral epiphysis, fracture of the humerus, and fracture of the clavicle are of fairly common occurrence. Dislocation of the shoulder may be a true congenital luxation or a traumatic displacement due to injury during birth; each is extremely rare. In the infant, cerebral palsy must also be excluded, and in older patients it is necessary to consider the possibility of an old, unrecognized poliomyelitis.

Prognosis. Improvement usually takes place during the months after birth; however, the prognosis varies with the type of paralysis, and perfect recovery is exceptional. In the upper arm paralysis, considerable return of power is to be expected; the lower arm type with paralysis of the intrinsic muscles of the hand has a relatively poor prognosis, and a claw deformity may develop; the palsy of the whole arm is especially likely to show only incomplete recovery. The paralysis is followed by the development of contractures. The typical deformity of internal rotation of the arm is maintained by firm fibrous contractures of the muscles about the shoulder. In extensive late cases the entire arm and shoulder girdle are underdeveloped, and secondary growth changes, such as abnormal prolongation of the acromion and of the coracoid process and subluxation of the elbow, take place. In such cases the functional loss is great.

Treatment. The early treatment of obstetric paralysis is conservative. After a week for rest of the acute nerve injury, a diligently applied passive exercise program should be carried out by the infant's parents. The specific exercises depend on the type of paralysis. For example, in the Erb-Duchenne or upper arm paralysis there is a great tendency for contractures to develop with the shoulder in adduction and internal rotation and the forearm in pronation. Exercises designed to combat these contractures consist of gentle passive stretching of the shoulder into abduction and external rotation along with turning the forearm into supination. Such exercises should be carried out several times whenever the infant is handled for feedings or diaper changes. Bracing of the extremity may be helpful but may itself lead to contractures. If a brace is used, it should be worn only at night in order not to interfere with the exercise program. Active and passive corrective exercises should be continued for a long while, and the child should be kept under periodic observation for signs of developing contractures.

Surgical repair of the brachial plexus for birth palsy is rarely indicated. In patients who do not recover spontaneously, intraneural fibrosis is usually so extensive that surgical resection and anastomosis are impossible. In the infant, contractures may be treated by passive exercises; in older patients, open release is preferable. The procedure described by Sever, comprising section of the contracted pectoralis major and subscapularis muscles near their insertion into the humerus, is often useful. It may be necessary also to section the coracobrachialis and short head of the biceps. If the elongated acromion is limiting shoulder abduction, it should be osteotomized at its base and bent upward. Late cases of internal rotation deformity may be treated by rotation osteotomy of the humerus or by tendon transferences favoring external rotation. These operations, although producing no increase of muscle power, may make motion of the arm less awkward. Occasionally in late cases with weakness in the hand and pronation deformity, tendon transferences in the forearm or rotation osteotomy of the radius may be indicated. After any of these surgical procedures the limb must be supported adequately in the appropriate position; subsequently a program of selected exercises should be prescribed.

Paralysis following dislocation of the shoulder

Paralyses complicating shoulder dislocations are fairly common and fall into two major groups: (1) *supraclavicular traction injuries of the plexus,* produced by the original trauma and independent of the dislocation, and (2) *infraclavicular nerve lesions,* resulting from displacement of the humeral head at the time of the original trauma or during its manipulative reposition.

Supraclavicular traction injuries are treated by measures similar to those used for the analogous obstetric paralysis. Some of these injuries are so severe that the limb is rendered flail and senseless, actually becoming a burden to the patient. Myelography done shortly after a severe injury may show leakage of dye through the torn root sleeves, indicating that the roots of the plexus have been avulsed from the spinal cord. In such cases the prognosis for return of function is very poor and amputation may prove advisable.

Infraclavicular nerve lesions complicating dislocation may involve chiefly the following: the *axillary nerve,* causing deltoid paralysis and hypesthesia on the lateral aspect of the shoulder and upper arm; the *axillary and the radial nerves,* causing in addition a widespread extensor paralysis; the *medial cord* with paralysis of the intrinsic muscles of the hand and anesthesia in the ulnar distribution; or the *lateral cord* with paralysis of the biceps, coracobrachialis, and flexor muscles of the fingers supplied by the median nerve. The extent of the paralysis depends in large degree upon the length of time during which the humeral head is allowed to remain displaced. The treatment is immediate reduction of the shoulder dislocation, which can usually be accomplished by a closed procedure, followed within a few days by splinting of the shoulder in moderate abduction and flexion and of the wrist in dorsiflexion. Daily physical therapy, including electrical stimulation of the muscles and motion of the involved joints, is helpful. Recovery is sometimes very slow and may continue for many months. Open operation for reduction of a late, untreated dislocation or for exploration of the nerves is occasionally indicated.

Axillary nerve

Etiology. The usual cause of injury is subcoracoid dislocation of the head of the humerus, but the paralysis may follow fracture of the surgical neck of the humerus, the incorrect use of an axillary crutch, therapeutic injections in the shoulder region, or traction injuries of the neck and shoulder.

Clinical picture. The deltoid muscle is paralyzed, causing a loss of strong abduction of the shoulder and subsequently an unsightly muscle atrophy, prominence of the acromion, and instability of the shoulder joint. The paralysis of the teres minor muscle is unimportant. There is a variable degree of sensory loss over the lateral aspect of the shoulder and upper arm.

Treatment. The treatment is usually nonsurgical. Surgical repair is difficult because of the short course of the nerve. After irreparable injury a muscle transplantation to provide abduction of the arm is sometimes done, or the shoulder may be arthrodesed in 50 to 75 degrees of abduction.

Long thoracic nerve

Etiology. The long thoracic nerve, supplying the serratus anterior muscle, may be injured by (1) the carrying of excessively heavy burdens upon the shoulder or (2) accidental division at operation. In some instances a definite cause for the paralysis has not been found.

Clinical picture. Paralysis of the serratus anterior muscle results in an inability to raise the arm above the shoulder level in front of the body, loss of forward pushing movements of the shoulder, and winging of the scapula. The deformity and disability are sometimes severe.

Treatment. Rest of the shoulder by a brace that limits scapular motion, and physical therapy are indicated. If recovery does not occur, excessive mobility of the scapula may be corrected by any of several surgical methods involving anchorage of its lower angle to the wall of the thorax.

Radial nerve

Etiology. The radial nerve with its long and exposed course about the humeral shaft is one of the most frequently injured nerves in the body. It is particularly liable to two types of mechanical injury: (1) laceration by the sharp edge of a bone fragment, in fractures of the humeral shaft or in supracondylar fractures, or involvement secondarily in a fibrous scar or in callus; and (2) compression from external objects, as in *crutch palsy,* or when the arm is allowed to hang for a long time against an

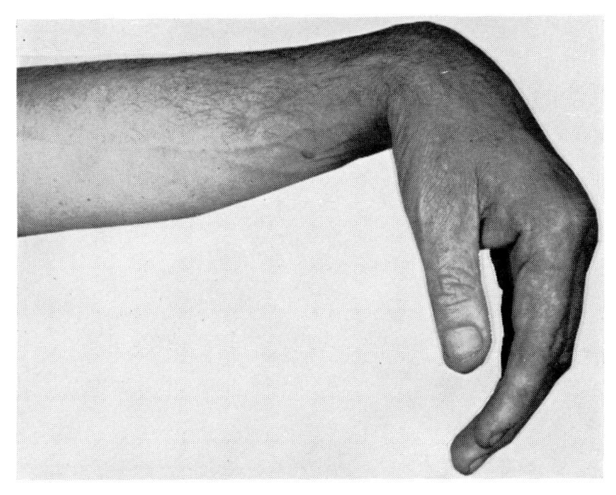

Fig. 6-13. Wrist drop in complete paralysis of the radial nerve.

object, causing local pressure. Even trivial injuries result usually in a complete rather than a partial motor paralysis.

Clinical picture. Radial palsy involves the extensor muscles of the elbow, wrist, and fingers and causes the characteristic sign of *wrist drop* (Fig. 6-13). If the lesion is in or below the middle third of the upper arm, paralysis of the triceps muscle is absent as its nerve supply is separated from the main trunk at a higher level. The sensory loss is less extensive than the wide distribution of the radial nerve would suggest; it consists usually of a small zone of anesthesia on the dorsal surface of the thumb and the adjoining portion of the hand. The dorsal interosseous branch of the radial nerve is occasionally injured in wounds of the elbow, with resulting paralysis of the extensor carpi ulnaris and the extensor muscles of the fingers and the thumb; since the extensor carpi radialis longus is not affected, the patient does not have wrist drop.

Treatment. Nonsurgical treatment of radial paralysis consists in the use of cock-up wrist splints, with elastic extension splinting for the digits, and of prolonged physical therapy. Surgical procedures upon the nerve itself—of which neurolysis and end-to-end suture are the most common—are often successful since the radial nerve is peculiarly capable of functional recovery. When the lesion is irreparable, however, tendon transferences about the wrist to restore lost function can do much to relieve the disability.

Ulnar nerve

Etiology. The ulnar nerve is frequently divided in lacerating wounds and may be injured by fractures in the region of the medial condyle and epicondyle, by dislocation of the elbow, and by secondary inclusion in scar tissue after elbow injury. It may be affected by postural pressure during a prolonged anesthesia. Lesions of the deep motor branch in the hand may result from prolonged use of certain vibrating tools. In addition, the ulnar nerve is subject to *late traumatic neuritis* from stretching or from friction against the posterior surface of the medial humeral condyle in valgus deformity of the elbow secondary to old fracture.

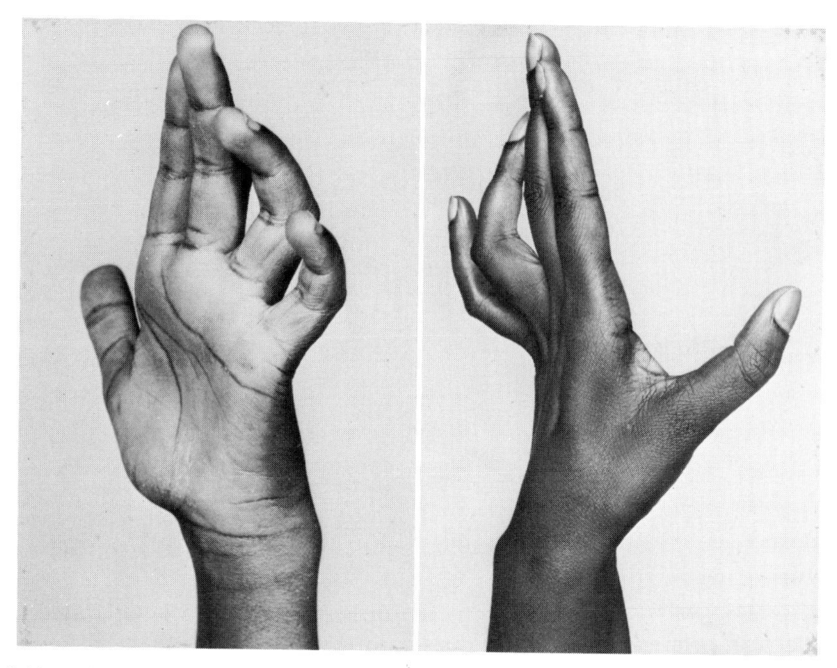

Fig. 6-14. Deformity of the hand in paralysis of the ulnar nerve. Active extension of the interphalangeal joints of the ring finger and the little finger is impossible because of paralysis of the two medial lumbrical muscles and the interossei. Atrophy of the first dorsal interosseous muscle is apparent.

Clinical picture. Lesions in the upper arm cause paralysis of the flexor carpi ulnaris, the ulnar half of the flexor digitorum profundus, the hypothenar muscles, interossei, two medial lumbricals, adductors of the thumb, and the deep head of the flexor pollicis brevis. In cases of long duration the flattening of the hypothenar eminence, interosseous atrophy, and ulnar clawhand deformity are characteristic (Fig. 6-14). A dry and atrophied appearance of the skin and nails due to vasomotor disturbances may be present. Sensory loss is usually considerable, even in incomplete division of the nerve, and is located over the ulnar border of the hand, the entire little finger, and the ulnar half of the ring finger. Protective sensation on the ulnar side of the hand is most important because without it the patient may unknowingly rest his hand on hot surfaces and sustain severe burns. Lesions in the middle third of the forearm cause motor and sensory losses in the hand, but power in the flexor carpi ulnaris and the flexor digitorum profundus is retained.

Late traumatic ulnar neuritis, or delayed ulnar palsy, may be seen in association with (1) recurrent dislocation of the nerve at the postcondylar groove and (2) old fracture of the lateral condyle with a cubitus valgus deformity. Various degrees of hypermobility of the nerve may occur developmentally or as the result of trauma, and occasionally there is slipping of the nerve trunk anterior to the medial epicondyle with each flexion of the elbow, which produces a typical friction neuritis. The nerve injury after old fractures of the lateral condyle is a result of stretching or friction due to the development of cubitus valgus (Fig. 8-17, *A*). Pain in the ulnar distribution is likely to be severe, and there may be sensory changes and atrophy of the interosseous muscles.

Treatment. Nonsurgical treatment consists of the use of splints and of physical therapy. When the nerve has been sectioned, the surgical procedure of primary suture is sometimes indicated. When length must be gained to overcome a gap, the nerve may with advantage be transposed to a position anterior to the epicondyle. The results of nerve suture are imperfect, since, as a rule, neither motor nor sensory losses are completely restored. In late traumatic neuritis, immobilization may produce satisfactory temporary improvement, but for permanent cure it is often necessary to perform an anterior transposition of the ulnar nerve at the elbow. The results of this operation are excellent in most cases.

Median nerve

Etiology. The commonest causes of median nerve paralysis are injuries from penetrating or lacerating wounds and primary or secondary involvement in association with supracondylar fractures of the humerus. In the forearm the median nerve may be compressed by scar tissue, as in Volkmann's ischemic contracture, or, rarely, by the pronator teres *(pronator syndrome)*. At the wrist a median paralysis sometimes follows traumatic lesions such as transverse laceration of the soft tissues, severely displaced Colles' fracture, or anterior dislocation of the lunate bone.

At the wrist the median nerve and the long flexor tendons traverse the carpal canal, formed by the carpal bones and closed anteriorly by the transverse carpal ligament. Compression of the nerve or its blood supply in this inelastic channel causes a common form of median neuropathy, termed the *carpal tunnel syndrome*. The compression may be produced by synovitis or fibrosis of the flexor tendon sheaths, by edema following Colles' fracture (especially if the wrist has been immobilized in the Cotton-Loder position of acute flexion and ulnar deviation), or by any other space-occupying lesion in the carpal canal.

Clinical picture. Lesions in the arm or at the elbow cause paralysis of most of the flexors of the wrist and fingers, of the pronators of the forearm, of the abductor-opponens group of the thumb, and of the two lateral lumbricals. Flattening of the thenar eminence is a conspicuous sign. The little finger and ring finger can be flexed through the ulnar supply to the profundus, but the patient cannot make a tight fist or pronate the forearm (Fig. 6-15). Power in the opponens pollicis muscle, which must be carefully differentiated from the adductor pollicis, is lost. In lesions at the wrist the loss of opponens and abductor brevis power is the most valuable diagnostic feature. The sensory loss is the same at whatever level the nerve is divided; it usually involves the thumb, the index finger, the middle finger, and the radial half of the ring finger. Loss of median nerve sensation compromises hand function, causing a severe disability. The nails may be atrophic and the skin shiny and dry. Incomplete division of the median nerve frequently results in causalgia.

Compression in the carpal tunnel usually causes incomplete interruption of the median nerve and spotty neurologic changes. Tingling pain and numbness over all or part of the median nerve distribution in the hand may be followed by atrophy of the thenar eminence. The pain is often worse at night. Tinel's sign may be positive.

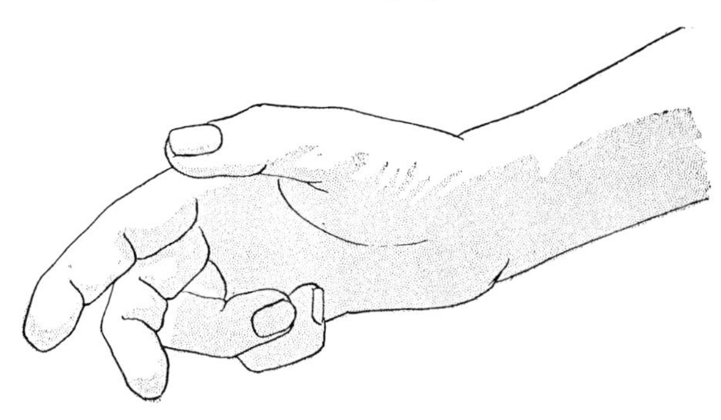

Fig. 6-15. Paralysis of the median nerve, showing inability to flex fully the index finger and the middle finger.

With volar flexion of the wrist, paresthesia may occur; it may be relieved by return to the neutral position. Determination of the rate of nerve conduction may be quite helpful in confirming the diagnosis. Sometimes evidences of a median nerve disturbance may develop several years after a wrist fracture has resulted in bony deformity; this condition is called *late median nerve palsy.*

Treatment. After clean lacerating wounds it is often possible to perform primary nerve suture. The technic must be scrupulous because of the likelihood of incomplete recovery and the development of causalgia. For traumatic neuritis, neurolysis is sometimes sufficient. Splinting, to prevent stretching of the weakened muscles, and physical therapy are important parts of the treatment. The results of suture of the median nerve are more favorable than those of ulnar nerve suture; however, the recovery is usually incomplete. Although early cases of the carpal tunnel syndrome may be relieved by simple splinting or hydrocortisone injection, complete section of the transverse carpal ligament is usually indicated. In irreparable lesions of the median nerve, tendon transplantations may be of benefit but persistent disability from sensory loss limits their value.

Combined ulnar and median nerve injury

Simultaneous injury of the ulnar and median nerves may occur at the elbow, in the forearm, or at the wrist. These lesions result in extreme crippling of the hand. Severe low injuries cause a total loss of sensation in the important palmar surface as well as loss of all intrinsic muscle power (Fig. 6-16). In more proximal lesions there is also paralysis of the long flexor muscles. Without treatment these losses lead to the development of a severe clawhand of intrinsic-minus type.

Treatment. The motor defect may be partially compensated by appropriate tendon transfers to the intrinsic muscles. Some restoration of sensation may follow free or pedicle nerve grafting.

Lumbosacral plexus and cauda equina

Lumbosacral plexus

The components of the lumbosacral plexus may be injured by penetrating wounds, fractures, infections, and tumors of the pelvic wall or viscera. The management of these conditions consists primarily of treating the causative lesions.

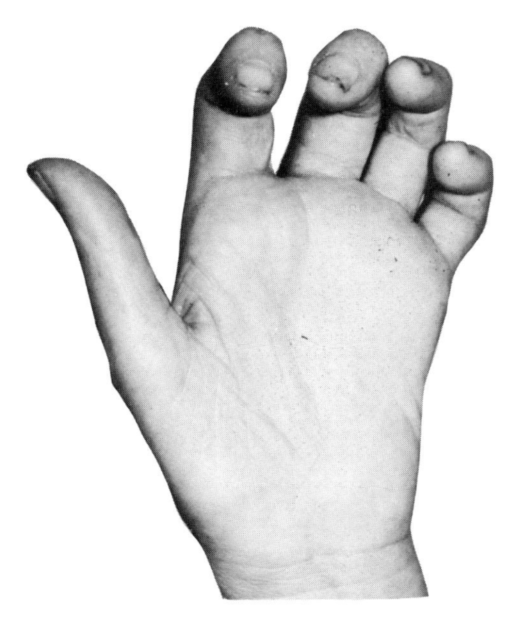

Fig. 6-16. Combined low ulnar and median nerve injury, showing atrophy of the thenar and hypothenar muscles, atrophy of the skin, and clawing of the fingers caused by paralysis of all intrinsic muscles.

Rarely compression by the fetal head is said to cause plexus symptoms. More common is the entity of *maternal obstetric palsy*, in which sciatic pain and sometimes peroneal paralysis and sensory changes are the result presumably of stretching of the lumbosacral trunks during a difficult labor. The prognosis for spontaneous recovery is good. Nonsurgical treatment, consisting of a foot-drop brace and physical therapy, should be employed. Very rarely, after a difficult birth, the infant may show temporary signs of a traction injury of the lumbosacral plexus analogous to the familiar stretching of the brachial plexus that produces obstetric paralysis of the upper limb.

Cauda equina and conus medullaris

These structures may be injured in penetrating wounds or by severe trauma with or without fracture of the lumbar vertebrae or sacrum. Motor and sensory losses may be slight or very extensive, depending upon the extent of the trauma and the level involved. The complete *cauda equina syndrome* includes a total flaccid paralysis of both lower limbs, paralysis of the sphincter muscles of the rectum and bladder, and anesthesia of the buttocks, the perineum, the entire posterior aspect of the legs, and the feet. The complete *conus medullaris syndrome*, resulting from involvement of the five fused sacral segments of the spinal cord, consists of paralysis of the sphincter muscles and anesthesia in a saddle-shaped area of the buttocks and perineum. In many of the cases the only practicable therapy is nonsurgical. Exploratory laminectomy may be indicated in injuries of the cauda equina.

In rare instances the cauda equina syndrome may result from the midline protrusion of a nucleus pulposus; in such cases early surgical intervention is important.

Sciatic nerve

Etiology. The sciatic nerve is involved commonly in gunshot wounds, and occasionally in displaced fractures of the pelvis or femur and dislocations of the hip. It is sometimes injured during the reduction of a traumatic or congenital dislocation of the hip or during a manipulation of the hip and low back for sciatic pain. In infants an occasional cause of sciatic injury is the accidental injection of medication into the nerve.

Clinical picture. Complete division causes a total paralysis of all muscles below the knee, and of the hamstring muscles also if the lesion is above the middle third of the thigh. In partial division the common peroneal fibers are most often injured. The outstanding motor sign of sciatic paralysis is *foot drop,* or inability to carry out active dorsiflexion of the ankle. Also lost is active dorsiflexion of the toes. The *steppage gait* of foot drop is characteristic. The paralyzed muscles develop severe atrophy. Sensation is lost in the lateral aspect of the leg and in the foot, except for its medial surface, which is supplied by the saphenous branch of the femoral nerve. Trophic ulcers are very likely to develop at points of pressure on the sole and may extend deeply to involve the bones. Incomplete division of the sciatic nerve sometimes leads to the syndrome of causalgia.

Treatment. Conservative treatment consists of physical therapy and of braces to prevent foot drop and contractures, such as a splint for use at night and an ankle brace, fitted with a spring or right-angle stop joint, for use when walking (Fig. 9-11). Surgical exploration and repair must sometimes be undertaken when satisfactory return of nerve function does not occur, but recovery is always slow and the results are usually disappointing. During and after operation, tension on the nerve may be relieved by keeping the hip extended and the knee flexed. At times advanced trophic changes and ulceration may make amputation of the limb necessary.

Common peroneal nerve

Etiology. As the common peroneal nerve winds superficially about the neck of the fibula, it may be injured by the following: (1) lacerating wounds; (2) sudden compression from a blow; (3) gradual compression from an ill-fitting cast; (4) laceration, edema, or cicatricial involvement associated with fracture or osteotomy of the upper end of the fibula, and sometimes of the tibia; or (5) stretching associated with the correction of a genu valgum or a knee flexion contracture of long duration.

Clinical picture. The motor signs consist of paralysis of the anterior tibial, peroneal, and toe extensor muscle groups with resulting foot drop. As in the case of the radial nerve, the usual result of trivial as well as of major injuries is complete motor paralysis. In high lesions sensation is lost over an area situated on the lateral aspect of the leg and the dorsum of the foot.

Treatment. Especial attention should be paid to the prevention or correction of foot drop. A brace is often used, and in most cases there is no indication for operation. When the signs of complete interruption are present after severe injuries, however, early exploration and suture are indicated. When the nerve lesion is irreparable, foot stabilization and tendon transplantation may improve the gait.

Tibial nerve

Injuries of the tibial nerve are uncommon; they are caused usually by penetrating wounds or by ischemia of the posterior compartment of the leg. Division causes paralysis of the plantar flexor and intrinsic muscles of the foot and a loss of sensation over the sole, the lateral surface of the heel, and the plantar surface of the toes. Vasomotor disturbances and trophic ulcers, particularly in the sole of the foot, may form serious complications. The treatment is based upon the usual principles.

Femoral nerve and obturator nerve

Division of either of these nerves is extremely rare. In patients with severe bleeding disorders such as hemophilia, intrapelvic hemorrhage may cause femoral nerve compression. Section of the femoral nerve causes paralysis of the quadriceps muscle, together with slight sensory losses in the thigh and leg. The power of extending the knee may be partially restored by transplantation of one or two of the hamstring tendons and the iliotibial band into the patella. Division of the obturator nerve causes an adductor paralysis without significant sensory changes.

NEURITIS (PERIPHERAL NEUROPATHY)

In addition to acute traumatic injuries, peripheral nerves are subject to a variety of chronic irritative disturbances. These disorders, although of interest to orthopaedic surgeons, are primarily of neurologic character and can be mentioned here only briefly and in their relationship to disease of the bones and joints.

Chronic irritative disorders of the peripheral nerves may be grouped conveniently into two classes: (1) traumatic neuritis, in which the causative factor is chronic mechanical injury, and (2) toxic neuritis, in which the nerves are affected by some deleterious chemical agent or toxin. Since such toxic substances are usually of systemic distribution, toxic neuritis is likely to involve more than one nerve, in which case it is known as *multiple neuritis,* or *polyneuritis.*

Traumatic neuritis

The symptoms and signs of traumatic neuritis vary with the degree of nerve irritation. There may be motor paralysis, as well as pain, sensory disturbances, tenderness along the nerve trunk, and trophic changes. Well-recognized forms of traumatic neuritis are those from prolonged hanging of the arm over the back of a chair or the edge of a bed or from the continuous pressure of a cervical rib. *Crutch palsy,* a partial paralysis of the radial nerve from pressure of a crutch in the axilla, is a common example. The sciatic symptoms often found in association with mechanical derangement in the low back can be interpreted on a similar basis. Irritation of the lateral femoral cutaneous nerve near the anterior superior iliac spine produces a burning discomfort and hypesthesia over the anterolateral aspect of the thigh; the condition is termed *meralgia paresthetica.* Paralysis of varying degree, resulting from pressure upon the common peroneal nerve as it winds about the head of the fibula, sometimes follows the application of a poorly padded or tightly fitting leg cast. The treatment of these lesions consists in removing the offending mechanical factor, in protecting the weakened muscles with braces when necessary, and in using daily physical therapy to prevent atrophy and contracture.

Toxic neuritis

Although the subject of toxic neuritis is primarily of medical interest, the orthopaedic surgeon is often called upon to aid in the prevention of deformity and the restoration of function. In most of these cases disturbances of sensation, tenderness along the nerve trunk, and motor losses are observed, and the distribution may be symmetric in the two corresponding limbs. Pathologically the chief change is a degeneration of the neuraxons, and the prognosis for recovery, usually good with adequate treatment, depends upon the extent to which this degeneration is followed by interstitial fibrosis. A common cause of toxic neuritis is poisoning by metals, such as lead or arsenic, or by organic compounds, such as methyl or ethyl alcohol or carbon monoxide. Other causative agents are the toxins of infectious diseases. The treatment of toxic neuritis is primarily medical and preventive, but when paralysis has developed, rest in supportive apparatus and subsequent physical therapy are indicated.

A rare form of toxic neuritis that causes paralysis about the shoulder girdle may follow the injection of prophylactic or therapeutic serum. The usual treatment is rest of the weakened muscles and physical therapy. If sufficient strength is not regained, surgical procedures of the types used for residual paralysis after poliomyelitis may be indicated.

GUILLAIN-BARRÉ SYNDROME (POSTINFECTIOUS POLYNEURITIS)

The Guillain-Barré syndrome is a form of polyneuritis characterized by a slowly progressive, symmetric, ascending motor weakness sometimes confused with poliomyelitis. The cerebrospinal fluid usually shows increased protein and a low or normal cell count; this combination is not found in poliomyelitis.

Etiology. The etiology is unknown. The syndrome frequently develops a few days or weeks after a mild infection of viral or bacterial nature. Some patients, however, have no recognized antecedent or associated illness.

Clinical picture. The onset is gradual with initial paresthesias. Pain may be present in the limbs, with muscle and nerve trunk tenderness; there may be mild signs of meningeal irritation. A symmetric flaccid paralysis gradually develops, reaching its maximum in 2 to 3 weeks. The paralysis usually but not always begins in the lower limbs and spreads to involve the trunk and upper limbs. Cranial nerve paralysis is common, the seventh nerve being most frequently involved. Sensory disturbances are much less severe than the motor paralysis and are usually limited to paresthesias and hypalgesia. The most serious aspect of the disease is paralysis of the respiratory muscles, which, if not promptly treated, may be fatal. When the paralysis has ceased to progress, it remains static for about 2 weeks and then improves slowly over the next 6 to 12 months.

Differential diagnosis. Differential diagnosis includes consideration of other types of polyneuritis, as well as poliomyelitis and acute ascending myelitis.

Treatment and prognosis. In the early stages, bed rest, an adequate nutritional intake, medication for pain, and proper bowel and bladder care are indicated. If the respiration is embarrassed, a respirator may be necessary, and possibly a tracheotomy. Moist or dry heat should be used over the tender and painful areas, and early, active exercises should be carried out daily through the range of painless joint motion. Recovery is usually complete but may take many months. If residual paralysis results, its treatment should be the same as that outlined for paralysis after poliomyelitis. Bracing may be required for many months.

HEREDITARY MUSCULAR ATROPHY OF PERONEAL TYPE (CHARCOT-MARIE-TOOTH DISEASE)

This type of progressive atrophy and paralysis, caused by degenerative changes in the peripheral nerves, begins clinically with involvement of the peroneal muscles. It usually makes its appearance between the ages of 5 and 10 years and affects boys more commonly than girls. It shows a familial tendency and may be inherited as an autosomal dominant or recessive trait. Often slight shortening of the heel cords is an early change. Equinovarus deformity associated with a steppage gait is characteristic. Stumbling and unsteadiness of gait become progressively worse. There may be cramps in the legs. The hands and forearms may become involved. Clawfoot (Fig. 14-7) or clawhand deformities often develop. The atrophy slowly progresses upward but never extends above the elbows or the middle of the thighs. The reflexes are hypoactive or absent, Babinski's sign is absent, and numbness is occasionally present. In affected motor nerves the conduction velocity is decreased.

Treatment. Surgical lengthening of the Achilles tendon may lessen the deformity of the ankle and foot and improve the gait for several years. The operation should be followed by systematic exercises. Occasionally it is advisable to support the feet and ankles with braces in order to prevent the development of severe equinovarus deformity. For patients in whom the disease is advancing only slowly, subtalar arthrodesis is sometimes indicated. With the arthrodesis it is sometimes advisable to transfer the anterior tibial tendon laterally or the posterior tibial tendon to the dorsum of the foot.

Involvement of muscles

PROGRESSIVE MUSCULAR DYSTROPHY

The term *progressive muscular dystrophy* is commonly used to include a number of genetically determined primary degenerative myopathies. Of these the most common is *pseudohypertrophic muscular dystrophy,* which was first described clearly in 1868 by Duchenne.

This affection appears most frequently at about the age of 5 years and is seen almost exclusively in boys. The usual transmission is that of a sex-linked recessive trait. Symmetric progressive muscular atrophy is accompanied by apparent hypertrophy of the muscles; the enlargement is usually greatest in the legs and forearms, whereas atrophy is more noticeable about the shoulder girdle. The increase in size is apparently due to actual hypertrophy of the muscle fibers as well as to fatty replacement of muscle. There are increases in the size and number of sarcolemmal nuclei, splitting of the fibers, increase of connective tissue, and deposition of fat.

Fig. 6-17. Progressive muscular dystrophy, showing the characteristic method of rising to a standing position (Gowers' sign). The arms are used to push the body erect. Note the increased lumbar lordosis, relaxed shoulder girdle, and enlarged calves.

Microscopically the muscle may look like a fatty tumor with little resemblance to muscle. The pathogenesis is unknown; it may be an intrinsic biochemical defect of the muscles.

Clinical picture. Often the first evidence of progressive muscular dystrophy is weakness of the legs and resultant fatigue. As the condition advances, the child stands with an obvious increase of the lumbar lordosis, walks with a peculiar waddling gait, has difficulty in climbing steps, and falls frequently. Weakness of the extensor muscles of the legs and trunk is characteristic. On attempting to get up from the floor the child usually climbs laboriously upon his legs and thighs (Gowers' sign, Fig. 6-17).

Weakness about the shoulder girdle is characteristic also; when it is severe, the affection is sometimes spoken of as *facioscapulohumeral dystrophy (Landouzy-Dejerine type of progressive muscular dystrophy).* When the affected child is lifted by the examiner with the examiner's hands under the armpits, the child may slip through because of weakness of chest and shoulder muscles. This type of dystrophy, which is inherited as an autosomal dominant, is as frequent in girls as in boys and appears usually about puberty. There may be facial weakness with inability to close the eyes or to whistle, a masklike facial expression, and winging of the scapulae. The reflexes may be hypoactive and later absent. Sensation is unimpaired.

Diagnosis. The diagnosis is usually made from the history, the characteristic clinical appearance, and the findings on physical examination. Because of the waddling gait, progressive muscular dystrophy may be confused occasionally with bilateral congenital dislocation of the hip, congenital coxa vara, or paralysis about the hips after poliomyelitis. Three laboratory tests are helpful in confirming the diagnosis. *Electromyography* shows a decrease in the amplitude of motor unit potentials on voluntary muscle contraction as well as some increase in polyphasic potentials. Fibrillations, characteristic of neurogenic processes, are absent in muscular dystrophy. Early in the disease there is a substantial elevation of the serum *creatine phosphokinase (CPK)*. The fact that this enzyme is often increased in female carriers of the disease is helpful in genetic counseling. *Muscle biopsy* frequently aids in diagnosis; care should be taken to obtain the specimen from a muscle that is moderately involved.

Prognosis. By the age of 12 years extreme disability is usually present, and later the patient dies of intercurrent pulmonary infection, respiratory failure, or cardiac failure. A common course is gradual progression of the disease in the 6 to 12 years after its onset, until the patient becomes severely crippled. Occasionally, for no obvious reason, the disease is arrested before the disability becomes severe. Arrest is more likely to occur in the facioscapulohumeral type than in the pseudohypertrophic type of dystrophy.

Treatment. There is no specific therapy. Treatment should be aimed at preventing deformity and keeping the patient ambulatory as long as possible. Progressive equinus deformities are frequently the factor that puts the patient in a wheelchair. They should be guarded against by the use of active and passive exercises, protective splinting, and bracing. When equinus deformity has developed, heel cord lengthening and occasionally anterior transplantation of the tibialis posterior tendon may be indicated. Care should be taken to get the patient walking immediately following operation because even short periods of immobilization lead to further weakness and disability. It is probably best to avoid use of a wheelchair until absolutely necessary. At that time a fitted chair with head rest, lap belt, and removable arms should be provided.

CHAPTER 7 Tumors

Tumors and tumorlike affections of bone

Bone tumors and lesions that simulate bone tumors clinically, pathologically, or in both respects, together make up a large and complex group of orthopaedic affections. Some of the individual entities are uncommon or rare. Some are of trivial import to the patient's welfare, whereas others are of the gravest significance. Accordingly the diagnosis, which is often difficult, must be made with the greatest care, and the treatment must be thoughtfully planned and executed.

When a patient's history and physical findings suggest the possibility of a bone tumor, roentgenograms should be made promptly. When, despite evaluation of the combined clinical and roentgenographic findings, the diagnosis of a malignant tumor remains uncertain, biopsy should be done promptly. It is often advisable to have the microscopic slides reviewed by a pathologist especially experienced in interpreting bone lesions. Only then is the orthopaedic surgeon ready to formulate and carry out the treatment that will be best for his patient. In the case of malignant bone neoplasms, the treatment must be started without unnecessary delay and with aggressiveness sufficient to eradicate all parts of the tumor.

Classification. Although several excellent classifications of the tumors and tumorlike affections of bone are available, none can be entirely satisfactory until the obscurities now surrounding the origin and nature of many of these lesions have been clarified. The following classification, based on that of Lichtenstein, groups the most important bone tumors according to presumptive tissue of origin and benign or malignant nature:

Tissue of origin	*Benign*	*Malignant*
1. Cartilaginous	Chondroma Osteochondroma Chondroblastoma Chondromyxoid fibroma	Chondrosarcoma
2. Osteoblastic	Osteoma Osteoid-osteoma Benign osteoblastoma	Osteogenic sarcoma Parosteal osteogenic sarcoma
3. Nonosteoblastic connective tissue	Nonosteogenic fibroma Benign giant cell tumor	Fibrosarcoma Malignant giant cell tumor
4. Mesenchymal connective tissue		Ewing's sarcoma

Tissue of origin	*Benign*	*Malignant*
5. Hematopoietic		Plasma cell myeloma Malignant lymphoma Reticulum cell sarcoma
6. Nerve	Neurofibroma Neurilemoma	
7. Vascular	Hemangioma Hemangiopericytoma	Hemangioendothelioma
8. Fat cell	Lipoma	Liposarcoma

Since limitation of space prohibits description of all of these entities, the more common tumors and tumorlike affections of bone will be discussed according to the following outline:

Benign

1. Chondroma
2. Benign chondroblastoma (Codman's tumor)
3. Osteochondroma
4. Osteoid-osteoma
5. Nonosteogenic fibroma (nonossifying fibroma, fibrous cortical defect)
6. Giant cell tumor (osteoclastoma)
7. Bone cyst
8. Aneurysmal bone cyst

Malignant

1. Osteogenic sarcoma (osteosarcoma)
2. Parosteal osteogenic sarcoma
3. Chondrosarcoma
4. Ewing's sarcoma
5. Reticulum cell sarcoma
6. Plasma cell myeloma (multiple myeloma)
7. Tumors metastasizing to bone

In addition to the entities enumerated in this simple classification, there are a number of less common new growths of bone that deserve mention. *Hemangioma* is common as a small, asymptomatic, innocuous lesion, of questionably neoplastic nature, in vertebral bodies; it occurs occasionally as an expansile tumor of a single vertebra that may cause paraplegia by posterior extension and collapse; involvement of the long bones, first evidenced by pain or pathologic fracture, is uncommon. *Benign osteoblastoma* is an uncommon, solitary, osteoblastic tumor found in long bones and vertebrae. *Chondromyxoid fibroma* is an uncommon tumor seen most often in the tibia in adolescents or young adults. Although the histologic findings in benign osteoblastoma and chondromyxoid fibroma may suggest malignancy, the clinical course of these tumors is benign; accordingly it is imperative that they be diagnosed accurately and treated by excision rather than amputation. *Adamantinoma,* a slowly growing tumor of obscure histogenesis, occurs rarely in the limb bones, where it is seen almost exclusively in the tibia.

Chordoma, a rare malignant tumor arising from remnants of the embryonic notochord, occurs about the base of the skull or in the sacrococcygeal region and is characterized by bone destruction, slow growth, and encroachment on the surrounding tissues. Chordoma of the sacrococcygeal area must be distinguished from tera-

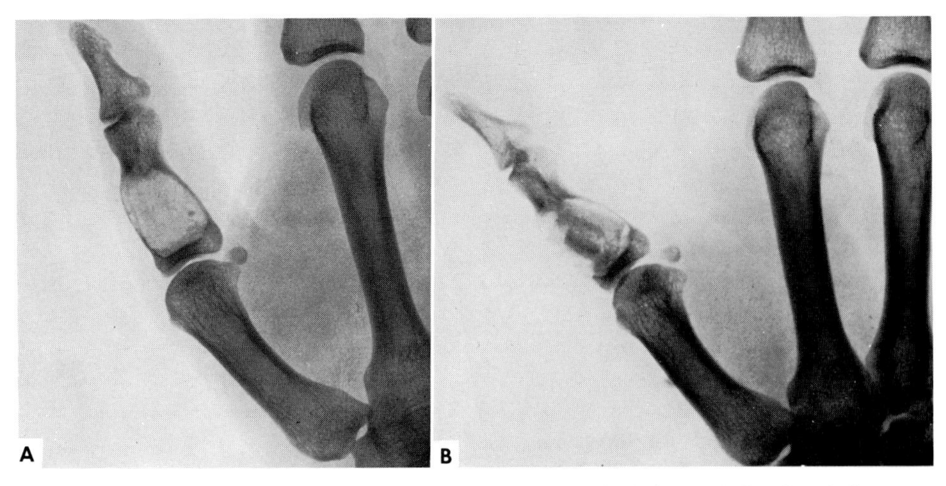

Fig. 7-1. Roentgenograms of enchondroma of the proximal phalanx of the thumb in a man 30 years of age. **A,** Note rounded radiolucent area with thinned overlying cortex. **B,** One week after excision of the tumor and packing of the cavity with small grafts from the ilium.

toma and from other tumors, primary or metastatic, which arise not infrequently in this location.

Benign tumors and tumorlike affections of bone

CHONDROMA

A benign tumor composed chiefly of cartilage is a chondroma. Usually these tumors arise within the medullary cavity and are termed *enchondromas.* Occasionally several bones are involved simultaneously by *multiple enchondromas.* Enchondromas are found most often in phalanges, metacarpals, and metatarsals and occasionally in the humerus, femur, and other bones. They occur most often in patients between the ages of 20 and 30 years. The clinical picture is that of a slowly growing tumor, producing as a rule only slight discomfort. Pathologic fracture may initiate the symptoms. The characteristic roentgenograms (Fig. 7-1) show a rounded area of decreased density with a smooth outline, often expanding beneath a narrow shell of cortical bone. Foci of calcification may be present with the lesion. Enchondromas of large tubular bones occasionally undergo malignant transformation to chondrosarcomas. The treatment of chondroma is thorough curettement or excision, after which it is usually advisable to pack the resulting cavity with small bone grafts.

BENIGN CHONDROBLASTOMA (CODMAN'S TUMOR)

Benign chondroblastoma is an uncommon new growth that typically arises near the end of adolescence in the epiphyseal region of a large long bone (Fig. 7-2). It occurs most frequently in the femur, tibia, or humerus. It is somewhat more common in males.

Pathology. Although benign chondroblastoma usually arises in an epiphysis, it often extends into the adjacent metaphysis. Histologically these tumors consist of

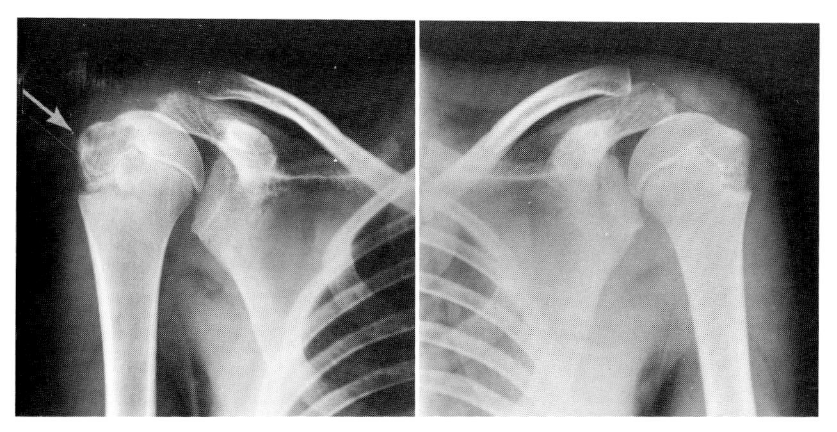

Fig. 7-2. Roentgenogram of benign chondroblastoma of upper end of the right humerus in a 13-year-old boy. He had had shoulder pain for 8 months.

closely packed round or polyhedral cells of somewhat variable size with relatively large nuclei. Multinuclear giant cells are usually present. There may be little chondroid. A characteristic finding is the presence of focal calcification throughout the tumor tissue, together with areas of cellular degeneration and necrosis.

Clinical picture. Most frequently the patient very gradually develops pain on motion of a major joint. Swelling, a palpable mass, and tenderness may be present; if a lower limb is affected, limping and muscle atrophy may be obvious.

Roentgenographic picture. The tumor often produces a round or ovoid, indistinctly mottled area of radiolucency surrounded by a thin margin of sclerotic bone and lying within, or predominantly within, an epiphyseal region of a large long bone.

Treatment and prognosis. Thorough curettement of the lesion and packing the resulting bony cavity with cancellous bone chips usually result in permanent cure.

OSTEOCHONDROMA

Osteochondromas are the commonest benign tumors of bone. Called by Jaffe and Lichtenstein *osteocartilaginous exostoses,* osteochondromas are hamartomas rather than true tumors, their growth usually ceasing simultaneously with that of the bones in which they have arisen. Multiple exostoses, often associated with skeletal deformity, form a distinct clinical entity of hereditary nature, which has been discussed (p. 61) under the subject of general affections of bone.

Osteochondromas occur usually in persons between the ages of 10 and 25 years and typically are situated near the ends of the long bones, where they form pedunculated bony overgrowths, the apices of which are covered by a layer of cartilage. The cartilaginous surface may be lobulated or roughened, and the protuberance is often covered, especially when in the vicinity of an overlying tendon, by a well-developed adventitious bursa containing fluid of synovial type. The symptoms produced by these growths are usually slight. Swelling is often the patient's only complaint, although there may be discomfort in the adjacent joint. The duration of

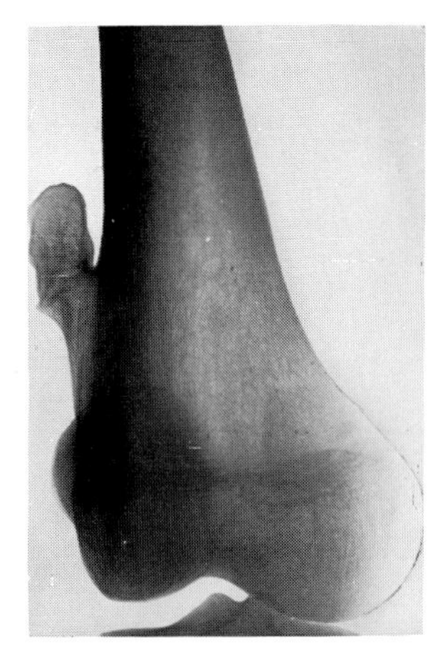

Fig. 7-3. Roentgenogram of osteochondroma of the lower end of the femur in a patient 12 years of age.

symptoms is usually long, and the tumor may eventually reach considerable size. The commonest location for these growths is near the adjacent ends of the femur and tibia or at the upper end of the humerus, although they are not uncommon in other sites. Of the flat bones the scapula is most frequently involved. Roentgenograms show the typical picture of an osseous base or pedicle of normal bone density, springing directly from the cortex of the underlying bone and capped by an expanded area that may show irregular calcification (Fig. 7-3). The marrow cavity of the osseous pedicle is continuous with the marrow cavity of the bone from which the lesion arises. Microscopically the tumors present bone of normal appearance covered by a layer of calcifying cartilage, which in turn may be bordered by a thin layer of fibrous tissue. During the growing period the tumor enlarges by enchondral ossification beneath the cartilage cap.

Treatment. The treatment is excision. Since osteochondromas rarely become malignant, operation is not always necessary; however, any large osteochondroma in a location susceptible to traumatization should be removed. If excision is not done, periodic roentgenographic examination should be made, particularly in the older age groups, to determine whether changes suggestive of malignancy have supervened.

OSTEOID-OSTEOMA

Osteoid-osteoma is a rather common small, solitary, benign lesion that occurs in either cancellous or cortical bone, or even in a subperiosteal location, and causes aching local pain that may persist for many months or years. Osteoid-osteoma is probably a reactive bone lesion of unknown cause rather than a neoplasm. It is seen most often in persons between the ages of 10 and 25 years, and more frequently in males than in females. It may arise in almost any bone; the femur and tibia are common sites.

Pathology. The typical lesion is a spherical nidus of osteoid not more than 1 cm. in diameter, surrounded by reactive bone that is often surprisingly extensive and dense, especially when the nidus is located in cortical bone. Early the nidus consists of vascular connective tissue packed with osteoblasts; later, osteoid trabec-

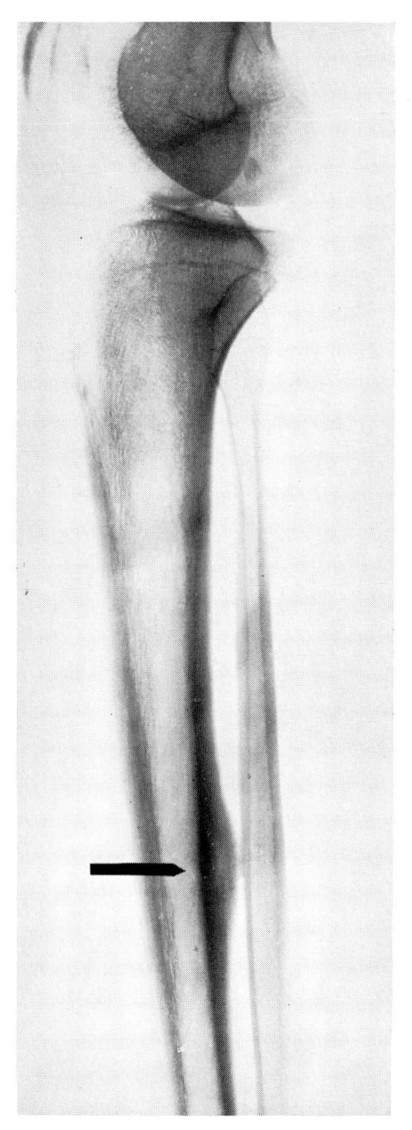

Fig. 7-4. Lateral roentgenogram showing an osteoid-osteoma of the tibial shaft. Note the small rarefied area surrounded by osteosclerosis and subperiosteal new bone.

ulae are formed among the osteoblasts; later still the osteoid gradually becomes converted into atypically dense bone with the microscopic appearance of an osteoma.

Clinical picture. The principal complaint is aching pain that has been present for several months. The pain is mild at first and then increases in intensity until it is severe enough to wake the patient at night; usually it can be relieved, however, by aspirin. There may be stiffness and weakness in the adjacent muscles, and a limp if the lesion is in a bone of the lower limb. Localized swelling and tenderness may be present; there is no elevation of body temperature.

Roentgenographic picture. The involved bone shows a small, radiolucent, well-circumscribed area, which is usually oval or round; in the center of this area there may be a minute shadow of increased density. This is surrounded by dense bone that, if the lesion is in the cortex, may extend several inches above or below the point of the original lesion (Fig. 7-4). Laminagrams may show clearly an otherwise obscure osteoid-osteoma. Later the entire lesion may become calcified.

Differential diagnosis. Osteoid-osteoma may be confused with Brodie's abscess, eosinophilic granuloma, and sclerosing osteogenic sarcoma.

Treatment and prognosis. The best treatment is complete excision of the lesion, which ordinarily is followed by prompt and permanent relief of pain. A small autogenous graft may be used to fill the resulting bone defect. If not excised, the lesion eventually regresses, but the pain may persist for a number of years.

NONOSTEOGENIC FIBROMA (NONOSSIFYING FIBROMA, FIBROUS CORTICAL DEFECT)

Nonosteogenic fibroma is a common, circumscribed, benign lesion seen near the ends of the diaphyses of long bones, most often in the lower limbs of children and

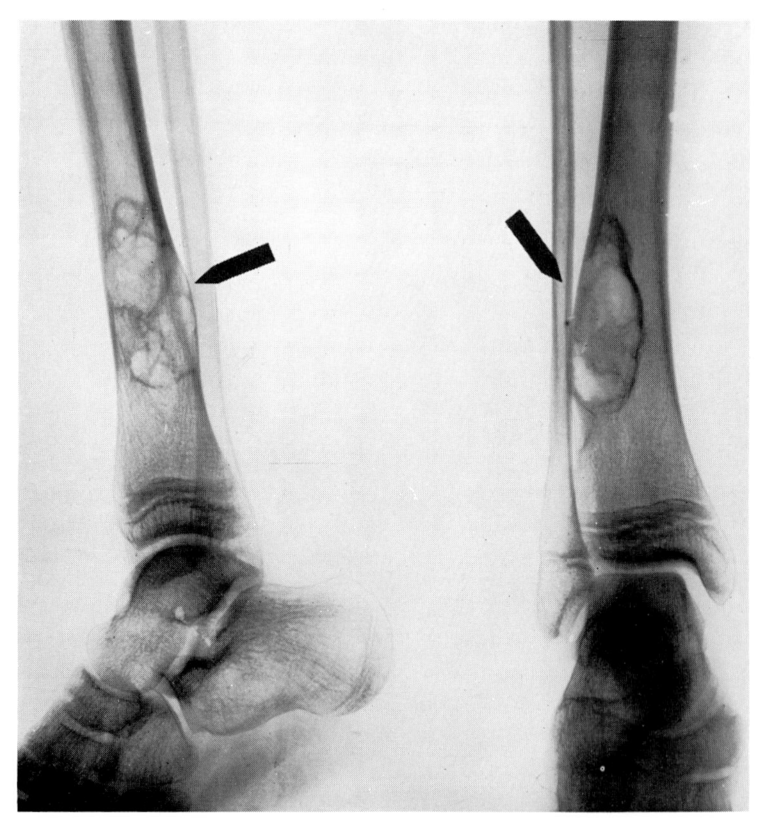

Fig. 7-5. Lateral and anteroposterior roentgenograms of a nonosteogenic fibroma of the tibial shaft in a 10-year-old child. The treatment consisted of curettement and packing with small bone grafts.

young adults. These lesions are usually innocuous, asymptomatic, and discovered incidentally on roentgenographic examination. Larger nonosteogenic fibromas, however, sometimes lead to pathologic fracture. Grossly these lesions consist of circumscribed, yellow-brown fibrous tissue underlying thinned cortex; microscopically they contain bundles of connective tissue cells, a few multinuclear giant cells, and frequently lipoid-containing foam cells. The typical roentgenographic appearance (Fig. 7-5) is a sharply demarcated, eccentric, loculated area of rarefaction. It is believed that many if not all of these lesions are obliterated in a few years by gradual ingrowth of bone from their periphery.

Differential diagnosis. Nonosteogenic fibroma must be distinguished from giant cell tumor, bone cyst, aneurysmal bone cyst, fibrous dysplasia, enchondroma, and eosinophilic granuloma. This can often be done without biopsy by consideration of the clinical and roentgenographic details.

Treatment. Asymptomatic, typical nonosteogenic fibromas require only extended observation. Unusually large, symptomatic, or questionable lesions should be treated by excision or curettement, after which the defect may be packed with small bone grafts.

GIANT CELL TUMOR (OSTEOCLASTOMA)

Giant cell tumor of bone is a solitary lesion, usually of slow rather than fast growth, seen most often in the epiphyseal region of a long bone in young adults. It causes swelling with little pain and appears in roentgenograms as an eccentric area of bone destruction. In recent years giant cell tumor has come to be recognized as a less common and more serious lesion than it was formerly considered. These tumors do not have a uniform clinical course. Most are considered benign, but many recur after curettement and some become frankly malignant.

Pathology. Giant cell tumor begins typically as a circumscribed growth occupying an eccentric position within the end of a long bone. The lower end of the femur, upper end of the tibia, and lower end of the radius are most commonly involved. As growth advances, the overlying cortex is expanded and thinned. Lines of relative thickening of the involved cortical bone, often continuous with fibrous septa extending into the tumor, may produce the roentgenographic appearance of trabeculation.

Grossly the tumor tissue is of brownish red appearance and friable consistency, with areas of fibrosis and of hemorrhagic softening. Microscopically the tumor consists of a vascularized network of fusiform and ovoid stromal cells, among which appear many multinuclear giant cells. The nuclei of the giant cells are rather uniform, tend to occupy the center of the cell, and may number fifty or more in a single cell.

Clinical picture. Giant cell tumor occurs usually in persons between the ages of 20 and 35 years. In many cases a history of local trauma can be obtained, but trauma has no proved importance in the etiology. Pain of moderate severity is usual and may be associated with slight tenderness. Increase in size is very gradual and may go long unnoticed. Occasionally pathologic fracture occurs.

Roentgenographic picture. Characteristically the giant cell tumor presents an area of radiolucency in the epiphyseal region of a long bone with expansion and

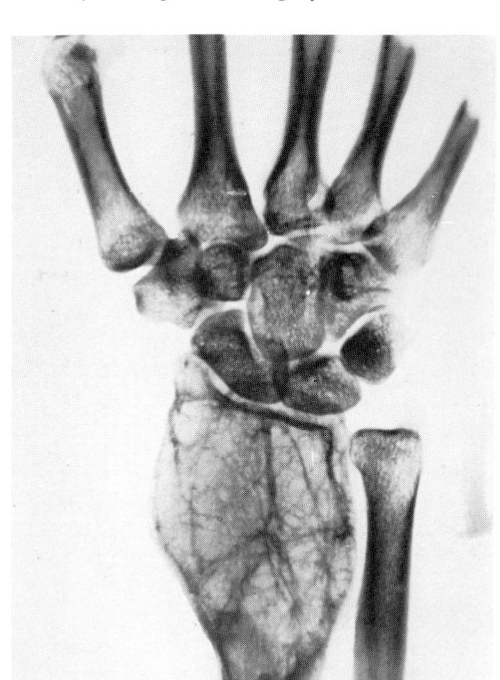

Fig. 7-6. Giant cell tumor of the distal end of the radius. Note rarefaction, trabeculated appearance, and the expansion and thinning of the cortex.

thinning of the overlying cortex (Fig. 7-6). The rarefaction often extends to, but does not invade, the articular cartilage of the adjacent joint. Evidence of subperiosteal new bone formation is minimal.

Treatment and prognosis. Biopsy is indicated to establish the diagnosis and disclose any stromal atypism suggestive of malignant tendency. As a rule the involved bone should be partially or completely excised and replaced, when necessary, by bone grafting, transplantation, or a prosthesis. In such cases the prognosis for cure is good. Curettement and packing with small bone grafts involves risk of recurrence. Roentgentherapy is inadvisable because it may be followed some years later by the development of sarcoma. Recurrent giant cell tumors should be treated less conservatively than primary ones; they sometimes require amputation.

BONE CYST

The simple or solitary bone cyst is not a tumor but is conveniently considered here because of its clinical similarity to certain bone tumors and its consequent importance in differential diagnosis. The typical bone cyst is a slowly growing, fluid-filled lesion occurring in childhood near one end of the diaphysis of a long bone. Its etiology is unknown.

Pathology. Most bone cysts occur in the proximal portion of the humeral shaft. They are seen also in the upper and lower ends of the femoral and tibial shafts and

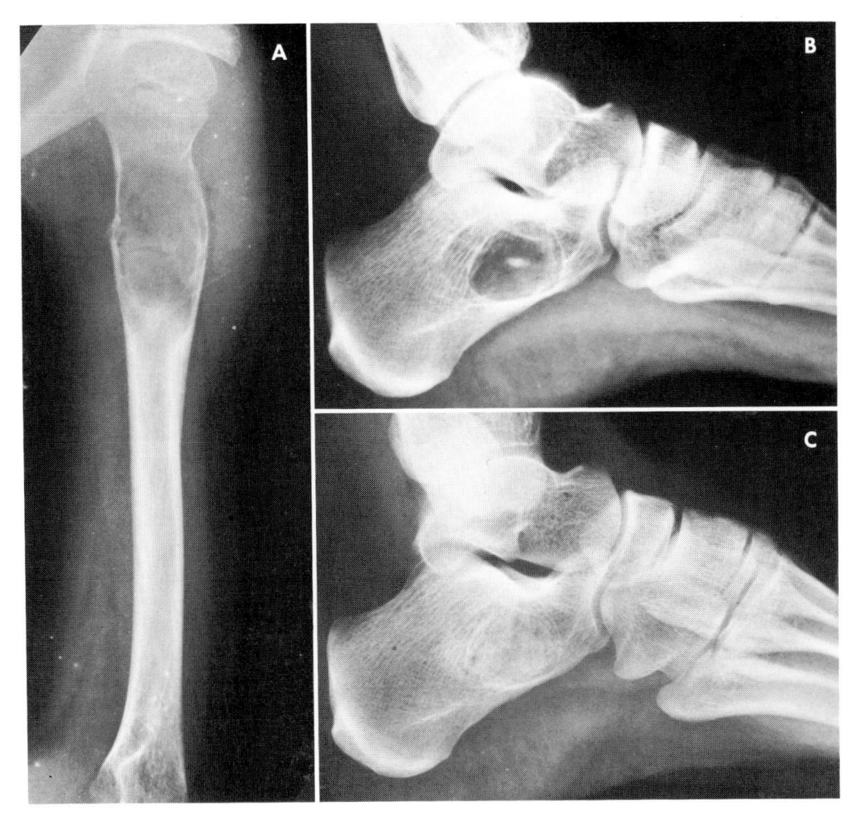

Fig. 7-7. Roentgenograms of solitary bone cysts. **A,** Cyst in upper part of the humeral shaft in a 7-year-old boy. Note large ovoid radiolucent area with thinned, bulging cortex and the transverse line of a partially healed pathologic fracture. **B,** Cyst of the calcaneus in a 33-year-old woman. **C,** The same cystic area 3 years after curettement and packing with bone chips from the ilium.

the upper end of the fibular shaft, and occasionally in other bones. Cortical bone overlying the cyst may be extremely thin and fragile. Although the cyst is unilocular, its wall may be traversed by narrow ridges of bone. The cyst commonly has a delicate connective-tissue lining; it usually contains an amber fluid. After recent fracture of the cyst wall the fluid may be discolored by blood and contain degenerating coagula.

Clinical picture. Bone cyst is a disease of late childhood, the majority of cases occurring in children between the ages of 10 and 15 years, although occasional cysts are found even in individuals over 20 years of age. Usually the first symptoms are occasioned by fracture, which takes place on slight traumatization of the thinned cortex.

Roentgenographic picture. The roentgenographic appearance of a single or, less typically, a multilocular central bone defect occurring in the diaphysis, with a smooth and intact outline formed by thinned and often slightly expanded cortical bone, is characteristic (Fig. 7-7).

Treatment and prognosis. When fracture has occurred through a bone cyst, treatment of the fracture is occasionally followed by spontaneous obliteration of the cyst. Ordinarily the treatment of bone cyst is thorough curettement and packing of the cavity with numerous small bone grafts, preferably from the ilium. Irradiation is ineffective and may be followed by sarcoma. With careful treatment the prognosis is good; however, the recurrence rate is about 25%. Cysts most likely to recur are those nearest the epiphyseal plate.

ANEURYSMAL BONE CYST

Aneurysmal bone cyst is an uncommon solitary, benign bone lesion of unknown etiology that contains distended channels or cavities filled with blood. Probably a disorder of the local vasculature of bone marrow rather than a tumor, it often requires differentiation from bone neoplasms.

Pathology. Not truly a cyst because the bony cavity is filled with soft tissue as well as fluid, aneurysmal bone cyst usually begins in cancellous tissue. The bone destruction may be the result of excessive blood supply to a localized area. Although any bone may be affected, aneurysmal bone cyst is seen most commonly in vertebrae, where it characteristically involves the posterior elements, or in long bones. Grossly it often appears as a rounded protrusion consisting of a thin shell of bone filled with a spongy mass of soft tissue, vascular channels, and bloody fluid. Histologic findings include small and large vascular spaces with connective-tissue septa containing many giant cells as well as plaques of reparative osteoid and bone.

Clinical picture. Aneurysmal bone cyst is commonest in adolescence but occurs also in childhood and adult years. The presenting symptom is most likely to be localized pain. Sometimes the first complaints concern an enlarging mass or a pathologic fracture. When located in a vertebra, an aneurysmal cyst may lead to paralytic symptoms from involvement of the spinal cord.

Roentgenographic picture. Typically the lesion appears as an ovoid radiolucent

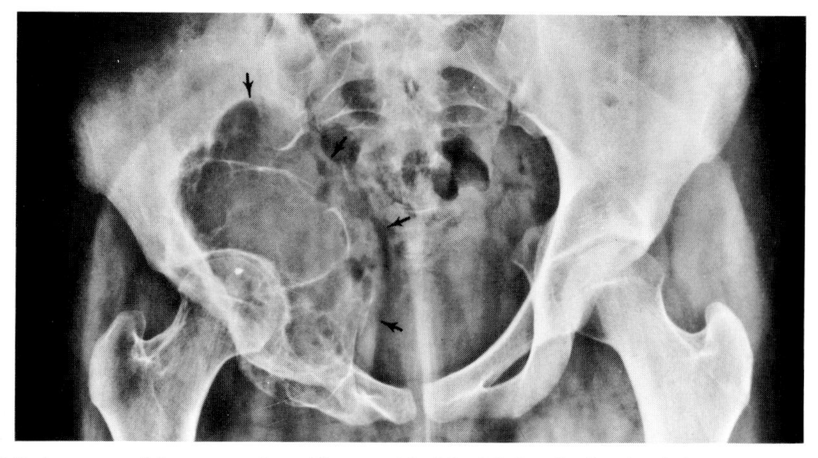

Fig. 7-8. Aneurysmal bone cyst in a 17-year-old girl with inguinal pain. A large mass outlined by a thin bony shell occupies almost half the pelvis.

area bulging eccentrically to produce the appearance of an aneurysm or blowout (Fig. 7-8). Fine or coarse trabeculation may be distinct and the overlying cortex extremely thin.

Prognosis and treatment. Untreated aneurysmal bone cysts usually continue to enlarge. Early recognition and treatment are usually curative. When surgically accessible, the lesion should if possible be excised; otherwise it should be thoroughly curetted and packed with small bone chips. For inaccessible lesions, irradiation is indicated.

Malignant tumors of bone

OSTEOGENIC SARCOMA (OSTEOSARCOMA)

Osteogenic sarcoma is one of the commonest and most malignant of the primary tumors of bone. Typically it occurs near one end of the diaphysis of a long bone in late childhood or early adult life (Fig. 7-9). It is characterized clinically by early and

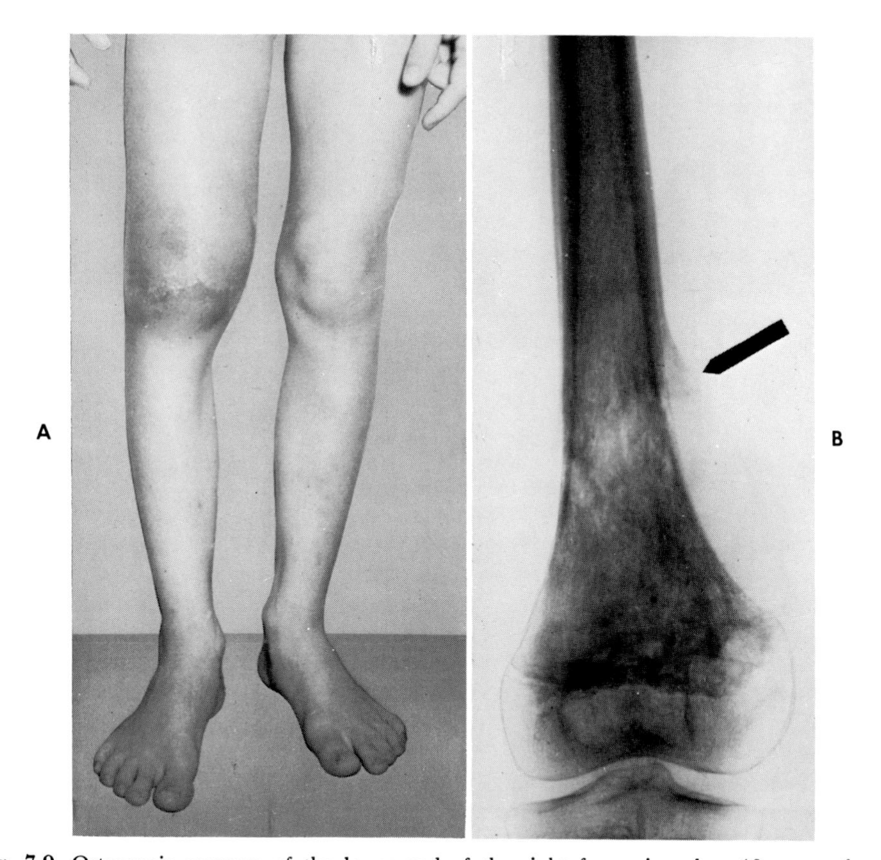

Fig. 7-9. Osteogenic sarcoma of the lower end of the right femur in a boy 10 years of age. **A,** Note enlargement of the right knee and thigh and discoloration of the skin about the knee. **B,** Roentgenogram of the femur showing moth-eaten appearance of the metaphysis, subperiosteal new bone, and Codman's triangle of reactive subperiosteal new bone (arrow).

severe pain, together with a rapidly increasing swelling. Radiographically both osteolysis and osteogenesis are usually present, although occasionally either may occur alone; invasion of the soft tissues may be conspicuous. Metastasis occurs early, and the prognosis for life, despite early and efficient treatment, is unfavorable.

Pathology. The upper tibia, lower femur, and upper humerus are the commonest sites of osteogenic sarcoma, although involvement of other bones, including those of the axial skeleton, is not infrequent. Pathologically there is considerable variation among the individual tumors of this large group. Microscopically the diagnostic findings are a frankly sarcomatous stroma and the production of osteoid and bone by the tumor cells. The typical cell of osteogenic sarcoma is spindle shaped, and in the more malignant tumors hyperchromatism, pleomorphism, mitotic figures,

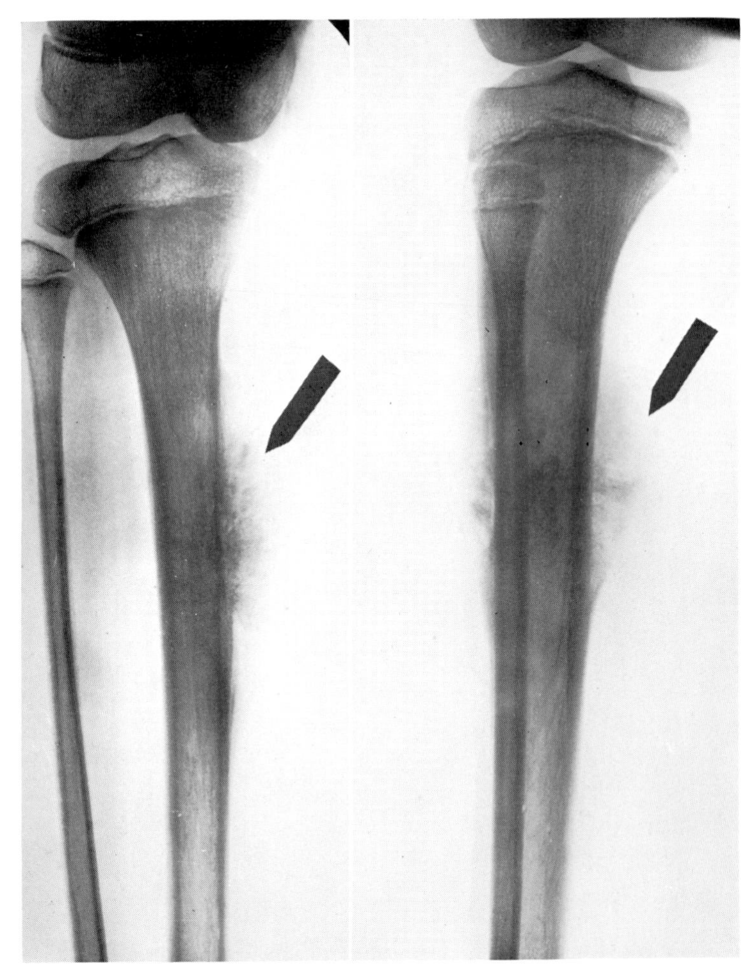

Fig. 7-10. Roentgenograms of osteogenic sarcoma of the tibia in a 10-year-old girl. In the proximal third of the tibia note indistinctly outlined new bone and changes in density of the shaft.

and tumor giant cells are frequent. Small islands of newly formed bone are common, together with areas of tumor cartilage in various stages of differentiation.

Clinical picture. Osteogenic sarcoma is commonest in children and young adults, and somewhat more frequent in males than females. It occurs occasionally in later adult years; in such cases an underlying osteitis deformans may be present. Pain is as a rule the first symptom. It is usually persistent even at rest. A history of local trauma is often obtained, but injury is thought to play no part in causing the tumor. Noticeable swelling and secondary limitation of motion in the adjacent joint soon follow. The more rapidly growing tumors are likely to show distended superficial veins and an increase in surface temperature, and to be accompanied by an elevation of the serum alkaline phosphatase. Other cases are less typical in onset and in their early stages are differentiated only with difficulty from the other bone tumors and from low-grade inflammatory conditions.

Roentgenographic picture. Areas both of bone destruction and of new bone formation are usually visible. Small spicules of bone at right angles to the shaft (Fig. 7-10) are sometimes seen; they are not specific for osteogenic sarcoma. In advanced cases there is evidence of extensive invasion of the surrounding soft tissues (Fig. 7-11).

Treatment and prognosis. Early biopsy-supported diagnosis and prompt treatment are important. The prognosis is always unfavorable because of early metastasis to the lungs, but amputation of the extremity, if performed early,

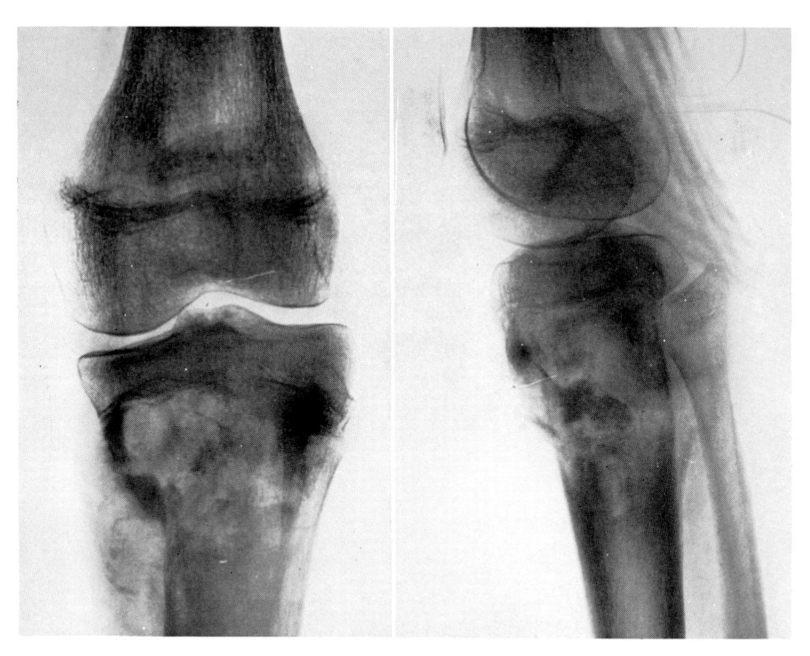

Fig. 7-11. Anteroposterior and oblique roentgenograms of osteogenic sarcoma of the tibia with pathologic fracture in a patient 13 years of age. Note areas of increased and decreased density in the upper tibia, new bone extending into the soft tissues, and slight angulation with break in continuity of the medial cortex of the tibia. Note also that the destructive process has not crossed the epiphyseal plate.

can be expected to result occasionally in cure. Even after the recognition of pulmonary metastases, amputation is often advisable in order to remove a painful lesion that may ultimately undergo pathologic fracture, ulceration, and hemorrhage. As a rule radiotherapy is considered to have no advantage over surgical ablation; it may, however, be used as an adjunct. In advanced cases roentgentherapy is usually advised, but it cannot be expected to accomplish more than a temporary retardation of the malignant process. Cytotoxic drugs have not proved effective. On rare occasions life may be prolonged by treating a solitary pulmonary metastasis of late appearance by lobectomy. The 5-year survival rate as reported in several series of cases is 10% to 15%.

PAROSTEAL OSTEOGENIC SARCOMA

A tumor of low-grade malignancy, parosteal osteogenic sarcoma is much less common than the usual form of osteogenic sarcoma and has a much better prognosis. This slowly growing lesion is characterized by the progressive proliferation of bone on the surface of a long bone, most often the femur. Radiographically it usually

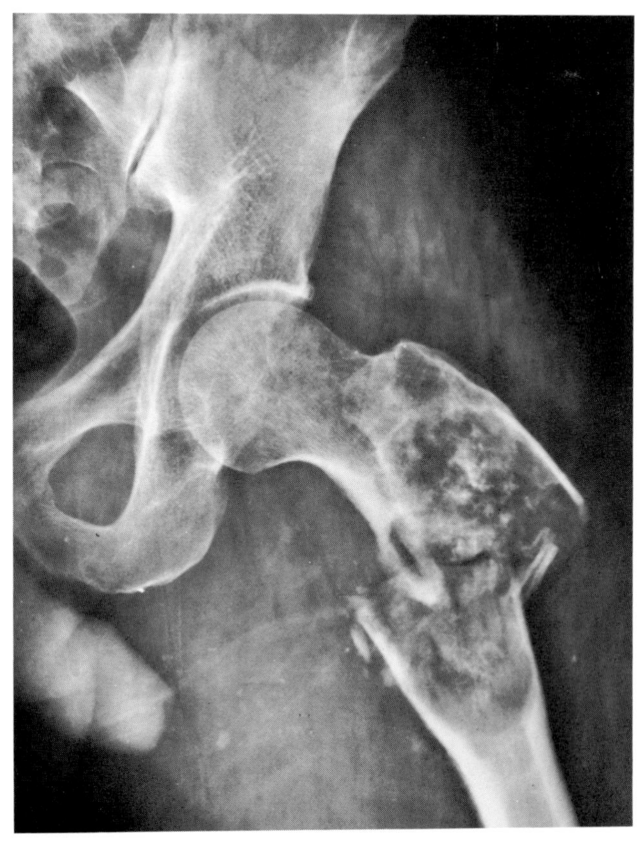

Fig. 7-12. Roentgenogram of chondrosarcoma of the femur in a 38-year-old man. Pathologic fracture has occurred through the radiolucent tumor area. Patchy calcification appears in the tumor tissue both within and outside the femur.

appears lobular, rather distinctly outlined, and very dense. The treatment is local resection or amputation.

CHONDROSARCOMA

Chondrosarcoma is a malignant tumor that develops from cartilage cells; it is about half as common as osteogenic sarcoma.

Pathology. *Central chondrosarcoma* arises in the interior of a bone, sometimes from malignant transformation of an enchondroma of a long tubular bone. In patients with Ollier's disease (p. 92), or enchondromatosis, one or more of the multiple cartilaginous lesions may undergo such malignant degeneration. *Peripheral chondrosarcoma,* developing on the surface of a bone, may arise from malignant change in an osteochondroma; this is more likely to occur in patients with multiple cartilaginous exostoses (p. 61) than in those with a solitary osteochondroma.

Grossly the tissue of a chondrosarcoma is usually recognizable as cartilage. Microscopically the diagnosis of malignancy may be very difficult, requiring generous biopsy material and experienced interpretation.

Clinical picture. Chondrosarcoma is a tumor of middle age. It is seen most commonly in the femur, humerus, tibia, pelvic bones, and scapula. Chondrosarcomas usually grow slowly, causing swelling and pain, and may reach considerable size.

Roentgenographic picture. Chondrosarcoma usually appears in roentgenograms as a bulky mass of soft tissue density, containing irregular, blotchy areas of more opaque calcification (Fig. 7-12). Uncalcified intraosseous areas of the tumor are relatively radiolucent.

Treatment and prognosis. Since chondrosarcoma is usually slow to metastasize, its prognosis ordinarily is far better than that of osteogenic sarcoma. The treatment is surgical, amputation or wide excision being indicated as soon as the histologic diagnosis has been made. Chondrosarcomas are not radiosensitive.

EWING'S SARCOMA

Ewing's sarcoma, called also *Ewing's tumor,* is an uncommon malignant tumor of medullary origin that often simulates a low-grade osteomyelitis, with an onset characterized by pain, tenderness, fever, and leukocytosis. It is primarily bone destructive, causing osteogenesis only by periosteal reaction. The tumor is at first most susceptible to irradiation—usually this response is only temporary, and metastasis and death soon supervene.

Pathology. The typical site of origin of Ewing's sarcoma is near the middle of the shaft of a long bone. The tumor occurs in the femur more commonly than in any other bone. It also occurs in the flat bones. Grossly the main mass of the tumor often lies subperiosteally but outside the thickened and eroded cortex. The histogenesis of these tumors is obscure. Microscopically the tumor may show areas of closely packed, small polyhedral cells with scant cytoplasm and dense round or oval nuclei. There is no intercellular stroma. Connective tissue septa form a network dividing the tumor tissue into rounded areas, and many large vascular spaces containing blood and lined by tumor cells may be present. Other highly cellular

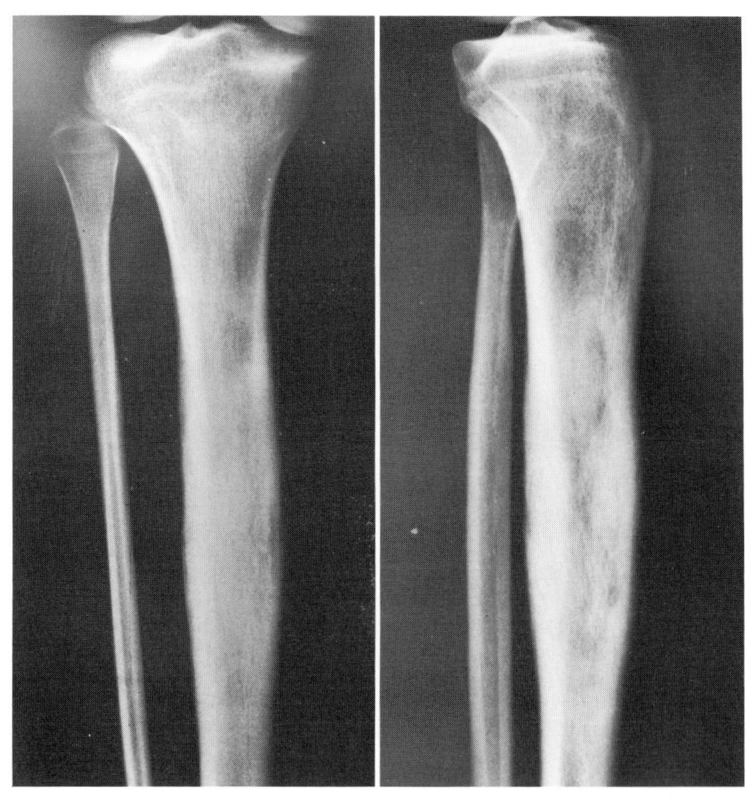

Fig. 7-13. Anteroposterior and lateral roentgenograms of Ewing's sarcoma of the tibial shaft in an 18-year-old girl with a history of increasing leg pain for 1 year. Note expanded segment of the diaphysis with areas of osteolysis, osteosclerosis, and subperiosteal new bone formation.

areas may show extensive necrosis. Differentiation from osteomyelitis, neuroblastoma, reticulum cell sarcoma, and anaplastic carcinoma may be extremely difficult.

Clinical picture. Ewing's sarcoma is most frequently encountered in patients between the ages of 10 and 25 years. A history of incidental trauma is often obtained. Pain is an early and distressing symptom, although there may be periods of complete freedom from discomfort. There is often a constitutional reaction suggesting a low-grade inflammatory process and probably caused by necrosis within the tumor. Palpation may reveal a smooth fusiform tumor continuous with the bone and of slight or extreme tenderness. Temporary subsidence of the pain and tenderness is believed to be due to relief of tension when the tumor has broken through the periosteum. In advanced cases there may be symptoms from metastases to the lungs or to bones, particularly the skull.

Roentgenographic picture. The roentgenographic appearance (Fig. 7-13) often is similar to that of a subacute or chronic osteomyelitis and must be carefully differentiated. In very early cases condensation of reacting bone about small areas of destruction may be visible. Some cases show widening of the shaft from the formation of reactive endosteal and subperiosteal bone in layers parallel to its long axis.

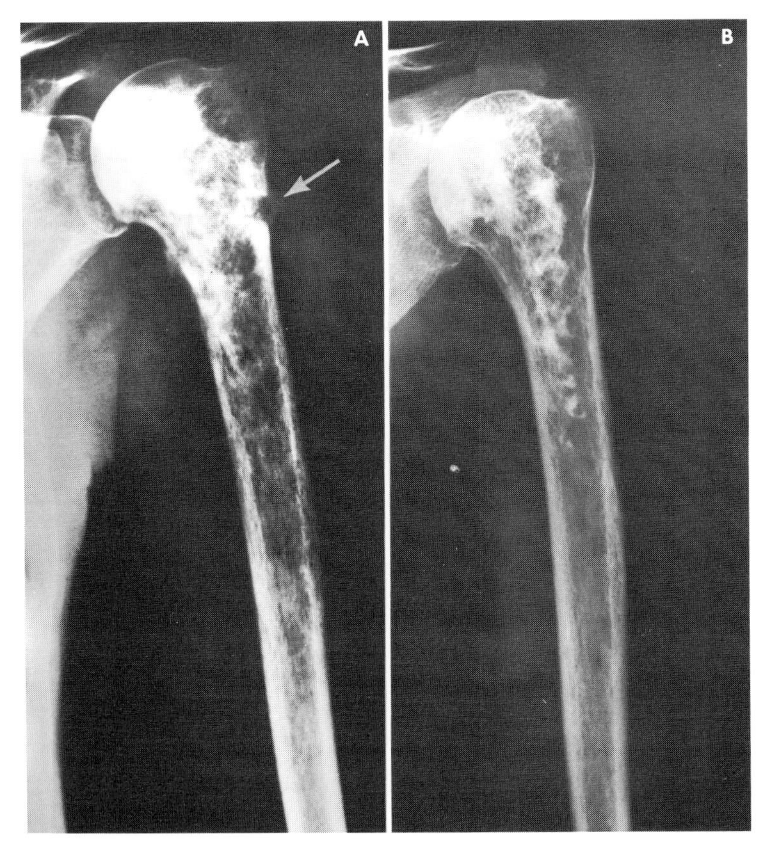

Fig. 7-14. Roentgenograms of reticulum cell sarcoma of the humerus in a 40-year-old woman. **A,** After biopsy (arrow) and before treatment. Note the mottled, moth-eaten appearance of the humeral head and shaft. **B,** Eight years after radiotherapy much bone healing has occurred and the patient is symptom free.

In later stages, rounded soft tissue masses as well as extensive areas of bone destruction may be evident.

Treatment and prognosis. Since Ewing's sarcoma is very sensitive to irradiation, roentgentherapy is often considered the best treatment. Even early radical operation offers little chance of cure. A combination of irradiation and surgery is considered by many authorities to be the treatment of choice. With any form of therapy, death is likely to occur not later than a very few years after the appearance of metastatic tumors. The survival rate in this disease is probably less than 5%. However, recent reports suggest that the prognosis may be improving with the use of supervoltage radiation and cytotoxic drugs.

RETICULUM CELL SARCOMA

An uncommon malignant tumor, reticulum cell sarcoma of bone is seen with about equal frequency in childhood and in adult years. The roentgenographic changes (Fig. 7-14) are predominantly those of patchy radiolucency from bone destruction, and pathologic fracture may occur. Histologically the lesion is composed of masses of small, fairly uniform cells with indistinct cytoplasmic boundaries. It may be very difficult to distinguish reticulum cell sarcoma from Ewing's sarcoma. Accurate diagnosis is important because of the much more favorable prognosis in reticulum cell sarcoma, which grows more slowly and metastasizes later than Ewing's sarcoma. Reticulum cell sarcoma is more vulnerable to irradiation than most tumors, and treatment either by irradiation or by amputation yields a 5-year survival rate of 40% to 50%. Cytotoxic drugs such as vincristine have also been reported to be useful in controlling reticulum cell sarcoma.

PLASMA CELL MYELOMA (MULTIPLE MYELOMA)

Plasma cell myeloma is a rather common, highly malignant neoplasm originating in the bone marrow of adults with a peak incidence between 50 and 60 years of age. Bones of the axial skeleton containing red marrow are most commonly affected. The lesions are entirely bone destructive and present roentgenographically a characteristic punched-out appearance. Being a neoplastic proliferation of cells that ordinarily produce immunoglobulins, plasma cell myeloma is attended as a rule by a hypergammaglobulinemia of abnormal type.

Pathology. Ribs, sternum, skull, and vertebrae are the commonest sites of plasma cell myeloma. Occasionally the first lesion to be discovered remains a solitary plasmacytoma, at least for some years. Many bones may be found to be involved simultaneously or in quick succession, and in advanced cases the distribution is widespread. The tumor begins in the medullary cavity and extends rapidly, replacing the normal bone marrow. The cortical bone is also invaded and usually shows many discrete areas of destruction. There is no bone formation other than a slight widening of the shaft from periosteal reaction, and pathologic fractures are frequent. The process may extend to perforation of the periosteum and invasion of surrounding soft tissues. In appearance the typical tumor cells closely resemble plasma cells.

Clinical picture. Unlike other bone sarcomas, plasma cell myeloma does not produce characteristic early symptoms. The onset is insidious, and vague pain, swelling, signs of local pressure, or pathologic fracture, commonly of a rib or vertebra, may be the first manifestation. Back pain is often the patient's first complaint. Anemia from widespread infiltration of the marrow by plasma cells is almost as common as bone destruction in producing the initial symptoms. Occasionally paraplegia results from myelomatous involvement of the spinal column. In later stages the pain may become more severe, and in advanced cases cachexia may be striking. Susceptibility to bacterial infections, related to decreased production of normal antibodies, is increased.

Laboratory findings. Sternal or iliac puncture may demonstrate myeloma cells. *Bence Jones proteinuria* is a helpful confirmatory finding but is absent in about half of the cases. The serum protein may be high because of an excess of globulin, and characteristically the albumin-globulin ratio is reversed. Serum electrophoretic studies show abnormal peaks in the gamma or beta globulin areas. Qualitative abnormalities in the globulins may be demonstrated by immunologic analysis. Hypercalcemia is a common finding in plasma cell myeloma, and occasionally the level of uric acid in the blood is elevated. Kidney function is often impaired.

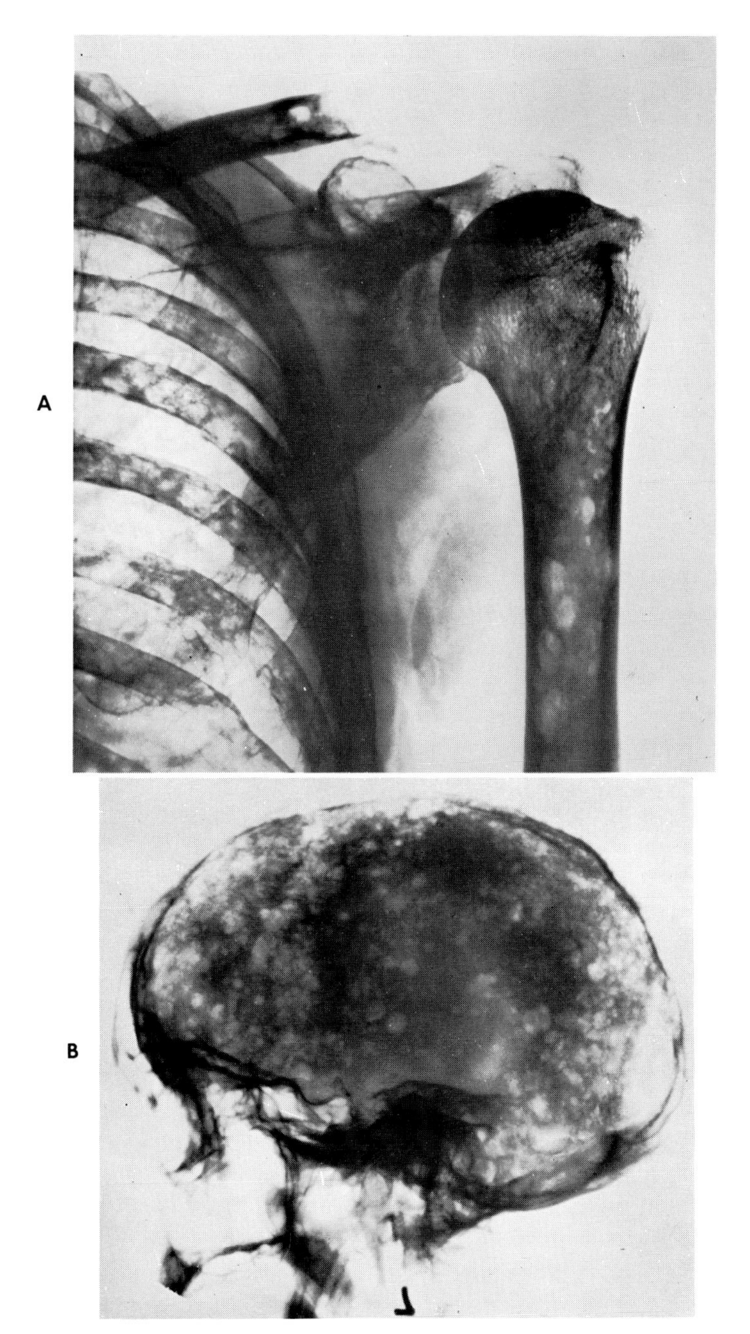

Fig. 7-15. Roentgenograms showing plasma cell myeloma. **A,** Note small punched-out areas in the ribs, scapula, and humerus and disappearance of the lateral end of the clavicle. **B,** Roentgenogram of the skull showing multiple punched-out areas.

Roentgenographic picture. Multiple punched-out areas appearing in the bones of the axial skeleton are characteristic (Fig 7-15). A generalized decrease of bone density is sometimes noticeable.

Differential diagnosis. Plasma cell myeloma may be confused clinically and roentgenographically with metastatic cancer of bone. When it involves primarily the vertebrae, plasma cell myeloma must be differentiated from senile osteoporosis, hyperparathyroidism, and other diseases that cause osteoporosis and compression fractures.

Treatment. The treatment of plasma cell myeloma is chiefly palliative. Irradiation of selected local areas, however, may effect dramatic relief of pain, and low-dose, whole body spray irradiation has also proved beneficial in cases with disseminated disease. In addition, a number of chemotherapeutic agents have proved effective in alleviating pain, improving anemia, and decreasing the production of abnormal globulins. Currently, cyclophosphamide and phenylalanine mustard are most widely used and are limited only by their toxic side effects on the bone marrow. In some cases treatment with sodium fluoride has seemed to improve both the roentgenographic appearance of the bones and the bone pain. Immobilization by brace, cast, or traction may be indicated in the prevention or treatment of pathologic fractures. Surgical treatment is limited to the relief of cord compression by laminectomy, the treatment of certain pathologic fractures, and rarely an amputation when severe involvement of a long bone is the only manifestation of the disease.

Prognosis. The outlook is almost uniformly unfavorable, the average duration of life being approximately 3 years, although instances of survival for as long as 10 years or more have been recorded.

TUMORS METASTASIZING TO BONE

In the diagnosis of bone tumors the possibility of involvement of bone by metastasis or by direct extension from a tumor that is primary elsewhere must always be kept in mind. Bone metastases are commonest in carcinomas of the breast, lungs, prostate gland, thyroid, and kidneys, and less frequent in cancers of the gastrointestinal tract and female genitalia. In infants and young children the commonest source of bone metastases is the highly malignant neuroblastoma (or sympathoblastoma), which usually arises in the adrenal medulla. Destructive changes in the bones may occur in the leukemias and in Hodgkin's disease.

Pathology. From their primary focus, malignant cells may reach bone via the bloodstream. The commonest bones to be involved by metastatic tumors are the vertebrae, pelvis (Fig. 7-16), upper end of the femur, ribs, skull, and humerus. Metastatic tumors are uncommon below the elbows and the knees. Metastases from lung and renal carcinomas are usually osteolytic; those from the breast are predominantly osteolytic, although they may present areas of bone formation as well. In cancer of the prostate gland, bone metastases of osteoblastic nature occurring in the pelvis and lumbar spine are typical. The histologic appearance of bone metastases depends upon the histogenesis of the primary tumor but is often not sufficiently characteristic to identify it.

Clinical picture. The clinical appearance of metastatic bone tumor is not uni-

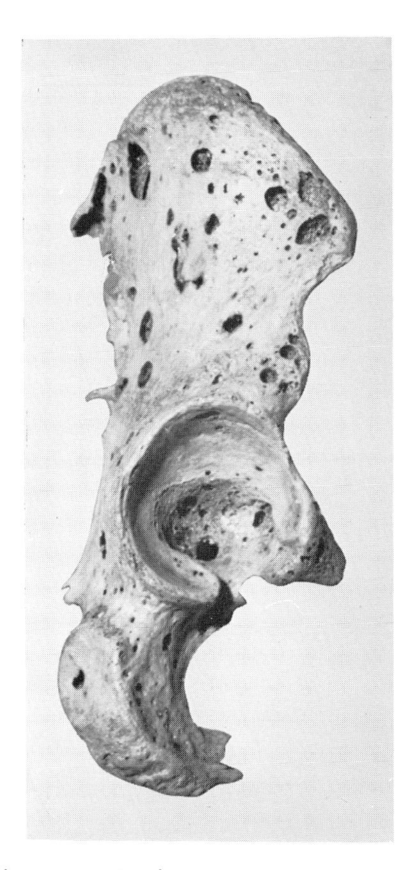

Fig. 7-16. Innominate bone showing multiple punched-out areas of metastasis.

form. The age of the patient, as a rule, is of aid in rendering primary bone sarcoma other than plasma cell myeloma unlikely. There is often a history of preceding operation upon breast or prostate gland. Pain may be an early symptom, often occurs before there is roentgenographic evidence of bone involvement, and may gradually become continuous and intense. In some cases, on the other hand, pathologic fracture is the first clinical sign (Fig. 7-17). This is commonest in metastases from cancer of the breast. If there is a pulsating mass with a bruit, the examiner should suspect metastasis from the thyroid gland. The presence of lung metastases is to be suspected and may be a valuable point in ruling out plasma cell myeloma. The serum alkaline phosphatase may be elevated because of the bone formation in osteoblastic lesions, or because of metastatic involvement of the liver. In bone metastases from prostatic cancer, there is also characteristic elevation of the serum acid phosphatase.

Roentgenographic picture. Osteolytic metastases must produce a severe degree of local bone destruction before they become visible on roentgenograms. The areas of involvement are often multiple. Their appearance varies with the type of the original tumor. Punched-out areas of bone destruction without new bone formation are frequent in the osteolytic tumors (Fig. 7-17), whereas in the case of metas-

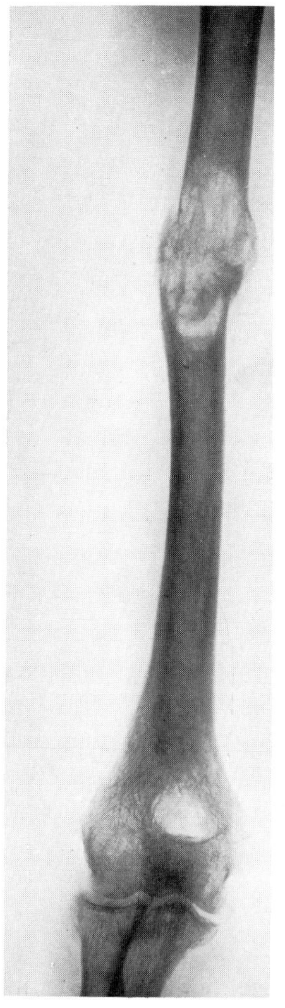

Fig. 7-17. Roentgenogram showing pathologic fracture of the humerus through a metastatic lesion from carcinoma of the breast. Note localized, well-demarcated area of bone destruction.

tases from the prostate gland a diffuse increase of density is the rule. When roentgenograms show no change in an area where the clinical findings strongly suggest metastatic bone disease, radioisotope scans may demonstrate the lesion.

Treatment and prognosis. Biopsy should usually be performed if the diagnosis is uncertain. Open biopsies yield information much more reliable than that obtainable by bone needling technics. Irradiation can do much to relieve the pain caused by metastatic tumors, and in the case of the more radiosensitive growths it may prolong life. Supervoltage radiation makes possible more effective dosage for the deeper tissues with less danger of skin damage. Cortisone sometimes affords relief of pain. Pain caused by bone metastases from carcinoma of the prostate gland can be relieved also by orchiectomy and the administration of estrogen. Bone metastases from breast cancer may be favorably influenced by castration, adrenalectomy, or the administration of hormones. Metastases from thyroid cancer may undergo prolonged remission, sometimes for years, with the use of suppressive doses of thyroid hormone and radioactive iodine. Cytotoxic chemotherapy is increasingly helpful as more effective drugs and technics are developed, and may produce long remissions. Traction, intramedullary nailing, cast, or brace may be indicated for a pathologic fracture. Although the long-term prognosis is uniformly unfavorable, compassionate management and carefully selected palliative measures can do much to relieve the patient's discomfort.

Tumors of joints, tendons, tendon sheaths, and bursae

True neoplasms of joint structures, of tendons and tendon sheaths, and of bursae are uncommon or rare. They may be simulated by several nonneoplastic entities that deserve mention.

Within the joints occur *lipomas* and *hemangiomas,* arising from fatty connective tissue and its blood vessels. *Synovial chondromatosis* (p. 364) is probably to be regarded as a self-limited metaplastic process rather than a neoplasm; it occurs most commonly in the knee but is also seen in other joints and gives rise to multiple loose bodies. *Pigmented villonodular synovitis* (p. 365) has been regarded by some observers as a neoplasm. So-called giant cell tumors of the synovium may be manifestations of pigmented villonodular synovitis. Cartilage cysts, such as those of the lateral meniscus of the knee (p. 357), probably arise through mucoid degeneration rather than neoplasia. A joint may be invaded by a malignant tumor of the articular end of one of its component bones. The only primary malignant tumor that arises within joints with any frequency is the *synovial sarcoma,* or synovioma. Synovial sarcomas are seen most often in or about the knee, are composed of a spindle cell stroma containing characteristic clefts, tend to recur locally after excision and eventually to metastasize, and carry a very grave prognosis.

Tendons may be involved by tumors of adjacent structures or by tumorlike granulomatous or degenerative processes but do not produce true neoplasms. Tendon sheaths and bursae, with lining membranes so similar to that of joints, are subject to the same tumors and tumorlike lesions that involve joints. So-called giant cell tumors occur frequently about tendon sheaths, especially in the hand and wrist; they are small, dense nodules whose nature is probably granulomatous rather than neoplastic. Ganglia of tendon sheaths and joints (p. 437) probably result from the myxoid degeneration of connective tissue rather than from neoplasia.

Diagnosis and treatment. Accurate diagnosis of tumors of the joints, tendons, tendon sheaths, and bursae is often impossible prior to exploration. The treatment is surgical and varies with the clinical and microscopic characteristics of the individual tumor. Most small, circumscribed growths should be removed with a margin of surrounding tissue. Synovectomy is sometimes indicated for diffuse involvement. Every patient with a joint tumor must be observed carefully for evidence of a recurrent lesion. When the diagnosis of malignancy is established, resort must usually be made to radical surgery, possibly combined with irradiation.

Tumors of muscles and fasciae

Although soft tissue tumors are encountered clinically with great frequency in muscular regions of the extremities and the trunk, these new growths usually arise from mesoblastic derivatives other than muscle. *Rhabdomyosarcoma,* a rare tumor derived from striated muscle, occurs in the extremities and about the face and neck. It is highly malignant. The treatment consists of wide excision or amputation supplemented by chemotherapy and irradiation. The great majority of the subcutaneous soft tissue tumors are derived from cells of connective tissue, of fatty, or of endothelial type. Of these, the benign varieties are *fibroma, lipoma,* and *angioma,* and the only malignant new growth of common occurrence is *fibrosarcoma.*

FIBROMA, LIPOMA, AND ANGIOMA

These benign tumors may arise at a subcutaneous level or deep within the musculature. They grow very slowly, cause few symptoms, and may be noted early as small, nontender nodules or only much later after having attained considerable size. *Fibromas* are encapsulated and are palpable as firm, circumscribed masses, often movable to some extent on the underlying tissue; they are particularly common in the hands and feet. *Lipomas* possess little or no capsule; they may be either circumscribed or diffuse and of irregular contour. *Angiomas* are of less frequent occurrence and have been described particularly as arising in the muscles of the extremities. Unlike fibromas and lipomas, they may extend locally and involve the surrounding tissues.

In diagnosis these tumors must be differentiated from masses of similar clinical appearance that may occur at corresponding sites; for instance, lipoma of the popliteal fossa may be distinguished only with difficulty from distention of the gastrocnemio-semimembranosus bursa. Their differentiation from the malignant fibrosarcoma is occasionally difficult, and the diagnosis is sometimes established only by surgical exploration.

The treatment of these benign tumors is excision. Recurrence need not be feared in the case of lipomas but must be kept in mind after the excision of fibromas. Any involvement of surrounding structures by an angioma necessitates wide excision and prolonged observation to rule out recurrence.

FIBROSARCOMA

Pathologically the fibrosarcomas of nonvisceral origin can be divided into several varieties according to the cellular morphology, but clinically they form a single important group with well-marked characteristics.

Fibrosarcoma may occur at any age but is commonest in the fourth and fifth decades. It is particularly likely to arise in the lower third of the thigh, in the forearm, or in the abdominal wall. A history of suddenly increased rapidity of growth or of recurrence after surgical removal suggests the diagnosis of malignancy. Fibrosarcomas are typically of firm consistency and, because of the involvement of surrounding fascial structures, are relatively immobile. The rate of tumor growth and time of metastasis vary greatly. Often metastases do not appear in the lungs until late in the course of the disease.

When malignancy is suspected, biopsy is always warranted. The treatment should consist of early, wide, and thorough excision of the tumor. The usual tendency is toward a too conservative removal, which favors recurrence. Irradiation and chemotherapy are usually ineffective. When the tumor is situated in an extremity, amputation is often the best treatment.

Tumors of peripheral nerves

Tumors of several types may arise from the cells that surround and support the nerve fibrils of peripheral nerves. *Neurilemoma* is a solitary, encapsulated, benign tumor that is thought to arise from the Schwann cells surrounding the axons. Its malignant counterpart, *malignant schwannoma,* is recognized by some authorities. From endoneural fibrous connective tissue arise the benign *neurofibroma* and the malignant *neurofibrosarcoma.*

NEUROFIBROMA AND NEUROFIBROSARCOMA

The typical neurofibroma, composed of masses of spindle cells, is a firm, nontender, slowly growing tumor attached to one of the larger nerves but not as a rule interfering with its function. When malignancy is present, growth is more rapid, the tumor mass becomes fixed, and nerve irritation or block may develop. Surgical removal of the tumor is indicated. In the presence of malignant changes wide excision should be carried out. In advanced cases of neurofibrosarcoma excision may be impossible, and amputation, as a palliative measure, may be indicated.

NEUROFIBROMATOSIS (VON RECKLINGHAUSEN'S DISEASE)

Neurofibromatosis, or von Recklinghausen's disease, which is thought to result from an inherited defect in tissues of ectodermal origin, is characterized by the

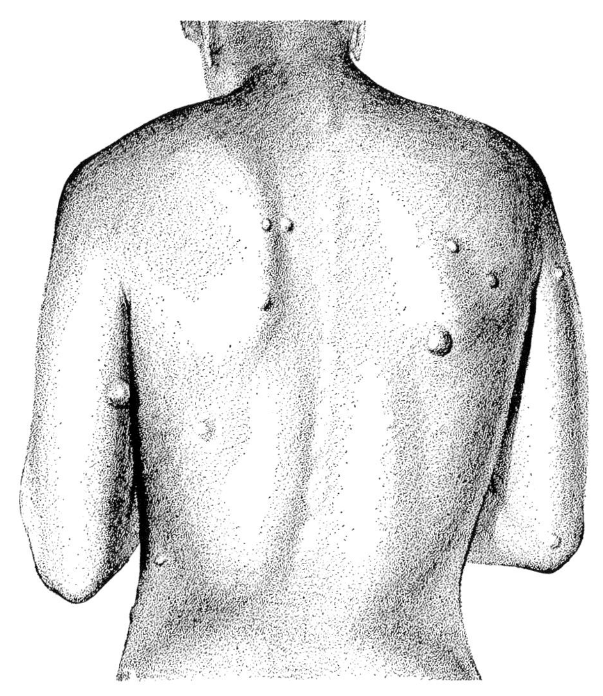

Fig. 7-18. Neurofibromatosis (von Recklinghausen's disease). Note multiple small tumor masses on the back and arms. These pedunculated soft tissue tumors are often associated with scattered areas of light brown skin pigmentation (café au lait spots), not shown in this drawing.

gradual development of numerous pedunculated soft tissue tumors (Fig. 7-18) and by the frequent association of small scattered areas of skin pigmentation with smooth edges, which are called café au lait spots. The tumors, most of which have the histologic characteristics of neurofibromas, vary greatly in number and size and lie in the subcutaneous tissue or skin of any or all parts of the body. A large diffuse mass may develop about the head or neck and form an unsightly, drooping fold of skin and soft tissue. Tumors attached to deeper nerve trunks, such as the sciatic nerve, as well as intraspinal tumors, are seen not infrequently. Bone changes, some of which seem to be secondary to involvement of the local nerves, are often observed; scoliosis and deformity of the limbs are frequent findings.

The tumors tend to progress slowly and to cause few symptoms except from local mechanical pressure. They possess, however, a definite tendency toward malignant change, and the ultimate outcome is often fatal. Excision of deforming or disabling growths must be done with the realization that removal is often followed by the appearance of recurrent tumors of heightened malignancy.

CHAPTER 8

Fracture principles, fracture healing

In a short text of this type there is insufficient space to present in detail the diagnosis and treatment of fractures. However, since fractures are an essential part of orthopaedic surgery and since the medical student often starts his study of orthopaedic diseases and fractures simultaneously, a brief description of the principles of fracture treatment and of the healing, normal and abnormal, of fractures may prove useful.

Fracture principles

DEFINITIONS

Certain terms are in common usage in the description of bone injuries. A *fracture* is defined as any break in the continuity of bone. The term *fractured bone* is generally used interchangeably with *broken bone*. A strong distinction is made between a fracture in which there is communication between the bone and the outside, an *open* or *compound fracture,* and a fracture in which there is no such communication, a *closed* or *simple fracture*. In the case of the open injury, infection, osteomyelitis, delayed union, and even nonunion of the fracture are frequent complications. *Pathologic fractures* occur in bones weakened by preexisting disease such as tumors, cysts, osteomyelitis, or osteoporosis.

Terms used to describe the configuration of fractures are of more than academic interest because they often convey information on the mechanism of injury and carry implications affecting the choice of treatment. Fractures are described as *transverse, oblique,* or *spiral* according to the direction taken by the line of breakage. *Transverse fractures* are usually caused by simple angulatory forces, whereas *spiral fractures* result from torsion. A fracture is said to be *comminuted* when the bone is broken into three or more fragments. Spiral, oblique, and comminuted fractures are often unstable; frequently their reduction cannot be maintained by simple cast fixation without danger of overriding of the fragments and shortening of the limb.

DIAGNOSIS

Establishing the diagnosis is the first step in the care of fractures. The ease of diagnosis varies with the severity and location of the fracture and with the degree of displacement and deformity. The presence of other serious injuries may obscure the fracture and make diagnosis more difficult. The clinician's awareness that certain types of injuries are likely to cause certain fractures and that some fractures may show little external evidence of deformity facilitates diagnosis. The diagnosis is based on data obtained from the history, physical examination, and roentgenograms.

History. A thorough history is important. Although difficult or impossible to obtain from a seriously injured patient, certain points in the history should always be sought. Details of the injury or accident should be carefully recorded since they may give helpful information regarding the severity and type of forces involved. A record of the time and place of the injury should be made. The age of the patient is important, since certain fractures predominate in particular age groups. A history of pain or deformity preceding fracture may be the first clue to the diagnosis of a pathologic fracture. In the case of open fractures it is important to learn whether the patient has had immunization to tetanus. Since the treatment of many fractures requires general anesthesia, one must often learn the time of the patient's last meal and what was eaten. A history of recent respiratory infection or of cardiac or renal difficulties may modify the treatment. Inquiry concerning allergies, particularly to medications used in connection with anesthesia and to antibiotics and antitoxins, is essential.

Physical examination. Immediate attention should be given to the patient's general condition. In *severely injured patients* the three most serious problems are (1) respiratory difficulties, (2) acute hemorrhage, and (3) shock. If any or all of these problems exist, they must be dealt with immediately.

Respiratory difficulties include obstruction due to edema from soft tissue injury about the face and neck, foreign bodies, and accumulated secretions in the respiratory tract. Tracheal intubation or emergency tracheotomy may be necessary. Tension pneumothorax, open wounds of the chest, and an unstable or flail chest are other acute respiratory problems requiring immediate attention.

Acute hemorrhage is obvious if external. Venous bleeding from a large wound can usually be controlled by direct pressure through a sterile dressing placed in the wound. Mild arterial bleeding can be controlled in a similar manner, but a bleeding larger artery may require direct clamping in the emergency room. Digital pressure over the arterial supply proximal to the bleeding area may be effective as a temporary expedient. Rarely is the use of a tourniquet to control hemorrhage warranted. If a tourniquet is used, the person applying it should be in constant attendance and responsible for release of the tourniquet within 30 to 60 minutes after its application. Internal hemorrhage within pleural, pericardial, or abdominal cavities may not be obvious. It should be suspected when shock persists despite adequate transfusion, and in the presence of injuries to chest or abdomen.

Shock, manifested by cold and clammy skin, rapid and thready pulse, and lowered blood pressure, is usually the result of blood loss. The amount of blood lost by a patient is difficult to determine and is frequently underestimated. A great deal of bleeding may have occurred from the relatively small wound of an open fracture or from scalp lacerations, but if the bleeding has ceased when the patient reaches the emergency room, extensive blood loss may not be suspected. A closed fracture of the femur in a large person may be followed by loss of a great quantity of blood in the soft tissues of the thigh. Intrapleural bleeding, or intra-abdominal bleeding from rupture of liver or spleen, may result in shock. Hemorrhagic shock is most effectively treated by whole blood transfusion. When compatible blood is not immediately available, the blood pressure may be temporarily sustained by giving intravenous dextrose in saline, or by giving plasma or a plasma expander such as dextran. The use of a balanced salt solution such as Ringer's lactate is advantageous. Severe shock frequently results in some degree of acidosis, for which the intravenous administration of a bicarbonate solution is advisable. Shock may result also from cardiorespiratory embarrassment with decreased oxygen supply to the tissues, vasodilatation, and circulatory collapse. Additional causes of shock include burns, crush injuries, overwhelming bacterial infections, and toxic conditions such as gas gangrene. Common to all forms of shock is a deficiency in the volume of circulating blood with respect to the effective capacity of the cardiovascular system. If the deficiency persists or increases, the outlook for survival is poor. In severe shock, central venous pressure, monitored continuously by means of a catheter in the superior vena cava, is helpful in evaluating the response to treatment. In addition to causing shock, crush injuries result in severe soft tissue damage and necrosis and may be associated with acute renal failure.

If none of these three grave emergencies exists or their treatment has already been started, one may proceed with a more detailed physical examination. This is best carried out with the patient completely undressed. Some of the clothing may have to be cut away. Much of the examination can be done with the patient supine on a table or stretcher. Observation of the entire body for lacerations, abrasions, swelling, or deformity can be carried out in seconds; unless this simple process of *looking* is done in a routine head-to-toe manner, it is surprising how many injuries may be overlooked. The second step in the examination is palpation of the entire skeletal system. Any tenderness, induration, or deformity is sought. Again, in a matter of minutes, every major bone in the body can be felt, the patient being moved very little during the process. The skull and jaw are palpated, the cervical spine next—and gently—and then the clavicles, sternoclavicular and acromioclavicular joints, shoulder, humerus, and elbow. The medial and lateral epicondyles, as well as the olecranon and radial head, are easily located. Palpation continues down the forearm to the small bones of the wrist and hand. Next the rib cage and sternum are carefully examined. The patient is then turned very slightly to enable the examiner to get his hand under the patient's back. Thoracic and lumbar spinous processes are palpated as well as the sacrum, sacroiliac joints, and ischial tuberosities. The skin of the back and buttocks may be inspected at this time. Palpation is continued anteriorly over the iliac crests, trochanters, pubis, thigh, patella, knee, tibia, ankle, foot, and toes. The circulation in the extremities is evaluated by noting temperature, color changes, and pulses. Sensation in the extremities may be briefly tested at this time. The patient is then asked to move fingers and toes.

A brief survey of the patient's cranial nerve function is followed by inspection of mouth, ears, and eyes. Auscultation of the chest and palpation of the abdomen may be done next.

The entire examination is a quick but careful survey that can be carried out rapidly with little disturbance to the patient. It will be modified by positive findings;

more detailed neurologic evaluation, for example, would be given a patient with evidence of head or spinal cord injury.

After this brief survey, attention is focused on the *injured area.* Other than distortion from swelling, there are three types of deformity that may be associated with fracture of a long bone: *angulation, shortening,* and *rotation.* Obvious angulation indicates fracture. One should be cautious, however, and by gentle, careful palpation seek possible associated fractures or dislocations above and below the angulation. When tenderness is not accompanied by deformity, the examiner may cautiously apply a slight bending force to the long bone, grasping it above and below the level of tenderness; if pain in the tender area results, a fracture is to be strongly suspected. Joint motion should be tested with the extremity well supported by the examiner's hands—it need not be tested beyond the point of pain.

When a fracture is suspected, the circulation and nerve supply distal to the injury should be carefully evaluated. In the lower extremity, simple tests for peroneal, tibial, and femoral nerve function should be done. Injuries to the blood supply manifest themselves by coolness, blanching or cyanosis, decreased sensation and motor function, and diminished or absent pulses distal to the injury. Vascular injuries are particularly common after elbow fractures; they demand immediate treatment. To evaluate possible nerve injury in the upper extremity the simple tests for radial, median, and ulnar function are quickly carried out, as discussed in Chapter 6. Where there is severe injury to the wrist or hand, these tests may be difficult, but with diligence a satisfactory evaluation can be made.

It should be noted that in emergency situations a thorough physical examination can be carried out in a very few minutes and that no equipment is required other than the hands and eyes of the examining physician.

When a fracture is suspected, the injured part should be splinted before the patient is moved to another location. Excessive motion at the fracture site not only is painful but also may increase soft tissue damage, injure blood supply and nerves, and even cause sharp fracture fragments to penetrate the skin, converting a closed injury into an open fracture. For ankle fractures an ordinary pillow wrapped about the leg and ankle and supported on the stretcher is quite effective. In the case of the humerus, having the patient hold the arm close to his body may be effective enough. Padded board splints are useful for forearm fractures. Prefabricated metal splints in common use for the leg and thigh require careful padding to prevent

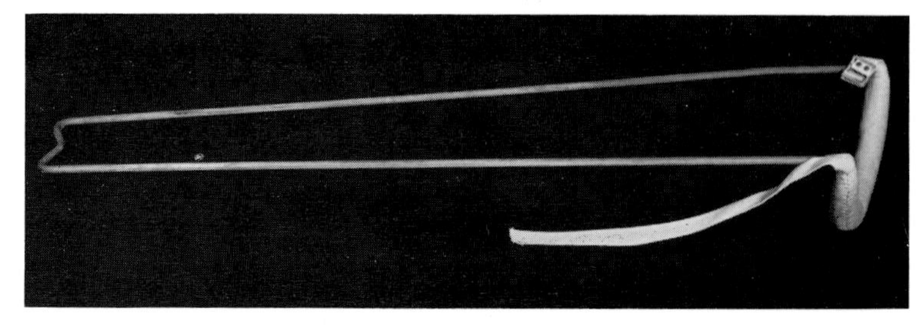

Fig. 8-1. The Thomas half-ring splint, used to provide support and traction for either upper or lower limb.

pressure sores about the heel, malleoli, and fibular head. For serious fractures of the leg, knee, and femur the Thomas splint (Fig. 8-1), which facilitates traction, remains an effective device; it should be available in every emergency room, and every physician should know how to use it.

Roentgenographic examination. After emergency treatment, physical examination, and splinting, roentgenograms of the fracture are made; they should be considered an integral part of the examination. The attending physician who has examined the patient can best tell the x-ray technician what views need be taken and where the x-ray tube should be centered. In severe injuries it is wise for the physician to accompany his patient and to aid in the taking of the films. He is best able to protect the injured area during x-ray positioning, and he will be assured that indicated views will be taken. In less severe injuries, adequate instructions should accompany the patient. A small ink mark on the patient's skin at the area of suspected fracture will often aid the technician in proper centering. Views of suspected areas should always be made in at least two planes, perpendicular to each other; additional oblique views are sometimes necessary. The fracture may be apparent on only one of several views. All necessary exposures should be made at the initial examination, any suspected areas being included. For early roentgenograms large films are best; many a fracture or dislocation has been overlooked by not being included on a small film.

PRINCIPLES OF FRACTURE TREATMENT

After interpretation of the roentgenograms the optimal treatment must be chosen. This will depend upon many factors, including the general condition of the patient, the presence of associated injuries, whether or not the fracture is open, and the location and displacement of the fracture. Fracture treatment includes three basic objectives:

1. *Reduction,* or replacement of the bone fragments to as near anatomic position as possible
2. *Maintenance of reduction* until healing is sufficient to prevent displacement
3. *Restoration of function* of the muscles, joints, and tendons

Reduction. Many fractures with little or no displacement require no reduction, only maintenance of position. Fractures with displacement may be reduced in several ways. Since reduction is painful and involves counteracting strong muscle pulls, general or regional anesthesia is usually necessary. Reduction by *manipulation* is the commonest method of restoring the alignment of fractured long bones. It may involve longitudinal traction to restore length, angulation to allow locked fragments to disengage and slide past one another, and manual pressure of the bone fragments into proper position. The details of manipulation vary with the individual fracture and the displacement of the fragments. A second method of fracture reduction is *traction* applied over a period of several hours or days. Reduction by sustained traction is generally used when the traction will also maintain reduction. It is used most commonly in the treatment of femoral shaft fractures and cervical spine injuries. A third method of fracture reduction is *open surgery.* Occasionally the fragments may be caught within the soft tissues in such a manner that reduction by manipulation or traction is either impossible or dangerous. In such instances reduction under direct vision is advisable. More frequently, however, open reduction is utilized because internal fixation is contemplated for maintenance of the reduction. It must always be remembered that open reduction is a hazardous procedure. It converts a closed fracture into an open one, and if unsuccessful or complicated by infection it may result in a much more serious situation than was present before operation. In children, with the exception of certain specific fractures about the elbow, it is quite rare that open reduction need be undertaken.

Maintenance of reduction. After a fracture has been reduced, the corrected alignment must be maintained until bone healing is well advanced. The three common methods by which fractures are held in position during the healing phase are *external fixation* by means of cast or splint, *traction,* and *internal fixation* by nail, plate, or screws. In many fractures the indications for one of these methods are fairly clear-cut, and it is used practically to the exclusion of the others. For example, in maintaining the reduction of fractures of the femoral neck, internal fixation has so many advantages that traction or external fixation is rarely used. On the other hand, many fractures may be treated satisfactorily by any of several methods.

External fixation. External fixation is the commonest method of maintaining reduction. It may be accomplished by means of a plaster cast or by splints of metal or wood. Such splints are of limited value since it is difficult to make them conform to the contour of the individual patient, and unless carefully applied with adequate padding, they cause pressure sores over bony prominences. Plastic materials have also been used for external fixation, but to date nothing else has been found quite so satisfactory as the plaster of Paris bandage first introduced by Mathijsen in 1852. Plaster of Paris is anhydrous calcium sulfate. In powder form it is incorporated in rolls of crinoline bandage that are dipped in water and then wrapped about the extremity. The fact that this wet bandage can be molded to the contour of the limb and to the requirements of the individual fracture accounts chiefly for its superiority to other methods of external fixation. Plaster of Paris sets in a matter of minutes and forms a strong and durable support for the injured limb. Plaster cast fixation is especially applicable where bones are close to the surface and can be held efficiently by the cast, as in fractures about the wrist, ankle, and tibia. Casts are less effective for immobilizing fractured bones that lie deep in muscle, such as the femur. In fractures of the long bones, casts are effective in controlling angulation and rotation, but another type of fixation is required when shortening is likely, as in oblique or comminuted fractures. Plaster cast fixation is used for most fractures in children.

Although the use of plaster of Paris is relatively simple, certain precautions must be observed. Since the hardened cast is quite rigid, bony prominences must be protected by adequate padding. Otherwise pressure sores may develop. Where major nerves pass between bone and skin, such as the ulnar nerve at the elbow and the common peroneal nerve around the neck of the fibula, protective padding is especially important. A common cause of foot drop is peroneal nerve injury from an improperly applied cast. It must be kept in mind that after the plaster has set, a circular cast cannot expand. If post-injury swelling takes places within a tight cast, circulatory embarrassment may produce serious consequences. Whenever there is doubt of the adequacy of the circulation in a limb, the cast should be split and spread immediately.

Traction. For fractures that cannot be immobilized efficiently by casts, maintenance of reduction by means of traction may be advisable. When traction is applied to an acutely injured limb, the muscles act as an internal splint to protect the fracture, and the patient usually experiences relief of pain. The principal disadvantage of treatment by traction is that it requires the patient to remain in bed and usually in hospital.

Traction may be applied in several ways. *Skin traction,* usually applied by means of adhesive strips and an encircling elastic bandage, may be used when not more than 5 or 6 pounds of pull is required. It is frequently used for longitudinal pull on the leg, as in *Buck's extension* (Fig. 8-2).

Another form of skin traction frequently used for children with femoral shaft fractures is *Russell's traction* (Fig. 8-3). For cervical spine injuries, traction may be applied to the head by means of a cloth head halter. Traction may be applied to the pelvis by means of a canvas girdle.

Skeletal traction, applied by drilling a wire or pin transversely through a bone, has several advantages over skin traction. It makes possible the stronger traction required in many cases. For example, a femoral shaft fracture in a muscular adult may need as much as 20 to 30 pounds of pull for reduction. Skeletal traction can be applied to distal areas, such as the ankle, where skin traction cannot be used. Abrasions, blisters, and reaction to adhesives, often troublesome with skin traction, are avoided by skeletal traction. In general, skeletal traction is comfortable and well tolerated by the patient.

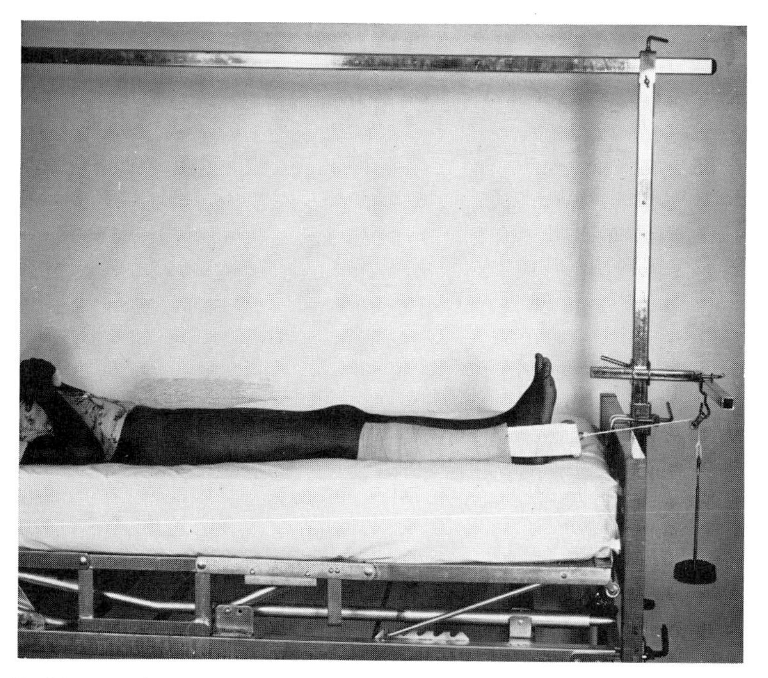

Fig. 8-2. Buck's extension, used to exert traction in the long axis of the lower limb with knee and hip in neutral position.

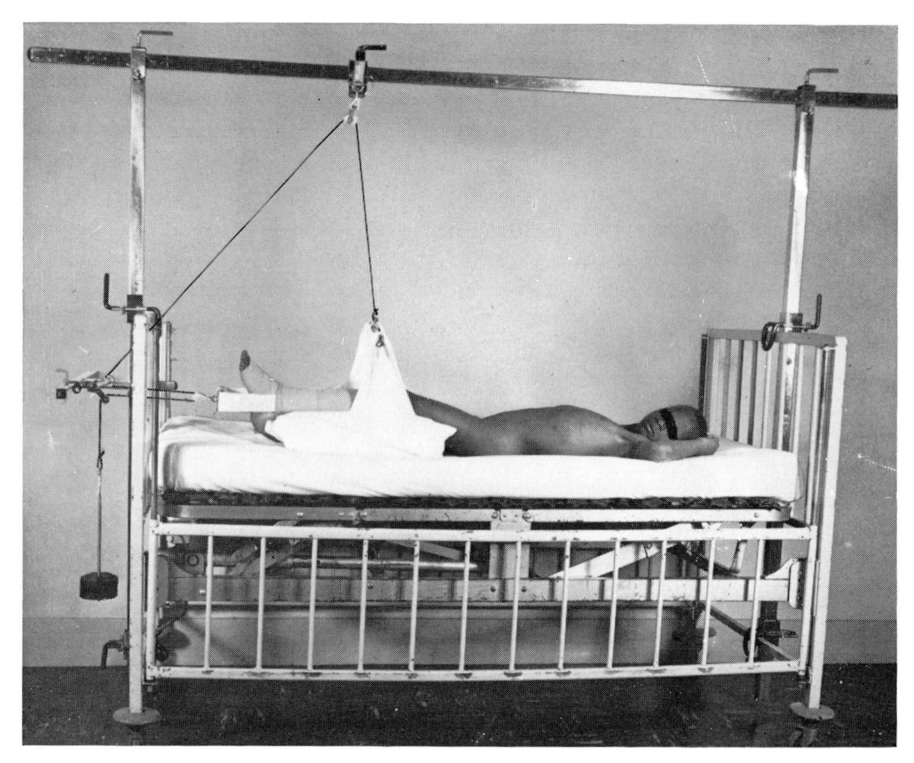

Fig. 8-3. Russell's traction, applied for a fracture of the left femoral shaft in a boy 9 years of age. This rope-and-pulley arrangement produces a traction force acting in the long axis of the femur that is approximately twice as great as the force suspending the knee. This type of skin traction is especially useful for children over the age of 3 years and for adolescents.

Skeletal traction is most commonly applied through the proximal part of the tibia, just distal to the tibial tubercle, in the treatment of femoral fractures. Among other areas that may be used are the distal part of the femur or tibia, the calcaneus, and, in the upper limb, the olecranon. Since the application of skeletal traction introduces the possibility of infecting the bone through which the pin or wire is inserted, strict aseptic technic is obligatory.

Skeletal traction is usually applied by means of a *Steinmann pin* or *Kirschner wire.* Either may be threaded to prevent slipping. The Steinmann pin, with a diameter of 1/8 inch or more, is relatively rigid and may be attached by a simple yoke to the rope and pulleys. The Kirschner wire has a much smaller diameter, is quite flexible, and must be supported by a special spreader that bowstrings the wire tightly to prevent bending. It has the advantage of requiring only a small opening in skin and bone. On the other hand, the Kirschner wire is so thin it may cut through osteoporotic bone.

The most frequent use of skeletal traction is in treating fractures of the femoral shaft in adults. While in traction, the limb may be supported in a Thomas splint. A *Pearson attachment,* clamped to the splint, allows knee flexion. Since the patient is to remain in this apparatus for weeks or months he must be allowed some

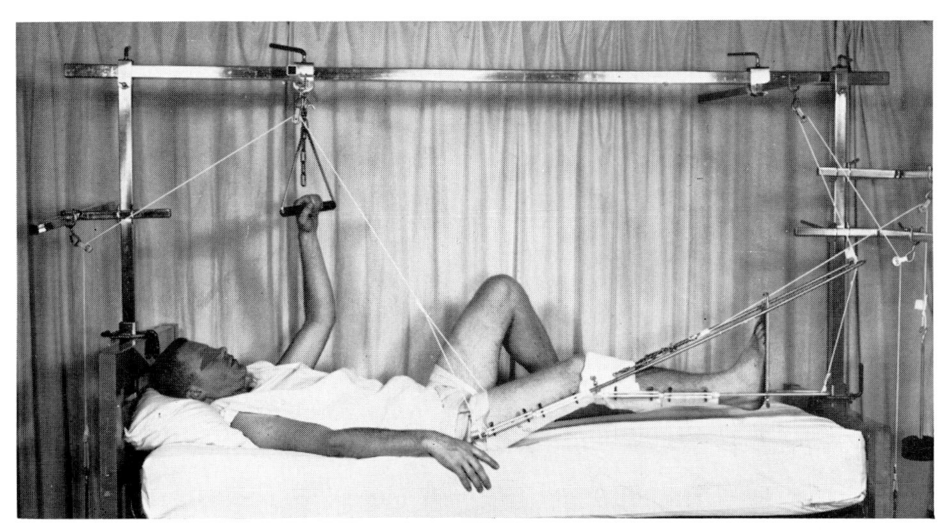

Fig. 8-4. Balanced skeletal traction. Traction in the long axis of the right thigh is applied by means of a Kirschner wire through the proximal portion of the tibia. The limb is supported by a Thomas splint beneath the thigh and a Pearson attachment beneath the leg. An additional attachment prevents foot drop. Weights apply countertraction to the upper end of the Thomas splint and suspend its lower end. By using his left arm and leg as shown, the patient can shift the position of his hips without change in the amount of the traction.

mobility for change of position in bed, but during such movements the pull upon his fractured femur must remain undisturbed. This is accomplished by so balancing the suspended Thomas splint by means of weights and pulleys that it will move up or down with the patient without disturbing the fracture. Such traction is a form of *balanced traction* (Fig. 8-4). Countertraction is provided by the splint and by the weight of the patient's trunk; when necessary, the countertraction may be increased by elevating the foot of the bed on blocks.

Skeletal traction is also used frequently in the treatment of dislocations and fractures of the cervical spine, being applied through the outer table of the skull. The most commonly used devices are *Crutchfield* or *Barton* tongs. Both are comfortable and well tolerated; they allow use of the strong traction (30 to 40 pounds) that is sometimes required for reducing dislocations of the cervical vertebrae.

Proper use of traction in the treatment of certain fractures yields excellent results, often superior to those of other methods. Skillful care and patience, however, are required. Traction apparatus must be inspected at least twice a day. Too little traction allows overriding of the fragments and shortening. Too much traction is worse; it separates the fragments and may result in delayed union or nonunion. Traction may be used throughout the period of fracture healing. More frequently, however, it is discontinued when union becomes strong enough to prevent angulation within a plaster cast. After the cast has been applied the patient may be discharged from hospital.

Internal fixation. Internal fixation is the third method of maintaining the reduction of a fracture. It is usually effected by means of metal plates, rods, or screws.

It is used when other methods of maintaining reduction are impracticable or unreliable. Among the disadvantages of internal fixation is the fact that it converts a closed fracture into an open one. Infection resulting from the surgical treatment of a closed fracture may be a major tragedy.

Although metallic implants had been used previously, it was not until early in the twentieth century that reasonably good results were obtained. A scrupulous "no-touch" technic developed by Sir Arbuthnot Lane lessened the danger of bacterial contamination during the insertion of metal plates. His success with metallic internal fixation stimulated its wider use by surgeons perhaps less skillful than Lane, however, with the result that many cases of osteomyelitis and nonunion occurred. It became apparent that some of the failures resulted from an unfavorable tissue reaction to the metal. In 1937 Venable and Stuck called attention to the electrolysis of metallic implants. They found that an alloy of cobalt, chromium, and molybdenum (Vitallium) was well tolerated by the tissues. With the development of Vitallium and certain stainless steels the feasibility of internal fixation increased. Attention then became focused on the design of devices for use in various types of fractures.

One of the simplest forms of internal fixation is the *transfixion screw.* This is applicable to oblique fractures, especially those of the tibial shaft. Fixation with transverse screws alone is not strong enough to maintain reduction. With the external support of a cast, however, transfixion screws can preserve the reduction of an oblique fracture that would override and shorten if treated by cast alone.

Another common method of internal fixation is the use of a *bone plate.* A metal plate is fastened to the surface of the fragments by at least two screws above and two below the fracture; the screws should be long enough to traverse both sides of the cortex. Plate fixation usually requires the additional support of a cast. Application of the plate involves periosteal stripping, which may tend to delay union by compromising the already impaired blood supply of the fragments. Some bone plates may also delay union by holding the fragments apart after slight resorption of the fractured surfaces has taken place early in the healing process. Fractures heal most readily when the pressure of contact between the injured surfaces is physiologic. Eggers stressed the importance of contact-compression in fracture healing and designed slotted plates which, affixed loosely to shaft fragments by screws, both maintain alignment and allow physiologic adjustment of the contact between the fractured surfaces. Of recent development is a technic for plating with firm compression at the fracture line.

A third type of internal fixation in common use is the *intramedullary rod* (Fig. 8-5). The fixation of femoral shaft fractures by inserting a long, inflexible rod down the medullary cavity was popularized in Germany by Küntscher during World War II. Treatment by this method provides excellent fixation of certain fractures, promotes contact-compression, obviates immobilization of the joints, and enables the patient to walk with crutches soon after the injury. Intramedullary fixation is also used for certain fractures of the shafts of other long bones.

One fracture deserves particular mention in connection with internal fixation. This is the hip fracture, or fracture of the femoral neck. Because of strong shearing force across the fracture line caused by muscle pull and because circulation to the

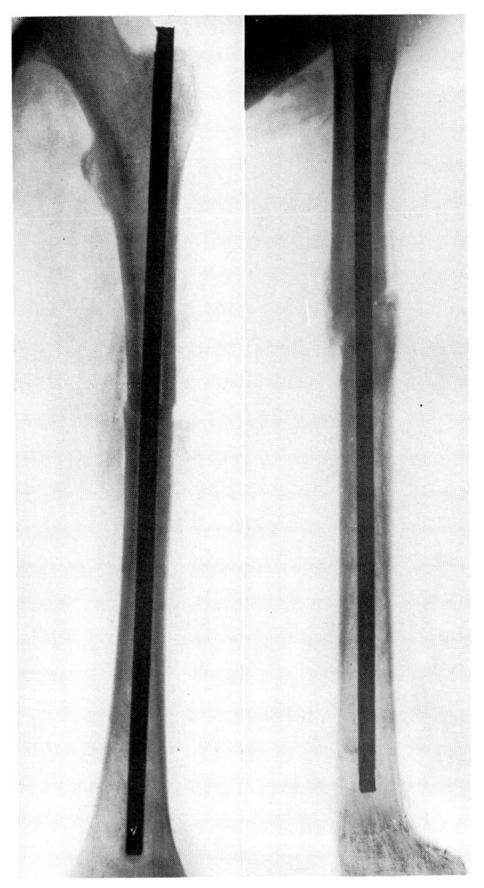

Fig. 8-5. Anteroposterior and lateral roentgenograms of Küntscher intramedullary rod, or nail, used to fix the fragments of a fracture of the middle third of the femoral shaft in an adult.

femoral head is most precarious, fractures of the femoral neck frequently fail to unite. In fact, at one time it was thought that these hip fractures never united. Royal Whitman demonstrated early in the twentieth century that many hip fractures united if held in abduction and internal rotation by a cast. However, since the old people who most often sustain this fracture tolerate prolonged immobilization poorly, the cast treatment left much to be desired. Internal fixation of femoral neck fractures by means of a three-flanged nail inserted just below the greater trochanter, up the femoral neck, and into the head (Fig. 8-6), which was introduced by Smith-Petersen and others in the late 1920's, enables these older patients to be up in a chair within a day or two after operation and has also resulted in a much higher rate of union. Instead of the nail, any of several other internal fixation devices, such as multiple pins, may be used. For hip fractures, early reduction and surgical fixation is the least hazardous method of treatment, since the mortality rate is lower than that in cases treated by cast immobilization or by traction.

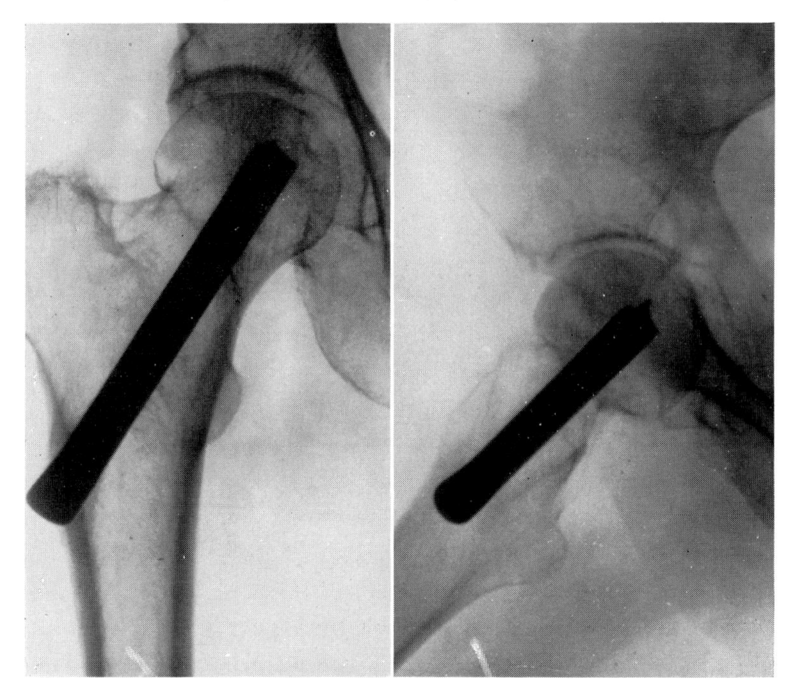

Fig. 8-6. Anteroposterior and lateral roentgenograms of a Smith-Petersen three-flanged nail used to fix the fragments of a fracture of the femoral neck.

Restoration of function. After adequate reduction and immobilization, most fractures progress to solid bony healing. In most instances, especially after cast immobilization, the patient develops some joint stiffness and muscle atrophy by the time the fracture has healed. The degree of stiffness and atrophy varies with many factors. A child develops little stiffness, even after long immobilization, and it usually clears up rapidly, whereas in older people some limitation of joint motion after fracture and immobilization may be permanent. The longer the period of immobilization the more severe will be the stiffness. The orthopaedic surgeon has the constant problem of being sure that immobilization is continued long enough for firm healing of the fracture but discontinued as soon as joint motion can be safely resumed. Another important cause of joint stiffness is soft tissue damage at the time of fracture, since torn muscles and ligaments are replaced in part by inelastic scar tissue. Infection also leads to scarring and loss of elasticity; an open fracture that becomes infected is likely to be followed by much stiffness in adjacent joints. Fractures near joints, and particularly fractures involving the articular surfaces, may not only disrupt the mechanics of joint function but may also lead to fibrous adhesions within the joint. Preexisting joint disease increases the tendency to stiffness after immobilization. A patient with chronic arthritis who sustains a fracture may develop ankylosis after a long period in a cast.

Measures to prevent or minimize joint stiffness should be started early in the period of fracture healing. Active movement of the joints above and below the cast should be carried out at frequent intervals. An exercise program is especially important in treatment of the common Colles' fracture of the distal end of the radius. This fracture, often seen in older people, may be followed by serious stiffness in the fingers and shoulder, but this can be prevented by seeing that the patient carries out a program of finger and shoulder exercises during the immobilization period. Setting, or isometric, exercises of the muscles covered by the cast can usually be started by the patient a day or two after application of the cast and increased as the fracture stabilizes. As a rule, joints that must be immobilized should be kept in a functional position. If, because of the nature of the fracture, the adjacent joint must be immobilized in an awkward position, it should be brought out of this position at the earliest possible moment. Prolonged casting with a foot in equinus, perhaps needlessly, may cause a permanent deformity. If ankylosis is to occur in a joint, it is far better that it take place in a position favorable for function of the limb. Thus if, following an olecranon fracture, an elbow should become stiff in full extension, the patient would be far worse off than if ankylosis had taken place with the joint in a position of about 90 degrees of flexion.

As soon as adequate fracture healing has taken place and the cast has been removed, an active and intensive program of restoring function should begin. Unfortunately, it is too easy for both the patient and the doctor, pleased with excellent fracture healing and the end of a long period of immobilization, to consider the battle won. The third principle of fracture treatment, restoration of function, is as important as the first two. Its application varies widely. The child, on the one hand, usually carries out his own program of exercises; once out of a cast, it is impossible to keep him from exercising the limb and restoring its function. At the other extreme is the older fracture patient suffering from chronic arthritis, who, after removal of the cast, has discomfort with every movement and will do little on his own to restore lost mobility. He must be instructed, begged, and coerced to carry out the exercise program.

Although many forms of physical treatments and devices have been promoted to help the patient regain the use of his injured limb, by far the most important factor in the restoration of function is active exercise of the patient's own muscles. The limb may be massaged and passively exercised, but unless the patient works actively at using his atrophic, contracted, and scarred muscles, little will be accomplished. Thus it is essential that the patient receive careful and detailed instruction in a series of graduated active exercises for the muscles that control motion in the stiffened joint. Early, the preliminary use of heat and gentle massage may facilitate the exercises. At first the exercises may be done most comfortably under warm water. The water supports the painful limb, and the heat promotes a certain amount of relaxation. Hydrotherapy can be carried out in a therapeutic pool, Hubbard tank, whirlpool bath, or ordinary bath tub. A stiffened joint that is slow in recovery may be helped along by gentle assistive passive exercise; this must be done with great care, because such exercise, if too vigorous, may do more harm than good. In difficult cases the active exercise induced by occupational therapy may be of great help. The patient is given interesting jobs to do, which are specifically designed to make him use the affected muscles and joints over and over again.

Other forms of physical therapy may supplement the active exercise program.

Heat can be applied in a number of ways. Moist heat in the form of hot packs may give relaxation and comfort, enabling the patient to move a painful joint more effectively. Heating of deeper tissues may be accomplished by means of shortwave or microwave diathermy or by ultrasonic therapy. The effect of such deep heat may be beneficial in the last phase of fracture treatment.

For the severely incapacitated patient, physical therapy must go much farther than simple exercise, heat, and massage. Instructions in ambulation with walker, crutches, or cane are often needed, and after these aids have been discarded, further gait training may be required. When multiple injuries have been sustained, instruction in the activities of daily living are important. Such patients may require special devices to help them carry out activities such as feeding and toilet care. Teaching the more seriously handicapped patients such common functions as getting in and out of a bed or a chair can tax the ingenuity of the physician and the therapist. With care and perseverance the solution to many such problems can be found and will provide the severely crippled individual with a wider range of activity and enhance his ultimate recovery.

OPEN (COMPOUND) FRACTURES

Open fractures are especially serious injuries from two standpoints:

1. They are contaminated. The degree of contamination varies from minimal in the case of a clean puncture wound to severe in a wound filled with dirt, grease, and other foreign material.

2. Compound fractures are usually associated with more soft tissue damage than are closed fractures. Muscles may be crushed, skin coverage lost, and vessels and nerves torn or severed. Severely injured soft tissue presents a favorable environment for the development of infection and may interfere with the process of bone healing. Osteomyelitis and nonunion are frequent complications of open fractures.

The principles of treating open fractures require emphasis. First is an attempt to minimize the chance of infection by early, thorough cleansing of the wound and by *debridement,* which is the careful removal of all nonviable tissue. Open fractures are acute emergencies. The longer definitive treatment is delayed the greater will be the incidence of serious infection. Only acute respiratory embarrassment, hemorrhage, and shock take precedence over treatment of the open fracture; as soon as these factors have been controlled, thorough cleansing and debridement should be carried out under appropriate anesthesia in the operating room. If the wound is clean and only a few hours have elapsed from the time of injury, the wound may be closed primarily. Most contaminated wounds are best left open and then closed secondarily 5 or 6 days later because if infection develops, an open wound allows ready escape of pus that otherwise might dissect along the tissue planes of the limb. Prophylactic antibiotic therapy is generally recommended in the case of compound fractures. Tetanus immune globulin or toxoid is usually indicated. After the soft tissue wound has been cared for, treatment of the fracture follows the general principles outlined previously. The immediate use of internal fixation devices in open fractures, however, is generally contraindicated. After the soft tissues have healed and all signs of infection are absent, open reduction and internal fixation may be carried out if necessary.

FRACTURES IN CHILDREN

Fractures in children differ in several ways from their counterparts in adults. The most striking difference is their tendency to heal rapidly and to undergo spontaneous correction of mild and even moderate degrees of angulation or overriding. With the exception of a very few fractures about the elbow or hip, almost all children's fractures should be treated by closed methods. Nonunion in childhood is quite rare unless the fracture has been exposed surgically or has become infected.

Certain types of fractures are found predominantly in children. Since the epiphyseal plates close with maturity, *epiphyseal fractures,* or *epiphyseal separations,* fall into this category. Fractures involving the epiphyseal plate occasionally cause premature fusion of the epiphysis, which may result in a progressive angulation or shortening of the affected limb; the degree of deformity is determined by the location and extent of the injury and the age at which it occurs. The *greenstick fracture* is peculiar to children; it is an incomplete fracture in which the angulating force bends the cortex on the compression side and breaks it on the distraction side. In children the *torus fracture* is frequently seen in the distal fourth of the radius, where mild angulation causes a buckling or bulging of the cortex on the compression side.

The bones of children are covered with a thick, active periosteum that accelerates the healing process. During the period of active bone growth, constant remodeling and replacement of normal bone structure takes place in accordance with Wolff's law (p. 56). Thus in a young child it is possible for a fracture that heals in a position of moderate deformity to be so completely remodeled during subsequent bone growth that the deformity becomes corrected. An interesting phenomenon of overgrowth is noted in complete fractures of the long bones of children. A displaced fracture of the femur of a young child, if reduced to anatomic position, will often result in a bone that is about ½ inch longer than the corresponding bone of the uninjured limb. Frequently, therefore, it is advisable to allow femoral shaft fractures in young children to heal with slight overriding.

Fracture healing

Fractures heal by physiologic processes similar to those that accomplish the repair of wounds of soft tissues. Fracture healing is conditioned by factors such as the rigidity of bone, the tendency for fracture fragments to displace, and the length of time required for newly formed bone to restore strong union of the fragments.

On occasion the process of fracture healing deviates from its normal course, that of prompt union in satisfactory alignment, to produce one of three late bony complications of fractures: (1) *delayed union,* when fracture healing is abnormally slow, (2) *nonunion,* when healing stops short of firm union, and (3) *malunion,* when healing takes place in unsatisfactory alignment.

REPAIR OF FRACTURES

Following fracture, blood extravasates into the spaces between the bone fragments, into surrounding soft tissue and the bone marrow, about the periosteum and endosteum, and into the haversian canals. As in the repair of all wounds, healing begins with the clotting of this extravasated blood. Organization of the hematoma begins within 24 hours. The blood clot is invaded by granulation tissue, which consists of a loose meshwork of capillaries and young fibroblasts. The importance of the clot in fracture healing has been the subject of much debate. The torn ends of periosteum and of endosteum and the bone marrow adjoining the fracture line supply cells that proliferate and differentiate into fibrous connective tissue, fibrocartilage, and hyaline cartilage, all of which take part in the formation of new bone. Undifferentiated cells of the marrow also contribute to the new bone formation.

New bone formation begins in young individuals on the inner and outer surfaces of the damaged bone within the first 48 hours. At the time of injury, elevation of the periosteum from the bone surface occurs and extends for a variable distance above and below the fracture. This tearing away of periosteum is a strong stimulus for proliferation of the cells in its deeper layers, and microscopic evidence of new bone formation by these cells may be seen within 2 days after injury (Fig. 8-7). Since this new bone formation is taking place early at a distance from the fracture site, cells in the immediate vicinity of the fracture may produce a cartilaginous or fibrocartilaginous tissue. The amount of callus varies with the type, location, duration, age, and treatment of the fracture. In an undisplaced linear fracture, callus

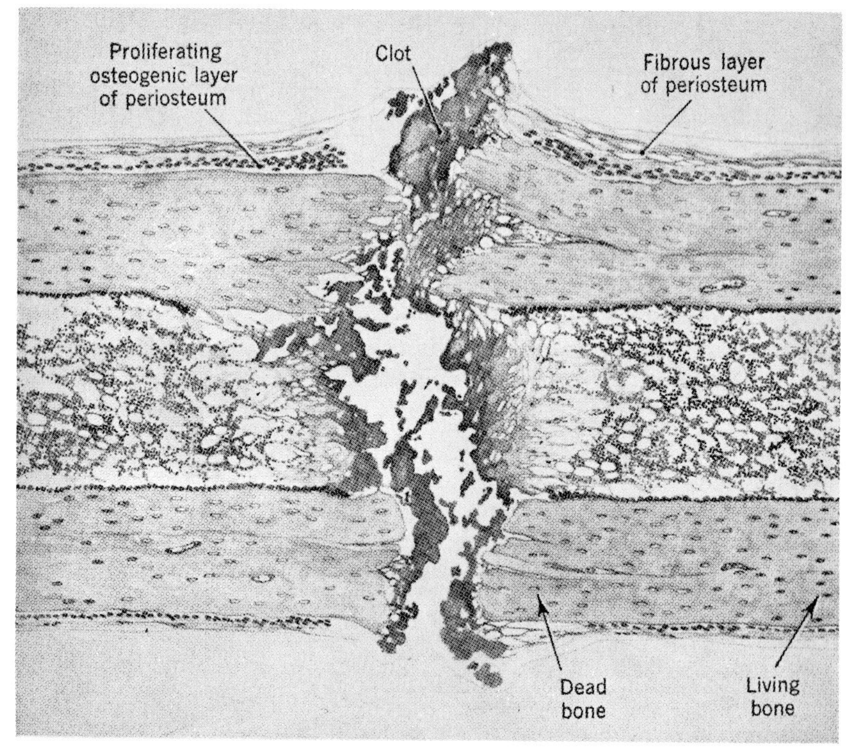

Fig. 8-7. Early fracture healing. Drawing of the microscopic appearance of a longitudinal section of a fractured rabbit rib in which healing had taken place for 48 hours. The drawing was made sufficiently out of perspective to permit both a considerable area and cell detail to be shown. (From Ham, A. W., and Harris, W. R.: In Bourne, G. H., editor: The biochemistry and physiology of bone, New York, 1956, Academic Press, Inc.)

may be minimal and is predominantly bone. Fractures of the shafts of young bones tend to produce an abundance of callus, especially if some movement takes place during healing. In early stages such callus may be predominantly cartilage. The presence of cartilage in the callus may result directly from motion at the fracture, or it may be secondary to the rapid and overabundant cellular proliferation resulting from such motion. When cellular proliferation is rapid, it may outstrip its blood supply. Cells of the osteogenic series tend to differentiate into bone when oxygen tension, which depends on the blood supply, is adequate and into cartilage when oxygen tension is reduced. Mature cartilage callus later undergoes changes (Fig. 8-8) similar to those seen in enchondral ossification at epiphyseal plates and is replaced by bone. Finally, the new bone becomes remodeled and adapted, according to Wolff's law (p. 56), to the stresses and strains to which it is subjected.

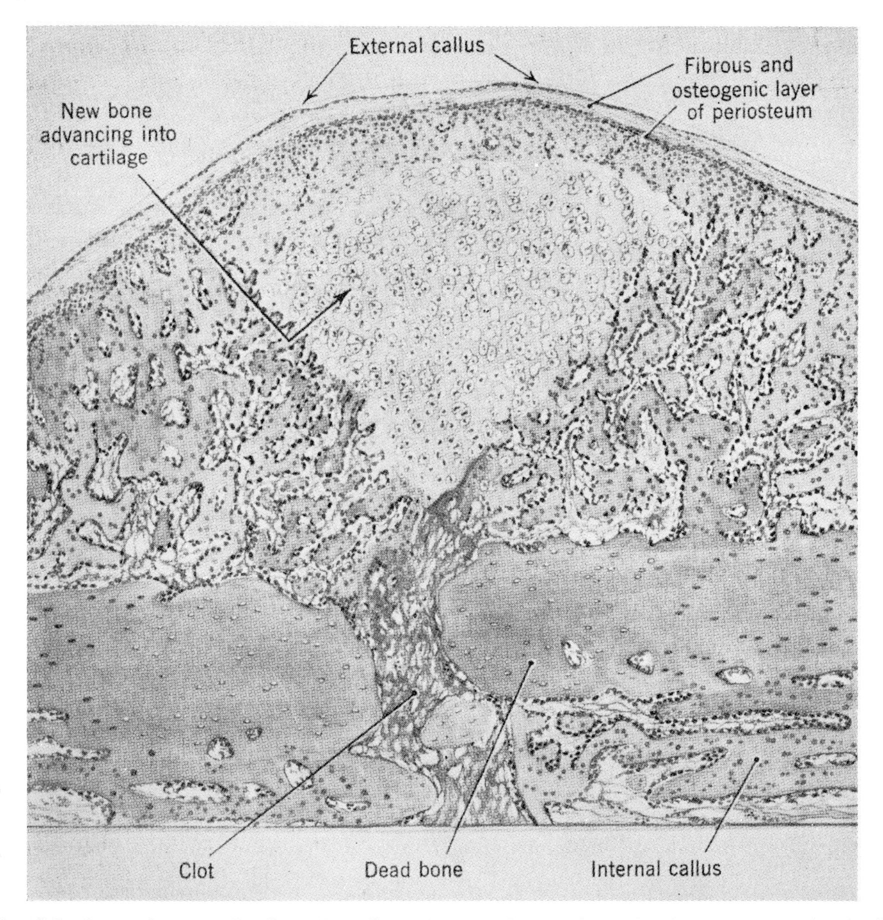

Fig. 8-8. Later fracture healing. Drawing of part of a section of a fractured rabbit rib that had healed for 2 weeks, illustrating the external callus to advantage and showing that the cartilage in it is being replaced by bone along a V-shaped line. Some clot, still unorganized, can be seen. (From Ham, A. W., and Harris, W. R.: In Bourne, G. H., editor: The biochemistry and physiology of bone, New York, 1956, Academic Press, Inc.)

The shape of the callus and the volume of tissue required to bridge a fracture depend upon the amount of bone damage and displacement. The healing time is directly proportional to the total volume of damaged bone and the breadth of the fracture defect. Some impacted fractures heal in a few weeks, whereas displaced fractures may require months or years for healing. With contact of the fractured bone ends, the humerus and forearm bones may unite in 3 months. The femur and tibia normally may require 6 months. Spiral fractures, with fragments of greater surface area, heal more rapidly than transverse fractures. The child generally produces more callus and heals a fracture faster than the adult. Beyond puberty, however, changes in age have much less influence on the rate of fracture healing.

Growing callus is calcified like cartilage and bone elsewhere in the skeleton. New bone ordinarily begins to calcify in the callus as soon as it is formed, provided that sufficient concentrations of calcium and phosphate ions are present in the blood plasma. In the process of calcification, minerals are transported to the area of ossification by the vascular system. They then pass to the extracellular fluid surrounding newly forming bone. The exact mechanism of calcium phosphate deposition within and about the collagen fibers of the bone matrix is poorly understood. In 1923 Robison noted the presence of alkaline phosphatase wherever bone formation takes place. This led to the hypothesis that the enzyme, by splitting phosphate esters, increases the local concentration of phosphate ions beyond the point of precipitation of calcium phosphate. This seemed to be a logical explanation of the calcification process, but analyses failed to show phosphate esters present in sufficient quantity to cause the precipitation. This and other objections have led to abandonment of Robison's theory of calcification. Current investigations point to some form of crystal seeding mechanism, apparently directed by specific molecular arrangements within the collagen fibers, as initiating the calcification process. For further study of this fascinating subject, reference should be made to the work of Neuman and of Glimcher.

The search for a substance that will accelerate the healing of fractures has motivated much research on the effects of minerals, vitamins, and hormones on callus formation. Thus far, no effective agent either to suppress or to stimulate bone repair has been found.

DELAYED UNION

Delayed union is said to be present when a fracture fails to consolidate in the time usually required for union to take place. In delayed union the processes of bone repair are retarded but are still going on and, with sufficient time, will of themselves produce firm union.

The period required for bone consolidation after fracture varies considerably in different individuals and under different circumstances. Delayed union is said to be present in the tibia if a closed fracture does not become clinically firm in 20 weeks; in the humerus, in 10 weeks; and in the femur, in 20 weeks. These are common sites of delayed union.

The causes of delayed union are the following:

1. Inaccurate reduction
2. Inadequate or interrupted immobilization

3. Severe local traumatization
4. Impairment of bone circulation following open operation
5. Infection, as in compound fractures
6. Loss of bone substance, such as might occur in compound fractures after free excision of devitalized fragments
7. Distraction or separation of fragments caused by excessive traction, a frequent sequel to too strong skeletal traction, or by improperly applied internal fixation

Treatment. Delayed union can best be avoided by (1) early, accurate, and gentle reduction and immobilization, uncomplicated by circulatory impairment from pressure of the splint or unnecessary surgical interference; (2) avoidance of the repeated traumatization of unnecessary attempts to perfect the alignment; (3) frequent observation of cases in skeletal traction to forestall separation of the fragments; (4) sound judgment and technic in the use of internal fixation; and (5) protection against undue strain on the fracture line.

The treatment of delayed union should be directed toward correction of recognizable factors that may be contributing to the delay. The type and quality of immobilization should be checked. An ill-fitting, inadequate cast should be replaced by a snug, well-applied cast including, in most instances, the joints above and below the fracture. Infection should be treated according to methods discussed in Chapter 4. Patients often ask whether they should drink milk or take calcium pills to speed healing. There is no evidence that supplementary dietary calcium has any effect on the healing rate of fractures. The body reserves of calcium are tremendous. The surfaces of bone crystals throughout the body are in constant equilibrium with the body fluids; this surface area has been estimated as greater than 200 square meters per gram of bone. Actually when a normally active person is immobilized in a cast, there is some generalized bone resorption, which creates an excess of mobilized calcium.

Delayed union calls for relatively long immobilization and for patience on the part of patient and physician. In delayed union of the tibia or femur, weight bearing in a walking cast or cast brace often seems to accelerate union.

NONUNION

Nonunion is present when the processes of bone repair, after having failed to produce firm union, have ceased completely. Unless this situation is changed radically by treatment, the nonunion will continue as a permanent and, in most instances, a severely disabling condition.

No fracture should be considered ununited until at least 6 months after the date of injury. Even after 8 months or more have elapsed, firm union will occasionally take place without surgical aid.

Etiology. Nonunion may result from the following:

1. Separation of the fragments
2. Loss of bone substance, resulting from extrusion or excision of fragments in open fractures
3. Inadequate fixation, allowing excessive motion at the site of fracture, such as might result from ineffective plating or nailing

4. Repeated manipulation to improve the position, causing injury to callus and its blood supply
5. Interposition of soft parts such as muscle and fascia between the fragments
6. Infection, which may develop after an open fracture or open reduction
7. Impairment of the circulation about the fracture by severe soft tissue injury, excessive periosteal stripping, and damage to the nutrient vessels at the time of the original injury or at open reduction

Pathology. In established nonunion the bone fragments may be connected by a fibrous or fibrocartilaginous tissue, or there may be a pseudarthrosis between the fragments with the formation of a thick bursal sac containing synovial fluid. The ends of the fragments usually consist of hard, sclerotic, eburnated bone; they may, on the other hand, become porous, atrophic, and cone shaped (Fig. 8-9). There may be excessive callus limited entirely to one fragment, usually the proximal.

Clinical picture. Mobility at the site of fracture varies widely. In some cases it is slight and hardly demonstrable, whereas in others, and particularly in pseudarthroses, there may be excessive movement in all directions. The pathologic mobility may be obvious when the patient attempts to move the limb. Occasionally motion elicits pain, but usually little discomfort is associated with the motion of a pseudarthrosis. In cases of nonunion in the lower limb there are often pain on weight bear-

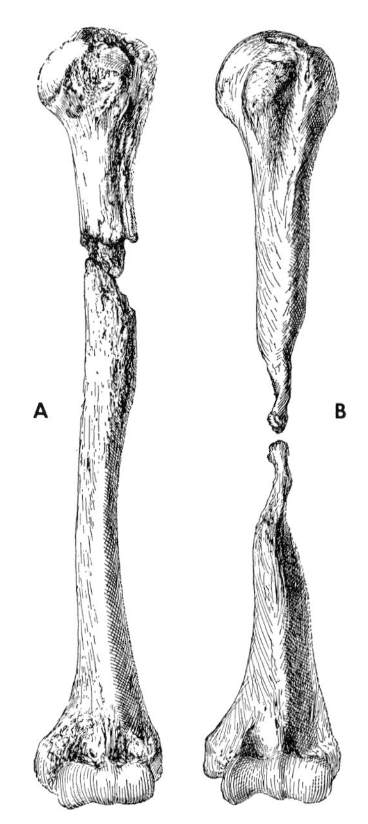

Fig. 8-9. Nonunion of fracture of the humerus. **A,** Common type with moderate bone absorption. **B,** Nonunion with extreme absorption. (Drawings from specimens.)

ing and some swelling after activity. Muscle atrophy is severe, and joint function may be limited. Especially great is the loss of mobility of the knee joint that may accompany an ununited fracture of the femur.

Treatment. The treatment consists of the use of braces or of surgery. Braces are entirely palliative and are used primarily to prevent deformity and to relieve strain and accompanying pain. In ununited fracture of the humerus a closely fitting laced leather cuff extending from the shoulder to the elbow will improve the function of the arm. A Thomas walking caliper splint fitted with a molded leather cuff for the thigh will brace an ununited fracture of the femur sufficiently to permit the patient to walk. A similar brace may be used for ununited fracture of the tibia; it should be fitted with a leather cuff that has been molded closely to the leg from knee to ankle.

When nonunion follows the frank infection of an open fracture, surgical treatment should usually be delayed 3 to 6 months after the healing of the wound since the development of infection in a wound containing transplanted bone often means complete failure of the surgical work. The use of antibiotics makes it possible to shorten the waiting period in some cases. It is sometimes advisable to perform the operation of bone grafting in two stages. In the first stage the potentially infected scar tissue should be thoroughly excised and the hard, eburnated ends of the bone fragments removed. Infected, scarred, or poorly nourished skin areas should be removed and replaced, preferably by a skin graft of pedicle, flap, or split-thickness type. If healing takes place without signs of infection, a bone graft operation may be performed several weeks later.

Most ununited fractures require a bone transplant before union can be obtained. When the bed is being prepared for the graft, all surrounding scar tissue and sclerotic bone should be removed until healthy bleeding occurs from the bone ends and from the surrounding soft tissues.

What happens to transplanted bone has been and continues to be extensively studied. In the case of autogenous cortical bone transplants it is known that the vast majority of the osteocytes die. Perhaps a very few of the transplanted cells survive, but new bone is formed by cells already near the fracture site. The graft becomes incorporated by this new bone and is invaded by blood vessels. Over a period of time the necrotic bone of the graft is resorbed. New bone may be formed as rapidly as the graft is removed, the fracture being bridged by this new bone as it replaces the graft. Under unfavorable circumstances graft resorption is not accompanied by new bone formation, and the nonunion persists. In transplants of autogenous cancellous bone it is possible that more of the transplanted bone cells survive because they are more accessible to the blood supply at the fracture site. Osteoblasts on the surfaces of the trabeculae of the cancellous graft may take some part in the new bone formation that bridges the fracture.

Autogenous bone is superior to homogenous bone for transplantation when satisfactory grafts can be obtained. However, in some instances insufficient autogenous bone is available and it is necessary to use another source. Bone banks are such a source. Bank bone is taken under sterile conditions from the cadaver, from amputated limbs, or from excess bone removed at operation; it is preserved by freezing or freeze-drying in sterile containers. Commercially processed heterogenous bone has also been used but with limited success.

Several types of bone graft are in common use. Choice of the most effective graft for the individual case depends largely upon the site of nonunion, the condition of the ends of the bone fragments, and the training and experience of the orthopaedic surgeon. For a *cortical graft,* bone is usually obtained from the tibia and may be fixed across the nonunion by means of screws above and below (Fig. 8-10). This type of graft adds stability at the fracture site. For a *cancellous graft,* bone is ordinarily taken from the ilium and packed in cancellous strips about the nonunion site; additional stability may be obtained through the use of a metal plate or intramedullary nail. The *subperiosteal onlay graft* of Phemister is excellent for certain cases with little displacement of the fragments. The periosteum is lifted on both sides of an ununited fracture, which has been approached through healthy tissue, and a rigid graft

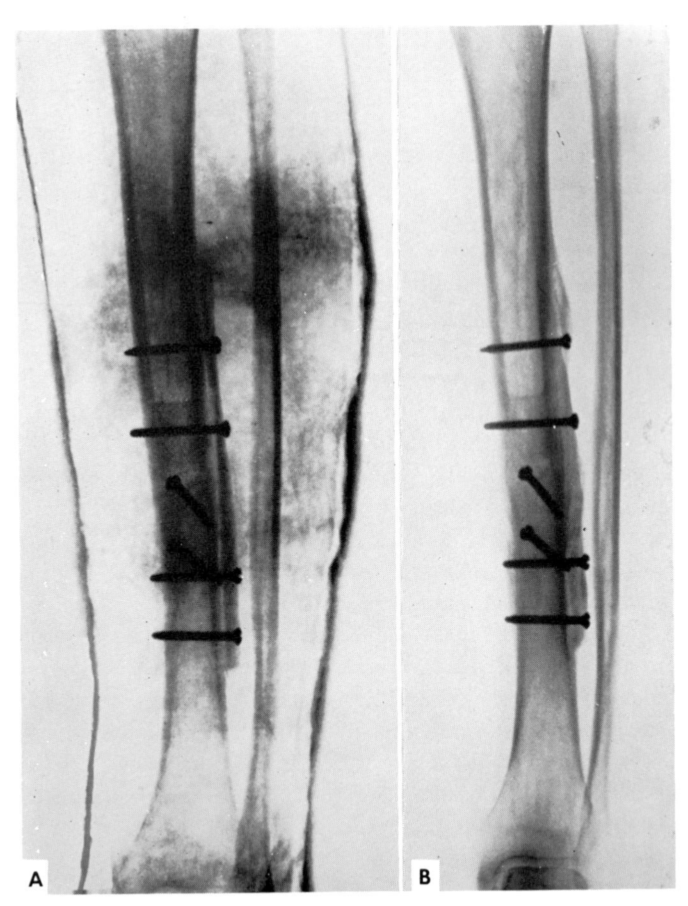

Fig. 8-10. Roentgenograms of onlay bone graft for ununited fracture of the tibial shaft. **A,** Massive cortical graft from the proximal half of the tibia has been fixed across the fracture site with two transverse screws in each fragment. The two middle screws transfix the oblique ends of the fracture fragments. Note cast, which is essential for further immobilization during the healing. **B,** Six months after operation. Note consolidation of the onlay graft with the tibial shaft.

7 to 12 cm. in length for an ununited shaft is inserted beneath the periosteum. The nonunion site is not disturbed, and no internal fixation is used. The *osteoperiosteal graft* is usually taken from the anteromedial surface of the tibia or from a rib in such manner as to include the periosteum and a thin layer of small adherent chips of cortical bone. The particular advantage of this graft is that it is pliable and can be molded to fit the contour of the fragments and the defect.

Many ununited fractures require metallic internal fixation at the time of bone grafting. In addition the fragments should usually be immobilized by means of ample and well-molded plaster splints or a cast. Bleeding from the denuded bone surfaces always occurs, and care must be taken postoperatively that this does not cause excessive pressure and ischemia. A properly padded cast, elevated promptly and continuously after operation, does not usually require splitting; if it must be cut to relieve pressure, complete temporary bivalving is the safest method. After grafting of a major bone of an extremity, a cast should be worn until there is roentgenographic evidence of enough bone formation to ensure union. This period varies with the site of the graft and the bone involved; it is usually from 12 to 24 weeks. The plaster should then be bivalved and active exercises started.

CONGENITAL FRACTURES AND NONUNION (CONGENITAL PSEUDARTHROSIS)

A defect in continuity resembling an ununited fracture or pseudarthrosis is occasionally seen roentgenographically in the bone of an infant immediately after

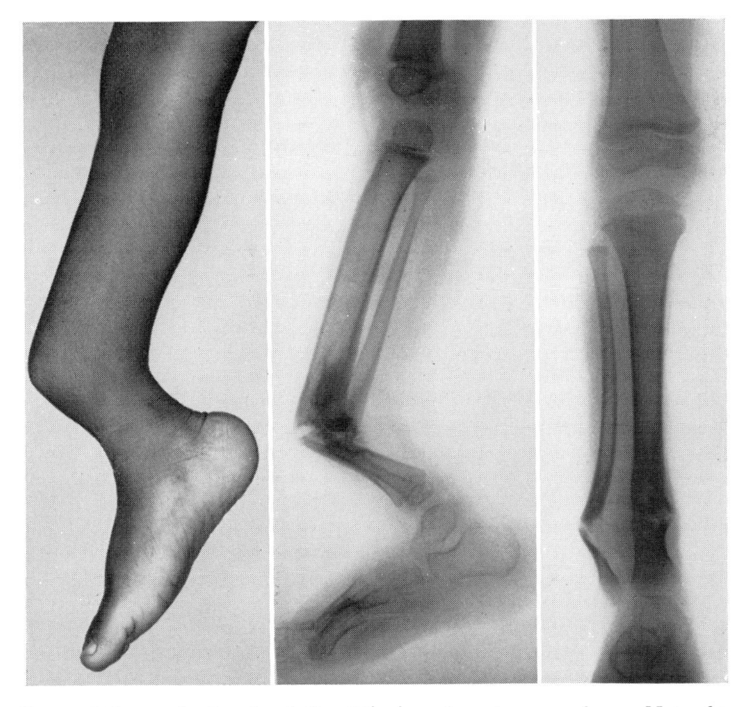

Fig. 8-11. Congenital pseudarthrosis of the tibia in a boy 4 years of age. Note sharp anterior angulation, changes in bone density at the nonunion site, and atrophy of the bones of the foot.

birth. Lesions of this type can be differentiated by means of roentgenograms from fractures occurring during delivery. Their etiology is unknown. They may result from a congenital deficiency of bone production. These fractures are rare, are seen in the tibia much more often than in other bones (Fig. 8-11), and unfortunately are very resistant to treatment. Poorly united congenital fractures are of two main types: (1) those that are quite unstable at birth and (2) those that present a weak union or malunion and are very easily refractured.

Attempts to induce union of these fractures, especially those of the first type, have often been unsuccessful. The dual allograft of Boyd, consisting of two onlay grafts spanning the defect on opposite surfaces of the fractured bone, allows early operation and occasionally has given good results. In some patients union has followed the operation of multiple osteotomies, reversal of the fragments, and intramedullary rod fixation advocated by Sofield. Sometimes, after repeated failure of bone-grafting procedures, the patient is left with a shortened, deformed leg that is best treated by below-knee amputation.

TREATMENT OF MASSIVE DEFECTS OF THE LONG BONES

Extensive loss of continuity in the shaft or at one end of a long bone is seen occasionally as the result of a severely comminuted open fracture, radical resection of bone for a malignant tumor, failure of bone regeneration in osteomyelitis, or a congenital anomaly. The replacement of a large defect in the continuity of the bone or the creation of a new bony structure to replace the absent end of the bone constitutes a problem much more difficult than the ordinary bone grafting for nonunion. In some instances replacement of the end of a long bone by a metal endoprosthesis may be preferable to bone transplantation. Such cases must be considered individually, and great care must be exercised in selecting the operation most suitable for improving the function of the disabled limb.

It is essential that certain conditions be fulfilled before the reconstructive bone operation is performed. Any infection that may be present must first be eliminated. Poorly nourished skin and scars must be replaced with healthy tissues by means of plastic surgical procedures. The nutrition, mobility, and strength of the limb must be restored as nearly as possible to normal by means of exercises, and the general physical condition and motivation of the patient must be good.

The technic of the reconstructive bone operation varies with details of the individual case. In the lower limb extensive defects may often be bridged successfully by a massive sliding onlay or inlay graft, by a massive graft from the tibia of the opposite leg, or by use of the shaft of the fibula as a graft. A dual graft is often indicated. In most instances the cortical graft should be supplemented by the addition of a considerable amount of cancellous bone; this may be taken conveniently from the ilium. The ribs also form a useful source of autogenous grafts. When the tibial shaft is involved, it is sometimes advisable to transfer the shaft of the fibula into the ends of the tibia, above and below (p. 267), without exposing the tibial defect.

In the upper limb, length is not an essential consideration. It is often advisable to approximate the ends of the fragments rather than attempt the spanning of a large defect in the humerus. The entire upper part of the fibula may be used to replace a lost portion of the humerus or radius. In rare instances of extensive defect of the shaft of the radius, a transference of the distal end of the radius to the shaft of the ulna may be indicated.

COMMON SITES OF NONUNION

Femoral neck

Although nonunion of femoral neck fractures is much less common than before the introduction of technics of internal fixation, it still occurs in approximately 15% of the cases (Figs. 8-12 and 8-13). The reasons for nonunion of the neck of the femur are (1) an anatomically meager blood supply and (2) difficulty in securing accurate approximation and rigid fixation of the fragments.

The outstanding manifestations of nonunion of fracture of the neck of the femur are (1) pain in the hip on weight bearing; (2) shortening and, as a rule, external rotation of the limb; and (3) grating in the hip on motion. Additional complicating factors that may be present include avascular necrosis of the femoral head, which appears roentgenographically as density changes and collapse of the head, and osteoarthritis, which may restrict the mobility of the head in the acetabulum.

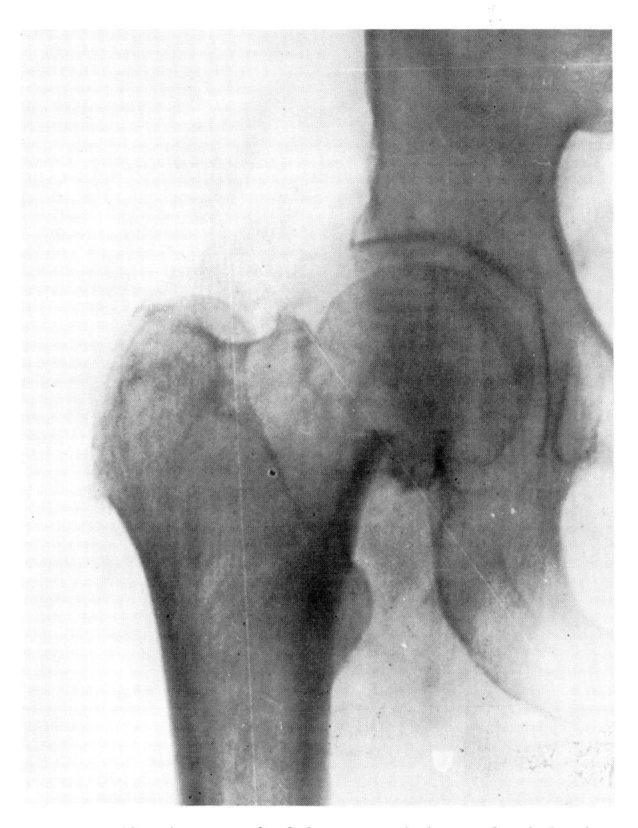

Fig. 8-12. Roentgenogram showing ununited fracture of the neck of the femur. Note upward displacement of the distal fragment. Similar density of the head and neck suggests at least partial viability of the head.

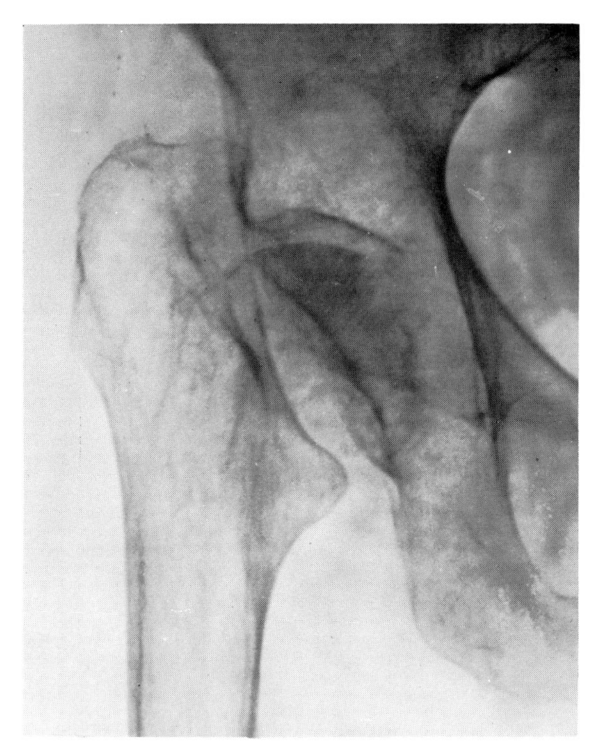

Fig. 8-13. Roentgenogram showing ununited fracture of the neck of the femur with marked upward displacement of the distal fragment 4 years after unsuccessful nailing of the fresh fracture in a patient 80 years of age. The femoral neck has disappeared. Relatively dense areas of the head suggest nonviability.

Treatment. Operation is required, and no one procedure is suitable for use in all cases. The patient's general condition and the details at the fracture site must be given careful consideration. In individuals under 60 years of age who are in good condition and who do not show severe avascular necrosis of the femoral head or significant absorption of the distal fragment, a *bone peg* operation may be followed by solid union. In the Albee type of bone peg operation the hip joint is opened, the fibrous tissue is removed, and the fragments are freshened on either side and approximated. A peg graft taken from the crest of the tibia is then inserted through the greater trochanter far into the neck and head. For additional strength a three-flange nail may be inserted alongside the graft. The results of this operation in selected cases have been satisfactory in leading to union and return of function.

When the fragments are not widely displaced and the femoral head is viable, an osteotomy just above the lesser trochanter with medial displacement of the distal fragment, according to technics described by McMurray, Dickson, Blount, and others, may be indicated. In many of the cases such osteotomies result in union of the hip fracture, relief of pain, and good hip function.

When the patient is over 70 years old, it may be advisable to abandon the effort to secure union and to replace the femoral head with a long-stemmed metal prosthesis (Fig. 5-20). This operation has the advantage of allowing the patient to return quickly to walking. However, normal hip function is usually not regained; the patient should be advised preoperatively that for most walking a cane or a crutch will be required.

Femoral shaft

Nonunion of the shaft of the femur is fairly common. If the fragments are overriding, it may be best to begin the treatment with a period of strong skeletal traction. When the fracture is near the middle of the shaft, the usual surgical treatment consists of trimming the fragment ends to bleeding bone, securing their approximation and fixation with an intramedullary nail, and surrounding the fracture site with multiple cancellous grafts taken from the ilium. For additional fixation a spica cast, extending down to the toes of the affected limb and to the knee on the other side, is often necessary.

Tibia

In the experience of many orthopaedists the lower third of the tibia is the commonest site of nonunion (Fig. 8-14). A contributing cause may be meager circulation over the subcutaneous medial surface of the tibia. In some cases the intact or healed fibula may prevent close approximation of the tibial fragments. A sliding inlay graft is often used at this site, or a large onlay graft may be placed on the posterior or lateral aspect of the tibia, which makes it possible to cover the transplant with healthy muscle tissue. In some cases of nonunion of the tibia, stabilization is obtained readily by means of a heavy compression plate. The freshened bone ends are forced tightly together and held securely by strong screws. In addition to the compression plating, cancellous bone grafts should be packed about the nonunion.

Surgical transference of the shaft of the fibula into the tibia above and below the level of the ununited fracture is sometimes preferable to the use of a free graft. This transference is particularly useful if the tibial defect is large or chronically infected as a sequel to open fracture or osteomyelitis. The operation is usually carried out in two stages. The postoperative cast, which is worn for several months, must often be followed by a brace to protect the fibular transplant until it has undergone sufficient hypertrophy to withstand the stresses normally placed upon the tibia.

Ununited fracture of the tibial malleolus should be treated surgically only if causing symptoms. If the malleolar fragment is quite small, it may be excised. If larger, it should be replaced and fixed by a screw; the addition of a small bone graft is often helpful.

Humerus

Ununited fractures of the middle third of the humerus are relatively common and are usually the result of inadequate initial approximation and immobilization of the fragments.

Nonunion of the shaft of the humerus is often treated by the *step-cut operation,* a splicing of the freshened and fitted ends of the fragments, plus internal fixation by means of screws, sutures, or an intramedullary nail. Some surgeons prefer simple osteotomies and internal fixation. The compression plate may be used. The addition

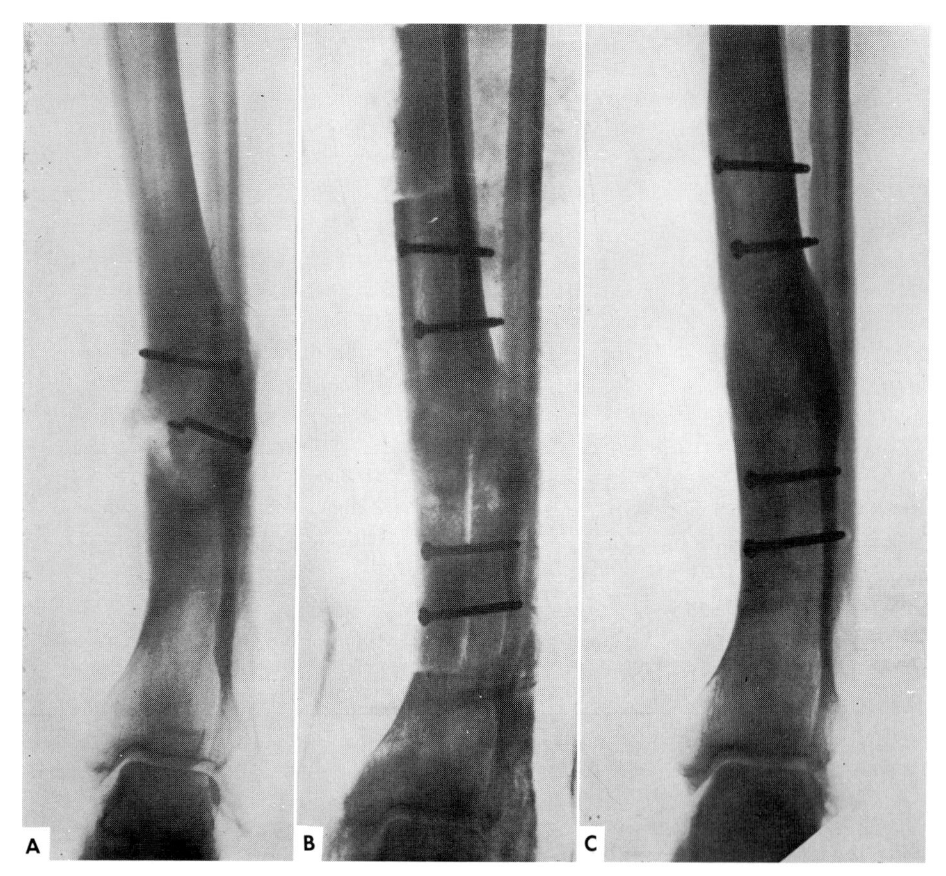

Fig. 8-14. Roentgenograms of an ununited fracture of the tibial shaft treated by a massive cortical bone graft. **A,** One year after fracture. Failure of previous treatment is indicated by nonunion, angulation, extensive sclerosis of fragment ends, and broken screws. **B,** Several days after graft operation, which consisted of taking a massive segment from the proximal part of the tibial shaft, placing it in a bed prepared across the old fracture site, and fixing it with two screws in each fragment. **C,** Eighteen months after graft operation. Note strong bone healing throughout the tibial shaft.

of autogenous iliac grafts is advisable. In difficult cases massive onlay grafts fixed by screws form an excellent means of inducing union.

Ununited fracture of the lateral condyle of the humerus, with resulting valgus deformity of the elbow, should be treated by replacing the condyle and fixing it with a screw, nail, or threaded Kirschner wire. Small osteoperiosteal grafts placed posteriorly are often advisable to facilitate bony union in this location. Small ununited fragments of the lateral or the medial condyle that cause pain and grating in the elbow should be excised.

Radius and ulna

For nonunion of both bones of the forearm the fixation of cortical onlay grafts, done through separate incisions to avoid radioulnar synostosis, is usually best effected with screws. Alternatively, alignment of the fragments can often be maintained conveniently by the insertion of an intramedullary nail or pin, after which cancellous strips of iliac bone are packed about the fracture sites. Ununited fracture of the radius or the ulna alone is more common. If the fracture is in the lower third of the radius, an inlay or onlay graft may be used, the distal end of the graft being sharpened and driven into the cancellous bone of the distal fragment. Ununited fracture of the lower third of the radius is often accompanied by an unsightly prominence of the lower end of the ulna, which may be resected.

Scaphoid bone

Ununited fracture of the scaphoid bone occurs frequently, resulting presumably from lack of immobilization and from poor blood supply; the scaphoid has three articulating surfaces, and only two small nonarticulating areas through which the blood vessels enter. Many wrists thought to be chronically sprained will on roentgenographic study show an ununited fracture of the scaphoid. With several months of immobilization by means of a plaster cast that includes the thumb and holds the wrist in slight dorsiflexion and radial deviation, many old fractures of the scaphoid bone will unite. When union cannot be obtained by immobilization alone, bone grafting may be indicated. In some instances excision of the small proximal fragment or excision of the radial styloid process leads to improved function and less pain in the wrist. If a nonunion of long duration is attended by pain and arthritic changes in the adjacent bones, arthrodesis of the wrist may be indicated.

MALUNION

Union in poor position is usually caused by (1) inability to secure accurate reduction or (2) inability to maintain effective immobilization for a sufficient length of time. Often malunion is preventable, but in some situations, such as the management of a patient with multiple severe injuries, it may be impossible or inadvisable to treat a particular fracture in an ideal manner. As a compromise, some degree of malunion must be accepted. The principal evidences of malunion are shortening of the limb due to overriding of the fragments (Fig. 8-15), deformity from angulation of the fragments, abnormal rotation of the fragments with relation to each other, and restriction of joint motion by a bone block.

COMMON SITES OF MALUNION

Malunited fracture may occur in any of the bones. The more common and clinically important malunions warrant individual description.

Femur

Malunion of intertrochanteric fractures of the femur is common. It may take place with the thigh externally rotated and adducted, with resultant coxa vara, retroversion of the femoral neck, and shortening. The deformity causes a limp and a tendency to tire quickly. Its correction requires an intertrochanteric osteotomy.

After fracture of the shaft of the femur, anterolateral angulation or rotation of the shaft may develop as a result of weight bearing before the callus has become firm. It is sometimes possible to correct this type of deformity by manipulation. If the fracture has become solid, an osteotomy is indicated. An intramedullary nail may be

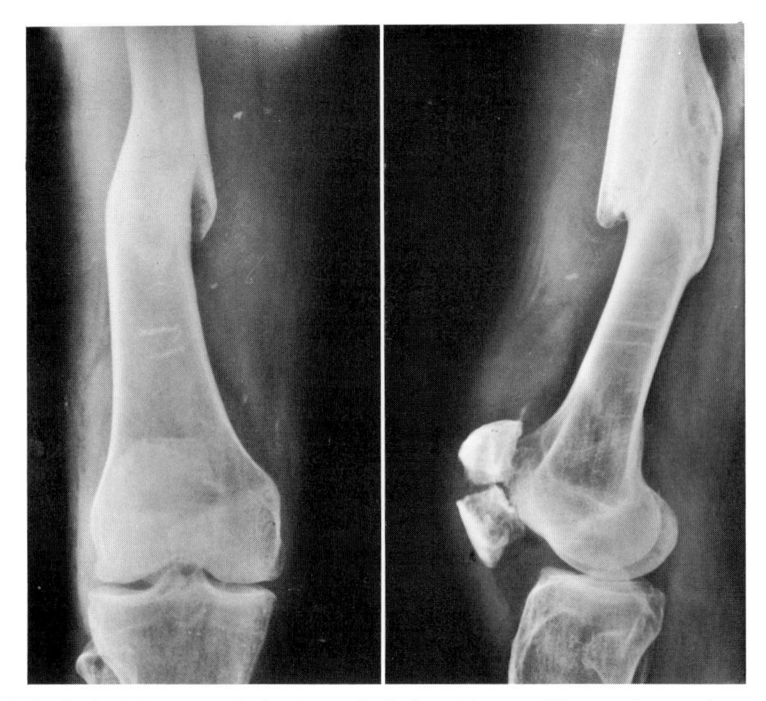

Fig. 8-15. Malunited fracture of the femoral shaft, with overriding and posterior angulation. Adherence of the quadriceps muscle at the fracture site severely limited this patient's knee motion, and when his knee was forcibly flexed in a second injury, the patella suffered this fresh transverse fracture.

used to maintain the corrected alignment, and cancellous grafts from the ilium are usually indicated. It is advisable to mobilize the stiffened joints before operation by the use of physical therapy. When overriding has caused appreciable shortening (Fig. 8-15), it may be best to perform an osteotomy at the old fracture line, use skeletal traction until the shortening has been overcome, and then carry out open reduction, internal fixation, and grafting.

After fracture of one of the condyles of the femur, a varus or valgus deformity of the knee may develop. In many instances this can be corrected satisfactorily by supracondylar osteotomy.

Tibia and fibula

Occasionally a depressed fracture of a tibial condyle leaves irregular or displaced joint surfaces and abnormal weight-bearing lines, giving rise to pain and disability on weight bearing. It may be possible to elevate and fix the depressed fragment and to remove the irregularities about the tibial plateau. After the fragment has been elevated, the space beneath it should be packed with bone chips. In less favorable cases arthrodesis or arthroplasty of the knee is indicated. The type of operation should depend largely upon the age, physical condition, temperament, and occupation of the patient.

Fractures of the tibial and fibular shafts may unite with unsightly and disabling

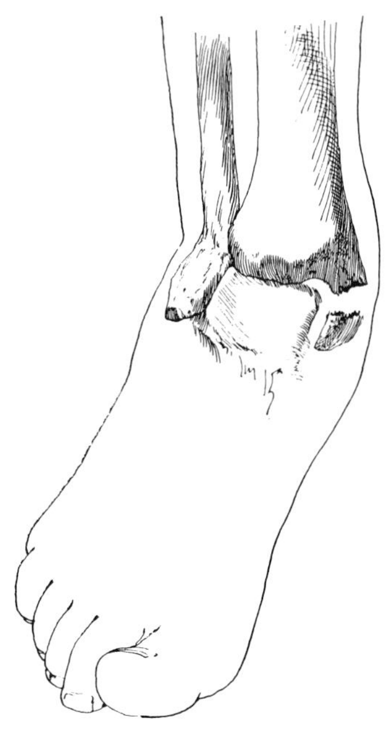

Fig. 8-16. Old fracture-dislocation of the ankle with malunion of fibular fracture, nonunion of fractured tibial malleolus, and abduction and lateral displacement of the foot (malunited Pott's fracture).

angulation. In order to correct the deformity, osteotomy may be performed through the site of angulation. After osteotomy of the middle or lower thirds, internal fixation and grafting are advisable. With moderate angulation a supramalleolar osteotomy to correct the angle of the ankle joint surface to the weight-bearing axis of the leg may be preferable. A rotation deformity of the foot and leg below the site of fracture sometimes follows failure to align the fragments accurately at the original reduction. Such rotation deformity can be corrected by osteotomy.

Ankle

Malunion of fractures of the lower end of the tibia and fibula results in severe disability. The foot usually becomes fixed in marked valgus with abnormal prominence of the medial malleolus (Fig. 8-16). The Achilles tendon may become contracted. Pain on weight bearing is severe. At the original reduction of a fracture of the lower end of the tibia and fibula, the talus should be so placed against the medial malleolus that the longitudinal axis of the tibia bisects at a right angle the transverse diameter of the upper articular surface of the talus. If this anatomic relationship can be maintained and there is minimal injury of the articular surfaces, permanent disability seldom results.

In cases of malunion with only slight displacement, the pain and disability may be relieved by a longitudinal arch support and ankle bandages. Such treatment, however, is only palliative. In adults most malunited fractures involving the lower articular surfaces of the tibia and fibula require open operation.

In cases of only a few weeks' standing it may be possible to break the beginning union and correct the deformity by manipulation. In cases of longer duration the fractured surfaces should be separated at open operation. Care must be taken to examine the posterior margin of the tibia (posterior malleolus), and if a large fragment is significantly displaced its position should be corrected. It is usually necessary to retain the corrected position of the fragments by means of a screw or nail.

When the malunion is less than 6 months old and traumatic arthritic changes have not yet become obvious, a reconstruction operation is sometimes indicated. It may be necessary to perform an osteotomy at the old fracture line in the fibula before the talus and foot can be mobilized and brought into alignment. For associated nonunion of a fracture of the tibial malleolus it may be necessary to excise the pseudarthrosis and fix the fragment to the tibia with a screw or nail. The Achilles tendon may require lengthening.

The persistence of severe pain in a patient over 45 years of age, due largely to traumatic arthritis, is an indication for arthrodesis of the ankle. After reconstruction operations there is always the danger of traumatic arthritis; arthrodesis, producing a stiff but painless ankle, is in many instances the preferable procedure. Arthrodesis involves removal of all joint surfaces, bone grafting, and immobilization of the ankle with the foot in functional position.

Calcaneus

Malunion of fractures of the calcaneus results in broadening of the heel, piling up of bone beneath both malleoli, especially the lateral, abduction deformity of the foot, and in many instances, flattening of the longitudinal arch. The patient seldom regains the normal range of inversion and eversion of the foot.

Longitudinal arch supports, Thomas heels elevated on the medial side, and ankle supports may afford partial symptomatic relief. Surgical treatment is more effective. Arthrodesis of the subtalar joint, with removal of enough bone to correct the weight-bearing alignment, is sometimes advisable. Subtalar motion will be lost, but the patient should walk with less pain. Occasionally pain resulting from malunion of a calcaneal fracture subsides gradually and spontaneously over a period of years without surgical treatment.

Humerus

Restriction of shoulder abduction caused by malunion of a fracture of the greater tuberosity of the humerus may be relieved by partial resection of the acromion, or *acromioplasty*. It is only occasionally necessary to perform an operation for malunion of the shaft of the humerus. An osteotomy may be indicated to correct an extreme angulation or rotation deformity.

The lower end of the humerus and the elbow are often involved in malunited fractures. Unreduced or incompletely reduced supracondylar fractures may restrict joint motion. In adults, bony spurs that project anteriorly into the joint and limit its range of motion may be resected. The operation should be followed by a program of supervised active exercise. The increase in range of motion is often slow.

In children, unreduced fracture of the lateral condyle is usually followed by the development of a severe valgus deformity of the elbow (Fig. 8-17, *A*). Delayed ulnar palsy, or late traumatic ulnar neuritis, from friction or stretching of the nerve (p. 204) frequently accompanies a cubitus valgus. For correction of the deformity a supracondylar wedge osteotomy of the humerus with replacement or removal of the detached lateral condyle is often necessary (Fig. 8-18). Varus deformity of the elbow (Fig. 8-17, *B*), which often follows an incompletely reduced supracondylar fracture, is sometimes an indication for a similar type of osteotomy.

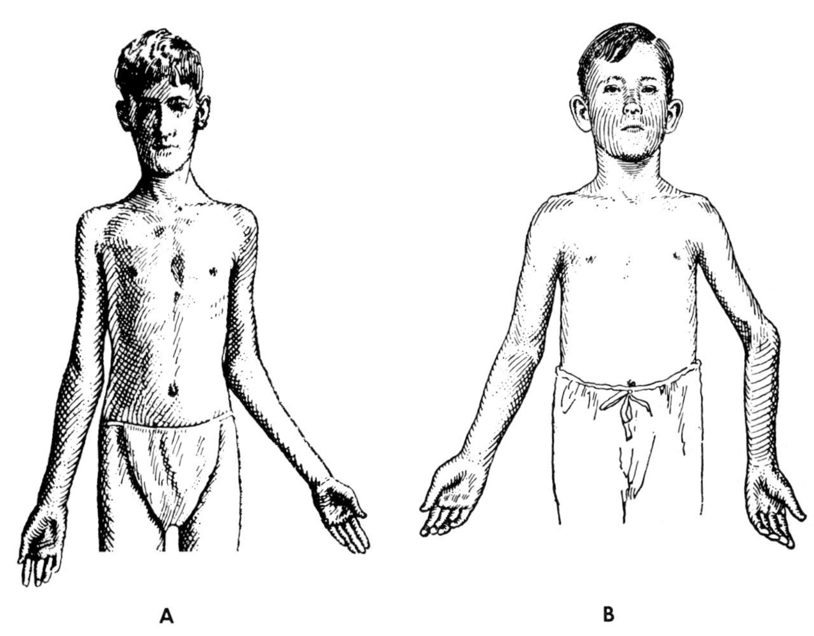

Fig. 8-17. A, Malunited fracture of the lower end of the left humerus, showing cubitus valgus. **B,** Malunited fracture of the lower end of the left humerus, showing cubitus varus.

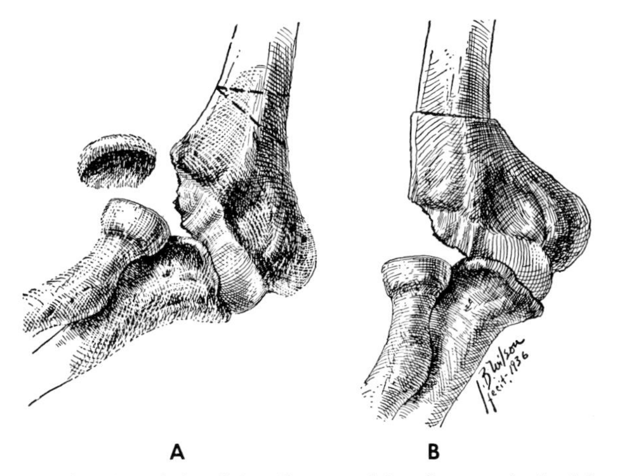

Fig. 8-18. A, Valgus deformity of the right elbow resulting from malunited fracture of the lower end of the humerus and ununited fracture of its lateral condyle. **B,** After operation, which included removal of the detached lateral condyle and a supracondylar wedge osteotomy.

Occasionally, after fracture about the elbow joint, a mass of new bone may form in the olecranon fossa and prevent normal extension of the elbow. Removal of the excess bone may be followed by improvement of mobility. Occasionally, after a severely comminuted fracture involving the joint surfaces, complete or partial ankylosis develops. For this type of lesion an arthroplasty or reconstruction operation may be indicated.

In association with fracture or dislocation about the elbow joint, a mass of bone may form in the brachialis muscle (*traumatic myositis ossificans,* p. 429). This bony mass may be directly connected with the periosteum of the lower end of the humerus. If spontaneous resorption fails to take place after prolonged rest and conservative therapy, the mass of new bone should be excised with as little traumatization of the surrounding tissues as possible. If the excess bone is removed too soon, it may recur; excision should not be attempted until at least 6 months after the injury.

After operations on the elbow the recovery of joint motion is notably slow.

Radius and ulna

Malunited fracture of the head of the radius may cause severe disability. Rotation of the forearm is restricted and painful, and extension of the elbow may be impaired. In such cases the head and neck of the radius should be completely excised. Small comminuted fragments of the head of the radius may have become lodged in the joint and at operation must be carefully removed. Because of the disturbance of forearm growth, the head of the radius should not be excised in children.

The combination of fracture of the upper third of the ulna and dislocation of the head of the radius, first described by Monteggia in 1814, is frequently seen. This type of injury may be followed by an unsightly deformity due to outward angulation of the upper third of the ulna. In such cases it is advisable to osteotomize the ulna, correct the alignment, and fix the fragments with a massive or dual onlay graft, an intramedullary pin, or a metal plate. The dislocation of the head of the radius should be reduced, especially if the patient is a child, and a new annular ligament should be constructed to maintain the reduction. When local conditions in the adult are unfavorable for the fashioning of such a ligament, the head of the radius may be excised.

Malunion or even nonunion of fractures of the olecranon usually results in little or no limitation of elbow function.

Fracture of both bones of the forearm may be followed by a radioulnar synostosis, which prevents rotation. In such cases the excess bone may be resected and fascia or fat placed between the radius and the ulna. In late cases, contractures and muscle atrophy jeopardize the postoperative result.

When fracture of both bones has been followed by severe angulation and overriding but firm union, osteotomies at the sites of angulation may be indicated for correction of the deformity. It is necessary to hold the fragments in apposition by means of massive onlay bone grafts, intramedullary pins, or metal plates.

Wrist

In the reduction of a Colles' fracture, whether recent or old, particular attention should be paid to the alignment of the joint surface of the radius. Normally, a line drawn anteroposteriorly across the articular surface of the lower end of the radius forms anteriorly an angle of from 70 to 80 degrees with the longitudinal axis of the radius, and a line drawn laterally across the articular surface of the lower end of the radius forms radially an angle of from 105 to 110 degrees with the longitudinal axis of the radius. These relationships must be restored before a Colles' fracture can be considered adequately reduced. The commonest cause of a poor functional result is failure to correct adequately the deformity of radial deviation.

Fibrosis and stiffness of the metacarpophalangeal and interphalangeal joints are often observed after immobilization of the wrist for a fresh Colles' fracture and may cause greater disability than the fracture itself. This is especially true in older people. It is essential to start motion of the fingers and thumb soon after the injury. In fractures about the wrist joint, and especially in Colles' fracture, it is important that the splints should not be carried posteriorly beyond the knuckles or anteriorly beyond the middle of the palm. If the fracture has been properly reduced and immobilized, the danger of slipping is not significantly increased by movement of the fingers and thumb. Colles' fracture in old people is commonly followed by stiffness of the elbow and shoulder. During the course of treatment these joints should be put through their full range of motion actively once or twice a day.

Unreduced Colles' fracture may cause an unsightly deformity. There is usually a silver-fork deformity, marked prominence of the lower end of the ulna, and radial deviation of the wrist (Fig. 8-19). The deformity may be accompanied by pain and disability.

If less than 5 weeks has elapsed since fracture, it may be possible to restore satisfactory alignment by closed reduction. When the deformity cannot be reduced easily by manipulation, a wedge of bone may be removed from the volar surface

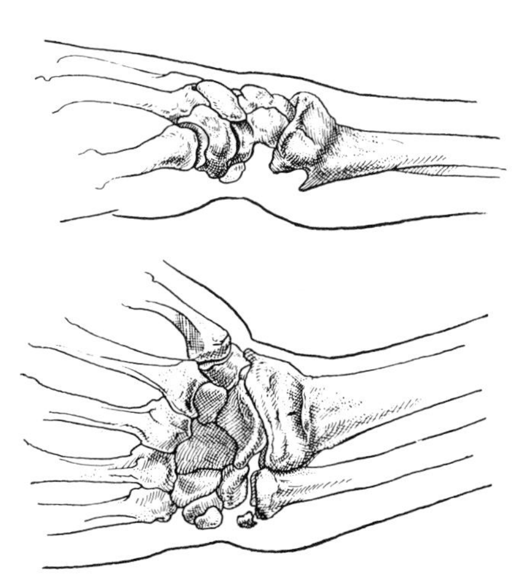

Fig. 8-19. Malunited fracture of the lower end of the radius (malunited Colles' fracture). Note dorsal displacement of the distal fragment, posterior tilting of the articular surface of the radius, radial deviation of the hand, prominence of the distal end of the ulna, and ununited fracture of the ulnar styloid process.

of the lower end of the radius and the alignment corrected. Internal fixation with wire or pin is useful to prevent postoperative slipping of the small distal fragment. When there is undue prominence of the lower end of the ulna and especially when it is associated with pain, restriction of pronation and supination, and shortening of the radius, the projecting bone should be removed at the same operation. In old, inactive persons, especially when the nondominant hand is involved, moderate deformity from malunited Colles' fracture causes little disability and requires no treatment.

RESIDUAL DISABILITY

Malunion of a fracture leaves the patient with some residual disability. If this is excessive, surgical correction may be necessary as noted earlier. In many instances it is necessary to accept a certain degree of permanent disability. Even a fracture that has healed in excellent position can be followed by some limitation of joint function. This is particularly true in older people. Following a fracture most functional return takes place within the first 6 months after bony union has occurred. After this period improvement is slow, but it may continue for a year or more. During this entire time the patient has a disability that may be considered either partial or total but one that is improving. After maximal functional recovery has been reached the patient may be left with a residual permanent disability. Today many injuries are financially compensable through liability insurance or workman's compensation insurance. It is the obligation of the patient's physician to evaluate the disability in order that the patient may be recompensed for his loss of function. Accordingly the physician who treats fractures must be familiar with conventional methods of evaluating disability. Excellent reference books and tables on disability evaluation are available through libraries and medical societies. Several are listed in the bibliography.

CHAPTER 9 Amputations, prostheses, braces

Familiarity with the closely related subjects of amputations, prostheses, and braces is important to the physician treating disorders of the musculoskeletal system. Injuries, infections, congenital deformities, diseases of the central nervous system or peripheral vascular system, neoplasms, and arthritis frequently cause disabilities that can be significantly lessened through the proper use of braces and prostheses. Gaining an understanding of the mechanical principles of these devices and knowledge of the normal and abnormal biomechanics of the skeletal system are essential parts of orthopaedic education.

AMPUTATIONS AND PROSTHESES

Amputations may be the result of congenital limb deficiencies (Fig. 9-1) or may be acquired. Acquired amputations are traumatic, ischemic, or surgical in origin. Surgical amputation is an ancient procedure. At first glance amputation may appear a purely destructive measure, incompatible with recognized orthopaedic principles of physical conservation and restoration. Usually, however, amputation is only the essential first step in a series of measures designed to achieve finally the maximum rehabilitation of the patient. The surgeon who plans an amputation should be prepared not only to execute the surgical operation but also to deal—early and late—with the patient's psychologic problems, to direct efficient postoperative care including physical therapy and preparation of the stump for use of the prosthesis, to supervise the proper selection and fitting of the prosthesis and adequate training in its use, to guide the patient's return to employment, and to carry out a medical follow-up that may last for many years. Such considerations emphasize the close relationship that exists between amputations and reconstructive orthopaedic surgery.

Indications

Surgical amputation should be done when in the judgment of physician and patient the patient's welfare will be significantly improved by the removal of an irreparably damaged, deformed, dangerous, painful, or useless part of the body. Seldom should an amputation be done without the supporting opinion of a consul-

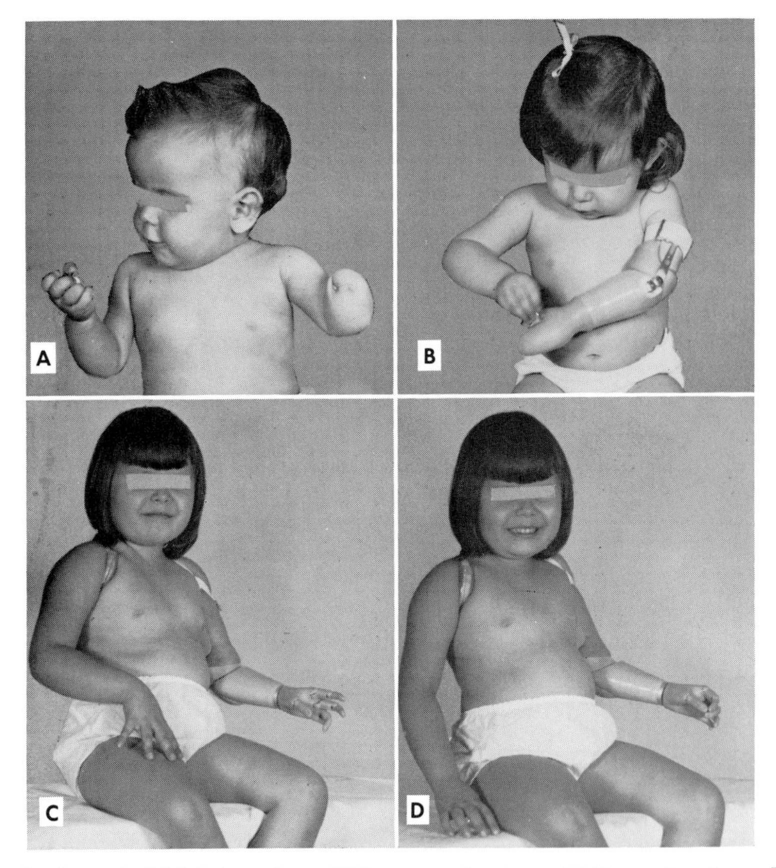

Fig. 9-1. A, Congenital left below-elbow (BE) amputation in a girl 11 months of age. **B,** Eight months later the patient had become accustomed to the use of a simple mitten prosthesis. **C,** At the age of 4 years the patient was fitted with a functional hand, shown here in open position. **D,** Patient with prosthetic hand in closed position.

tant. When the blood supply of a limb has been lost and cannot be restored, amputation is always necessary. In permanent, irreparable loss of nerve supply, amputation is occasionally indicated to remove a limb or part of a limb that may be useless, unsightly, and subject to chronic ulceration and other trophic changes. The commonest reasons for amputation are the following: (1) vascular disease or accident, (2) trauma, (3) infection, (4) tumor, (5) thermal, chemical, or electrical injury, and (6) congenital anomaly.

Vascular accident or disease. Local blood supply, especially in the lower limbs, may be destroyed suddenly by thrombosis or embolism, or gradually by peripheral vascular disease such as arteriosclerosis or thromboangiitis obliterans. When an adequate blood supply cannot be maintained or restored by treatment, gangrene ensues and amputation becomes necessary. The level of amputation is determined by the adequacy of the remaining circulation.

Trauma. Amputation is indicated when severe trauma has so destroyed the blood supply or so damaged the tissues of a limb that gangrene is inevitable or useful reconstruction impossible. For the evaluation of such wounds, careful examination and mature judgment are essential.

Infection. In acute fulminating infections that endanger life by extending proximally and that cannot be controlled by less radical means, amputation is indicated. The commonest of these infections is gas gangrene of high virulence.

In chronic infection, such as a draining osteomyelitis of long standing that has not responded to medical and ordinary surgical treatment, amputation may be indicated because of either local or systemic sequelae. Systemic complications, which include amyloidosis, glomerular nephritis, and bacterial endocarditis, may prove fatal unless amputation is done. Local sequelae of chronic infection that may make amputation and the fitting of a prosthesis advisable include mutilating loss of muscle or bone, deforming contractures and ankylosis, chronic ulceration, painful scars, ischemic changes, persistent sinuses, and the development of carcinoma about sinuses. In adults, extensive tuberculous or mycotic infection of the foot or ankle with multiple sinuses is often best treated by amputation.

Amputations for infection and its sequelae are becoming less frequent as improved methods for the control of sepsis are devised and adopted.

Tumor. Amputation is frequently indicated when a limb is involved by a primary malignant tumor, of which osteogenic sarcoma is a classic example. Amputation should not be done, however, until it has been ascertained, usually by biopsy, that the tumor is definitely malignant and that it cannot be treated satisfactorily by measures such as resection, irradiation, or chemotherapy with or without perfusion. If metastasis has not occurred, amputation may be curative. When the tumor has already metastasized, amputation may still be the treatment of choice as a palliative measure to relieve pain, improve systemic status, and obviate ulceration, infection, hemorrhage, and pathologic fracture.

For certain late, extensive benign tumors of the limbs, amputation may be indicated when excision of the tumor would cause severe loss of function of the limb.

Congenital anomaly. Several types of congenital anomaly are best treated by amputation. Supernumerary fingers or toes form a clear example: amputation is indicated when they impair appearance or function. Congenital absence of distal parts of a limb may require amputation for the modeling of a stump satisfactory for a prosthesis.

Thermal, chemical, or electrical injury. Extensive, severe tissue damage from excessive heat or cold or from chemical or electrical burns may necessitate amputation.

Types of surgical amputation

Amputation performed through a joint is called *disarticulation.* Amputation in which the surface of the wound is not covered with skin but left unclosed is termed *open amputation.* This is a temporary amputation, used for the control of actual or potential infection, and often must be followed by a secondary surgical closure. *Closed amputation* is usually a final or definitive amputation performed to create a stump that can be used effectively with a prosthesis.

Amputations through or distal to the metacarpus or the metatarsus are called *minor amputations.* At these levels an artificial limb will not be worn, and the stump should be so planned that as much as possible of its function is retained or recreated. All amputations proximal to the metacarpal or metatarsal bones are termed *major amputations.* They are designed primarily to produce a stump suitable for an artificial limb.

Amputations of the upper limb

Improvements in artificial limbs have made it possible to fit properly any level of the upper limb, provided that the condition of the stump is satisfactory. The site of election has become the most distal point at which sound surgical principles permit the formation of a satisfactory amputation stump. The ideal levels of amputation are those that have proved most functional, without a prosthesis in the case of minor amputations and with a prosthesis in the case of major ones.

In amputations of the upper limb the basic objective is to maintain function or to substitute for lost function. At the level of the fingers and metacarpal bones, amputations are designed to retain the most useful elements for grasp, pinch, and hook. At higher levels the surgeon attempts to create a stump suitable for supporting and activating a prosthesis comfortably and capable of withstanding the trauma occasioned by its use.

In amputations through the fingers, as much as possible of their length and useful mobility should be retained. When a fingertip is lost, the bone end should be covered with a well-cushioned tactile pad by means of a full-thickness skin flap or graft. In amputations through the distal half of the terminal phalanx the fingernail should be preserved, if possible. It is extremely important that no part of the thumb be sacrificed unless absolutely necessary. Total loss of the thumb, the loss of multiple digits, and amputations in the metacarpus and carpus are individual problems requiring careful selection of the procedure best suited for restoring the complex function of the hand.

Above the metacarpals the entire carpus should be retained when possible, since it will be extremely useful to the patient with or without a prosthesis. Disarticulation at the wrist has the advantages that the prosthesis need not include the elbow joint and that pronation and supination are retained. In amputations through the forearm as much length as possible should be preserved. The usual prosthesis is hinged at the elbow and includes a forearm socket with a wrist unit to which a prosthetic hand or a hook may be attached interchangeably. The hook is more useful than the hand. It can be opened by the pull of a cable attached to a harness about the patient's opposite shoulder and closed by rubber bands about its base. Several types of hooks are available, some designed specifically for certain occupations.

Disarticulation at the elbow is uncommon. Above-elbow amputations (Fig. 9-2) are most satisfactory at the supracondylar level; above this, functional efficiency becomes less as the shoulder is approached, and at least 2 inches of bone stump should remain below the anterior axillary fold. In amputations at the shoulder the head and neck of the humerus should whenever possible be preserved to minimize disfigurement. Interscapulothoracic or forequarter amputation, a severely deforming procedure including removal of the scapula and most of the clavicle, is sometimes required in the treatment of malignant disease. After these extremely high

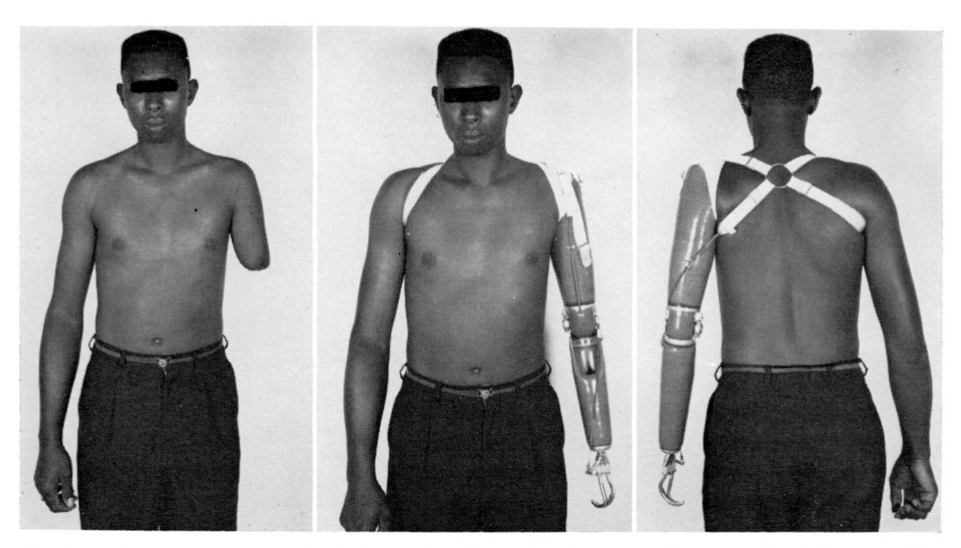

Fig. 9-2. Amputee and prosthesis for above-elbow (AE) stump. Dorrance hook is opened by a cable from the shoulder harness and closed by an elastic band.

amputations of the upper limb, effective use of a prosthesis is unusual except in the case of bilateral high amputees, where it becomes a necessity.

In *cineplastic amputations,* designed for use in the upper limb, the power of one or more of the patient's muscles, transmitted by means of a small peg traversing a muscle tunnel lined by skin, is used to activate the hand mechanism of the prosthesis. The commonest example is use of the biceps as a motor after short below-elbow amputation. Successful cineplasties require a careful selection of patients and the combined, skillful efforts of surgeon, prosthetist, and physical or occupational therapist. Cineplastic amputations are rarely used today.

In the *Krukenberg operation* the forearm stump, after a below-elbow amputation, is converted into a crude pinching mechanism by separating the lower ends of radius and ulna and covering them with soft tissues. No prosthesis is used. Because of its unsightliness the Krukenberg stump has not become popular. However, the fact that it possesses both tactile sensation and pinching function makes it the best expedient in certain situations, such as that of the blind, bilateral forearm amputee.

Despite many recent advances in the design and manufacture of prostheses, lack of tactile sensation in the prosthetic limb continues an unsolved and difficult problem for the upper extremity amputee. Another problem is that of providing power to move the terminal device. A hook or artificial hand is usually activated by means of a shoulder harness (Fig. 9-2). The more proximal the amputation the greater becomes the problem of controlling the terminal device. In the bilateral upper extremity amputee the situation is indeed critical. In recent years much research has been carried out in the field of prostheses powered pneumatically or electrically. Although still at the prototype stage, they are expected to become increasingly available in the future.

Amputations of the lower limb

Amputations of the lower limb are about three times as common as those of the upper limb. The most important requirement of the lower extremity stump is that it be able to bear weight in standing and walking. Depending upon the level of amputation, weight may be borne on the end of the stump (end-bearing), the metaphyseal flare of the tibia (side-bearing), the patellar tendon (PTB), or the ischial tuberosity. It is desirable that the weight-bearing area be as large as possible and covered with skin that is free from scar and capable of withstanding pressure.

The commonest levels for amputation in the lower limb are shown in Fig. 9-3.

Amputation of the great toe does not significantly impair standing or unhurried walking. Disarticulation of the second toe leads to hallux valgus. Loss of the three smaller toes does not cause appreciable disability. In transmetatarsal amputations as much length as possible should be preserved. These amputations require no prosthesis; the end of the shoe may be filled with sponge rubber on a removable insole, and for better weight distribution a metatarsal bar may be added to the shoe.

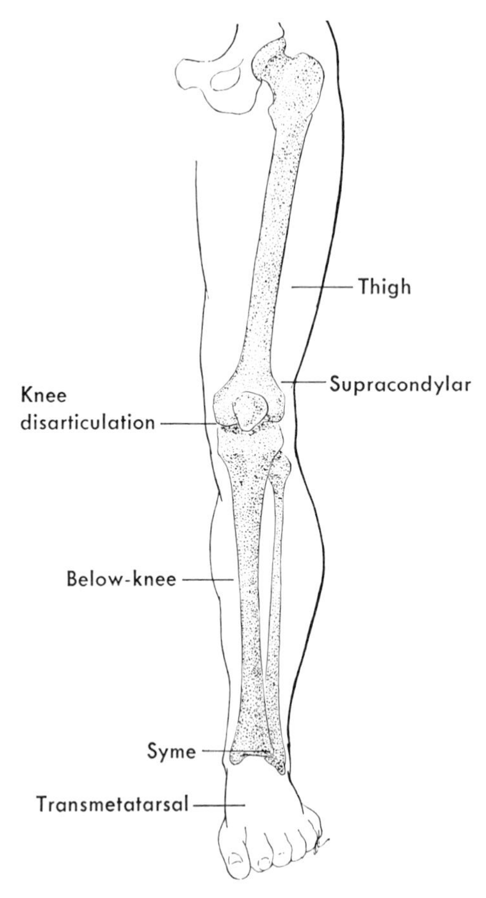

Fig. 9-3. Common levels for amputation of the lower limb. These levels have proved relatively favorable for subsequent function.

Amputation through the tarsal bones is usually unsatisfactory. Lisfranc's disarticulation at the tarsometatarsal level and Chopart's amputation through the talonavicular and calcaneocuboid joints have become obsolete because by distorting the muscular balance of the foot they produce intractable deformity. The amputations of Boyd and of Pirogoff include tibiocalcaneal fusion. Above the transmetatarsal level and below the middle third of the leg there is only one generally accepted and widely used amputation, the Syme amputation. This is an end-bearing amputation in which the tough plantar skin of the heel is used to cover the end of the tibia after it has been severed ¼ inch above the ankle joint. The Syme prosthesis includes only a below-knee molded plastic socket and a modified solid-ankle, cushion-heel (SACH) foot. Since the stump is bulbous the prosthesis is enlarged at the ankle; thus for cosmetic reasons the Syme amputation is a better procedure for men than for women.

Above the Syme level the below-knee stump is side bearing and should be about 6 inches in length. Longer stumps fit prostheses poorly and are liable to circulatory difficulties. The fibula should be ½ to 1 inch shorter than the tibia. The older below-knee prosthesis has consisted of a leather thigh corset laced in front, lateral hinges at the knee, and a shin piece simulating the normal leg, enclosing the stump-fitting socket and connected to the ankle mechanism and foot. The currently used patellar tendon-bearing (PTB), cuff-suspension prosthesis (Fig. 9-4) eliminates in most

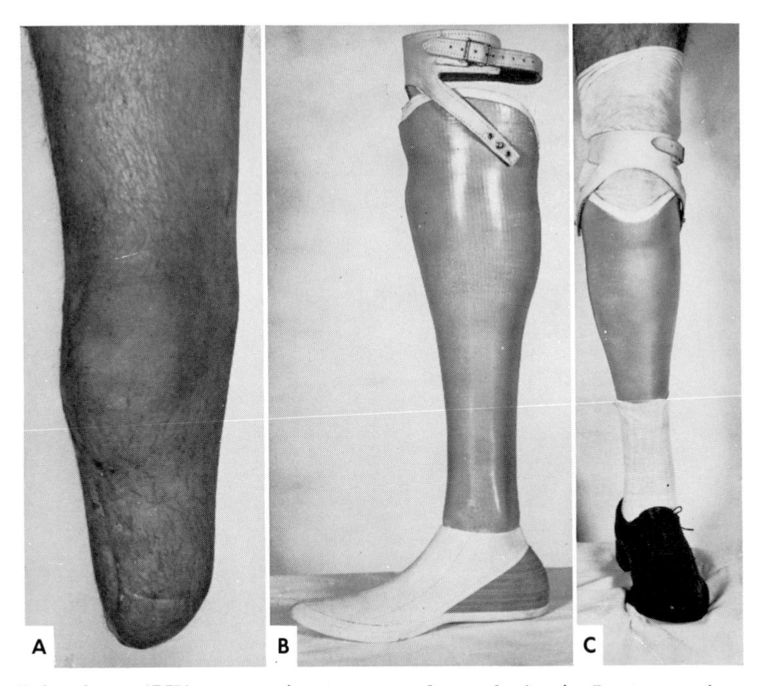

Fig. 9-4. Below-knee (BK) amputation stump and prosthesis. **A,** Posttraumatic amputation stump of adequate length. **B,** Plastic patellar tendon-bearing (PTB) prosthesis with leather cuff and solid-ankle, cushion-heel (SACH) foot. **C,** Prosthesis is suspended by cuff over patella at the supracondylar level of the thigh.

cases the need of thigh corset and knee hinges. This prosthesis, in which a portion of the weight is borne by the skin over the patellar tendon, is simple to put on and take off, is light, and permits great freedom of movement. An agile, healthy person with an adequate stump and a well-fitted PTB prosthesis can run, jump, and climb, and suffers little restriction of physical activity or job capability. Such freedom of activity is impossible in the case of higher amputations, for without active knee control, walking with a prosthesis becomes much more difficult. For this reason it is important that amputations for peripheral vascular disease be done at the below-knee rather than the above-knee level when there is any chance that the lower amputation will succeed.

Disarticulation at the knee and supracondylar amputation may produce stumps that can tolerate some end bearing. These amputations have not been widely used, however, because of technical difficulties in making a satisfactory prosthetic knee joint for stumps at these levels. When an amputation must be proximal to the knee, the level generally most acceptable is 4 to 5 inches above the joint line. This leaves a stump 10 to 12 inches in length that provides a long lever arm for the prosthesis but

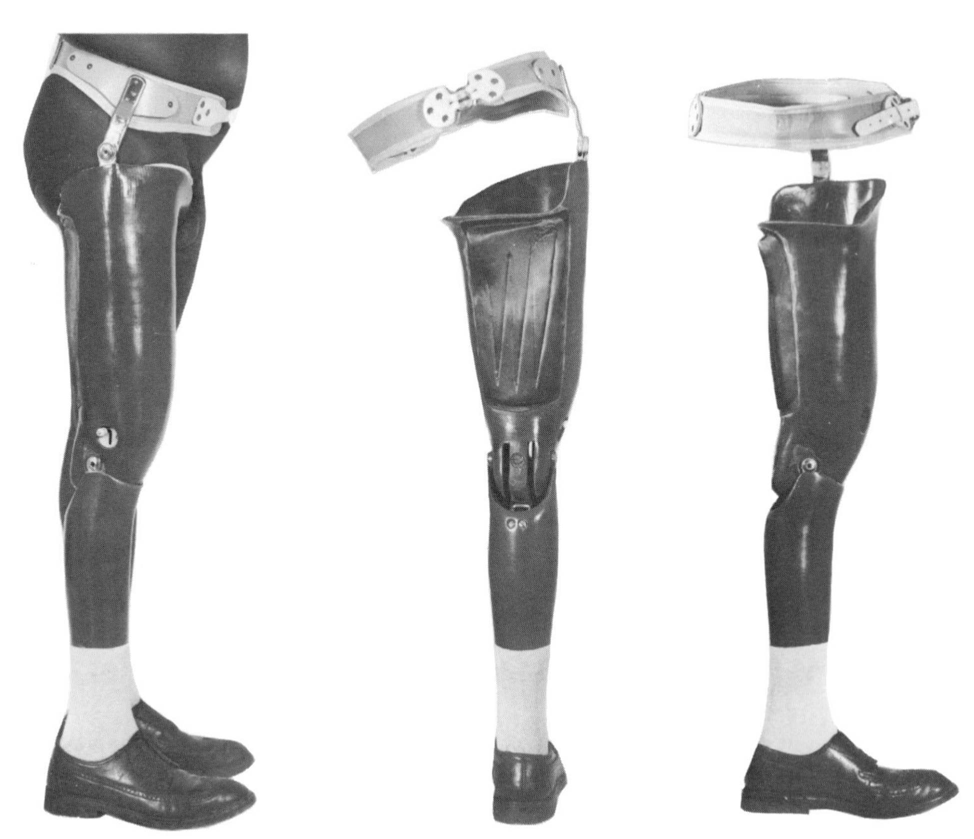

Fig. 9-5. Above-knee (AK) prosthesis with pelvic band suspension and quadrilateral socket. The patient's weight is borne on the ischial tuberosity.

also allows room for its knee mechanism. Above this level the difficulty of fitting the prosthesis increases as the stump length decreases. Since up-and-down motion takes place within the prosthesis with each step, it is important that the skin over the end of the stump be neither tight nor redundant. The conventional above-knee prosthesis (Fig. 9-5) includes an ischial weight-bearing thigh socket and is held to the body by a pelvic band with metal hip joint, or in the case of longer, stable stumps, by a cloth belt (Silesian bandage). In agile patients with suitable stumps the suction socket prosthesis may be used, which eliminates the need for belt or other suspension device. A valve in the lower part of the socket permits air to escape during the stance phase of gait, and the suction thus created holds the prosthesis in place during the swing phase.

The upper part of the above-knee socket is quadrilateral in shape with high anterior and lateral walls and low posterior and medial walls to accommodate the ischial tuberosity and the adductor muscles. A great difference between the below-knee and the above-knee amputee is that the latter lacks active knee control. The knee joint and leg portion of the above-knee prosthesis function as a pendulum, which is propelled forward during the swing phase of gait by flexion of the femoral stump within the socket. The axis of the prosthetic knee joint must be carefully positioned so that the knee is stable in full extension at the time of heel contact (the beginning of the stance phase); otherwise the knee may buckle and the patient fall. Friction mechanisms must be incorporated in the knee joint to allow the leg section to swing forward at a speed similar to that of the normal knee to create a normal gait, without limp. A great variety of prosthetic knee units is available, and the physician who is caring for amputees must decide just which type of knee is best for each of his patients.

Disarticulation at the hip, or the even more mutilating hindquarter amputation in which half of the pelvis is removed, is sometimes indicated in the presence of malignancy such as osteogenic sarcoma or chondrosarcoma of the proximal part of the femur. Even after such extensive losses, satisfactory ambulation can be carried out with the help of a Canadian-type hip-disarticulation prosthesis or a hemipelvectomy prosthesis.

Amputations in children

Of the entire group of orthopaedically crippled children amputees make up an important and increasing segment with great rehabilitation potential. Most of these limb deficiencies are either congenital or traumatic in origin. In addition certain other congenital anomalies of the limbs, such as severe unilateral shortening, are frequently best treated by the fitting of a prosthesis, with or without prior surgical amputation.

In children the object of amputation and the fitting of a prosthesis is to facilitate early function, to enhance appearance, and at the same time to produce an optimal stump for maturity. It must be kept in mind that children's bones are growing; not only should length be preserved, but every effort should be made to save the epiphyseal growth centers.

A major postoperative complication of amputations in children is overgrowth of bone as compared with the soft tissues of the stump. This may lead to perforation of the skin and secondary infection of low grade. Such overgrowth occurs most commonly in the humerus, fibula, or tibia and is of unknown cause. It appears to be of endosteal origin; attempts to prevent or correct it by the surgical arrest of epiphyseal growth are not indicated. The usual treatment is revision of the stump. The stump complications of shortness, scarring of the skin, bony spurs, and neuromata, which so often lead to disabling symptoms in adults, cause little difficulty in children, and phantom limb sensations are temporary and painless.

Prostheses for children are of standard types with minor technical variations from those for adults. It is important that they be comfortable and simple. Early fitting tends to minimize psychologic difficulties. The upper limb amputee may be fitted with a simple mitten prosthesis (Fig. 9-1, *B*) as early as 6 months of age; the lower limb amputee should begin to use a prosthesis at normal standing age. The prosthetic training of children is often less time consuming than that of adults. The child amputee and his prosthesis should be reexamined at frequent intervals, and the prosthesis should be changed as the child grows.

Preparation for the prosthesis

The period between sound healing of the surgical wound and the fitting of the prosthesis is occupied by several measures; some of these are begun immediately after the operation, and all are of the greatest importance in the final rehabilitation of the amputee. They include: explanation and encouragement; proper positioning of the limb to prevent contractures; massage of the stump to mobilize the scar, decrease tenderness, and improve vascularity; active exercise to strengthen muscles and mobilize joints; and firm, smooth elastic bandaging to prevent edema and shrink the stump to contours that will remain relatively constant.

In recent years immediate or early postsurgical fitting of prostheses has been utilized effectively in many cases. Instead of the usual soft postoperative dressing, an accurately fitting, rigid plaster cast socket is used. To this is added an adjustable prosthetic unit, and the patient is permitted to stand and to bear some weight on the extremity within a few days after amputation. Earlier rehabilitation and fewer postamputation complications such as flexion contractures and stump edema are benefits of this method.

Selection of the prosthesis

The prosthesis should be chosen carefully with due regard not only to the stump but also to the personal and occupational needs of the individual patient. This always calls for the closest cooperation between orthopaedic surgeon, patient, and prosthetist. The assistance of a physical therapist is often desirable, and in amputee centers the consulting team may include also an occupational therapist, vocational counselor, psychologist, and social worker.

Training in use of the prosthesis

Without sound prosthetic training, few persons who have had major amputations will acquire satisfactory use of the prosthesis, and none is likely to achieve an optimal functional result. Longer periods of training are required for upper limb than for lower limb amputees, for bilateral than for unilateral amputees, and for adults than for children. The above-knee amputee requires much more training, time, and effort than the below-knee amputee. The amputee must learn efficient control of stump and prosthesis as a unit in the activities of his daily living and in the demands of his occupation. He is best taught, under the orthopaedic surgeon's supervision, by physical therapists and occupational therapists experienced in work with amputees and prostheses.

Evaluation of the prosthetic gait

To evaluate the patient's effectiveness in using his prosthesis and to be aware of technical problems in the fit and alignment of an artificial leg, the physician must have knowledge of the normal and abnormal mechanics of gait. He must be able to analyze the various components of the gait of the amputee on his new limb.

The normal walking gait (Fig. 9-6) is divided into two parts, the *stance* or weight-bearing phase and the *swing* or nonweight-bearing phase. The stance phase is longer than the swing phase, taking up about 60% of the walking cycle. Thus in walking there is always a period of double support when both feet are on the ground at the same time. Walking differs from running in that in the latter there is no period of double support and in some period of the running cycle both feet are off the ground.

The stance phase of gait begins with heel strike, or contact of the heel with the ground. This occurs with the knee in or nearly in full extension. As the trunk moves forward over the weight-bearing limb, the entire foot contacts the floor and full weight is assumed by the limb. The large gluteal muscles stabilizing the hip and pelvis come into play during this early stance phase, as do the muscles of the quadriceps. As the trunk moves forward over the limb, the body weight shifts to the ball of the foot and the heel comes off the ground while the gastrocnemius and other foot flexors contract for push-off. The swing phase begins as the toe leaves the ground and is characterized by foot dorsiflexion and strong activity of the iliopsoas to bring the femur forward again. During the swing phase, the knee comes from a position of more than 60 degrees of flexion to full extension, helped

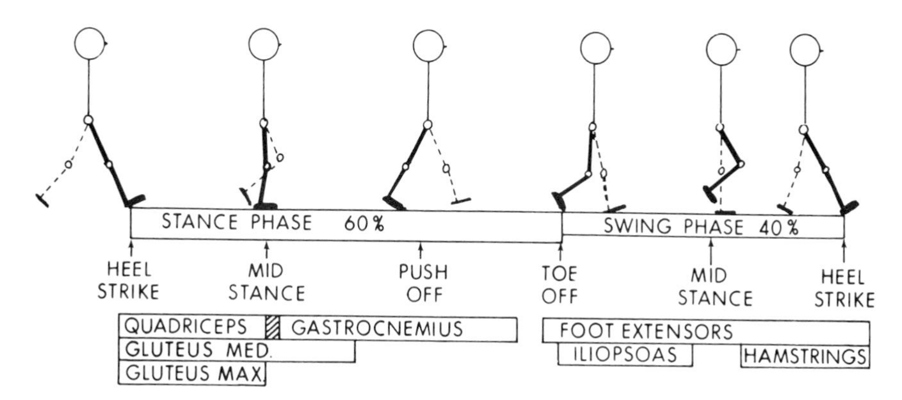

Fig. 9-6. Stance and swing phases of the normal gait cycle.

in the very beginning by a short burst of activity in the rectus femoris. In the latter part of the swing phase the hamstring muscles exert a braking action to stop knee extension at 0 degrees at the time of heel strike. In addition to these movements of the limb, vertical and horizontal motions of the trunk play an important part in gait efficiency.

Any deviation from the normal gait pattern results in an additional expenditure of energy by the patient when walking. The energy expended by an elderly above-knee amputee when walking with his prosthesis in the most efficient manner may be more than 50% greater than the energy he would consume if he had no amputation. It is therefore important that the fitting, alignment, and component parts of the prosthesis be so accurate that the patient will walk with a gait as nearly normal as possible.

Disabilities of the amputation stump

Amputation stumps are subject to many early complications and late disabilities. Some of these difficulties are attributable to recognized surgical or prosthetic factors; others are of unknown pathogenesis. Frequent among late disabilities of the stump are ulcerations, contractures, infections, and noninfectious conditions characterized by pain or tenderness.

Ulceration of the stump is usually associated with local ischemia and secondary infection; it often develops along the surgical scar. Ulceration may arise from several causes, including excessive postoperative scarring and the friction, tension, and pressure exerted by the prosthesis. The ulcer should usually be treated by local rest, elevation, and hot compresses until it is clean and the skin about it appears healthy. The ulcer, a narrow rim of skin surrounding it, and its entire scarred base should then be excised; if necessary, the underlying bony stump is shortened to allow closure of healthy soft tissues without tension. Modification of the prosthetic socket in the old area of contact with the ulcer may be necessary.

The unfavorable circulatory and mechanical influences present in amputation stumps predispose them to the development of *infection.* Early infection of low grade frequently produces pain in the stump before clinical signs of sepsis appear. Common forms of infection include cellulitis, abscesses, sinus tracts, and osteomyelitis. These lesions are treated according to established surgical principles, the final step usually consisting of plastic closure of the stump after all diseased or redundant tissue has been removed.

The normally proliferating end of a sectioned nerve, especially when it becomes involved in scar tissue and subjected to tensions and pressures from the prosthesis, may become a *painful amputation neuroma.* Surgical excision of neuroma and scar is indicated. The nerve should be resected at a higher level and its end allowed to retract into normal tissue. In some cases it may be advisable simply to interrupt the nerve well above the level of amputation, with care to produce no anesthesia of the stump.

Phantom limb sensations follow almost all amputations. Usually the phantom limb is painless and temporary or intermittent. The *painful phantom limb* is a serious sequel of amputation. Its pathogenesis has not been established, and its treatment is unsatisfactory. Resection of a neuroma, procaine blocks, sympathectomy, and chordotomy are among the therapeutic measures frequently but often unsuccessfully used.

Hyperesthesia of the stump, unassociated with the phantom limb phenomenon, is another recognized complication of unknown cause. It is not relieved by local treatment of the stump; after reamputation it recurs at the higher level. Sympathetic blocks are sometimes helpful.

BRACES

Orthopaedic braces are mechanical supports used to provide stability, relieve pain, improve motor function, and prevent or correct deformity. Since every brace must meet needs peculiar to an individual patient, braces are necessarily of many different types and degrees of complexity. To all varieties, however, certain principles apply.

Principles of bracing

The chief function of most braces is to relieve weakened structures of undesirable stresses by passively restricting motion, tension, or pressure. Thus a back brace (Figs. 4-14 and 11-3), by restricting spinal mobility, may relieve pain arising from a damaged intervertebral disk; a shoulder brace (Fig. 9-7), by restricting adduction, may provide relaxation for a temporarily paralyzed deltoid muscle; and an ischial-bearing brace (Fig. 9-8), by taking resting body weight off the lower limb, may relieve it of considerable stress. In disabilities of the lower limb a brace is often used also to provide stability. When the lower limb is completely paralyzed, or flail, a long leg brace with the knee locked in extension keeps the knee from buckling and stabilizes the foot and ankle, enabling the patient to walk. Release of the lock allows knee flexion when the patient sits.

Varying degrees of protection from stress are afforded by braces of differing design. Thus a neck brace with four turnbuckles (Fig. 9-9) is required for adequate immobilization and support of the severely injured or dangerously unstable cervical spine, whereas a cylindric Thomas collar (Fig. 9-10) will provide sufficient restriction of motion for a less serious or less acute disability.

In more complex braces the principle of an active corrective force is often utilized. This is usually done by incorporating a metal spring or elastic band to produce the desired self-adjusting tension or pressure. A wrist brace fitted with wires and rubber bands is useful for stretching early flexion or extension contractures of the fingers. An ankle brace fitted with a spring mechanism favoring dorsiflexion (Fig. 9-11) is often indicated for paralytic foot drop. Braces that incorporate an active corrective force are helpful not only in overcoming contractures but also in providing power to balance that of unopposed muscles, facilitating therapeutic exercise of the involved structures. Bunnell pioneered the development of extremely useful devices of this type, suited to the needs of various disabilities of the hand and its digits. Such dynamic or physiologic splinting is especially useful in preoperative and postoperative treatment of the disabled hand.

In recent years the application of these principles has been further developed in the field of *functional bracing* of the permanently paralyzed upper limb. Power can be supplied from an external source of pneumatic or electrical type. When

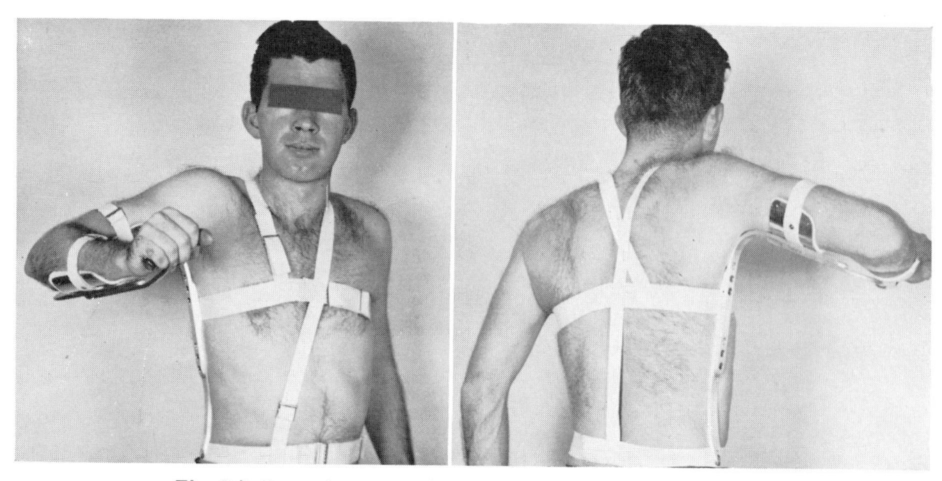

Fig. 9-7. Brace for supporting the right shoulder in abduction.

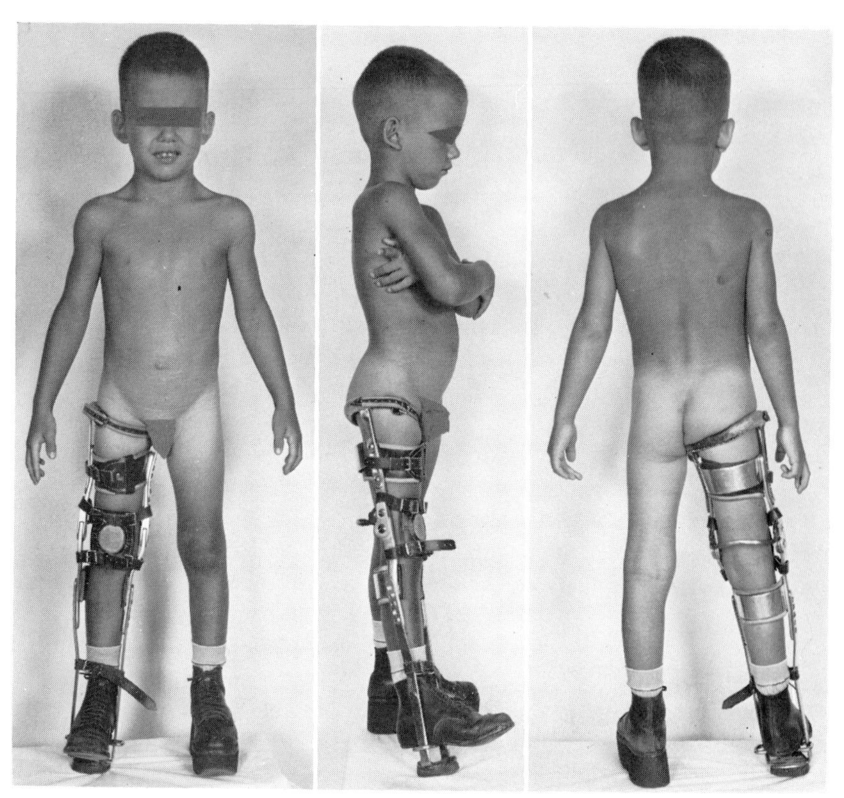

Fig. 9-8. Ischial weight-bearing brace with knee lock and patten bottom to allow standing and walking without bearing weight on the right leg. To keep the pelvis level, the left shoe is elevated.

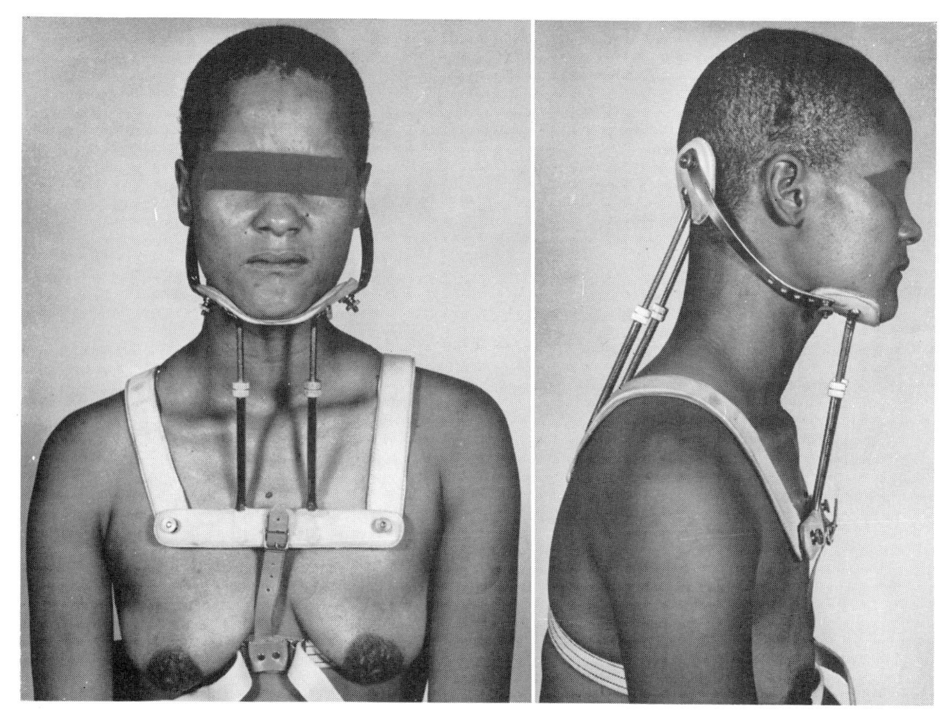

Fig. 9-9. Turnbuckle brace for the neck. Adjustment of the four turnbuckles facilitates effective support and slight distraction. A brace of this type should be available in hospital emergency rooms for early protection of the patient with injury of the cervical spine.

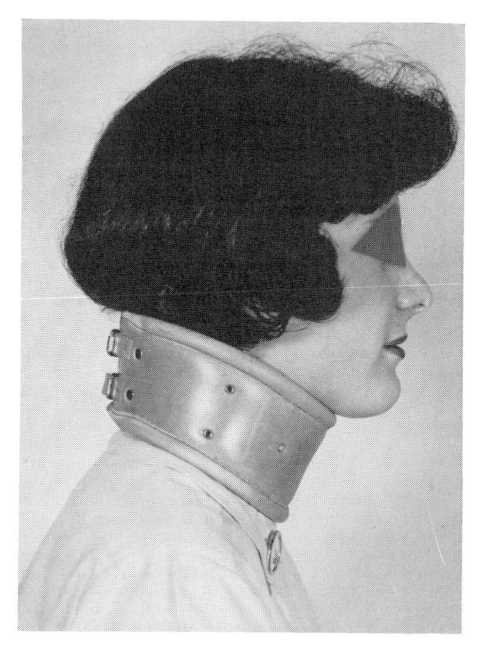

Fig. 9-10. Thomas collar (modified). This light, inexpensive collar provides adequate support in minor disabilities of the neck.

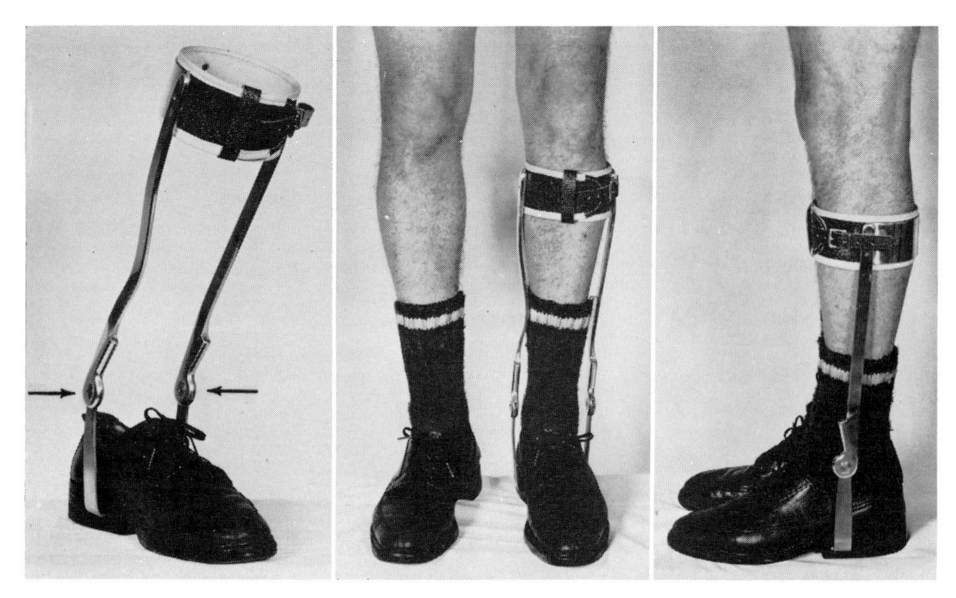

Fig. 9-11. Spring foot-drop brace. When weak dorsiflexor muscles are overbalanced by stronger plantar flexors, the adjustable spring at the ankle hinge of each upright (Klenzak joint, Pope Foundation, Inc.) is used to supply passive dorsiflexion and thus prevent foot drop and an equinus limp.

sufficient experience and facilities are available, it is possible by means of individually designed bracing to restore useful function even in the presence of severe weakness and disability.

The prescription of braces

Having determined the details of his patient's need of a brace, the physician selects from available types that which best combines utility, comfort, strength, durability, lightness, cleanliness, and economy. Many bracing problems call for the closest cooperation between orthopaedist, patient, and the brace-maker, or orthotist. On delivery the brace frequently requires minor adjustments for the individual patient, who also must be instructed in its application and use. If the brace is to be worn for a long period, arrangement for subsequent examinations, adjustments, and repairs must be made.

CHAPTER 10 Affections of the spine and thorax

In this chapter abnormal curvatures of the spine and deformities of the chest are discussed. Affections of the lumbosacral region are so common and so important that they have been placed in a separate chapter, Chapter 11, "Affections of the low back."

Affections of the spine

SCOLIOSIS

In scoliosis, or lateral curvature of the spine, a series of vertebrae is persistently deviated from the normal spinal axis; the lateral deviation is invariably accompanied by some degree of rotation of the vertebrae. Scoliosis is a deformity rather than a disease and is often secondary to pathologic changes outside the spine.

Incidence. Scoliosis is a common deformity. It usually becomes apparent during the years of growth. The factors that result in scoliosis are often established long before the child reaches school age, and definite deformity is usually recognizable before the age of 14 years. In early school ages scoliosis is slightly more common in boys, but during adolescence it occurs three to five times as often in girls. A survey in Delaware disclosed scoliosis of mild to severe degree in 1.9% of 50,000 chest roentgenograms in a population over 14 years of age.

Classification and etiology. For purposes of description, cases of scoliosis can be conveniently classified in two ways: (1) according to the causative agent and (2) according to the shape and level of the curvature.

The etiologic classification is necessarily imperfect since our knowledge of the causes of scoliosis is incomplete. One large group is made up of patients whose spinal curvature is a result of temporary postural influences. Such *postural* or *nonstructural scoliosis* (Fig. 10-1) is not accompanied by asymmetric changes in the individual structures of the spine and is relatively easy to correct. *Structural scoliosis* (Fig. 10-2), on the other hand, is characterized by definite morphologic abnormalities, and it is with this group of cases that therapeutic effort is most concerned. It is possible that some cases of nonstructural scoliosis in children may progress, develop morphologic changes, and so undergo transition into the structural type.

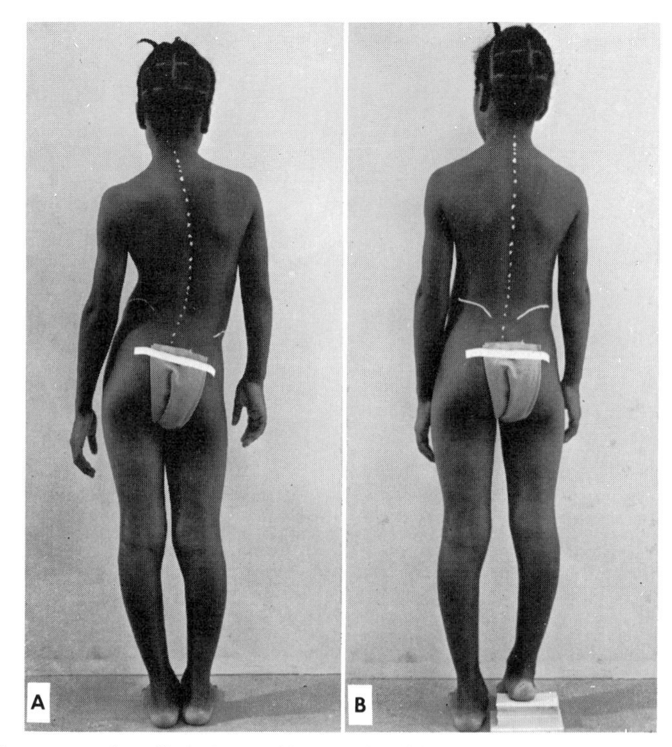

Fig. 10-1. Nonstructural scoliosis in an 11-year-old girl with congenital hemihypertrophy, the left limbs being longer than the right. **A,** Note right C-shaped scoliosis secondary to inequality of leg length. **B,** Note correction of nonstructural scoliosis when shortness of the right leg is compensated by elevation of the right foot.

The causes of structural scoliosis are numerous. *Congenital scoliosis* is usually caused by defective embryologic development. A clear example of this type is scoliosis associated with the presence of a hemivertebra or with the asymmetric fusion of two or more vertebrae (Fig. 10-3). Another common cause of structural scoliosis is asymmetric paralysis of the trunk muscles, which destroys the normal balance of the supporting musculature. Such *paralytic scoliosis* is seen after poliomyelitis and in other neurologic affections, such as infantile spinal muscular atrophy, progressive muscular dystrophy, Friedreich's ataxia, syringomyelia, and spastic paralysis. Structural scoliosis occurs also in neurofibromatosis, in skeletal diseases such as osteogenesis imperfecta and osteomalacia, and, especially in children, after unilateral thoracic conditions such as thoracoplasty and chronic empyema. In the great majority of patients with structural scoliosis, however, the pathogenesis is unknown. In such cases, called *idiopathic scoliosis,* the underlying factor may be a slight unilateral weakness of the musculature of the trunk or asymmetric growth at the epiphyseal plates of the vertebrae. Idiopathic scoliosis usually becomes manifest shortly before or during adolescence. It may, however, appear in infancy or between 3 and 6 years of age.

In the regional classification of scoliosis are encountered several types of curva-

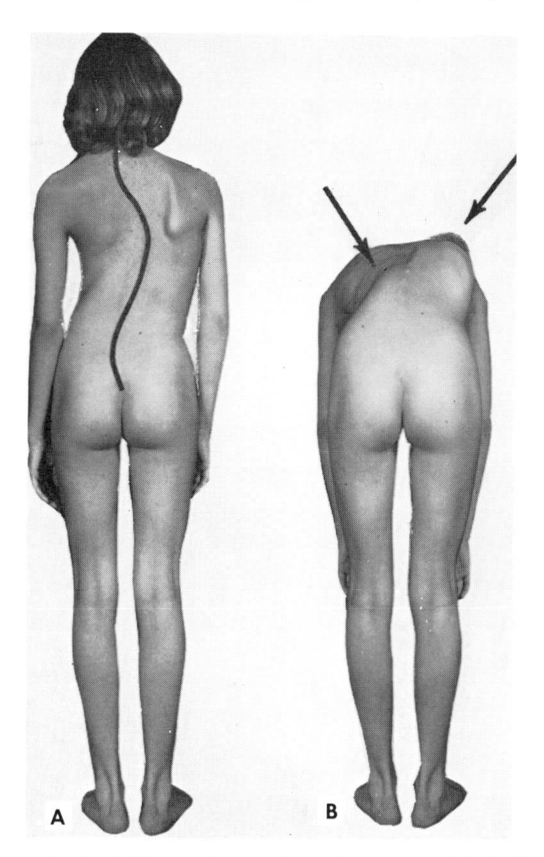

Fig. 10-2. Structural scoliosis of idiopathic type in a 13-year-old girl. **A,** Note the right thoracic, left lumbar curvature and the prominence of the lower part of the right scapula. **B,** The posterior prominences of the back, associated with vertebral rotation, are accentuated when the spine is flexed.

ture whose prognosis varies with etiology as well as location. A scoliotic curve is designated by the side of its convexity. An S-shaped compound curve, consisting usually of a major right dorsal curve and compensatory left cervical and left lumbar curves, is often idiopathic in type. Sometimes two curves seem to participate equally in the deformity, which is termed a double major curve. A very long C-shaped curve involving most of the lumbar and dorsal areas is usually seen in scoliosis of paralytic origin. In general, curves in the dorsal and cervicodorsal regions become fixed earlier and cause more deformity than do curves in the lumbar area.

Pathology. The extent of the pathologic changes varies with the degree of lateral curvature. All the structures of the concave side are compressed or shortened, whereas those of the convex side may remain normal or become lengthened. The apical vertebra, situated at the middle of the curve, shows the greatest change, being wedge shaped and most rotated. The vertebrae above and below it undergo similar but less significant changes. The intervertebral disks are compressed on the side of the concavity and may bulge on the opposite side as a result of the pressure;

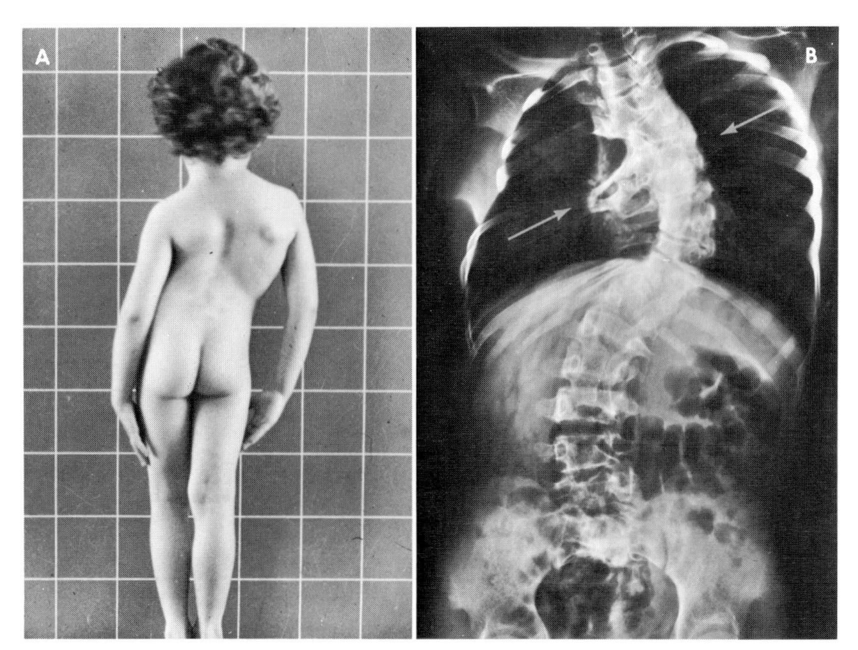

Fig. 10-3. Congenital scoliosis. **A,** Clinical appearance of deformities in a 4-year-old girl. **B,** Roentgenogram showing multiple anomalies of the ribs and vertebrae associated with the curvatures in a 10-year-old girl.

the nucleus pulposus migrates toward the convex side. The anterior longitudinal ligament is thickened on the concave side and thinned on the convex side. As the deformity increases, proliferation of bone takes place; later, ossification of the ligaments may occur. Malalignment of the spinal joints leads to degenerative arthritic changes. Rarely such changes may progress until at the level of greatest distortion the spine becomes ankylosed by the fusion of two or three vertebrae. In severe, late cases the muscles may become atrophic and on section exhibit fatty and fibrous degeneration. Because of rotation of the vertebrae, which is invariably present, the ribs on the side of the convexity are usually thrust backward with an increase in their angularity. When the curve affects the lower part of the spine, the pelvis may be distorted. Changes in the anterior chest wall include a flattening and diminution of thoracic capacity on the convex side of the curve and a prominence on the concave side. All of the organs of the chest and abdomen may be distorted and their function compromised because of abnormal stresses created by changes in the shape of the thorax (Fig. 10-4).

Clinical picture. As a rule there is no complaint until deformity of the back is noticed, and since the deformity is of very gradual development, it may reach considerable proportions before its presence is observed. The patient may be brought to the doctor because of a high shoulder, a prominent hip, or a projecting shoulder blade. When the patient is a young girl, the fitting of a dress may first call attention to the deformity. Occasionally the child may complain of fatigue and backache before a deformity is noted. As the scoliosis increases, the discomfort may become

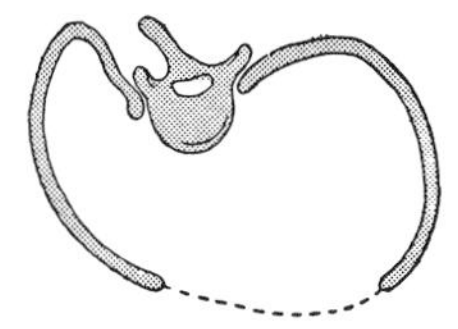

Fig. 10-4. Cross section of the thorax in scoliosis, showing distortion of the ribs associated with rotations of the vertebra. (After Hoffa.)

greater. In severe cases, pain developing in the lumbar region may be caused by pressure of the ribs on the crest of the ilium. There may be shortness of breath, from diminished respiratory capacity, and gastrointestinal disturbances from crowding of the abdominal organs. In patients with only slight deformity, however, symptoms usually do not occur until middle age or later.

Physical findings that may throw light on the cause of the scoliosis include inequality of leg length, asymmetric weakness of the abdominal muscles, and the café au lait spots of neurofibromatosis.

The physical examination should include careful note of the following points: (1) inspection of the natural, relaxed standing posture; (2) estimation of the shape and degree of spinal curvature, the number of vertebrae involved, and the level and extent of the maximum deviation; (3) estimation of the degree of rotation or twisting of the vertebrae; and (4) estimation of the amount of spinal flexibility. Each of these findings is important in determining the treatment.

In the majority of cases of structural scoliosis the backward rotation is on the side of the convexity, the prominence of the ribs posteriorly causing a corresponding asymmetry of the thorax (Fig. 10-2, *B*). The shoulder on the side of the convexity is elevated and anterior with respect to the other shoulder. In cases of purely postural character the curvature can be straightened voluntarily and undergoes complete spontaneous correction during recumbency.

The compound or S-shaped type of curve is found in most cases of idiopathic scoliosis and in approximately 30% of the entire group of scoliosis patients. The deformity consists most often of a major dorsal curve convex to the right and of compensatory cervical and lumbar curves convex to the left. The curvature cannot be corrected voluntarily and in recumbency may be less pronounced but does not disappear. The asymmetry and rotation may be severe. The rotation deformity is best observed when the patient leans sharply forward.

Idiopathic lumbar scoliosis is most commonly convex to the left with its apex at the level of the second lumbar vertebra. The iliac crest is more prominent on the side of the concavity, whereas there is fullness of the back on the side of the convexity. Since the ribs are not involved in lumbar scoliosis it tends to cause less deformity than does thoracic scoliosis. Single dorsal curves tend to develop accentuation of the normal dorsal kyphosis, a rotation deformity, and compensatory curves in the cervical and lumbar regions.

Diagnosis. Since scoliosis is not a primary disease of the spine but the resultant of certain mechanical forces exerted upon it, the examiner must attempt to determine the basic etiologic factor. Careful history and physical examination, to-

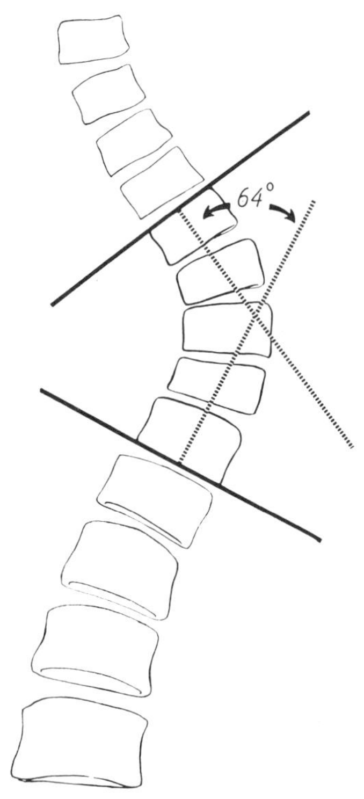

Fig. 10-5. Measurement of scoliotic curve by method of John R. Cobb. Highest vertebra is that with the disk below it wider on the convex side and the disk above it wider on the concave side; the lowest vertebra has the disk above it wider on the convex side and the disk below it wider on the concave side. Intersecting perpendiculars determine the angle of curvature. The Ferguson method of measuring scoliotic curvature utilizes lines drawn from the centers of the bodies of the highest and lowest vertebrae of the curve to the center of the body of the middle vertebra of the curve.

gether with good roentgenograms, are essential. The roentgenograms may show the character of the curve, the degree of distortion of individual vertebrae, and the presence of bony defects. The degree of curvature should be measured on the roentgenograms by a standard method and recorded (Fig. 10-5). Similar roentgenograms and measurements, made several months later, should be compared with those of the initial examination to demonstrate whether the curvature is increasing or is static.

Prognosis. In most cases of structural scoliosis, significant progression of the curvature ceases spontaneously with the cessation of vertebral growth at the age of 15 years in girls and 17 years in boys. A useful roentgenographic guide, described by Risser, is the fact that idiopathic scoliosis increases little after the age at which the iliac crests have been fully capped by their apophyses. After skeletal maturity, structural curvatures may very gradually increase because of degenerative changes in the spine; this is often especially noticeable in paralytic scoliosis.

During childhood, however, the deformities of structural scoliosis often progress rapidly, constituting a serious therapeutic problem until arrest of advancing curvature can be assured. Every case demands careful attention and supervision over a long period of time. In adult scoliosis pain is a common symptom, appearing in about two thirds of the cases, very often in the fourth decade of life and in association with lumbar curves.

Without treatment the nonstructural curvatures may remain stationary, but the structural curvatures in growing children, including those of congenital type, often increase to cause severe deformity and disability.

With treatment, scoliosis of the nonstructural type may be completely cured. The structural type in young children may be prevented from becoming worse and often may be greatly improved; rarely, however, is complete correction of the curvature obtained. The earlier the treatment is started, the better the prognosis. The C-shaped curvatures are more amenable to treatment than are the compound or S-shaped curves. Thoracic curvatures become more deforming than lumbar curvatures.

Treatment. With these prognostic considerations in mind the treatment of scoliosis must be suited to the individual patient. Many persons with scoliosis will not require definitive treatment. Cosmetic considerations usually govern the decision. As a general rule, thoracic curves that exceed 35 degrees as measured by the Ferguson technic, in the supine position, are cosmetically unacceptable. However, the appearance of any given patient varies not only with the angular measurement but also with the spinal segment involved, the length of the curve, the patient's physique, and the degree of vertebral rotation.

In treating scoliosis the basic aims are to recognize curvature early, to evaluate accurately its chance of progressing, and insofar as possible to correct it and maintain its correction.

Recognition of the earliest stage of scoliosis is important; it is usually concerned with the preschool child rather than the older girl or boy. Any recognizable underlying cause of scoliosis should be treated. A leg length discrepancy of significant degree should be equalized. Unilateral defects of sight or hearing may play an etiologic role and should be corrected. The child who has had poliomyelitis or other neurologic disorder must be observed at intervals for a number of years to rule out the development of paralytic scoliosis. Exercises may be prescribed to improve posture. When an increase of any curvature is considered likely, the patient should be examined at frequent intervals.

Nonsurgical measures. At the start of therapy for established scoliosis nonsurgical treatment is used and, in the majority of patients, need not be followed by surgical measures. Four types of nonsurgical treatment have been used, alone or in combination: bed rest, exercise, traction, and plaster cast or brace. Both bed rest and traction reduce the deformity but do nothing to maintain the correction. Exercises are useful to help maintain flexibility of the curvature, but they cannot be relied upon to reduce the curve or to slow its progression.

Gradual straightening of the curvature can be accomplished by utilizing the principle of a hinged jacket. For this the turnbuckle plaster jacket of Risser has been much used. It consists of a body cast that incorporates one or both thighs and the head and neck and that is provided with anterior and posterior hinges and a turnbuckle on the side of the concavity; after correction has gradually been effected, stabilization can be secured by means of surgical fusion. More rapid improvement of the curvature can be secured, however, by combining longitudinal traction with pressure over the apex of the convexity. The localizer body cast of Risser is applied with the patient on a special frame provided with attachments for exerting head traction, pelvic traction, and localized pressure posterolaterally over the rib angulation.

Of the many and diverse types of braces that have been devised for the treatment of scoliosis, that in most effective use is the Milwaukee or Blount brace (Fig. 10-6). It incorporates both active distraction, encouraged by adjustable uprights extending from head to pelvis, and adjustable posterolateral pressure over the thoracic prominence. In it the patient can lie down, sit, walk, carry out prescribed exercises, and

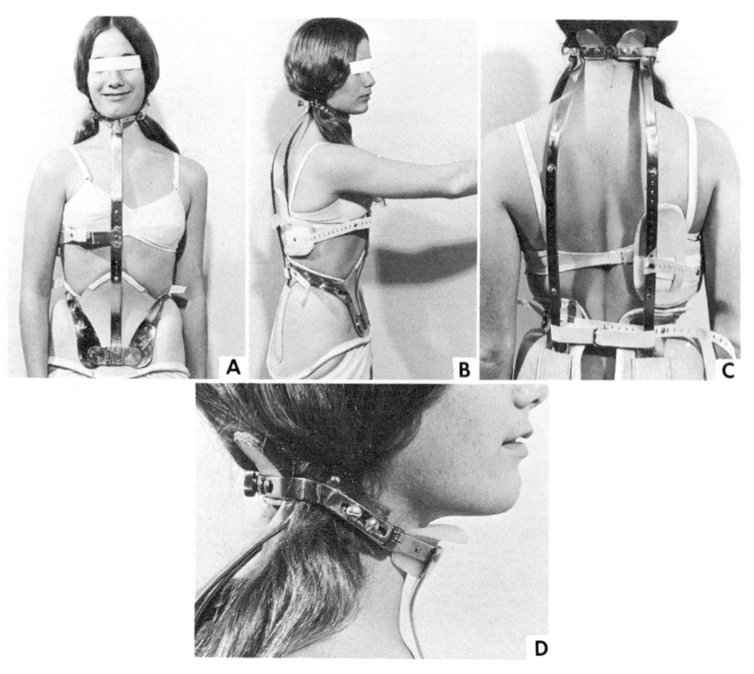

Fig. 10-6. Milwaukee brace for scoliosis. **A,** Girdle is carefully molded above the waistline and in front of the anterior superior iliac spines. The abdomen is well supported by the leather apron. An outrigger controls the direction of the holding force of the thoracic pad and avoids pressure on the breast and chest wall during inspiration. The throat mold is scarcely visible when the gaze is level. **B,** Uprights follow the body contour. The strap for the thoracic pad is advanced anteriorly as the deformity is actively overcome. The girdle extends low in back to assist the apron in maintaining pelvic tilt. The throat mold maintains the position of the head over the occiput pads, which furnish distraction. **C,** Right thoracic pad engages the ribs caudad to the apex of the major thoracic curve. The occiput pads fit under the base of the skull. The lumbar pad holds against the transverse processes of second and third lumbar vertebrae. **D,** When the chin is elevated, the occipital pads actively distract. The tonguelike projection of the throat mold is a reminder to keep the gaze level. (From clinical material of Dr. Walter P. Blount, Milwaukee, Wis.)

engage in some sports. After initial passive lessening of the spinal deformity by the brace, further improvement is obtained by such activities and by special exercises taken under the direction of a physical therapist. As further correction occurs, the pads are advanced and the brace lengthened.

In early progressive but flexible curves of young children, the Milwaukee brace can be used to correct and maintain correction of the deformity. During the period of spinal growth it may, by alleviating pressure on the concave side of the vertebral epiphyses, prevent and to some extent correct fixed deformity of the spine. The brace must be worn until all tendency toward increase of the curvature has ceased. Thus bracing must usually be continued until skeletal maturity, demonstrated by closure of the vertebral and iliac apophyses, has been reached. If satisfactory correction can be maintained without need for extensive surgery, the effort is very worthwhile. Special training on the part of the physician and the brace maker in the manufacture and adjustment of this brace is advisable. Since possible complications of wearing the brace include skin allergies, pressure sores, and emotional disturbances, the patient should be seen at frequent intervals.

Surgical measures. Surgical treatment is indicated when curvatures of unacceptable degree cannot be satisfactorily improved, or their improvement satisfactorily maintained, by nonsurgical measures. In many clinics operation is advocated for all

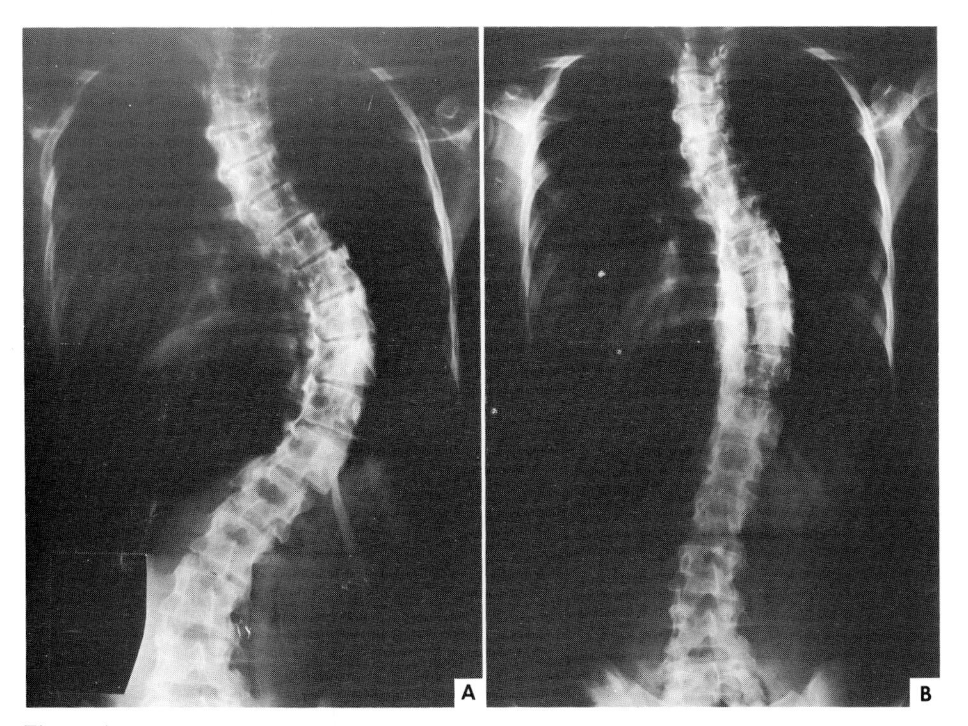

Fig. 10-7. Roentgenograms of idiopathic scoliosis in a 15-year-old girl. **A,** Before treatment. Angulation of right dorsal curve, as measured by the method of Cobb, is approximately 85 degrees. **B,** Two years after treatment in turnbuckle jacket and arthrodesis from fourth thoracic to second lumbar vertebra. Angulation by Cobb measurement is 40 degrees.

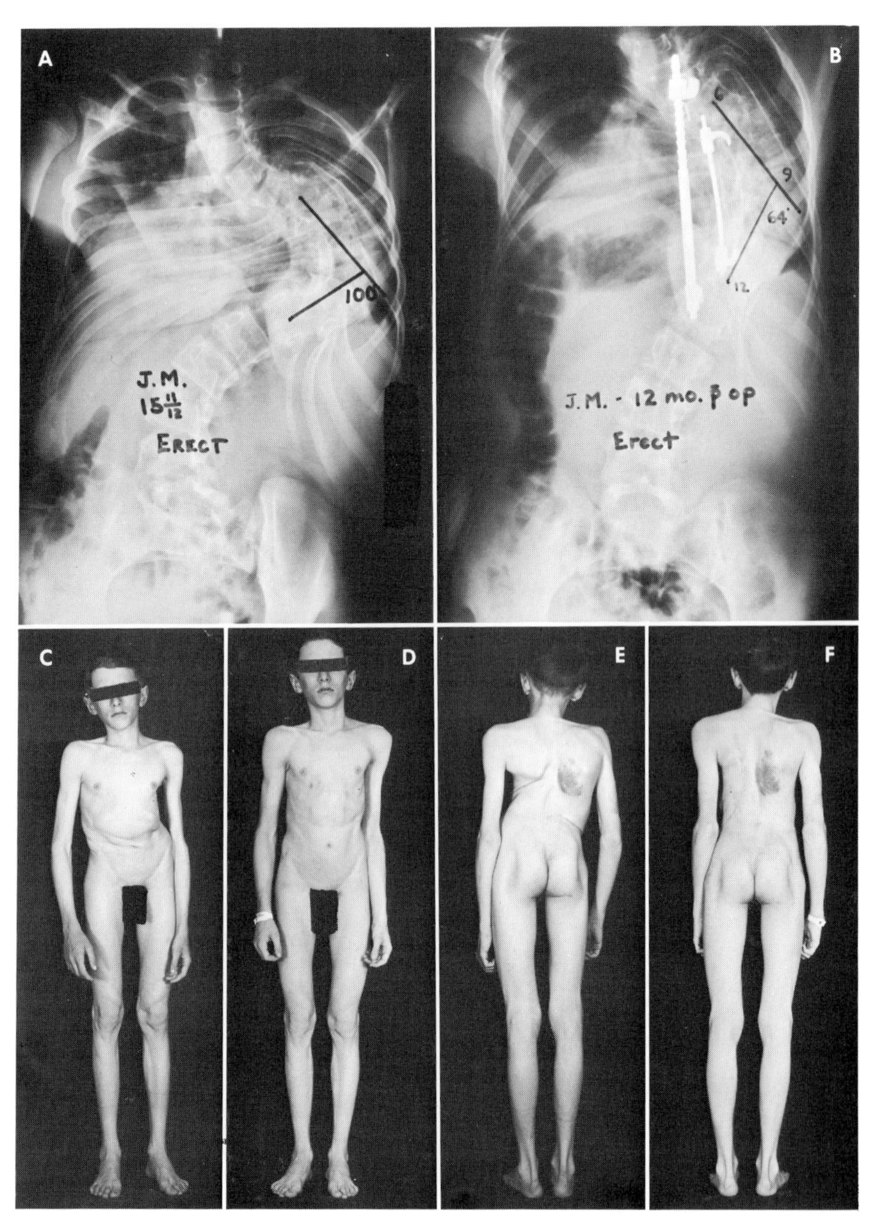

Fig. 10-8. Severe structural scoliosis treated by Harrington rod instrumentation and spinal fusion. **A,** Preoperative roentgenogram at age 15 years 11 months. **B,** Roentgenogram a year after operation. Curvature has been improved from the measured angle of 100 degrees to that of 64 degrees. **C,** Front view before operation. **D,** Front view after operation. **E,** Back view before operation. **F,** Back view after operation. In **E** and **F** note also the large café au lait spot of neurofibromatosis.

forms of paralytic scoliosis, for rapidly increasing idiopathic scoliosis, and for any other type that is accompanied by severe deformity in adolescence. As a rule, however, only about 5% of the cases are severe enough to require fusion.

Spinal arthrodesis, applied in the treatment of scoliosis by Hibbs in 1914, is the most effective means of permanently maintaining correction of the curvature (Fig. 10-7). It may be done in one or more stages after maximum correction of the curvature by brace or cast has been obtained. Alternatively the arthrodesis may precede correction since straightening of the curvature is facilitated by the temporary increase in spinal mobility after operation. Surgical fusion of a long segment of the spine, the level depending upon the details of the individual curvature, may be carried out according to the Hibbs technic (Fig. 4-11) or according to modifications of this method with reinforcement by autogenous iliac bone. The surgical correction of severe curvatures may be facilitated by inserting the metal distraction rods devised by Harrington (Fig. 10-8); in addition to these devices, however, spinal fusion is necessary. Resection of the proximal portion of the ribs on the concave side of a thoracic curve, advocated by Flinchum, may improve the correction of resistant curves.

Most spinal fusions for scoliosis should be followed by rest in bed in a plaster cast or Milwaukee brace, for a period of from 3 to 6 months, after which an ambulatory plaster jacket or strong spinal brace should be worn for an additional period of 6 months or more, until roentgenograms show a solid column of bone sufficiently strong to maintain permanent correction of the curvature. The patient must then be followed at intervals for a number of years, during which any recurrent increase of curvature suggests that there is a pseudarthrosis or that the extent of the fusion is inadequate. In such cases the pseudarthrosis should be repaired or the fusion extended.

KYPHOSIS (ROUND BACK)

Anteroposterior curvature of the spine in which the convexity is directed posteriorly is called *kyphosis.* The thoracic and sacral levels of the normal spine exhibit this type of curvature. Posterior convexity of abnormal degree results from pathologic changes located primarily in the vertebral bodies, the intervertebral disks, or the supporting musculature.

Kyphotic deformity occurs not uncommonly in children and young adults. The most frequent cause is faulty posture. Kyphosis resulting from diseases such as chronic arthritis or tuberculosis has been described in preceding chapters. Other entities that may cause kyphosis in youth include *Scheuermann's disease* and *vertebral osteochondritis.*

Kyphotic deformity seen in middle or late age groups has been called *adult round back.* Its causes include postural influences and common bone and joint diseases, as well as degenerative spinal lesions peculiar to adult years.

Adolescent kyphosis (Scheuermann's disease, vertebral epiphysitis)

The term *adolescent kyphosis* has been applied to a chronic affection of the epiphyses of the vertebral bodies that is evidenced clinically by fatigue, backache, and kyphosis of gradual development (Fig. 10-9, *A*). The process always involves

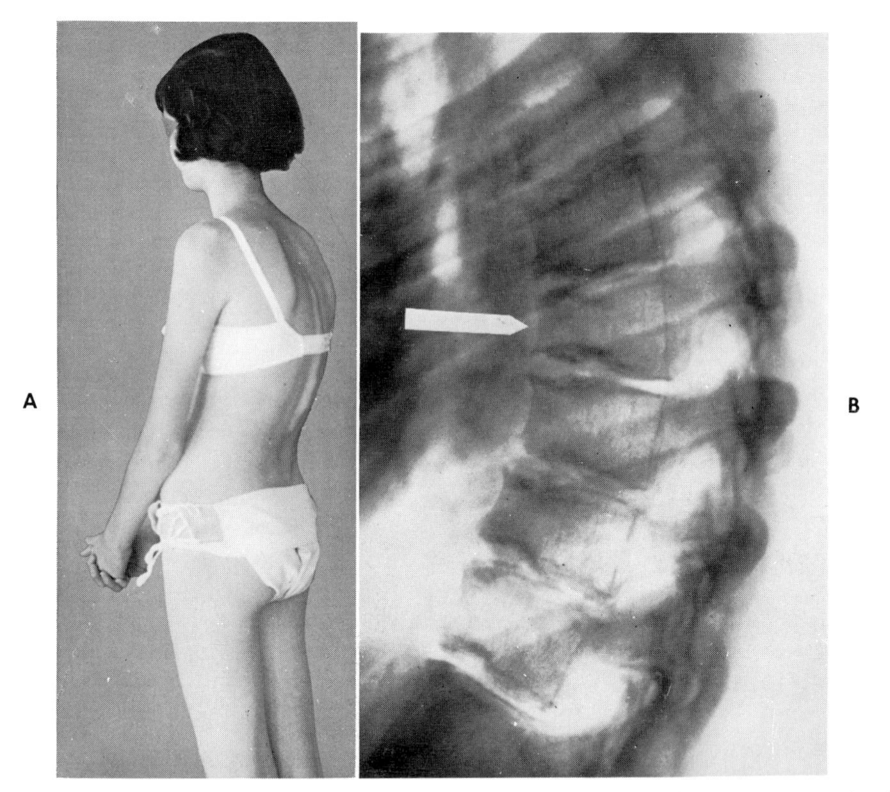

Fig. 10-9. A, Adolescent kyphosis showing pronounced dorsal kyphosis and poor posture. **B,** Lateral roentgenogram showing changes of adolescent kyphosis. Note the narrowing or wedging of the anterior portion of some of the vertebral bodies and the irregularity of their superior and inferior surfaces.

a number of contiguous vertebrae and usually is most advanced in the lower or middle portion of the dorsal spine. Adolescent kyphosis makes its appearance at puberty and is self-limited, its active course lasting usually for 2 or 3 years. It is more common in boys than girls.

Etiology. Scheuermann believed that the cause is a growth disturbance of the vertebral epiphyses. It is considered by some to be a vascular disturbance similar to that seen in coxa plana. Schmorl noted the frequent finding of irregular thickness in the vertebral end plates and protrusion of the nucleus pulposus into the vertebral bodies *(Schmorl's nodes)* in this condition. He believed that the cause is congenital deficiency in the thickness of the vertebral plates, and that partial loss of disk substance causes excessive pressure on the anterior portion of the vertebral epiphyses, which in turn results in localized growth disturbance and wedging of the vertebral bodies.

Roentgenographic picture. In early cases the bone edges above and below the intervertebral spaces are ill defined and of uneven density. The epiphyses are irregular in outline, particularly at their anterior edges, and may appear fragmented. As healing takes place, the fragmentation disappears, and the bone outlines become relatively more distinct but remain irregular. The vertebral bodies are wedged anteriorly, and the interbody spaces are distorted (Fig. 10-9, *B*). Localized indentations of the spongiosa, resulting from protrusions of nuclear material through defects in the cartilage plates, may be seen. There is often a persistent vascular groove, appearing as a horizontal cleft in the middle portion of the vertebral body and extending from the anterior margin posteriorly for about two thirds of the depth of the body. In most normal children such clefts disappear before the tenth year of age.

Clinical picture. The symptoms usually begin between the ages of 12 and 16 years. The first subjective evidence may be fatigue and pain in the back. Undue prominence of the spinous processes of the vertebrae may be noticed, especially at the lower dorsal and upper lumbar levels, and gradual development of kyphosis takes place. Compensatory increase of lumbar lordosis may be noticeable. The deformity persists in recumbency. Stiffness and tenderness may be present throughout the spine. Expansion of the chest is abnormally restricted. The affection sometimes progresses to the point of severe disability, whereas in other cases it causes few or no symptoms. In later years osteoarthritic changes and backache may develop.

Diagnosis. Adolescent kyphosis is to be distinguished from tuberculosis of the spine. In tuberculosis there is significant spinal rigidity, which is never striking in epiphysitis, and the roentgenograms show localized bone destruction and an angular rather than a rounded kyphosis. Adolescent kyphosis is also to be differentiated from osteochondritis, which develops at an earlier age and is localized to a single vertebral body, and from rheumatoid and Strümpell-Marie arthritis.

Treatment. The therapy depends upon the severity of the symptoms. In milder cases it may be advisable to have the patient limit activities that put stresses on his spine, use a fracture board without pillow, and carry out kyphosis exercises. It is sometimes necessary to support the spine by means of a brace or plaster jacket. Occasionally, recumbency upon a Bradford or a Whitman frame, followed by the use of a brace or plaster jacket, may be advisable. In progressive cases the Milwaukee brace (Fig. 10-6) has been found highly effective. In addition to the chin and pelvic portions of this brace, anterior and posterior pressure pads are used instead of the lateral pads required in the treatment of scoliosis. In an occasional case, persistent pain in later years is an indication for spinal arthrodesis.

Vertebral osteochondritis (vertebra plana, Calvé's disease)

Vertebral osteochondritis is an uncommon affection, occurring usually in the dorsal spine of children between 2 and 12 years of age and characterized by pathologic changes localized in a single vertebral body. It has been considered an ischemic necrosis somewhat analogous to that of the femoral head in coxa plana (p. 336); however, in recent years histologic examinations have shown that in many if not all of these cases the vertebral lesion may be a pathologic fracture secondary to a benign destructive process such as eosinophilic granuloma.

Roentgenographic picture. The affected vertebral body may appear fragmented or eroded; it becomes uniformly flattened or wedge shaped and abnormally dense. Adjacent disk spaces are normal or thickened. With healing the affected body becomes normal in density; in a young child it may regain considerable height as growth progresses.

Clinical picture. Pain, fatigue, and a mild angular kyphosis are characteristic. Muscle spasm and tenderness may also be present.

Diagnosis. The differential diagnosis should include eosinophilic granuloma, tuberculosis, tumor, compression fracture, and congenital anomaly. Lesions in bones other than the affected vertebra should be excluded by roentgenograms. Biopsy may be advisable.

Treatment. Rest in recumbency is indicated until the diagnosis has been made and the pain has subsided. A brace or plaster jacket should then be used until roentgenograms show the lesion healed.

Adult round back

Etiology. Adult round back may be due to faulty posture alone or to an occupation that requires constant flexion of the spine, to old age with associated degeneration of the intervertebral disks *(senile kyphosis),* to atrophy and collapse of the vertebral bodies *(postmenopausal* and *senile osteoporosis),* or to a pathologic entity such as chronic arthritis, osteitis deformans, poliomyelitis, fracture, metastatic tumor, plasma cell myeloma, tuberculosis, or other disease affecting the spine. Osteoporosis due to endocrine abnormalities such as hyperparathyroidism, Cushing's disease, and prolonged steroid therapy may result in vertebral collapse. Degeneration of the intervertebral disks sometimes takes place in middle life; it may progress to cause a single, long kyphosis with flattening of the lumbar and cervical portions of the spine and forward projection of the head.

Pathology. In adult round back due to lesions of the intervertebral fibrocartilages, the pathologic changes that occur in the disks are characteristic. The normal disk, as demonstrated by Schmorl, comprises a central portion, the *nucleus pulposus,* consisting of a semifluid medium under pressure, and a peripheral portion, the *annulus fibrosus,* which forms a fibrocartilaginous capsule. Two thin plates of hyaline cartilage separate the disk from the bones above and below; these cartilage plates are in contact with the inferior and superior surfaces of the vertebral bodies. In 30% of all adult spines, localized protrusions of nuclear material through the cartilage plates and into the spongy bone of the vertebral bodies *(Schmorl's nodes)* have been found. These lesions may occur in any case of abnormal anteroposterior spinal curvature, whether in a child or in an adult. They are also seen in the absence of clinical deformity.

Gradual thinning or collapse of the intervertebral disks allows adjacent vertebral bodies to become approximated. Altered local stresses lead to new bone formation about the edges of the vertebral bodies. If in addition the anterior portion of the cartilage plates is destroyed, bridges of new bone may develop across the intervertebral spaces, resulting in ankylosis.

In adult round back from senile osteoporosis the disks remain relatively normal, but the spongy bone within the vertebral body becomes atrophic and the cortex becomes thinned. Disuse atrophy may play a part. In the thoracic spine the vertebral bodies may become wedge shaped, whereas in the lumbar region they may assume a biconcave or hourglass contour as seen in lateral roentgenograms. The osteoporosis may lead to pathologic compression fractures of the vertebral bodies (Fig. 3-33), which may further increase the kyphosis.

Clinical picture. The deformity of kyphosis is of characteristic appearance and may be associated with considerable pain, weakness of the back, and general fatigue. The aching and tiring of the back usually occur below the apex of the kyphosis. Seldom is there any localized tenderness; it is usually present, however, in senile osteoporosis with recent compression fracture.

Diagnosis. Since adult round back is not a disease entity but a deformity that may result from many different conditions, accurate diagnosis is important. Determination of the serum calcium, phosphorus, and alkaline phosphatase and of the protein fractions is often helpful. A search for other foci of bone disease and for primary sites of malignancy or infection may lead to the proper diagnosis.

Treatment. Attempt should be made to maintain correct posture; exercises to strengthen the muscles of the back and abdomen and to expand the chest will sometimes aid in accomplishing this. In more advanced cases it may be necessary to apply a light spinal brace or corset. If the symptoms are located in the cervical spine, a Thomas collar (Fig. 9-10) may be used to support the head and so relieve the constant dragging sensation. Daily periods of rest on fracture board and sponge rubber mattress are helpful. In postmenopausal and senile osteoporosis the administration of sex hormones will usually relieve the back symptoms but may induce undesirable side effects.

LORDOSIS (HOLLOW BACK)

Anteroposterior curvature of the spine in which the concavity is directed posteriorly is termed *lordosis*. Cervical and lumbar levels of the spine normally exhibit lordosis. Abnormal degrees of lordosis are usually secondary to other deformity of the spine or to deformities of the lower limbs. Lordosis is often found in association with flexion contracture of the hip, congenital dislocation of the hip, progressive muscular dystrophy, paralysis from poliomyelitis, obesity of the abdomen, dorsal kyphosis, and shortening of the Achilles tendons. Some cases are idiopathic, no cause being recognizable.

Clinical picture. Constant aching throughout the lower part of the back, protrusion of the abdomen, and generalized fatigue are often present. Asymptomatic lordosis is fairly common in young children.

Treatment and prognosis. Therapy should include general postural exercises with particular effort to strengthen the abdominal and gluteal muscles, since these groups act to decrease the forward tilt of the pelvis and so straighten the lumbar spine. In adults a strong support for the back with a lower abdominal pad may be helpful. The result of treatment is often complete relief of the backache, but usually the improvement will not be permanent unless the pathologic factor underlying the deformity can be identified and remedied.

Deformities of the thorax

Slight deformities and asymmetries of the thorax are common, cause no disability, and often require no treatment. Such deformities, including flat chest associated with round back, can be made less apparent if exercises are prescribed to improve the posture and to build up the pectoral muscles and shoulder girdle.

The more serious deformities of the chest are almost certainly congenital in origin and are probably interrelated. They are always associated with overgrowth of the ribs. Often several members of a family, more commonly the males, are affected; they may have chest deformities of differing types.

PIGEON BREAST (PECTUS CARINATUM)

In the chest deformity known as pigeon breast, the sternum projects forward and downward like the keel of a boat (Fig. 10-10). This increases the anteroposterior diameter of the thorax, impairs the effectiveness of coughing, and restricts the volume of ventilation. Premature development of emphysema or cor pulmonale sometimes results from severe degrees of pigeon breast.

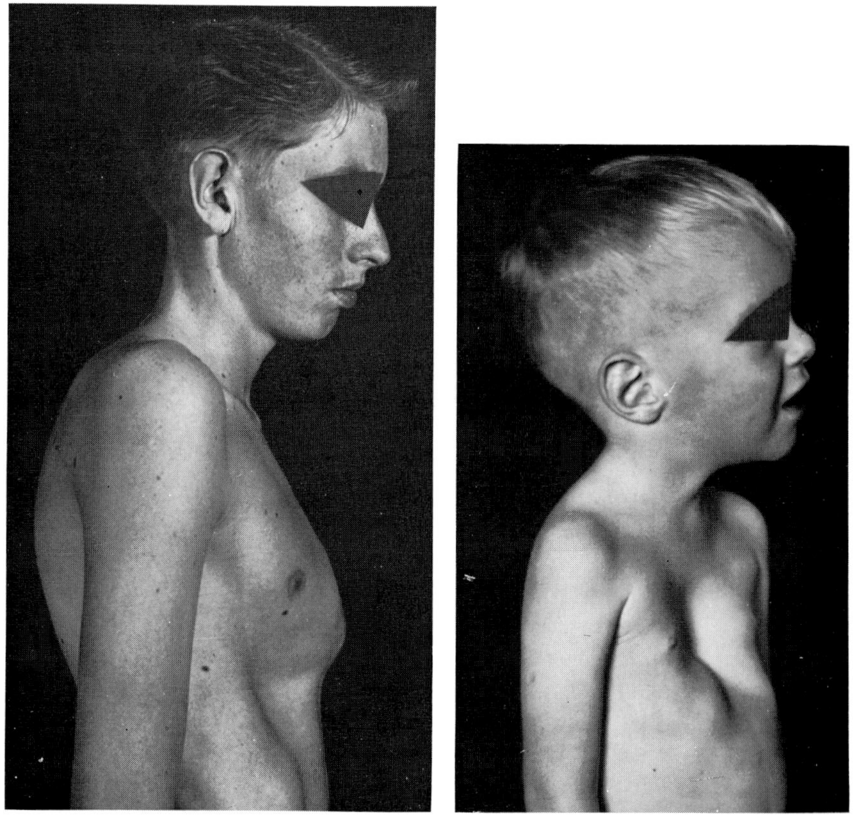

Fig. 10-10 **Fig. 10-11**

Fig. 10-10. Pigeon breast (pectus carinatum) in a 19-year-old boy. Note projection of lower portion of sternum and deformity of ribs.

Fig. 10-11. Funnel chest (pectus excavatum) in a 3-year-old boy. Note depression of sternum and deformity of ribs.

Treatment. Mild deformities can be made less noticeable by exercises that increase the strength and size of the pectoral muscles. The more severe deformities require thoracic surgery, the affected costal cartilages and the xiphoid process being resected and the sternum replaced in normal position. The cosmetic and functional results are gratifying.

FUNNEL CHEST (PECTUS EXCAVATUM)

Funnel chest (Fig. 10-11) is the mirror image of pigeon breast, the sternum being pushed posteriorly by overgrowth of the ribs. The anteroposterior diameter of the thorax is decreased. The heart is often displaced into the left side of the chest. Shortening of the central tendon of the diaphragm has sometimes been considered the cause of the deformity, but there is little evidence to support this concept. Severe forms of funnel chest can result in kyphosis, repeated respiratory infection, wheezing respiration, cardiac arrhythmia, and premature emphysema. Funnel chest may produce considerable disfigurement in both boys and girls; it may cause the youth to avoid circumstances in which he must remove his shirt, and the girl to shun normal social contacts.

Treatment. In mild cases exercises to improve posture and build up the shoulder girdle and pectoral muscles will greatly improve the patient's appearance. Swimming is especially helpful. More severe cases require a thoracic surgical procedure similar to that for pectus carinatum. The deformed costal cartilages and xiphoid process are resected, and the sternum is restored to proper position. In the seriously disabled patient the improvement after this operation is often dramatic.

CHAPTER 11 Affections of the low back

The term *affections of the low back* is used in this chapter to include ligamentous and muscular strains, spondylolisthesis, intervertebral disk lesions, and several less serious disorders involving the lower part of the back.

The frequency with which affections of the low back are encountered brings them constantly to the attention of every practitioner of medicine. Patients whose low back pain is caused by visceral rather than musculoskeletal lesions or in whom it is a hysterical conversion symptom associated with tension, nervousness, and chronic fatigue are problems for the internist and the psychiatrist, respectively, rather than the orthopaedist. However, low back pain, with or without sciatic pain, is the commonest complaint with which orthopaedic surgeons are confronted in adults. The variety of bone, joint, muscle, ligament, and nerve lesions that may cause low back pain and the difficulties encountered in their treatment make these affections formidable orthopaedic problems. In each case the diagnosis and treatment must be based upon a careful evaluation of the clinical and roentgenographic findings in the light of available information on the various recognized types of low back lesions.

Classification. For convenience of description, cases characterized by low back pain may be discussed according to the following outline:

1. Lesions peculiar to the low back:
   a. Ligamentous and muscular strains of the normal low back. Of these, the syndrome of lumbosacral strain resulting from acute or chronic traumatic influences is the most important.
   b. Abnormalities of the bony structure of the low back. Among these are congenital anomalies, spondylolisthesis, and prespondylolisthesis.
   c. Lesions of the lumbar intervertebral disks.
   d. Other lesions of muscle and fascia.
2. Common osteoarticular lesions of infectious, neoplastic, or traumatic nature affecting the low back.
3. Visceral lesions that may cause low back pain.

LIGAMENTOUS AND MUSCULAR STRAINS OF THE NORMAL LOW BACK

Most cases of low back pain of orthopaedic nature fall into the group of ligamentous and muscular strains. It has been reported that 80% of low back pain is caused by mechanical strain, but undoubtedly in many of these cases the strain is superimposed upon preexisting degenerative disk disease, which may or may not be recognized. The mechanism of injury is either an episode of trauma or a continued mechanical strain of postural or occupational type. In many cases the lesion is presumably a tear or stretching of ligaments that support and limit the mobility of a joint. In some cases the lesion is a tear of the posterior longitudinal ligament, a tear of the annulus fibrosus, or traumatic changes within the disk substance. Muscle injury may be associated.

The lumbosacral joint, situated at the critical level between movable and immovable portions of the spine, is particularly liable to injury from forces applied in an obliquely anteroposterior direction, as in violent flexion or hyperextension of the spine, falls upon the buttocks, and gravitational stresses associated with excessive lordosis. In addition the lower lumbar area is susceptible to rotational injuries. Lumbosacral sprains are therefore very frequently encountered. The sacroiliac joints, on the other hand, are large in area and protected by strong ligamentous structures; little motion can be demonstrated in them (5 degrees in young women). They may nevertheless be injured occasionally by violence applied to the low back in a rotary or obliquely lateral manner. During pregnancy, temporary relaxation of the ligaments of the symphysis pubis and the sacroiliac joints occurs normally and sometimes may be of sufficient degree to play a part in causing back pain.

Clinical picture. The symptoms and physical signs are very similar to those of other affections of the low back and warrant detailed analysis. Many of these symptoms and signs are identical with those of rupture of an intervertebral disk (p. 325).

Localized pain. Pain in the low back is the outstanding symptom. The discomfort is often localized to a single area. At times it is generalized across the whole back, however, and the area of maximum intensity may shift from one side to the other. If pain in the low back is precipitated or aggravated by motion, it will in most instances be relieved by recumbency.

Radiating pain. Radiation of pain into the lower limbs is a frequent symptom in low back affections. Such pain may be (1) radicular or (2) somatic. *Radicular pain* results from nerve root irritation. The commonest cause of root irritation is rupture of an intervertebral disk, which will be discussed later in this chapter. The root may also be irritated mechanically in a narrow, irregular intervertebral foramen, or by local inflammatory or neoplastic processes. Radicular pain tends to be sharp, shooting, and fairly well localized to a dermatomal pattern. *Somatic pain* arises within ligamentous and muscular tissue. It is of diffuse, deep, aching character. Such pain may be produced by injecting small amounts of hypertonic salt solution into the lumbar paravertebral muscles, fairly constant areas of radiation to the lower limbs being demonstrable. Deep segmental pain commonly extends into the gluteal region and into the posterior or lateral aspect of the thigh. Radiation below the knee is less common. It is probable that the occasional good results from injecting tender areas in the back with a local anesthetic are produced by blockage of this pain mechanism.

Tenderness. There is usually localized pain on firm pressure over the affected area. When the lumbosacral joint is involved, the tenderness may be most acute over the lumbosacral supraspinous ligament, but it may be present over the ilio-lumbar ligament of either side or generalized across the whole low back. When the sacroiliac structures are affected, tenderness may be present over one sacro-iliac joint.

Muscle spasm. In patients who are having low back pain there is often severe spasm of the lumbar muscles. The spasm may be equal on the two sides; it is often asymmetric, however, causing curvature of the spinal column or sciatic scoliosis (Fig. 11-1). Spasm may be present also in the hamstring muscles, making it painful to flex the hip while the knee is extended.

Abnormal limitation of mobility. Usually motion is much limited by muscle spasm, especially in acute cases. Tests of mobility dependent upon the localization of the muscle spasm are of value in differential diagnosis and should be carried out with the patient in the standing, sitting, and lying positions. In the standing position, flexion and hyperextension may be sharply limited. Rotation of the lower portion of the spine is often restricted, the patient being able to turn slightly more toward the normal side than toward the affected side. Flexion in the sitting position may be definitely restricted. In acute low back affections the lumbar spine is often flattened.

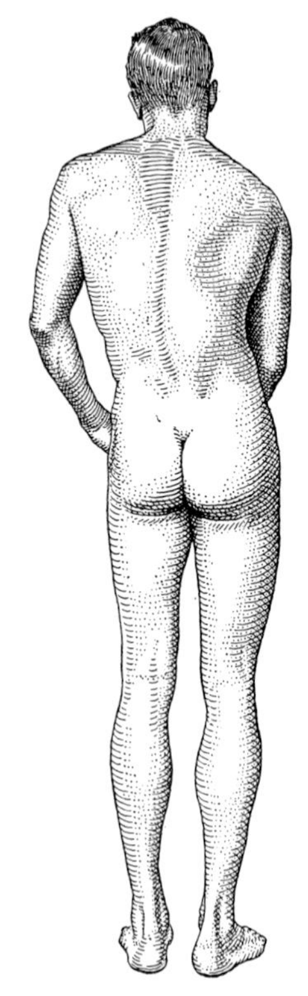

Fig. 11-1. Lateral curvature of the lumbar spine in acute affection of the low back (sciatic scoliosis).

Special tests of passive mobility. A number of passive mobility tests have been devised or popularized by various observers and are of value in the clinical analysis of low back affections. Among tests in common use are the following:

1. Extreme flexion of both thighs upon the abdomen, with the knees flexed, places the lumbosacral joint under flexion stress while causing no unilateral rotary stress of the sacroiliac joints or stretching of the sciatic nerves. Lumbosacral pain produced by this maneuver is therefore suggestive of a disorder of the lumbosacral joint.

2. Straight leg raising, that is, flexion of the hip while the knee is held in extension (Fig. 11-2), stretches the sciatic nerve and places a rotary and flexion stress upon the low back structures. It causes pain when sciatic irritation is present.

3. When the hip of the affected side is flexed, extension of the knee will cause pain in the affected side of the low back; this pain is accentuated by dorsiflexion of the foot (Lasègue's sign). The mechanism is similar to that of the preceding test.

4. Forceful and prolonged lateral compression of the iliac crests puts no stress on the lumbosacral joint but sometimes produces pain in an affected sacroiliac joint.

Roentgenographic examination. Roentgenograms are often of little help toward making a positive diagnosis. They are of great assistance, however, in ruling out conditions such as neoplasms, pyogenic infections, or tuberculosis. The presence of bony anomalies, such as sacralization of the last lumbar vertebra (Figs. 11-4 and 11-5), should be looked for. In lumbosacral affections the plane and smoothness of the articular facets and the width of the intervertebral disk spaces are to be noted, as well as the degree of the lumbosacral angle (in lateral roentgenograms, the angle between the upper sacral surface and the horizontal plane, with the patient stand-

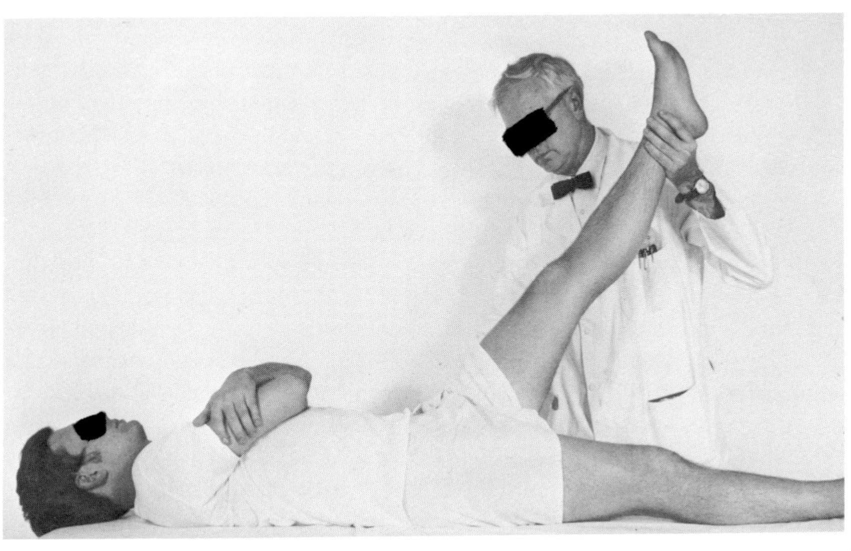

Fig. 11-2. Straight leg-raising test. The restriction of hip flexion by pain is a useful index of the degree of sciatic irritation.

ing). Arthritic spurring and bridging may be seen. The roentgenograms should include anteroposterior, lateral, and oblique views. A search for defects of the isthmus or pars interarticularis should be made, and any forward displacement of a vertebral body should be noted. The examiner should also look for destructive changes and increased bone density about the sacroiliac joints.

Differential diagnosis. Ligamentous and muscular disorders of the low back must frequently be differentiated from (1) arthritis of the lumbar spine or sacroiliac joints, (2) fracture of the body or processes of a lumbar vertebra, (3) spondylolisthesis and prespondylolisthesis, (4) injury of an intervertebral disk, (5) tuberculosis and other infections, (6) malignancy, (7) tender fatty and fibrous nodules overlying ligaments and muscles, (8) hysteria, and (9) malingering. When the diagnosis is being made, the visceral diseases that may cause low back pain must be considered. A history of aggravation of the pain by exercise and of relief during recumbency suggests that the symptoms are of osteoarticular rather than visceral origin.

Treatment. In the majority of cases nonsurgical treatment of ambulatory or recumbent type is indicated. In certain cases, however, surgical treatment is needed.

Nonsurgical measures. The principles of the treatment of ligament and of muscle injuries of the low back are essentially the same. The primary requisite is rest and support. In the acute phase uninterrupted recumbency is most helpful; the patient should lie supine in bed upon a firm mattress or on a thin mattress that is separated from the bedsprings by a fracture board. Frequently it is advisable to apply traction to one or both legs in order to relieve muscle spasm and pain. Traction may be exerted by applying straps to a pelvic belt; 15 to 30 pounds can be applied in this manner with the foot of the bed elevated 4 to 6 inches. Recumbency may sometimes be made more effective by placing a pillow under the knees or under the painful portion of the back. Recumbency should continue until pain and muscle spasm subside. It is often advisable to accompany the rest in bed with local heat and massage, maintaining constant warmth over the painful area by means of a heating pad. Aspirin may be helpful. Acute cases attended by muscle spasm may be relieved by muscle relaxant drugs. If there is a point of acute tenderness, its injection with a local anesthetic may be very helpful. In moderately acute cases in which the symptoms improve rapidly with recumbency, it is usually possible to allow gradual return to walking in 1 or 2 weeks, at which time the back may be firmly strapped with adhesive plaster or supported with a snugly fitting, low back brace or corset (Fig. 11-3). In lumbosacral affections the support should splint the back from gluteal folds to the level of the tenth thoracic vertebra; in the much less common sacroiliac affections the support should exert lateral compression upon the ilia and need not extend above their crests. Occasionally it is desirable to obtain still stronger support by the application of a light plaster cast. Such a cast should be well molded to the patient's torso and should extend well down over the iliac crests.

An acute low backache may become chronic if weight bearing without proper support of the back is allowed too soon. It is sometimes desirable to have the patient use crutches when he first returns to walking. Flexion exercises to strengthen the abdominal and gluteus maximus muscles, and to stretch their antagonists, have been advocated by Williams. By decreasing the anterior tilt of the pelvis, they tend to

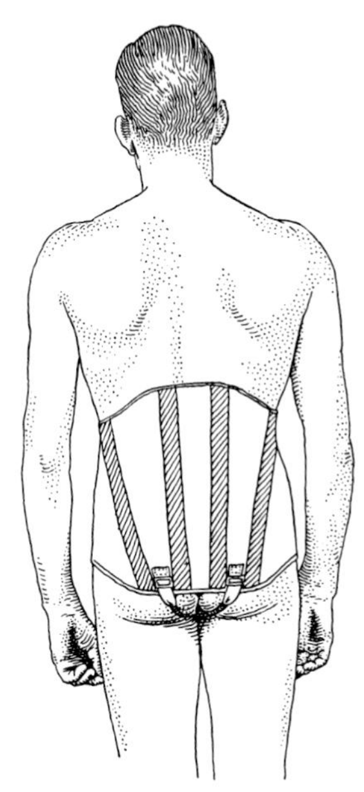

Fig. 11-3. Lumbosacral support made of heavy canvas with stays of spring steel. This wide belt may be used for low back affections that do not require rigid bracing.

straighten the lumbar lordosis and improve postural relationships at the lumbosacral angle and the intervertebral foramina. A program of such exercises should be started before the patient begins to walk and should be increased as he becomes more active. Too much emphasis cannot be placed on the necessity of warning the patient to be careful about lifting and bending. He should not be allowed to return to strenuous work until the prescribed exercises can be performed without causing pain. If a contributing factor such as inequality of leg length, hip flexion contracture, or obesity is present, special attention should be given to its correction.

Cases of only moderate severity may be treated without the initial period of recumbency.

Surgical measures. When a thorough trial of conservative treatment fails to produce lasting relief of chronic recurring and incapacitating low back pain, lumbosacral arthrodesis may be considered. Seldom is it indicated, however, in the absence of demonstrable abnormality in the low back structures, and the patient for whom arthrodesis is contemplated must be very carefully evaluated. Surgical treatment is ordinarily contraindicated unless the patient is psychologically sound, uninfluenced by compensation or insurance expectations, and strongly motivated toward recovery and return to activity. Applying a plaster body jacket and observing the patient's response while he wears it, ambulatory, for perhaps 6 weeks will sometimes facilitate the evaluation of pertinent physical and psychologic factors. The orthopaedist must also keep in mind the facts that arthrodesis does not always produce solid fusion and that solid fusion does not produce a normal back. These considerations should be weighed and discussed preoperatively with the patient.

Arthrodesis may be done posteriorly by a modification of the Hibbs technic (Fig. 4-11); the addition of an iliac prop-graft at the lumbosacral level, as advocated by Bosworth, may be helpful. Posterolateral arthrodesis with iliac bone grafts between the sacrum and transverse processes of the fifth and sometimes the fourth lumbar vertebra is an effective method. Anterior interbody fusion is sometimes done, but the incidence of pseudarthrosis is probably high with this type of fusion. Postoperatively it is essential to provide a period of recumbency, external support, and prohibition of strenuous activity sufficiently long for strong bony union to take place —6 to 12 months of such protection of the back may be required.

ABNORMALITIES OF THE BONY STRUCTURE OF THE LOW BACK

A great many of the patients with low back pain present roentgenographic evidence of anomalous bony development. In interpreting the relationship of these abnormalities to the clinical symptoms, the examiner must remember that similar anomalies are often present in individuals who are symptom free. Nevertheless, some statistical studies support the natural inference that certain anomalies of the

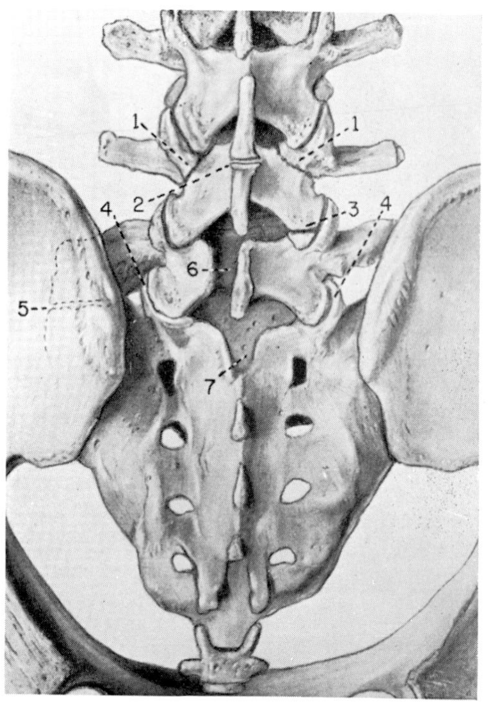

Fig. 11-4. Common anomalies of the lumbosacral region: **1,** bilateral isthmus defect; **2,** impingement of third and fourth lumbar spinous processes; **3,** laminar fissure at articular facet; **4,** abnormal plane of articular facets; **5,** sacralization of fifth lumbar transverse process with pseudarthrosis; **6,** spina bifida occulta of fifth lumbar vertebra; **7,** spina bifida occulta of first sacral segment. (From Freiberg, J. A.: In Bancroft, F. W., and Marble, H. C., editors: Surgical treatment of the motor-skeletal system, Philadelphia, 1951, J. B. Lippincott Co.)

bony structure of the low back render it somewhat less stable than the normal back and more inclined through faulty mechanics to develop the ligamentous and muscular affections that have been described.

The commonest bony anomalies of the low back are shown in Fig. 11-4. They include the following: sacralization of the last lumbar vertebra; elongation of the transverse process of the last lumbar vertebra; defects of the laminae (spina bifida occulta); variations of the spinous processes, the lumbosacral angle, and the articular facets; constitutional variations; and isthmus defects. Isthmus defects are an essential factor in spondylolisthesis, which is an important cause of low back pain.

Sacralization of the last lumbar vertebra

A bony conformation in which one or both of the transverse processes of the last lumbar vertebra are long and wing shaped and articulate with the sacrum, the ilium, or both (Fig. 11-5) is termed sacralization of the last lumbar vertebra. The disk space between sacralized vertebra and sacrum is usually narrow. The condition

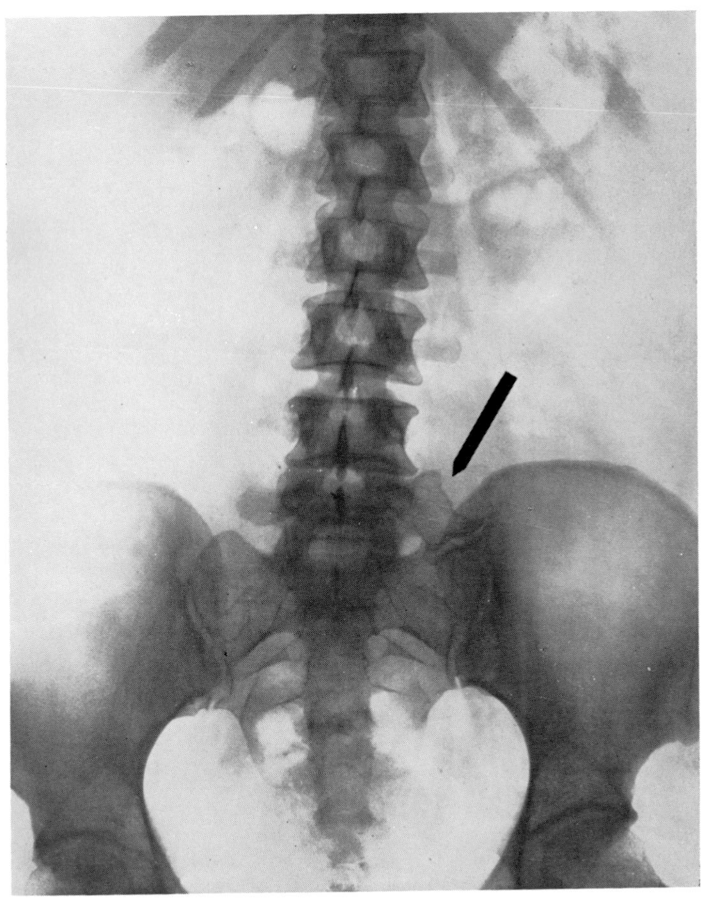

Fig. 11-5. Roentgenogram showing congenital anomalies of the lumbar spine. The left transverse process of the lowest of six lumbar vertebrae is sacralized.

is more often bilateral than unilateral; it is found in 3.5% of all individuals. A pseudarthrosis between the long transverse process and the ilium may occur, and on rare occasions this is a source of back pain and even a cause of root irritation. In most instances, however, it is an incidental roentgenographic finding and the actual cause of the patient's back pain is degenerative disk disease at a higher level.

Defects of the laminae (spina bifida occulta)

Clefts are frequently found in the vertebral laminae (Fig. 11-6) and may be associated with underdevelopment of the supporting ligaments. Partial or complete lack of fusion between the laminae of the last lumbar vertebra or of the first sacral segment is common; this imperfection has been found in 5% or more of all spines examined by roentgenograms. There is little evidence of any precise relationship between mild degrees of spina bifida and mechanical low back pain. A patient with low back pain whose roentgenograms show such an anomaly should be treated according to the principles discussed under back pain of ligamentous or muscular origin.

Variations of the spinous processes

When the neural arch has failed to fuse, the spinous process may be attached to only one lamina. In other instances a spinous process associated with normal

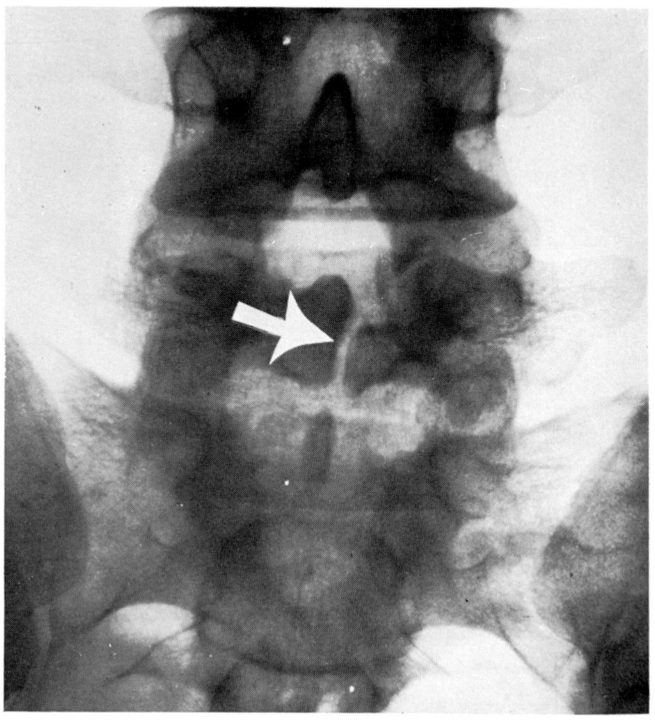

Fig. 11-6. Roentgenogram showing spina bifida occulta of the fifth lumbar vertebra. Note cleft in the laminal arch.

laminae may be so large and elongated that it touches the spinous process of the vertebra above or below. Increased lumbar lordosis and narrowed intervertebral disks tend to increase such contact between the spinous processes, which may lead to the formation of a painful bursa. The symptoms can usually be relieved by flexion exercises and without resort to surgical treatment.

Variations of the lumbosacral angle

Wide variations of the angle between the upper surface of the sacrum and the horizontal plane when the patient is standing are not uncommon. Increase of the lumbosacral angle is often associated with the so-called horizontal sacrum and excessive lordosis. In such backs the lumbosacral joint is susceptible to injury from strain, with subsequent formation of small bony spurs about the margins of the vertebral bodies. The treatment is the same as that of lumbosacral sprain.

Variations of the articular facets

Variations in the size and plane of these small but important joint surfaces are common (Fig. 11-4). The facets between the last lumbar vertebra and the sacrum are particularly susceptible to such asymmetries. Their significance as a cause of low back pain has not been proved.

Constitutional variations

The structure of the spine varies with the development of the body as a whole. In tall, thin persons the presence of a sixth lumbar vertebra is not uncommon, whereas short, stocky individuals sometimes have only four lumbar vertebrae. Low back pain is common in individuals with long, narrow backs or with flat backs. The treatment is that of lumbosacral sprain.

Isthmus defects

Absence of bony continuity at the pars interarticularis, or isthmus, of the fifth lumbar vertebra is found in about 5% of adult skeletons. It is called *spondylolysis;* when bilateral, it is also termed *prespondylolisthesis.* The defect is more often bilateral than unilateral. It may lead to low back pain and spondylolisthesis. Its treatment is the same as that of spondylolisthesis.

SPONDYLOLISTHESIS

Spondylolisthesis is a forward displacement of the fifth lumbar vertebra and spinal column upon the sacrum (Fig. 11-7) or, less commonly, of the fourth lumbar vertebra and spinal column upon the fifth lumbar vertebra. In approximately 85% of the cases the fifth lumbar vertebra is displaced on the sacrum. Apparent posterior projection of the last lumbar vertebra on the sacrum is sometimes demonstrable in lateral roentgenograms but of no clinical significance. Slight actual posterior displacement, constituting the reverse of the usual spondylolisthesis, is probably related to settling of the vertebra in the presence of degenerative disk disease.

Etiology. The underlying cause of spondylolisthesis is an absence of bony continuity at each isthmus, the narrowest part of the neural arch. The pathogenesis of isthmus defects has not been established; they are believed to result from fracture,

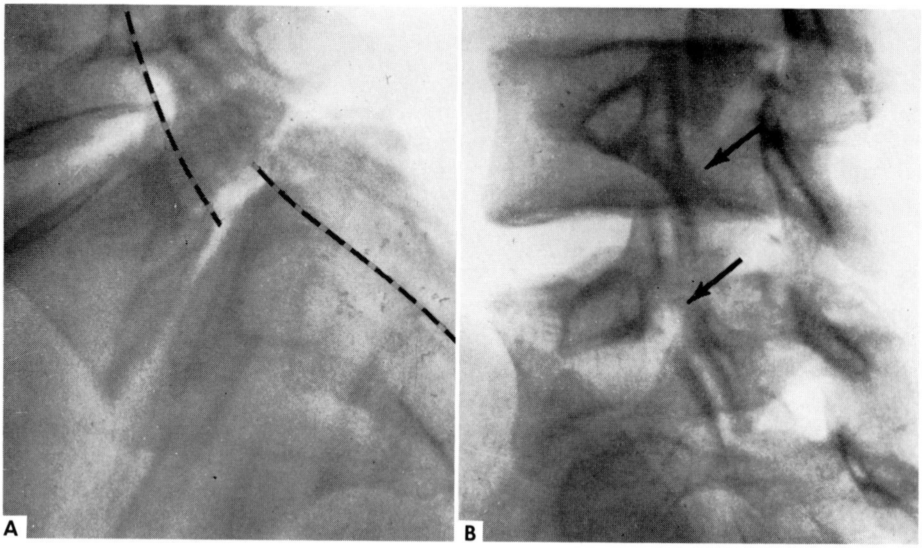

Fig. 11-7. Spondylolisthesis. **A,** Lateral roentgenogram. Note forward slipping of the fifth lumbar vertebra upon the sacrum (shown by interrupted line). **B,** Oblique roentgenogram. Note intact isthmus of the fourth lumbar vertebra and defect in isthmus of the fifth lumbar vertebra.

possibly at birth. However, these defects are much more common in adults than in children, and they may represent ununited fatigue fractures. Bilateral isthmus defects divide the vertebra into two separate pieces of bone. The anterior portion is made up of the body, pedicles, transverse processes, and superior articular processes. The posterior portion, often referred to as a separate neural arch, consists of the laminae, the inferior articular processes, and the spinous process. When a bilateral fifth lumbar isthmus defect is present, the bony anchorage of the vertebral column to its base on the sacrum is lost and its stability is altogether dependent upon the ligaments. The presence of a bilateral isthmus defect without forward displacement of the vertebral body, called *prespondylolisthesis* or *spondylolysis,* is thought to be a cause of low back pain.

Although the slipping or listhesis may be caused by a single severe injury, chronic postural strains are probably far more often responsible. Contributing mechanical factors are the mobility of the lumbosacral joint and the inclination of the upper sacral surface. Progressive increase of slipping in adults is uncommon; it is most likely to occur in adolescence. The symptoms are associated with narrowing of the intervertebral disk, degenerative changes in the disk, and proliferative bone changes about the intervertebral foramina.

Rarely the body of a vertebra together with its intact neural arch may become displaced forward when articular facet lesions are present; this condition has been termed *pseudospondylolisthesis* or *articular spondylolisthesis.* The displacement may produce symptoms of pressure on the roots of the cauda equina.

Clinical picture. Pain may be severe, slight, or entirely absent. It is often well localized in the lumbosacral joint region but may radiate down one or both legs along the course of the sciatic nerves and especially into the distribution of the peroneal nerves. There is often complaint of stiffness of the back, and all of the symptoms become worse with exercise and strain. Prior to the onset of frank symptoms there may have been recurrent episodes of low back fatigue and weakness. Upon examination the low back appears lordotic. The spinous process of the detached neural arch is prominent on palpation and often is evident on inspection. There is a tender depression just above it. The sacrum is tilted nearer the vertical plane. In cases of extensive slipping the torso is shortened, the ribs may rest on the iliac crests, and the abdomen may protrude. In severe cases the pelvic inclination is decreased and the body is swayed backward. There may be great limitation of anteroposterior motion in the affected area and spasm of the erector spinae and hamstring muscles. When the condition is severe, the gait is sometimes awkward and waddling.

In women, spondylolisthesis may be an important complication of pregnancy.

Diagnosis. It may be impossible to differentiate clinically between spondylolisthesis of slight degree and a ruptured intervertebral disk as cause of the symptoms. In spondylolisthesis the roentgenograms show the pathognomonic finding, however, of isthmus defects, seen best in oblique views, and the unmistakable forward slipping of the body of one vertebra on another. When surgical treatment is contemplated, it may be advisable to carry out myelography to investigate the possibility of coexisting disk rupture and spondylolisthesis.

Treatment. For cases in which the symptoms are trivial, no treatment is indicated other than periodic x-ray follow-up in children. When moderate symptoms are present, immobilization of the spine in a flexed position, by means of a plaster cast extending from the lower part of the thighs to above the costal margins, will relieve most of the acute pain. The cast should be followed by a back brace. In the milder cases an ambulatory plaster jacket or brace, together with exercises to decrease pelvic tilt and lumbar lordosis, may be adequate treatment. The majority of cases of spondylolisthesis can be treated by conservative means of this type.

In patients whose pain has continued despite conservative treatment and in children with severe or progressive slipping, surgical fusion of the last two or three lumbar vertebrae to the sacrum is indicated. It should be followed by from 6 to 12 weeks of recumbency and then by a brace for a period of from 6 months to a year after the date of operation. The functional results are usually satisfactory. In the rare cases with evidence of pressure on the cauda equina, laminectomy should be carried out.

LESIONS OF THE LUMBAR INTERVERTEBRAL DISKS

In the lower lumbar region, degenerative changes in the intervertebral disks are a frequent cause of back pain, and rupture of disk material posteriorly is the commonest cause of sciatic pain, or sciatica. Although these lesions do not necessarily affect the general health of the patient, they often result in prolonged disability. As an economic problem, disk disease is enormous, because it frequently affects the wage earner and results in the loss of many man-hours of work. In addition, in some instances a person with a ruptured disk finds himself unable to return to his previous occupation and must seek training in another field.

Etiology and pathology. Before considering the clinical picture referred to as ruptured disk, one should review the anatomy and the pathologic changes associated with aging and with trauma. The intervertebral disk consists of the nucleus pulposus, surrounded by the annulus fibrosus and enclosed above and below by the cartilaginous end-plates of the vertebral bodies (Fig. 11-8). The disk is reinforced by the strong anterior longitudinal ligament in front and by the posterior longitudinal ligament in back. These structures, along with the adjacent margins of the vertebral bodies, undergo characteristic, progressive changes from the beginning to the end of life. It is often impossible to distinguish sharply between what should be considered the normal aging process and what should be called pathologic.

In early life the nucleus pulposus is a gelatinous material with strong water-binding properties. Cells of notochordal origin are present in very young disks but gradually disappear. At its periphery the nucleus is more fibrous, blending with the annulus. Centrally its ultrastructure has been shown to consist of a network of fine collagen fibrils surrounded by mucopolysaccharides (principally chondroitin sulfates), which account for the water-binding property; at birth the nucleus pulposus is 88% water. As age increases, the water content decreases; this change is associated with and is probably the result of a decrease in mucopolysaccharide. The collagen content increases at the expense of the mucopolysaccharide. Thus as age progresses, the nucleus becomes more fibrous and loses some of its hydrodynamic properties. The young, healthy nucleus acts as a hydraulic shock absorber between two verte-

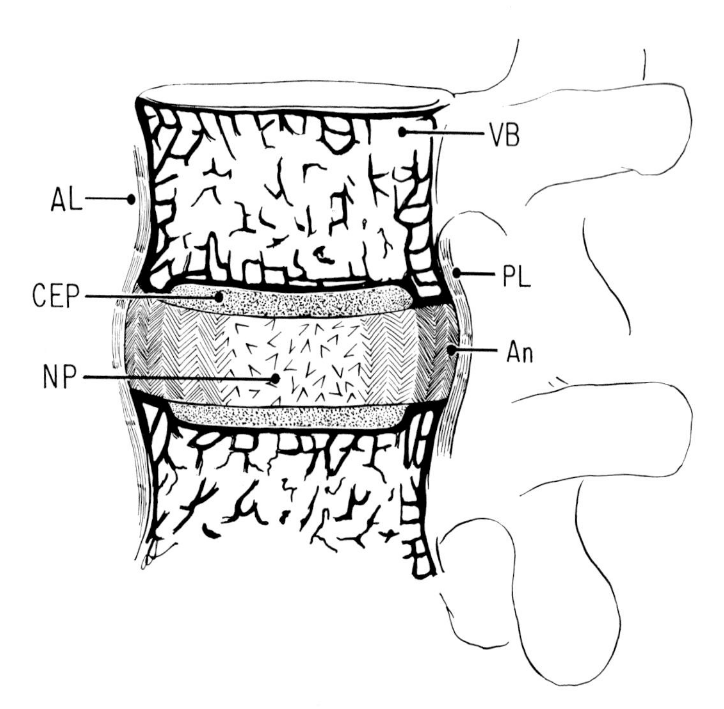

Fig. 11-8. Diagram of normal disk and adjacent vertebrae. **AL,** Anterior longitudinal ligament; **CEP,** cartilage end-plate; **NP,** nucleus pulposus; **VB,** vertebal body; **PL,** posterior longitudinal ligament; **An,** annulus fibrosus.

bral bodies. Since it is essentially a fluid, stresses placed upon it are equally distributed in all directions to the fibers of the annulus and to the cartilaginous end-plates, regardless of the position of adjacent vertebrae in relation to each other. Thus the danger of excess stress in any one area is minimized. As the disk loses some of these fluid properties with age, excessive localized stresses may occur and result in pathologic changes.

The annulus fibrosus surrounds the nucleus pulposus as a series of concentric fibrocartilaginous lamellae. The fibers of the lamellae run obliquely from one vertebra to the other, alternating in direction in adjacent lamellae. The peripheral fibers insert directly into the bony rim of the vertebral body as *Sharpey's fibers.* The more central fibers insert into the cartilaginous end-plates. The fibers of the annulus blend in front with the strong anterior longitudinal ligament and behind with the weaker and narrower posterior longitudinal ligament. Age changes become apparent in the annulus during the third decade of life, at which time the fibers become coarse and fissures appear in the lamellae. Thinning of the disks occurs as age advances. The thinning may be a very gradual process or may be abrupt. The causes of decrease in disk material include progressive dehydration and fibrosis of the nucleus pulposus, and rupture of the annulus fibrosus with extrusion of disk material. In osteoporosis, nuclear material may compress the cartilaginous end-plates, causing the centra to become biconcave. Actual ruptures of the cartilaginous end-plates may occur. A discrete amount of nuclear material may protrude into the vertebral body as a *Schmorl node.*

Associated with thinning of the disks and the aging changes in the nucleus and annulus, reactive changes occur about the margins of the vertebral bodies (Fig. 11-9). Bony spurring or lipping develops. This is similar to the lipping seen in

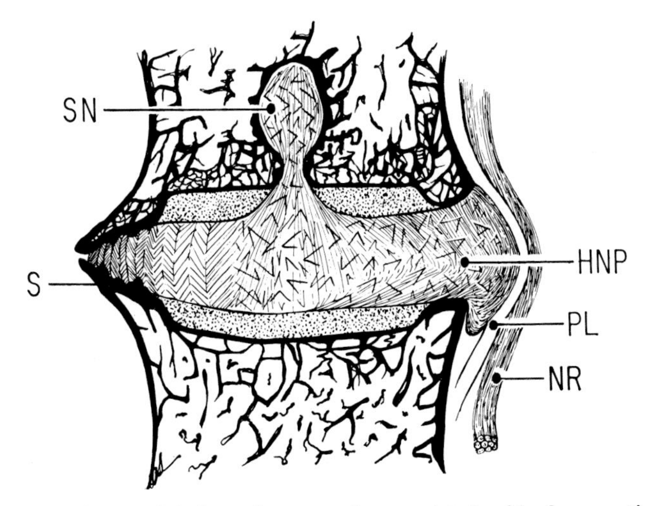

Fig. 11-9. Diagram of several lesions that may be associated with degenerative disk disease: marginal lipping, demonstrated by anterior bony spur, **S**; herniation of nuclear material through the cartilage end-plate into the vertebral body, forming a Schmorl node, **SN**; posterior herniation of nucleus pulposus, **HNP**, beneath the posterior longitudinal ligament, **PL**, displacing the nerve root, **NR.**

osteoarthritis, but since synovial joints are not involved some authors prefer to refer to this process as *spondylosis* or *spondylophytosis*. The entire complex of changes noted here is sometimes referred to as *degenerative disk disease*. Although these changes are related to aging and to wear and tear, other factors are probably involved. Why some individuals show extensive degenerative disk disease at an early age whereas others fail to do so has not been answered satisfactorily. The role of trauma in disk rupture is sometimes difficult to ascertain. Rupture of a normal disk is probably uncommon. Acute disk rupture is usually the result of a combination of acute trauma and preexisting degenerative changes. When the degenerative changes are minimal, one may assume that relatively severe trauma is required to cause rupture. When degenerative changes in the annulus are advanced, minimal trauma such as simple forward bending may cause rupture.

As degenerative changes in the disk occur, and provided that an acute rupture does not cause nerve root pressure, the patient may experience only mild or moderate back pain and stiffness. It is probable that many cases classified under lumbosacral sprain are actually manifestations of early degenerative disk disease. In most instances the degenerative changes cause few symptoms unless the back is put under abnormal stress. It is rare, however, for an individual to pass through adult life without having experienced some back discomfort. As the degenerative changes become more advanced, the disks thinner, and the marginal lipping more distinct, a certain amount of stability results from fibrosis and the approximation of adjacent vertebrae. At this stage, back symptoms may be minimal or absent; one often sees individuals in late adult life with extensive spondylosis and disk narrowing but no back complaints. These people have rather stiff backs but experience little discomfort on ordinary activity.

Herniation of nuclear material often occurs after the onset of degenerative

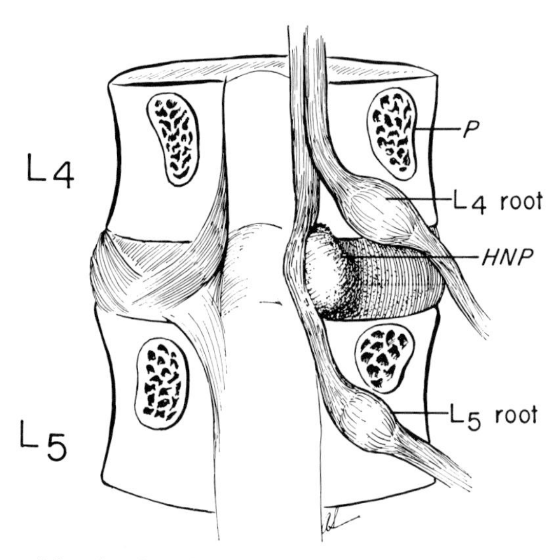

Fig. 11-10. Diagram of herniated nucleus pulposus, **HNP**, as seen from the back with spinous processes and laminae removed from the pedicles, **P.** Note that the disk protrusion between the fourth and fifth lumbar vertebrae impinges on the fifth lumbar nerve root.

changes but before they reach an advanced stage. Most disk ruptures occur in the third and fourth decades, which represent the most active period of adult life. The rupture may take place through the cartilaginous end-plate or through the annulus. Ruptures through the annulus usually occur in its thin posterior portion. Disk herniation that does not involve a nerve root may result in back pain without sciatica. From the clinical standpoint, however, diagnosis of ruptured disk is made only when nerve root symptoms are present.

The herniation of nuclear material usually occurs through a small rent or tear in the annulus just lateral to the posterior midline (Fig. 11-10). Posterior midline ruptures are less common because the posterior longitudinal ligament is thickest in its midportion. The nuclear material, in the form of a bulge 5 to 15 mm. in diameter beneath the fibers of the posterior longitudinal ligament, pushes and stretches the overlying nerve root to produce sciatic pain. Most disk herniations in the lumbar spine occur at the two lowest interspaces. Disk ruptures at the L_4-L_5 level usually involve the fifth lumbar root, which emerges between L_5 and the sacrum; ruptures at the lumbosacral level frequently involve the first sacral root, which leaves the spine between the first and second sacral segments.

Clinical picture. The patient's primary complaint is pain in the low back and leg. The pain is usually first noted in the back, and the onset may be either abrupt or gradual. Frequently the patient gives a history of previous attacks of non-radiating back pain diagnosed as lumbosacral sprain. In some cases, but by no means the majority, the onset is associated with significant trauma. Such trauma may occur when the patient is helping to lift a heavy object and the entire load is suddenly forced upon him. He may experience an immediate snapping sensation in the low back, presumably from the acute rupture of an annulus. Other patients may date the onset to a catching sensation in the back while stooping; still others cannot date the onset but have become aware of gradually increasing low back discomfort. The leg pain usually appears from a few days to several weeks after the onset of back pain, but in some instances the two are associated from the beginning, and occasionally the patient complains of leg pain alone. The limb pain is often of two types: (1) a deep, aching pain extending to the thigh and leg and (2) a sharp, lancinating pain shooting down the limb to the lateral side of the leg and ankle and even into certain toes. Both back pain and leg pain are aggravated by spinal motion and by activities such as coughing or sneezing that increase the intracranial pressure. The patient may complain also of numbness in his foot or toes, and occasionally of weakness and instability of his foot and ankle.

On examination the back is found to be held stiffly. The lumbar lordosis is flattened and often reversed. The lumbar spine may be shifted to the left or right (*sciatic scoliosis,* Fig. 11-1). The scoliosis may be toward or away from the painful side, the direction apparently being determined by the relationship of the protruded disk to the nerve root. If the protrusion is lateral to the root, the curvature will be to the side opposite the lesion. If the protrusion lies medial to the root, the curvature is to the same side as the lesion. Thus the back assumes the curvature most favorable to the relief of root pressure. Palpation of the back discloses tightness of the lumbar muscles from spasm, and tenderness usually localized to the interspinous ligaments of the lowest one or two lumbar segments.

Abnormalities in the lower limbs result from irritation and injury of the nerve root. Since as a rule the root is not completely destroyed by the disk protrusion, the findings are variable. Damage to the nerve root is manifested by muscle weakness and atrophy, reflex changes, and hypesthesia and other sensory disturbances. Since the first sacral root usually supplies sensation to the lateral aspect of the leg and foot, hypesthesia in this area suggests a disk protrusion at the lumbosacral level. The Achilles reflex is also supplied at this level; accordingly, an absent or diminished ankle jerk is frequently found with lumbosacral disk lesions. Involvement of the fifth lumbar root by a protrusion at the L_4-L_5 level may produce hypesthesia of the dorsum of the foot and first two toes, as well as weakness of the extensor of the great toe. Motor impairment is quite variable, however, and may be so slight that it can be detected only by special tests such as electromyography. Disk herniations above the fourth lumbar vertebra, which are uncommon, may involve roots making up the femoral nerve; accordingly they may be associated with anterior thigh pain and alterations in the knee jerk. Although disk herniations may be noted at several levels simultaneously and although a large single herniation may affect more than one root, neurologic changes involving multiple roots are suggestive of other lesions, such as spinal cord tumors.

Signs of nerve root irritation may be demonstrated by any method of increasing tension on the root. Of these the straight leg-raising test (Fig. 11-2) is the most obvious. The sacral plexus roots making up the sciatic nerve join within the pelvis to emerge through the greater sciatic foramen posteriorly. The sciatic nerve passes down the posterior aspect of the thigh to continue as the tibial nerve and terminate as its medial and lateral plantar branches. As the limb, with the knee fully extended, is flexed at the hip, tension on the nerve is increased; actual movement of the roots

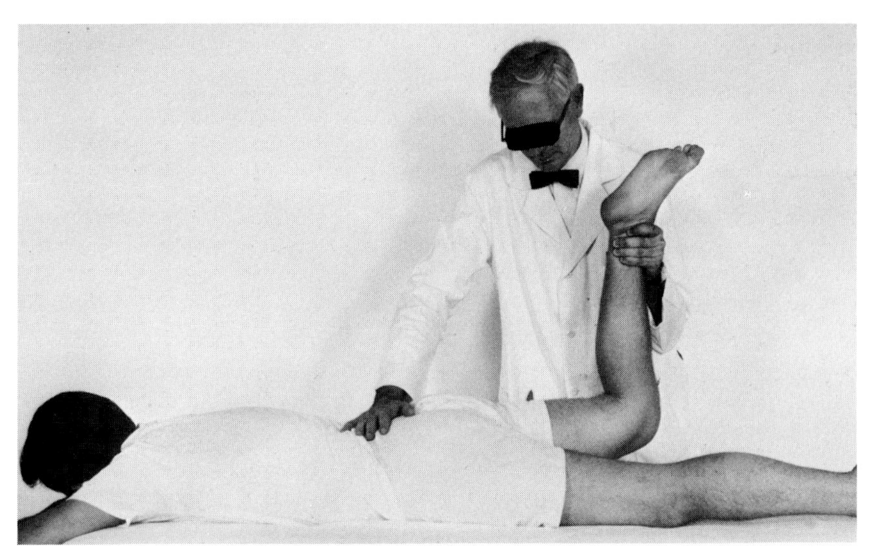

Fig. 11-11. Reverse leg-raising test. Hyperextension of the hip with the knee flexed places tension on the femoral nerve and its roots, causing pain roughly proportional to root irritation.

within the foramen may be demonstrated in anatomic preparations, and a small manometer placed beneath the root will register a positive pressure. Thus straight leg raising stretches the already taut root over the ruptured disk and causes an increase of pain. There are numerous variations of this confirmatory test. When the limb has been raised to the point where pain is first noted, the patient may be asked to strain, cough, or flex his neck, or his foot may be passively dorsiflexed by the examiner, and increased sciatic pain will result. If the hip and knee are flexed and the knee then extended to the point of discomfort, the tibial nerve can often be felt as a tight band in the middle of the popliteal space. A gentle flick of the nerve by the examiner's thumb may then send pain shooting up to the buttock or down to the foot. A bilaterally positive straight leg-raising test suggests the presence of a large disk protrusion. For lesions involving the third or fourth lumbar roots the reverse leg-raising test (Fig. 11-11) is helpful. This test, done with the patient in the prone position and the knee flexed, places tension on the femoral nerve and its roots.

Diagnosis. Presumptive diagnosis of intervertebral disk rupture is made on the history and physical findings. It must be remembered, however, that most of the findings are based only upon the presence of nerve root irritation. Although disk herniation is the causative lesion in the great majority of cases of root irritation at the lowest two lumbar segments, the list of conditions that have been found to

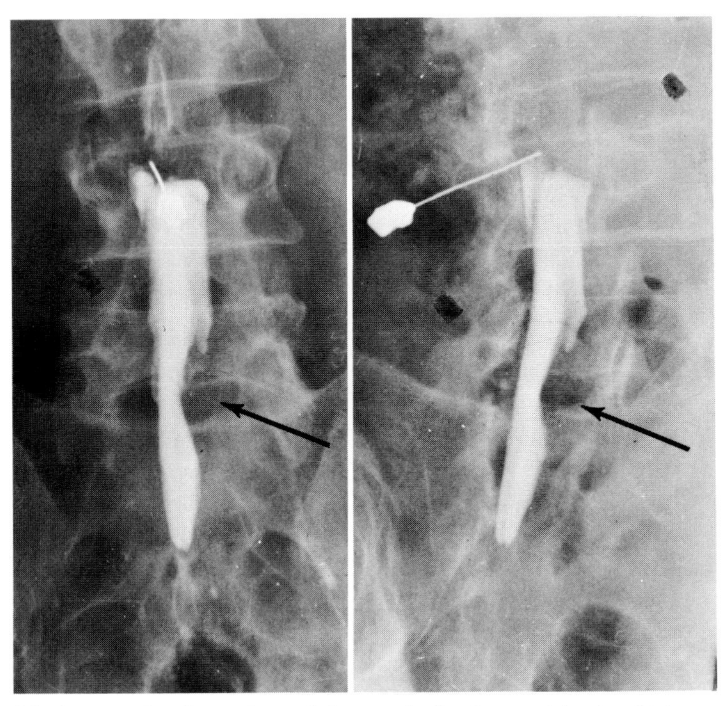

Fig. 11-12. Myelogram showing a large defect at the lumbosacral level, which at surgery was proved to have been caused by a herniated disk.

mimic disk rupture is quite long. Primary or metastatic tumors of the spinal cord, cauda equina, spinal nerves, lumbosacral plexus, or vertebrae have all been found to produce symptoms and signs very similar to those of a ruptured disk. Inflammatory conditions such as spinal tuberculosis, vertebral osteomyelitis, and Strümpell-Marie arthritis can cause irritation of a single nerve root. Pain mimicking that of nerve root compression may result from arterial insufficiency or from herpes zoster. Many of these affections can be brought to light by means of a careful history and physical examination. The history of a mastectomy several years before, or of a little blood in the sputum, should stand out as a red flag because it may be the physician's only warning that this ruptured disk is actually metastatic carcinoma from breast or lung. Some of these conditions may be ruled in or out on the basis of routine spinal roentgenograms. In other instances, particularly when the history and physical findings are atypical or suggest involvement of more than one nerve root, myelography (Fig. 11-12) may be necessary. Electromyography may be helpful in establishing the diagnosis and in localizing the lesion to a particular nerve root.

Treatment and prognosis. Since disk herniation is an acute lesion superimposed upon a degenerative process common in greater or lesser degree to all adults, one cannot expect treatment to produce the strong, painless, flexible back of a teen-ager. Accordingly the aim of treatment is to provide relief of pain and a back strong enough to allow the patient to resume his normal occupation. Occasional recurrent symptoms should be expected, and it is possible that disk rupture may occur later at another level. Nevertheless, in most cases satisfactory relief of the symptoms can be obtained by conservative means. Some of these have been described on p. 314. In the acute stage, rest in bed is the treatment of choice. The majority of patients will be most comfortable in a semi-Fowler position with spine, hips, and knees slightly flexed. Traction applied to the legs or pelvis may help. Many patients can be treated conservatively at home. Those who fail to respond to home treatment can often be relieved successfully by the stricter rest program accomplished in hospital. As the patient's condition improves, exercises to strengthen the abdominal muscles and decrease the lumbar lordosis are quite helpful.

Manipulation has been used in the treatment of the ruptured disk by both qualified and unqualified practitioners. Various forms of manipulation have been advocated, including bilateral forced flexion of the hips and knees on the pelvis with an accompanying rotary movement, bilateral or unilateral straight leg raising, and hyperextension of the spine or hips. The movements are gross, the lesion small. Nevertheless, in some instances sudden and dramatic relief of pain has been noted. Such relief may have resulted from a slight shift in the position of the nuclear material, or the stretch may have severed the nerve root. Forceful manipulation in patients with osteoporosis, arthritis, or destructive lesions of the vertebrae may cause serious damage.

Certain patients, probably less than one fourth of those with ruptured lumbar disks, require surgical treatment. The indications for disk surgery are (1) severe symptoms that fail to improve on an adequate conservative program, (2) progressive neurologic involvement, and (3) moderately severe symptoms with failure of conservative therapy when economic or other factors make it difficult for the patient to continue nonsurgical treatment. It is unfortunate that many patients believe that surgical removal of the disk, like an appendectomy, eliminates the offending condition. This, of course, is not true. Degenerative disk disease remains, and symptoms arising at other levels are always possible. Surgical treatment usually consists of a partial laminectomy, exposure of the protruding mass, and excision of the mass as well as all nuclear material remaining in the disk. When back symptoms have been a prominent feature, spinal arthrodesis may be carried out at the time of disk surgery. In most instances, however, a satisfactory result can be obtained without arthrodesis.

OTHER LESIONS OF MUSCLE AND FASCIA

Myofascitis

Localized areas of tenderness and sometimes induration are occasionally found in the lumbodorsal fascia and its underlying muscle, as well as along the attachments of these structures to the iliac crests. The pathologic nature of these areas is unknown; it seems likely that a number of different factors are involved. The terms *myositis* and *fibrositis* are sometimes used to describe such conditions, but the presence of a microscopically demonstrable inflammatory reaction in the designated tissue is questionable—perhaps the term *myofascial pain,* used by Wiles, is sufficient. Most of these affections respond to the use of local heat and massage. When the lesion is well localized, its infiltration with a local anesthetic may bring relief.

Herniation of fascial fat

Lobules of fat may herniate through small defects in the lumbodorsal fascia. In many instances such nodules are painless; they are frequently palpated as an incidental finding in patients with low back pain of other cause. At times, however, such a nodule is quite tender and is associated with diffuse low back pain. Injection with a local anesthetic relieves both the pain and the tenderness. If the symptoms recur after several such injections, excision of the nodule is recommended.

COMMON OSTEOARTICULAR LESIONS OF INFECTIOUS, NEOPLASTIC, OR TRAUMATIC NATURE

Specific pathologic conditions of the bones and joints, not limited necessarily to the lumbosacral or sacroiliac joints, often cause low back pain directly or lead to an impaired mechanical resistance that may augment the incidence of sprain. Low-grade infections, neoplastic processes, and old fractures are examples. Spinal tuberculosis or carcinoma of the lung that has metastasized to the lower lumbar or sacral area may be first manifested as low back pain, sometimes with nerve root signs which mimic a herniated disk. In young adults Strümpell-Marie arthritis is a frequent cause of low back pain. In stout individuals over 40 years of age, with proliferative bone changes about the joint margins, susceptibility to low back pain following acute or chronic strain is greatly increased, and the symptoms are likely to persist for long periods. Arthritic changes in the lower lumbar and lumbosacral articular facets may cause low back pain, sometimes accompanied by pain radiating to the legs.

Low back pain is occasionally associated with a localized increase of density in the ilia adjacent to the sacroiliac joints; this condition is usually found in women and has been called *osteitis condensans ilii.* In adolescents sacroiliac symptoms are sometimes associated with roentgenographic changes in the sacroiliac joint surfaces, resembling bone destruction and suggestive of osteochondritis.

VISCERAL LESIONS THAT MAY CAUSE LOW BACK PAIN

Low back pain unassociated with local osteoarticular changes may be caused by any of a great variety of visceral diseases. This pain, although located definitely over a bone or joint by the patient, may be entirely of referred type. A carefully performed history and physical examination will usually reveal the source of the pain.

Among such visceral causes of low back pain are various disorders of the gastrointestinal tract. The diagnosis is not easily confused since the pain is not usually aggravated by mechanical factors such as bending or lifting. Chronic pancreatitis or posterior perforation of an ulcer may give rise to lumbar pain.

Chronic progressive occlusion of the aorta or the common iliac arteries may cause back, buttock, or lower limb pain. Such pain may simulate that of disk disorders. It differs, however, in that it is not aggravated by single motions such as forward bending or coughing but, rather, is brought on by more prolonged activity such as walking. Changes in the peripheral pulses and the presence of calcification in roentgenograms of the aorta or iliac arteries are helpful in suggesting the diagnosis, which may be confirmed by arteriography.

Tumors of the spinal cord are often accompanied by low back pain. Tumors of the cauda equina are likely to be evidenced by pain radiating into one or both legs and by spasm of the hamstring muscles. Syringomyelia, lateral sclerosis, tabes, and similar affections occasionally give rise to pain in the low back.

Tumors of the female pelvic organs, retroversion and retroflexion of the uterus, and endometriosis may lead to referred pain in the low back. Overemphasis has sometimes been placed on these conditions, however, with the result that gynecologic operations perhaps have been done in the hope of relieving low back pain that was actually of musculoskeletal origin. In rare instances the sacral plexus has been involved by endometriosis, causing typical sciatica during menstruation.

Most of the diseases of the genitourinary tract may at times cause moderate or severe low back pain. This is particularly true of infections and tumors of the prostate gland, as well as lesions of the kidneys and other retroperitoneal structures. Involvement of the lumbar or lumbosacral plexus may lead to pain radiating into the legs. Febrile illnesses, such as influenza, are often accompanied by generalized muscular pain that may be most noticeable in the low back.

COCCYGODYNIA

Pain about the coccyx and lower part of the sacrum, or coccygodynia, is in many cases a result of sprain of the sacrococcygeal ligaments and often follows a direct blow or fall, such as may be occasioned by sitting down suddenly or sliding down a flight of stairs. It is sometimes caused by sitting for a long period on a hard surface. The pain is not always experienced immediately after the injury but may develop gradually later. Coccygodynia sometimes begins insidiously without evident trauma. It is occasionally seen after severe weight loss, when thinning of the buttocks leaves the coccyx relatively unprotected and vulnerable to trauma.

Clinical picture. There is usually an aching, nagging pain located near the sacrococcygeal junction. At times there may be shooting pains in the buttocks or even down into the legs. These pains are especially severe when the patient is sitting on a hard surface or when arising after sitting for any length of time. There may be considerable pain on defecation. When the condition occurs in an individual with a psychoneurotic personality, the symptoms may be greatly exaggerated. Examination shows a tender coccyx, which on rectal examination is sometimes rough and angulated. It may be abnormally mobile or fixed.

Diagnosis. The diagnosis should be made with caution since pain in the coccygeal region is often associated with arthritis of the low back or with psychoneurosis. Roentgenograms occasionally show deviation of the coccyx forward or to one side as a result either of injury or of developmental variation, but identical changes occur in many individuals who have no symptoms.

Treatment. In acute traumatic cases showing displacement of the coccyx, the deformity may be reduced by gentle manipulation, especially if the coccyx has been angulated forward and its motion appears to be restricted. Hot baths, strapping across the buttocks and the low back, and massage may afford considerable relief. The patient should be instructed to sit on cushions or soft chairs. Occassionally, however, sitting on cushions that are very soft may cause the buttocks to spread and may thus increase the pain. The use of a large rubber ring in sometimes indicated to relieve the pressure occasioned by sitting. An alternative method is to have the patient sit on a firm cushion, which is placed forward in the chair.

If the pain cannot be relieved by nonsurgical measures, excision of the coccyx, which can be carried out very simply, will often relieve the patient of all symptoms. Before resort is made to operation, however, great care should be taken to remove all contributing causes of the discomfort, since in occasional cases the diagnosis is somewhat uncertain and the relief of pain after operation is incomplete. Any psychoneurotic tendency contraindicates coccygectomy.

COMMENT

Current treatment of patients with low back pain is not uniformly satisfactory. In some cases the symptoms may be prolonged because of insufficient rest of the back in the acute stage, inadequate support of the back, or too early resumption of activity. With conservative treatment that is efficient and thorough, the great majority of mechanical disorders of the back can be relieved without resort to surgery. On the other hand, when a patient is unable to afford a prolonged period of such care, surgical treatment may be desirable for economic reasons to enable him to resume his occupation. One cannot be too cautious, however, in operating on the back of a patient who has a pending compensation or insurance claim or who shows any evidence of an abnormal psychosomatic state.

Many unorthodox practitioners and cultists thrive on the too frequent failure of physicians to appreciate or employ optimal treatment of the common affections that cause low back pain. Most of the cultists' activity involves manipulation to break adhesions and so relieve pain. The patient may be told that a dislocated joint or bone has been put back into place—a statement that makes a profound impression on the lay mind. Much misunderstanding on the part of the patient will be avoided if the physician can recognize and explain carefully to the patient the true nature of the problems involved in the treatment of low back pain.

CHAPTER 12 Affections of the hip

The ball-and-socket structure of the hip joint allows a wide range of motion that is exceeded in no other joint of the body except the shoulder. At the same time a remarkable degree of stability is provided by the close fit of the femoral head into the acetabulum and its deepening lip, the glenoid labrum, and by the support of the strongest capsular ligaments and the thickest musculature of the body. Of all the joints, the hip is most deeply situated. This relative inaccessibility increases the difficulty of diagnosing hip lesions, renders thorough operative exposure of the joint arduous, and so plays a large part in making affections of the hip one of the most challenging parts of orthopaedic surgery.

The blood supply to the femoral head is of crucial importance in disorders of the hip, and especially in relation to hip affections of childhood and adolescence, hip fractures, and hip surgery. Injection studies by Trueta and others have shown that the sources of blood for the femoral head vary with age (Fig. 12-1). They include the following: (1) retinacular arteries, which arise from the medial circumflex artery, extend up the femoral neck between the bone and its synovial covering, and form the main vascular source throughout life and usually the only source in young children; (2) the artery of the ligamentum teres, arising from the obturator or medial femoral circumflex artery, which is inconstant and supplies at most only a limited segment of the head; and (3) the nutrient artery or arteries of the femoral shaft, which arise from perforating branches of the deep femoral artery, contribute little or no circulation to the epiphysis in children, and may be a supplementary source in some adults. The posterior retinacular arteries, sometimes inaccurately termed capsular arteries, are the chief source of circulation to the femoral head.

The especial affections of the hip that are discussed in this chapter can be grouped conveniently according to the location of the lesion. *Avascular necrosis of the femoral head, coxa plana,* and *slipping of the upper femoral epiphysis* are common and important entities characterized by pathologic changes in the femoral head or at the epiphyseal plate. *Congenital coxa vara* and *coxa valga,* marked by alterations in the angle of femoral neck to femoral shaft, are less common. *Pathologic dislocation of the hip* and *intrapelvic protrusion of the acetabulum* are characterized by a changed relationship between femur and pelvis. To be remembered also in the diagnosis of hip affections are lesions of the soft tissues, which include *synovitis, bursitis,* and *snapping hip.*

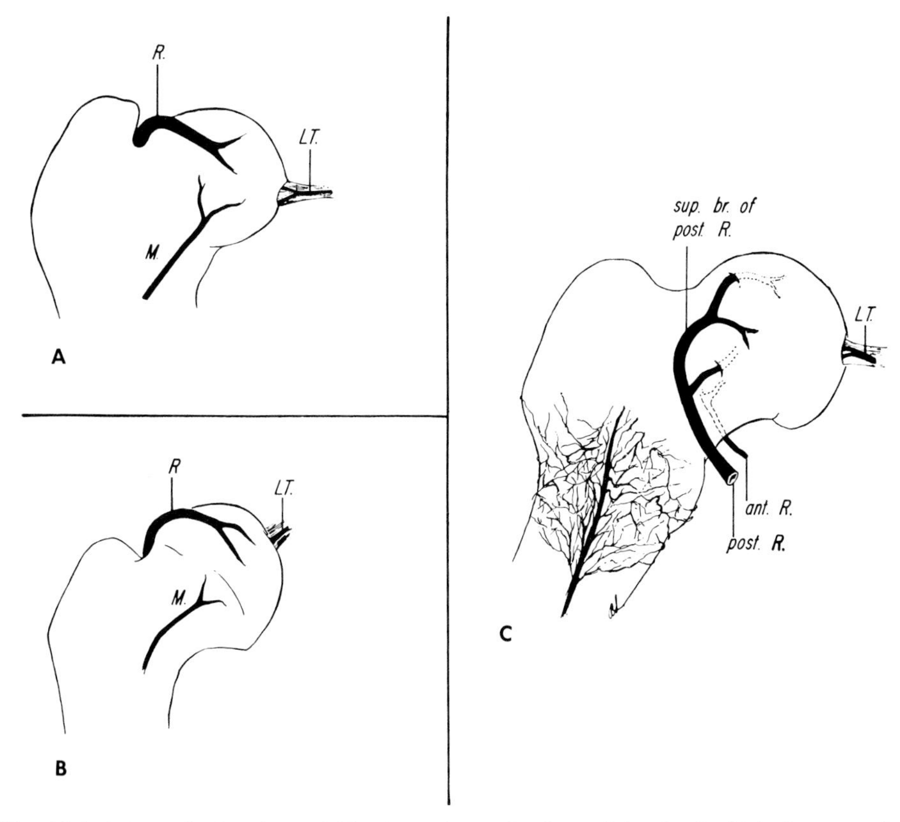

Fig. 12-1. Schematic drawings of blood supply to the femoral head, of which there are three sources:

1. Retinacular arteries, **R**, ascend the neck between synovium and bone. Their posterior group, arising from the medial femoral circumflex, includes a superior branch (the lateral epiphyseal artery of Trueta) that is the main vascular supply to the head in adults and usually the only supply in children prior to penetration of the cartilage of the femoral head by the artery of the ligamentum teres. The anterior retinacular artery, from the lateral femoral circumflex, appears to be inconstant and relatively unimportant.

2. Artery of ligamentum teres, **LT**, arising from obturator, medial femoral circumflex, or both, is present in only 33% of children but over 70% of adults. Never the chief vascular supply to the head, it may be supplementary to the vital supply from the retinacular arteries.

3. Superior branch of nutrient artery of femoral shaft, **M**, ascends in medullary cavity, traverses metaphysis, and anastomoses with cervical branches of the retinacular arteries. Not crossing the epiphyseal plate before age 13 years, it is unimportant as concerns the capital epiphysis in children but a supplementary source for the head in some adults.

A, At birth (anterior view) the retinacular and metaphyseal supplies are constant and that via the liagmentum teres is inconstant. **B**, At ages 4 to 7 years (anterior view) the retinacular arteries are usually the only significant supply to the bony epiphysis, since the ligamentum teres and metaphyseal vessels meet cartilaginous barriers. In preadolescence the artery of the ligamentum teres begins to enter the bony epiphysis from above, and in adolescence the metaphyseal arteries begin to enter it from below. **C**, In the adult (posterior view) the head is supplied chiefly by the retinacular arteries, the ligamentum teres and metaphyseal vessels being inconstant or meager contributors.

AVASCULAR NECROSIS OF THE FEMORAL HEAD (ASEPTIC NECROSIS)

Necrosis of the femoral head as the result of impairment of its blood supply is the common complication of a variety of affections spanning an age range from early childhood to late adult life. The term *aseptic necrosis* was originally used to differentiate this condition from necrosis associated with pyogenic and tuberculous infections of the hip. The designation *avascular necrosis,* implicating a disturbance of the circulation, is preferred by many.

Etiology. There are a number of entities in which trauma to the major blood vessels supplying the hip seems to be the cause of necrosis. The commonest of these is fracture of the neck of the femur. In such injuries, tearing of the retinacular vessels is probably the cause of the necrosis, which occurs in 20% to 30% of all femoral neck fractures. Traumatic dislocation of the hip, in which the ligamentum teres as well as the retinacular vessels may be damaged, is also a cause of avascular necrosis. In the treatment of congenital dislocation of the hip, forceful manipulation or wringing-out of the joint capsule by fixing the hip in a position of extreme rotation may result in necrosis of the femoral head. A similar complication may be caused by the forced manipulation of a slipped upper femoral epiphysis.

In another group of affections impairment of the circulation seems to occur in the small vessels and sinusoids of the femoral head. Avascular necrosis may complicate sickle cell disease or other hemoglobinopathies, caisson disease (decompression sickness), Gaucher's disease, gout, systemic lupus erythematosus, and renal transplantations. It may occur after any prolonged steroid therapy and in such cases has been suspected of resulting from fat embolism.

In other instances avascular necrosis arises in patients in whom there is no history of significant trauma or underlying disease. One of these, Legg-Calvé-Perthes disease, is discussed in the following section. Another is *idiopathic avascular necrosis,* an affection found most often in men between the ages of 30 and 60 years. Bilateral involvement is frequent. Clotting defects have been found in many of these patients, and there is often a history of chronic alcoholism.

Pathology and roentgenographic picture. Infarction of the femoral head may be total or incomplete. If incomplete, necrosis may be limited to one segment of the femoral head or may be spotty in distribution (Fig. 12-2). Infarction results in death of the marrow elements, mainly fat in the case of the femoral head, and in death of the cancellous bone. This is manifested by the degeneration and disappearance of osteocytes from their lacunae within the bone trabeculae. Necrosis of the femoral head, like that in other areas of the body, results in marked hyperemia of the tissues adjacent to the infarct. This is followed by invasion of the infarcted area by young connective tissue and new blood vessels. As revascularization proceeds, both osteoclastic resorption of dead trabeculae and osteoblastic repair with new bone take place. Either of these cellular activities may predominate. There is great variation in the rapidity and degree of the cellular reaction. New bone may be laid down on a core of dead trabeculae, resulting in an increase in the total amount of bone. On the other hand, if resorption predominates there is less bone in a given area. Resorption may be so extensive in the subchondral areas that collapse of the articular surface occurs, resulting in flattening of the femoral head. The process of removal of dead

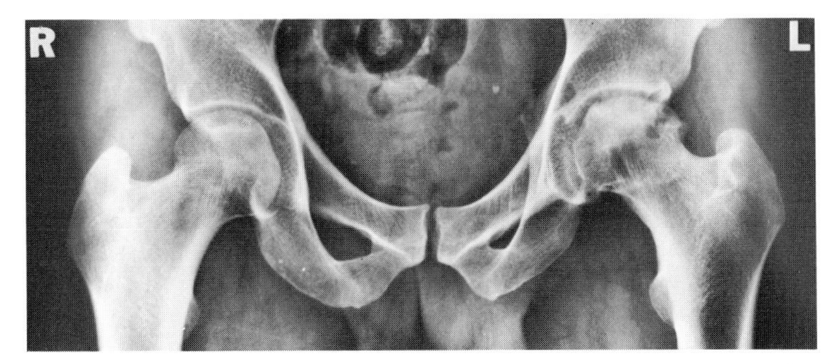

Fig. 12-2. Idiopathic avascular necrosis of the femoral heads. On the right, the early change is a radiolucent area underlying the weight-bearing surface. On the left, advanced changes include partial collapse of the femoral head from pathologic fracture of its necrotic segment, and secondary alterations in the acetabular roof. The patient was a 34-year-old man who had had mild left hip pain for 3 months.

bone and replacement by new bone has been referred to by Phemister as *creeping substitution.* In the femoral head it may take 2 to 4 years to be completed, and in the process the head may become severely deformed. In some patients one or both humeral heads also may show the changes of avascular necrosis.

During the course of avascular necrosis, the head of the femur, or portions of it, may appear quite dense on the roentgenogram. Such increase in density may be apparent or real. Apparent increase in density may be the result of osteoporosis of the surrounding bone due either to hyperemia associated with the infarction or to disuse atrophy. The avascular portion of bone retains its density, since bone resorption does not occur in the absence of a blood supply. Real or absolute increase in density of the femoral head may result from one or more of several changes. Collapse of the femoral head increases the amount of bone in the collapsed area, thus increasing its density. In the healing process if new bone is deposited on existing dead trabeculae, the roentgenographic density is increased. Finally, as noted by Johnson, the saponification of necrotic fat can lead to precipitation of calcium soaps in the marrow spaces.

Clinical picture. The symptoms and signs of avascular necrosis of the femoral head vary with the underlying cause and the age of the patient. Onset of the symptoms may either precede or follow the first appearance of roentgenographic changes. In children a limp and slight spasm about the hip are often the first clinical manifestations, followed by pain present on weight bearing and often referred to thigh or knee. In adults pain in the groin is usually the first symptom. It may be referred to the thigh or the knee, may be sudden in onset, and is usually worse on standing and walking and relieved by rest. Spasm about the hip is an early sign. In later stages muscle atrophy and restriction of abduction and internal rotation may be noticeable.

Treatment. The treatment of avascular necrosis varies with the underlying disease process and with the age of the patient. In the child, treatment is usually conservative protection of the hip joint in abduction for a prolonged period until reconstitution of the femoral head is complete; this is discussed in the next section. In the adult, surgical treatment is usually needed. Of a variety of useful procedures, those most often indicated are intramedullary bone grafting, osteotomy, interposition or replacement arthroplasty, and arthrodesis.

COXA PLANA (LEGG-CALVÉ-PERTHES DISEASE, AVASCULAR NECROSIS OF THE CAPITAL FEMORAL EPIPHYSIS)

Coxa plana is a flattening of the epiphysis of the head of the femur due to osseous changes without primary involvement of the articular cartilage. It was

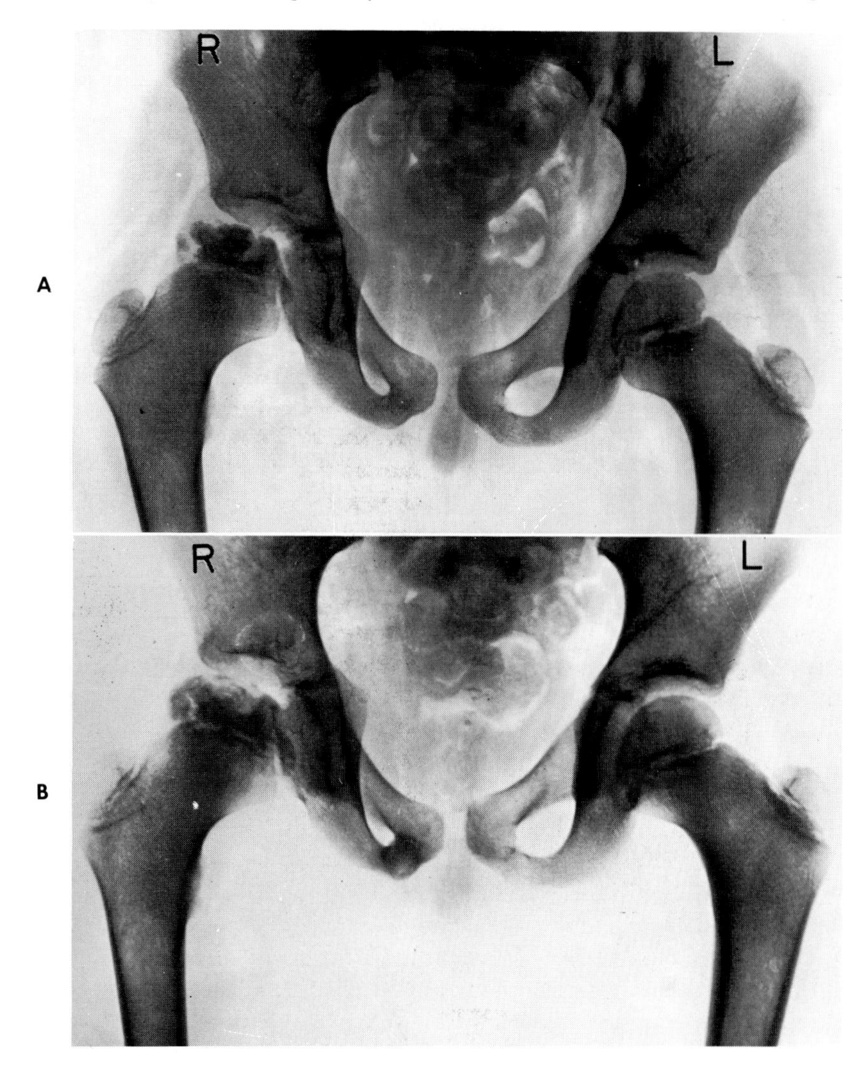

Fig. 12-3. Roentgenograms of the pelvis showing coxa plana of the right hip. **A,** Note fragmented appearance of the right capital epiphysis with areas of increased density and flattening. The widened capital epiphysis is not covered well by the roof of the acetabulum. The patient is a boy 8½ years of age. **B,** Nine months later. Note reshaping of the right capital epiphysis with new bone formation. Fragmented appearance is still present. **C,** Five years later. Note well-rounded right femoral head, which is slightly larger than the left. The treatment, which lasted for 3 years, consisted of traction in recumbency followed by crutches and a hip sling.

described in 1909 by Legg in Boston and by Waldenström in Sweden, and in 1910 by Calvé in France and Perthes in Germany. Before the affection was recognized as an independent entity, these cases were believed to represent a mild form of tuberculosis of the hip.

Coxa plana occurs in children between the ages of 3 and 12 years and is found much more commonly in boys than in girls. Unilateral involvement is more common than bilateral.

Etiology. The cause of the avascular necrosis occurring in coxa plana has not been established. Injury or disease of the blood vessels supplying the femoral head, or within the head itself, may be responsible. Increased intraarticular pressure as the result of disease or trauma has been suggested as a cause of ischemia of the femoral head.

Pathology. Massive subchondral necrosis of the bone and marrow of the epiphysis of the femoral head takes place. It may involve the entire epiphysis or only a part of it. This is followed by slow resorption of the dead bone and its replacement by new, living bone. Collapse of the head may occur, producing a characteristic flattening. As new bone is formed, healing takes place, and the fragmented areas become reorganized into a smooth, regular mass of bone that rarely, however, regains the shape and appearance of a completely normal epiphysis. The regeneration and return to the normal shape usually become more nearly complete in the younger than in the older child. The articular cartilage remains smooth; it continues to grow. The synovial membrane and capsule become thickened.

Clinical picture. The most constant early sign is a limp, which is associated with muscular spasm and may or may not be accompanied by pain. The pain is often referred to the inner side of the thigh or knee. As the disease progresses, the limp persists and may become more severe. The hip suffers slight limitation of all motions, and particularly of internal rotation and abduction. In rare instances all of the symptoms may be exaggerated and the disability may be severe. A moderate flexion and adduction deformity with prominence of the greater trochanter may

C

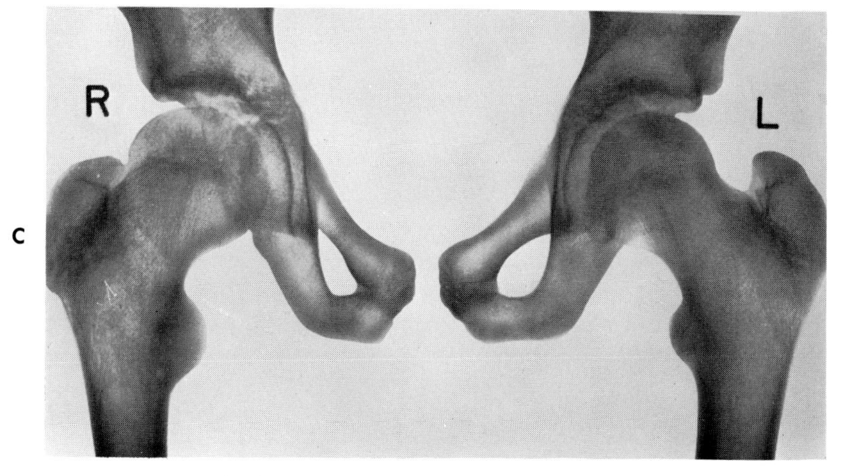

Fig. 12-3, cont'd. For legend see opposite page.

develop. There is atrophy of the muscles of the hip, thigh, and leg. Thickening of the joint capsule is sometimes palpable, giving a boggy feel, and usually slight shortening of the leg is demonstrable. During the stage of repair the signs and symptoms become less severe, and the hip may appear almost normal. Later in life, however, the hip may show evidence of a degenerative process that has the appearance of osteoarthritis (p. 150).

Roentgenographic picture. The stages of degeneration and repair can be visualized by x-ray examinations (Fig. 12-3). Small areas of resorption appear in the femoral neck adjacent to the epiphyseal plate. In early cases the epiphysis shows increased density, thinning or flattening, and in many instances an appearance suggesting fragmentation. Within the head appear dense areas that become less conspicuous as healing takes place. The whole head, or only a major part of the head, may be involved. The neck becomes broad and short and may present a coxa vara. The acetabulum changes in shape to correspond with the deformity of the femoral head. Flattening and broadening of the head may cause it to protrude laterally from the acetabulum. Shands has observed in a high percentage of the cases an increase in the normal anterior torsion of the femur (anteversion), as determined by roentgenograms made with a special positioning technic.

Diagnosis. Coxa plana is to be differentiated from transient synovitis, rheumatoid arthritis, early tuberculosis, slipping of the upper femoral epiphysis, congenital dysplasia of the hip, and congenital coxa vara.

Bilateral coxa plana must also be distinguished from *multiple epiphyseal dysplasia,* an uncommon developmental affection, often familial, which is usually manifested by hip pain and stiffness, as well as roentgenographic changes in the femoral head epiphyses. Commonly the child with multiple epiphyseal dysplasia also shows changes in the knees, stubby digits, and shortness of stature.

Prognosis. Coxa plana is a self-limited disease with a constant tendency toward spontaneous recovery as the cycle of avascular necrosis, revascularization, resorption of necrotic bone, and its replacement by living bone proceeds. The prognosis in partial involvement of the femoral head is better than that when the entire head is affected. It is more favorable in younger than in older children. It has been demonstrated repeatedly that relieving the strain of weight bearing by means of traction, recumbency in plaster or splints, or crutches during the stages of degeneration and repair minimizes the progressive distortion of the head. The symptoms may disappear, but further flattening of the head is likely to develop and persist. The functional end result when weight bearing is prevented is usually very good, but on the other hand, there may be a permanent limp and restriction of hip motion. Without early treatment the femoral head and neck may become greatly deformed (Fig. 12-4), which is likely to lead to severe disability of the hip.

Treatment. The primary consideration is protection of the softened femoral head from mechanical stresses leading to deformity and, in later years, probably to degenerative arthritis from incongruity of the joint surfaces. As far as possible the deforming forces of weight bearing, muscular tension, and subluxation should be avoided. Containment of the femoral head within the concave hemisphere of the acetabulum is an important factor in preventing deformity of the head.

When pain, muscle spasm, or clinical deformity is present, rest in bed with skin

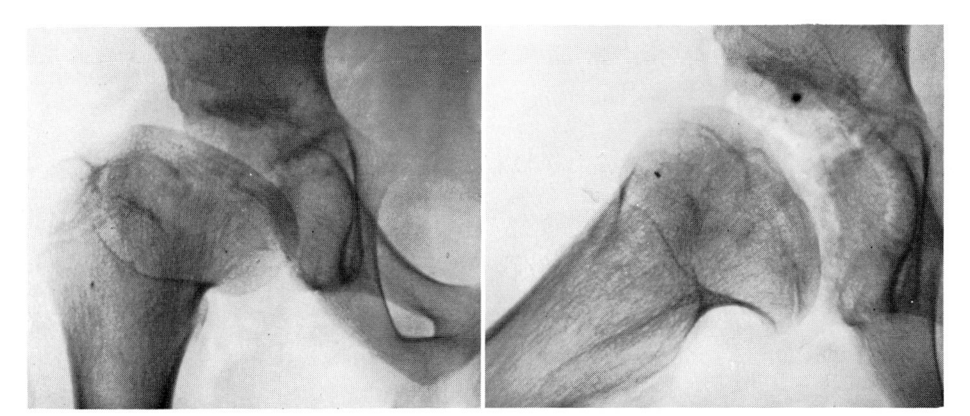

Fig. 12-4. Anteroposterior and lateral roentgenograms of coxa plana in a 12-year-old boy who had had a limp for 3 years before being brought to a physician for treatment. Note the flat, wide femoral head, the short, thick femoral neck, and secondary changes in the acetabulum.

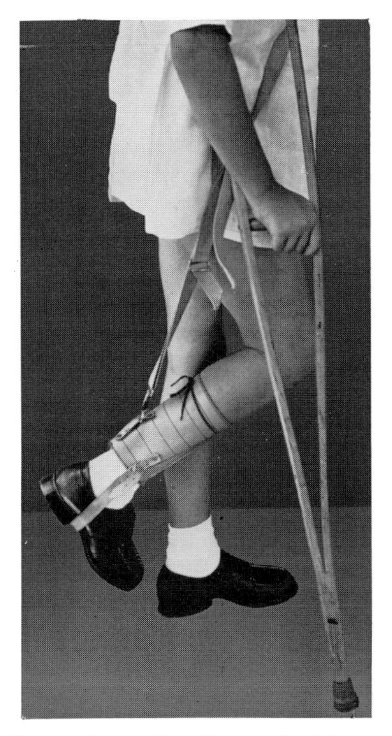

Fig. 12-5. Snyder-Fort hip sling and crutches. Leg cuff with shoe strap is attached to a strap that circles the opposite shoulder to suspend the hip and knee in flexion and prevent weight bearing on the affected hip.

traction, with the hips partially abducted, is indicated and should be continued until there is no pain or muscle spasm on motion. In some children's hospitals skin traction in bed is continued until roentgenographic evidence of nearly complete repair is present, which may be as long as 1 to 2 years.

In the early days of bone regeneration, however, keeping the femoral head deeply seated and covered in the acetabulum by means of a brace or cast applied with the hip in abduction may be preferable to recumbency and traction. Some orthopaedists allow weight bearing when the hips are widely abducted in casts or brace, believing that in abduction the femoral head, deep in the acetabulum, is unlikely to undergo subluxation or become misshapen. Use of a brace holding the affected hip in wide abduction and internal rotation, with the knee flexed, and allowing crutch walking with weight on the normal leg has given good results. The hip may be protected, less effectively, by the use of crutches accompanied by a high-sole shoe on the sound leg or by crutches and a sling from the lower leg of the affected side to the shoulder of the opposite side (Fig. 12-5). Strenuous activity involving weight bearing on the affected hip should be avoided until the roentgenograms indicate that an end stage of repair has been reached.

Innominate osteotomy may be used in older children, before the head becomes deformed, to cover it more effectively by the acetabulum and so minimize subluxation and deformity.

Children under 4 years of age with only partial involvement of the femoral head have a favorable prognosis and may require little treatment other than periodic observation.

COXA VARA AND COXA VALGA

Coxa vara is an abnormality of the upper end of the femur, consisting of a decrease in the neck-shaft angle (angle of inclination) and hence resulting in a shortening of the limb (Fig. 12-6). Cases of coxa vara can be classified according to whether the primary anatomic change is in the neck *(cervical),* in the head *(epiphyseal),* or in both neck and head *(cervico-epiphyseal).*

Coxa valga, an increase of the neck-shaft angle, is normal in infants before

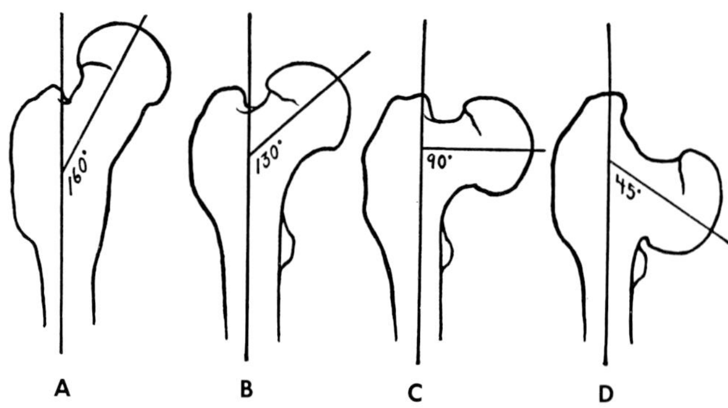

Fig. 12-6. Angles of femoral neck to shaft. **A,** Coxa valga; **B,** normal; **C,** moderate coxa vara; **D,** severe coxa vara.

weight bearing, persists in children unable to stand because of chronic disability such as severe paralytic disease, and is seen as a secondary deformity in congenital dislocation of the hip (Fig. 2-12) and other affections. Ordinarily coxa valga is not of itself clinically significant.

Etiology. Coxa vara may be congenital or acquired. The acquired form is by far the more common, developing as a result of rickets, osteomalacia, osteitis deformans, or other bone disease. It may occur also in the end stages of coxa plana or of congenital dislocation of the hip with or without reduction; it is also seen after slipping of the capital femoral epiphysis, intertrochanteric fracture of the femur that has not been properly reduced and immobilized, and greenstick fracture of the neck of the femur in childhood.

Congenital coxa vara and slipping of the upper femoral epiphysis are of especial interest and will be discussed as separate entities.

CONGENITAL COXA VARA

Congenital coxa vara is usually first observed in early childhood but may not be recognized until later in life. It is of the cervical type and often bilateral. It is sometimes associated with other congenital defects, especially in the femur.

Etiology. The cause of congenital coxa vara is thought to be a congenital or growth defect of the femoral neck.

Clinical picture. Often the first evidences of the affection appear after an episode of slight trauma. The complaint may be weakness and stiffness in the leg, together with an awkward gait or limp. The range of abduction and internal rotation is usually restricted, whereas that of adduction and external rotation may be

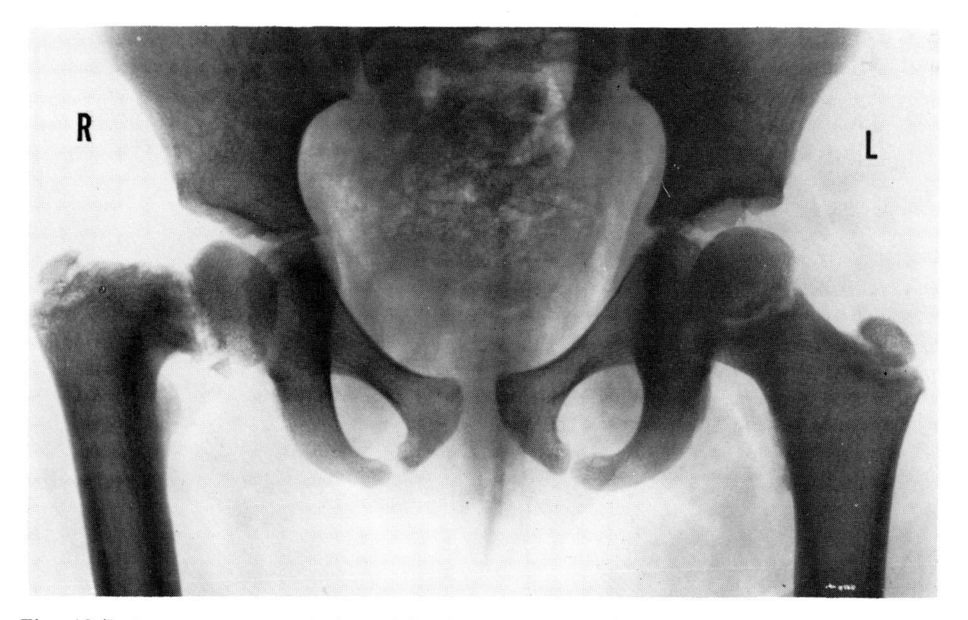

Fig. 12-7. Roentgenogram of the pelvis of a 4-year-old girl showing right congenital coxa vara. Note decreased neck-shaft angle and irregular ossification at epiphyseal line.

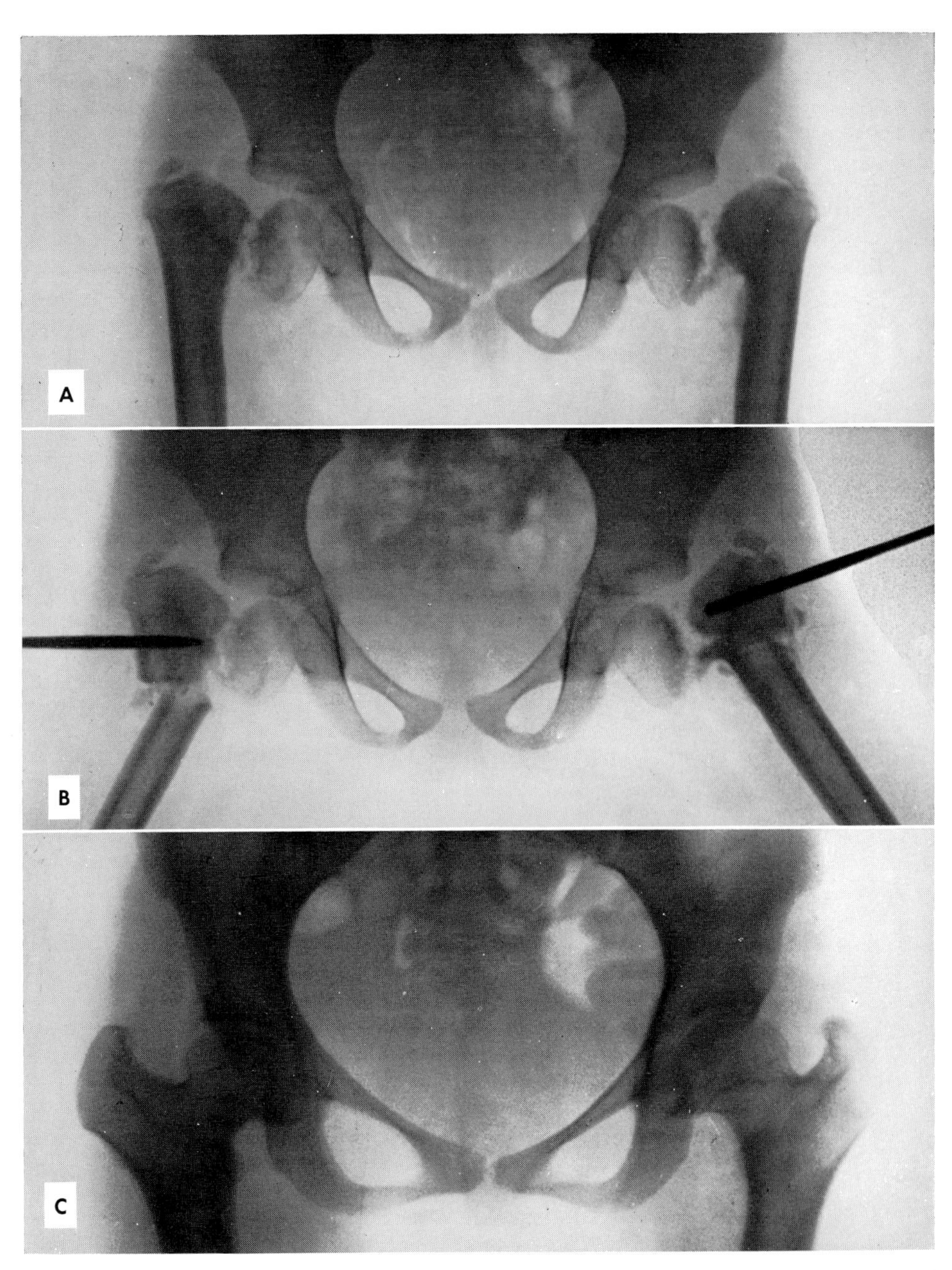

Fig. 12-8. Roentgenograms of bilateral congenital coxa vara. **A,** Note severe varus of femoral necks and widened, ragged epiphyseal lines. The patient is a girl 5 years of age. **B,** At operation. Note lines of subtrochanteric osteotomies, corrective valgus angulation at osteotomy sites, and Steinmann pins to control position of proximal fragments and to be incorporated in postoperative cast. **C,** Thirteen years after operation. The patient has no hip symptoms, only slight restriction of hip abduction, and a normal gait.

increased. A deformity characterized by prominence of the greater trochanter and external rotation of the limb may develop. The amount of actual shortening depends upon the extent of the depression of head and neck of the femur on the shaft; it is usually from 2 to 4 cm. Atrophy of the muscles is always great. There is usually little or no associated pain. At times all of the discomfort is referred to the thigh or knee.

In bilateral coxa vara there may be a duck-waddle gait, in which the body sways from side to side; this is similar to the gait seen in bilateral congenital dislocation of the hip.

Roentgenographic picture. In the young child, in addition to the decreased neck-shaft angle, a frequent roentgenographic change is the demarcation of a triangular area of bone in the lower side of the femoral neck close to the head; the epiphyseal line lies on one side of this area, and across the neck on the other side is an abnormal band of decreased density. The head of the femur is situated low in an acetabulum that is often shallow and sloping and that may appear abnormally wide (Fig. 12-7).

In older cases the depression of the head and neck on the shaft becomes greater, and the triangular area of bone fuses to the neck to assume the appearance of a dependent lip. There is usually a considerable upward prominence of the greater trochanter and upper part of the shaft.

Diagnosis. The diagnosis is made from the clinical picture together with roentgenograms showing decrease in the neck-shaft angle of the femur.

Congenital coxa vara should be differentiated from coxa plana, congenital dislocation or subluxation of the hip, slipping of the capital femoral epiphysis, rheumatoid arthritis, and tuberculosis.

Treatment. When there is little deformity, protection of the hip against unusual weight-bearing stresses constitutes adequate treatment. Fixed adduction deformity should be corrected by an abduction osteotomy through the trochanteric or subtrochanteric area (Fig. 12-8). When the patient is a child with unilateral involvement, the affected femur, even after osteotomy, may be retarded in length growth; such a patient may later require an operation to equalize leg length. In severe cases in adults, a reconstruction operation or arthrodesis may be indicated for relief of pain and disability.

SLIPPING OF THE CAPITAL FEMORAL EPIPHYSIS (EPIPHYSEAL OR ADOLESCENT COXA VARA)

Slipping of the upper femoral epiphysis is most often observed in children between 10 and 16 years of age and is more common in boys than in girls. In about 40% of the cases both hips are affected. In most cases a history of trauma or strain can be elicited, but the traumatic episode is often trivial. Slipping of the upper femoral epiphysis (Fig. 12-9) occurs predominantly in two types of children: (1) very tall, thin children and (2) obese children with undeveloped sexual characteristics. Many of these patients give a history of rapid skeletal growth immediately preceding displacement of the epiphysis.

The underlying cause of the epiphyseal slipping is unknown. Perhaps, as pointed out by Howorth, it is a combination of rapid growth, obliquity of the epiphyseal

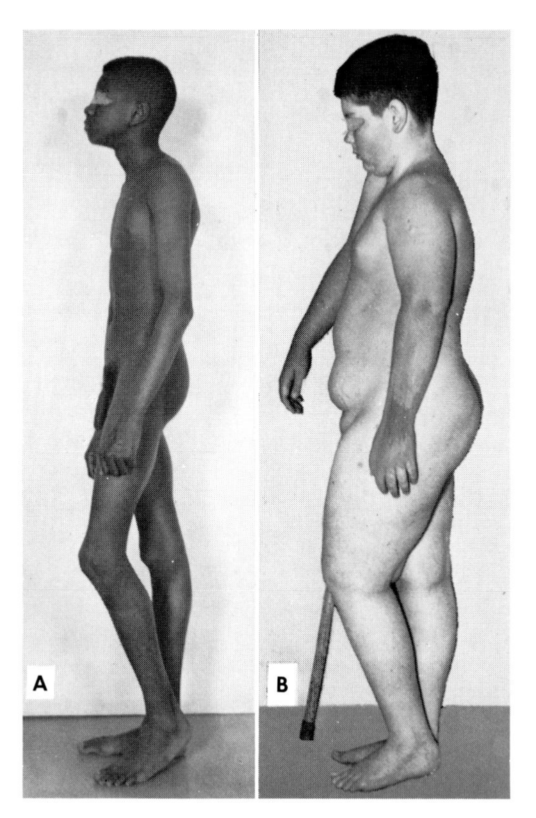

Fig. 12-9. Body types seen frequently in slipping of the upper femoral epiphysis. Both of these patients had slipping of the head of the left femur and were photographed before treatment. **A,** Tall, thin boy of 14 years; **B,** obese boy of 12 years.

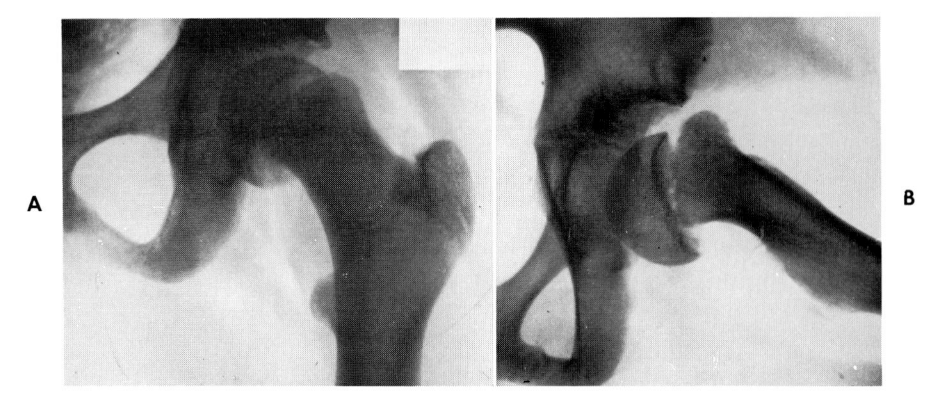

Fig. 12-10. Anteroposterior, **A,** and lateral, **B,** roentgenograms showing slipping of the upper femoral epiphysis in a girl 11 years of age. Note upward slipping of the neck of the femur. In relation to the neck the head has become displaced downward and backward.

plate, and minor traumas. Endocrine factors may play a part, but no specific endocrine abnormality has yet been proved. The head of the femur slips at its epiphyseal plate to become displaced downward and backward on the neck (Fig. 12-10), which retains its normal angle but which later may show roentgenographic evidence of posterior torsion.

Clinical picture. When there is no history of trauma, the onset is usually very gradual, the early symptoms being fatigue after walking or standing and later a slight amount of pain and stiffness, and a limp. After trauma the symptoms may either develop immediately or come on gradually. The degree of trauma apparently has little effect on the extent of the disability. The diagnosis should be suspected in any adolescent who has a limp accompanied by hip or knee discomfort and shows slight restriction of internal rotation of the hip; roentgenograms should be made at once to confirm the diagnosis, since prevention of lifetime crippling may depend upon early recognition and prompt surgical treatment.

The affected limb gradually becomes shorter and smaller, and the range of hip motion, especially of internal rotation and abduction, becomes restricted. Hyperextension may be more easily carried out on the affected side than on the normal side. When the affected hip is flexed, it tends to abduct and rotate externally. At first the discomfort may consist of referred pain in the knee, but later the pain is located in the hip. There is usually pain about the anterior aspect of the hip when the limits of its range of motion are reached. Tenderness may be present about the anterior and lateral aspects of the hip.

Roentgenographic picture. Early slipping can be recognized in the lateral roentgenogram before it becomes apparent in the anteroposterior view. When further slipping has taken place, the anteroposterior view may show a displacement of the neck upward and forward on the head of the femur so that the lowest point of the proximal end of the neck projects as a beaklike process (Fig. 12-11, *A*). If the displacement is not corrected, the upper part of the neck gradually becomes lengthened, the lower part shortened, and the whole neck bowed, exhibiting coxa vara. New bone may form between the lower border of the neck and the overhanging head. In advanced stages the femoral head appears quite atrophic, especially in its lower half, and the neck is very thick and short. In early cases after severe trauma the roentgenogram may demonstrate complete separation of the head from the neck of the femur.

Diagnosis. Slipping of the upper femoral epiphysis is to be differentiated from coxa plana, congenital coxa vara, rheumatoid arthritis, subacute infectious processes including tuberculosis, and fracture. Roentgenograms are necessary to establish the diagnosis; a lateral view is essential, in addition to the anteroposterior film.

Treatment. The object of treatment is to correct the displacement, when possible with a minimum of trauma, and to maintain the correction until bony union between neck and epiphysis has taken place. For the individual case the most suitable type of treatment depends upon the degree and duration of the slipping. In some early cases traction, preferably skeletal, in a position of slight abduction and internal rotation for approximately 2 weeks will reduce the displacement.

The corrected alignment is best maintained by inserting across the epiphyseal plate the multiple pins of Moore or Knowles (Fig. 12-12). Thereafter, to provide freedom from weight bearing, crutches should be used for several months.

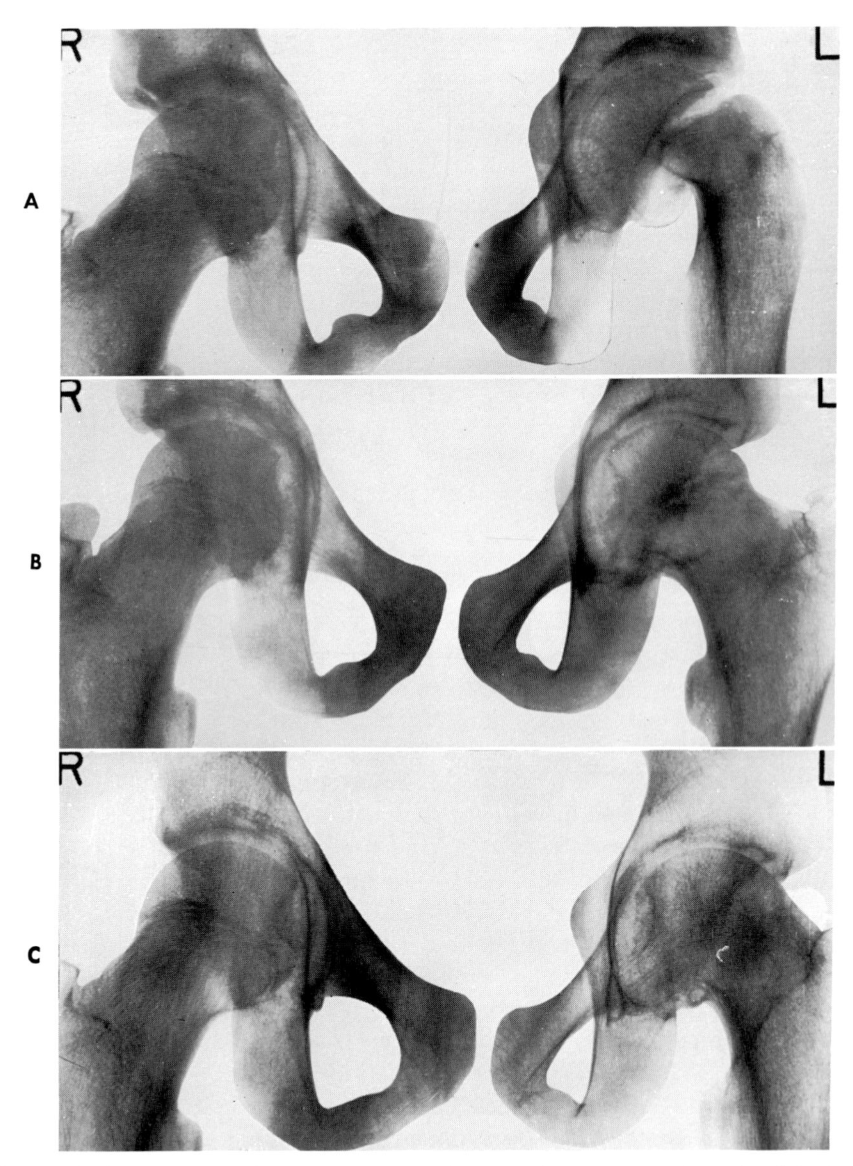

Fig. 12-11. A, Anteroposterior roentgenogram showing slipped capital femoral epiphysis of the left hip in a boy 14 years of age. Note upward displacement of the neck of the femur. **B,** Two months after reduction by skeletal traction and closed manipulation. Note almost normal relationship between the capital epiphysis and the femoral neck. **C,** Eighteen months after closed reduction. In the left hip the epiphyseal line has become obliterated, the femoral head is of uneven density, the head and neck are slightly deformed, and slight secondary changes are apparent in the acetabulum. At this time the mobility of the left hip was about half that of the normal right hip.

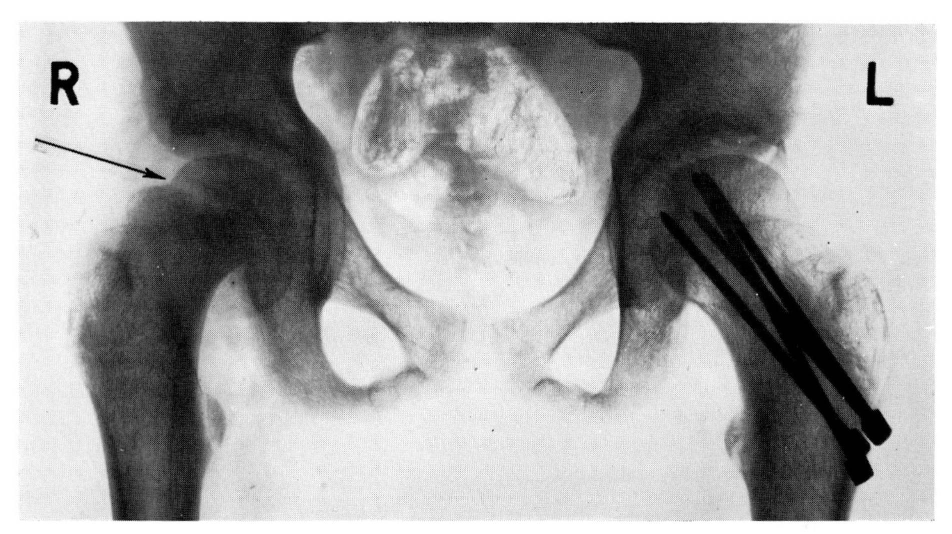

Fig. 12-12. Roentgenogram of slipped capital femoral epiphyses in a 13-year-old boy. Seven months earlier, slipping of the left epiphysis had been reduced by traction and fixed by four Knowles pins. Note slipping of the right epiphysis, which was associated with hip pain of 2 weeks' duration.

Patients with only slight displacement (less than 1 cm. in both anteroposterior and lateral roentgenograms) that is not corrected by a trial period of traction should also be treated by internal fixation with no attempt at forceful correction, since slight malalignment is compatible with satisfactory hip function.

In early cases with moderate or severe displacement of the epiphysis (Fig. 12-11), reduction may sometimes be accomplished by the maneuver used for reducing a fractured femoral neck (Leadbetter maneuver), which consists of exerting traction, with the hip and knee flexed, followed by gradual extension, internal rotation, and abduction of the hip. Forceful manipulation is always contraindicated, however, because of the danger of causing avascular necrosis of the femoral head by interrupting its blood supply.

In late cases also, manipulation or open reduction is inadvisable because of the danger of interference with the blood supply of the epiphysis. The results of open operations on the hip, however well done, are sometimes disappointing. If the deformity is severe, the function of the hip can often be improved by resection of protruding bone, as suggested by Heyman, or by subtrochanteric osteotomy. In adult years, arthrodesis or an arthroplasty may be indicated.

If the patient is obese, his weight should be reduced. In all degrees of slipping, rest of the hip and protection from weight bearing are important. The opposite hip should be watched for any clinical or roentgenographic sign of early slipping since bilateral involvement is common (Fig. 12-12).

Prognosis. Cases with slight slipping, treated early, have a favorable prognosis. Late cases with severe displacement often leave a permanent disability despite any form of treatment. Slipped epiphysis complicated by avascular necrosis or by acute cartilage necrosis, which appears roentgenographically as severe narrowing of the cartilage space, has a relatively unfavorable prognosis for hip function.

PATHOLOGIC DISLOCATION OF THE HIP

Pathologic dislocation is a late complication of many affections of the hip. It develops as a result of (1) erosion of bone about the acetabulum, the femoral head, or both, or (2) paralysis of the muscles and relaxation of other soft tissues around the hip joint. In most cases the head of the femur becomes displaced upward and posteriorly.

Etiology. Pathologic dislocation often follows pyogenic arthritis of the hip or osteomyelitis of the upper end of the femur. Tuberculosis of the hip, in which gradual erosion of the acetabulum and destruction of the head of the femur may take place, may result in dislocation of the hip. In these conditions the dislocation is partly due to relaxation of the ligamentous supports following distention of the joint capsule by fluid; the relaxed ligaments allow displacement of the head of the femur to the dorsum of the ilium. Pathologic dislocation is seen also in cerebral palsy, poliomyelitis, and spina bifida with myelomeningocele. It may follow the general muscular atrophy of prolonged febrile illness. Occasionally it may accompany rheumatoid arthritis or neuropathic arthropathy. Flexion and adduction of the hip, accompanied by pain and frequently by spasm, are important factors leading to dislocation.

Treatment. Preventive therapy is imperative. The possibility of dislocation should be anticipated and should be guarded against by preventing flexion and adduction. If there is likelihood of displacement, the leg should be kept in traction in an abducted position. As soon as a dislocation is recognized, it should be corrected.

Before reduction in late cases can be accomplished, it may be necessary to remove by open operation fibrous material that has filled the acetabulum. The depth

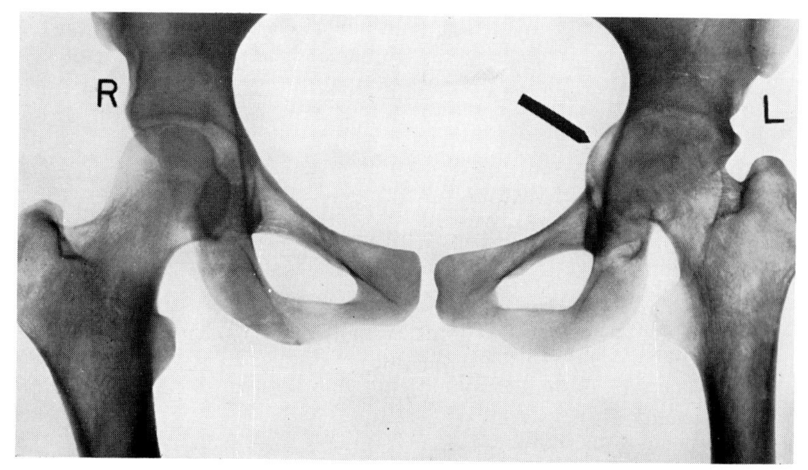

Fig. 12-13. Roentgenogram showing intrapelvic protrusion of the acetabulum in the left hip of a girl 14 years of age. (Courtesy Children's Seashore House, Atlantic City, N. J.)

of the hip socket can be increased by innominate osteotomy (Fig. 2-11, *A*), pericapsular iliac osteotomy (Fig. 2-11, *B*), or an iliac shelf operation. In paralytic dislocations, transference of the iliopsoas tendon through an opening made in the ilium to the greater trochanter, to provide an additional stabilizing factor, must be considered. Occasionally subtrochanteric osteotomy or hip arthrodesis may be the procedure of choice.

INTRAPELVIC PROTRUSION OF THE ACETABULUM

Intrapelvic protrusion of the acetabulum is an uncommon affection of undetermined etiology. In some cases it is of congenital or familial origin. In other instances it is probably secondary to common affections such as rheumatoid or low-grade pyogenic arthritis. It is characterized by a deepening or inward protrusion of the acetabulum, which allows the head of the femur to project farther into the pelvis than normally (Fig. 12-13). The roentgenograms usually show thinning of the walls of the acetabulum, but occasionally there is evidence of increased bone formation; there may be narrowing of the cartilage space. Usually little change occurs in the head of the femur, but occasionally it is irregular in contour or enlarged. The affection is found more often in females than in males and may be either bilateral or unilateral.

Clinical picture. Discomfort and limitation of motion develop gradually over a period of years. Abduction and rotation are especially restricted. There may be little pain until osteoarthritic changes are superimposed. In the end stage, ankylosis of the hip usually results.

Treatment. Little treatment is indicated in unilateral cases unless the deformity is accompanied by pain, in which event rest and traction for several weeks, followed by the use of crutches, may result in symptomatic improvement. Night traction may be helpful for a long period. In persistent disabling pain and in many bilateral cases with stiffness, arthroplasty may be indicated.

TRANSIENT SYNOVITIS OF THE HIP

Hip pain and spasm, developing without obvious cause and subsiding after a week or two, may be attributed to a transient synovitis from mild trauma or low-grade, short-lived infection. This obscure affection is almost always unilateral and is seen mostly in boys between 4 and 10 years of age.

Clinical picture. The child complains of pain in the hip, thigh, or knee, and there is usually tenderness over the hip joint. Restriction of passive hip mobility by muscle spasm is a constant finding. A limp is usually associated. The hip is often held in a flexed and abducted position. Systemic signs of infection are slight or absent.

Diagnosis. Synovitis of the hip joint is to be distinguished from coxa plana, osteomyelitis, pyogenic arthritis, tuberculosis, and early slipping of the upper femoral epiphysis. Roentgenograms that are negative except for showing distention of the joint capsule support the diagnosis of transient synovitis. If the symptoms persist, additional roentgenograms should be made after an interval of several weeks.

Treatment and prognosis. Rest and hot applications are indicated. When the pain is severe, a short period of traction, perhaps followed briefly by a plaster cast, may be advisable. The usual outcome is quick, permanent recovery, but recurrences sometimes take place. Prolonged follow-up is indicated since on rare occasions what seems to be transient synovitis is followed within a year or two by the much more serious disorder of coxa plana.

BURSITIS ABOUT THE HIP

Eighteen or more bursae about the hip have been described, but of these only four are of clinical importance. These are the iliopectineal or iliopsoas, the deep trochanteric, the superficial trochanteric, and the ischiogluteal bursae (Fig. 12-14).

Iliopectineal or iliopsoas bursa

The iliopectineal or iliopsoas bursa is located between the iliopsoas muscle and the iliopectineal eminence, on the anterior surface of the hip joint capsule, and frequently communicates with the joint cavity.

Clinical picture. When the iliopectineal bursa is inflamed, tenderness is usually present over the anterior aspect of the hip at about the middle of the inguinal ligament. Pain caused by pressure on the femoral nerve in this area may radiate down the front of the leg. The hip is usually held in flexion, abduction, and external rotation. Pain is elicited upon attempting to extend, adduct, or internally rotate the hip.

Diagnosis. Femoral hernia, psoas abscess pointing in the groin, synovitis, and infection of the hip joint are to be considered in the differential diagnosis.

Treatment. The patient should be placed at rest in bed with traction applied to the lower limb. Hot applications should be placed over the area of tenderness on the anterior aspect of the hip. If cellulitis or frank infection is present, antibiotic therapy should be used. The bursitis usually subsides completely within several weeks.

Deep trochanteric bursa

The deep trochanteric bursa is located behind the greater trochanter and in front of the tendinous portion of the gluteus maximus muscle.

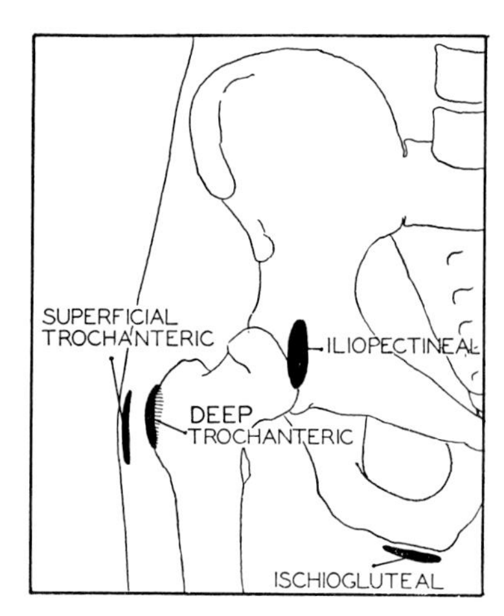

Fig. 12-14. Most commonly affected bursae about the hip.

When the deep trochanteric bursa is enlarged, the normal depression behind the greater trochanter is obliterated. At this point there may be considerable tenderness. The leg is usually held in an abducted and externally rotated position, which relaxes the tension on the gluteus maximus muscle and the bursa. Pain may radiate down the back of the thigh, and any motion of the hip joint may cause discomfort.

Deep trochanteric bursitis is to be differentiated from infection of the hip joint and from osteomyelitis of the upper end of the femur.

Rest and heat usually constitute adequate treatment. When a pyogenic infection is present, treatment with an appropriate antibiotic is indicated, and drainage must be considered. For tuberculous infection complete excision of the bursa is advisable.

Superficial trochanteric bursa

The superficial trochanteric bursa is located between the greater trochanter and the skin and subcutaneous tissue.

Tenderness and swelling may be present over the inflamed bursa, and there is pain on extreme adduction of the hip.

The therapy is similar to that outlined for deep trochanteric bursitis.

Ischiogluteal bursa

The ischiogluteal bursa is located superficial to the tuberosity of the ischium. Chronic ischiogluteal bursitis, called *weaver's bottom,* develops in tailors, boatmen, and other individuals whose occupation necessitates prolonged sitting upon hard surfaces.

Tenderness over the tuberosity of the ischium is characteristic, and pain radiating down the back of the thigh along the course of the hamstring muscles may mimic that of a herniated intervertebral disk.

If the inflammation is not severe, heat and rest will usually cause the symptoms to subside. The use of a pillow or cushioned seat may aid in preventing recurrence. Procaine and hydrocortisone injections may be helpful. Persistence of pain may indicate need for excision of the bursa.

SNAPPING HIP

Snapping hip is an uncommon affection, seen as a rule in young women. With the knee flexed, active internal rotation of the hip will sometimes give rise to a snapping noise. It may be present with every step. It is sometimes caused by the slipping to and fro over the greater trochanter of the iliotibial band or of a fibrous thickening on the deep surface of the gluteus maximus muscle. The snapping can seldom be heard with passive motion. It is annoying rather than painful.

Treatment. As a rule no treatment is indicated, other than explaining to the patient that the snapping is harmless.

CHAPTER 13 Affections of the knee

The knee is the largest joint in the body. In general it is hingelike in character, and its stability is dependent upon (1) an intricate group of strong ligaments and (2) the supporting muscles and tendons. The motions of the knee joint are an extensive anteroposterior, rotary gliding movement and a very limited axial rotation of the tibia upon the femur. The tibia undergoes slight external rotation on the femur as the knee is fully extended and slight internal rotation as the knee is flexed. Flexion and extension of the knee take place actually between the semilunar cartilages and the femur. Rotation has been shown to take place between the semilunar cartilages and the tibia.

The knee joint is formed by the articulation of the two rounded condyles of the femur with two shallow depressions in the tibia, which are also called condyles. Between the condyles of the tibia is the tibial spine, which divides to form medial and lateral tubercles. Immediately anterior to the tibial spine is a flattened triangular area on the anterosuperior aspect of the tibia, at the base of which lies the tibial tuberosity. On either side of the joint are strong collateral ligaments that prevent lateral angulation and displacement, and posteriorly the joint capsule forms the floor of the popliteal space. Between the corresponding condyles of femur and tibia are semilunar cartilages or menisci, each of which acts as a wedge-shaped cushion between the tibia and the femur and helps to maintain tension in the two cruciate ligaments. The cruciate ligaments extend from the intercondylar notch of the femur to the upper surface of the tibia. The anterior cruciate ligament prevents forward displacement of the tibia, and the posterior cruciate ligament prevents backward displacement. Synovial membrane lines the joint and is continued above into a large anterior pocket called the suprapatellar bursa. Behind the patellar ligament and projecting into the anterior portion of the joint is the infrapatellar fat pad, which is extrasynovial and changes shape with every movement of the joint. This pad is connected with the intercondylar notch of the femur by the ligamentum mucosum.

INTERNAL DERANGEMENTS OF THE KNEE JOINT

Internal derangements is a term commonly applied to those intra-articular and extra-articular affections of the knee, most often of traumatic origin, which are the result of lesions of the semilunar cartilages, the joint surfaces, the ligaments, the fat pads, or the synovial membrane. The more common derangements include (1) lesions of the semilunar cartilages, (2) rupture of the tibial and fibular collateral ligaments, (3) rupture of the cruciate ligaments, (4) injury of the tibial spine, (5) loose bodies, as in osteochondritis dissecans and synovial chondromatosis, (6) hypertrophy and pinching of the infrapatellar fat pad and synovial membrane, and (7) exostoses.

LESIONS OF THE SEMILUNAR CARTILAGES

Displacements and tears of the semilunar cartilages, or menisci, constitute by far the most frequent type of internal derangement of the knee joint. The medial cartilage is injured about five times as frequently as the lateral cartilage; the reasons for this are (1) a difference in structure, the medial cartilage being longer and bifurcated at its anterior pole, and (2) a difference in etiology, the mechanism of injury which causes damage of the medial cartilage being more common than that which affects the lateral cartilage. It is convenient to consider the etiology and symptomatology of injuries of each meniscus separately.

Injuries of the medial semilunar cartilage

Etiology. Derangement of the medial semilunar cartilage is usually caused by a sudden internal rotation of the femur upon the fixed tibia while the knee is abducted and flexed. The damage occurs most frequently about the anterior portion of the cartilage. A greater degree of knee flexion at the instant of injury causes the tear to be situated nearer the posterior end of the cartilage. If the anterior portion of the cartilage is torn and slips into the joint, it may lodge between the joint surfaces, prevent complete extension, and give rise to so-called locking. If the rotary strain is severe, the connection between the cartilage and the deep portion of the tibial collateral ligament may be torn, allowing the cartilage to slip into the joint. With extension of the knee the free border of the cartilage may be caught between the

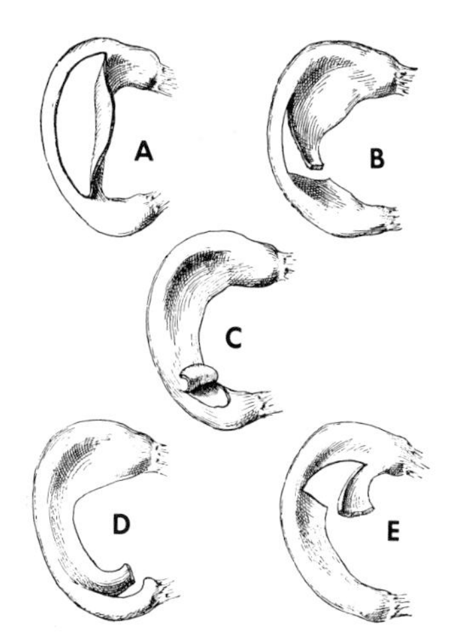

Fig. 13-1. Types of semilunar cartilage injury. **A,** Longitudinal splitting (bucket-handle type). **B,** Tear of middle third. **C,** Tear of anterior tip. **D,** Longitudinal splitting of anterior third. **E,** Tear of posterior third. (After Henderson.)

condyles and be split longitudinally. This produces the bucket-handle lesion, in which the inner portion of the meniscus may be easily displaced centrally between the joint surfaces and cause locking. When the joint is unlocked, the cartilage becomes dislodged from between the articular surfaces.

Pathology. The lesion (Fig. 13-1) may consist of the following: (1) a tearing of the anterior or posterior part of the cartilage with or without displacement; (2) a transverse tear through any portion of the cartilage; (3) a longitudinal splitting with or without the displacement characteristic of the bucket-handle cartilage; or (4) a simple loosening of the cartilage at its peripheral attachment, allowing it to slip into and out of the joint. Any one of these lesions may give rise to the symptoms of a joint derangement. When the cartilage is partially torn from its peripheral attachment, it may heal without difficulty if further displacement is prevented. Tears limited to the central rim of the cartilage probably never heal. At operation the bucket-handle lesions are as common as all the other lesions combined.

Clinical picture. An accurate history is most important, since often the diagnosis is based more upon the history than upon the physical findings. A typical history is that of an athlete whose partially flexed knee is suddenly twisted inward (Fig. 13-2). Acute pain on the inner side of the knee accompanies the injury, and the athlete is likely to fall to the ground. His knee cannot be straightened, and the joint may swell

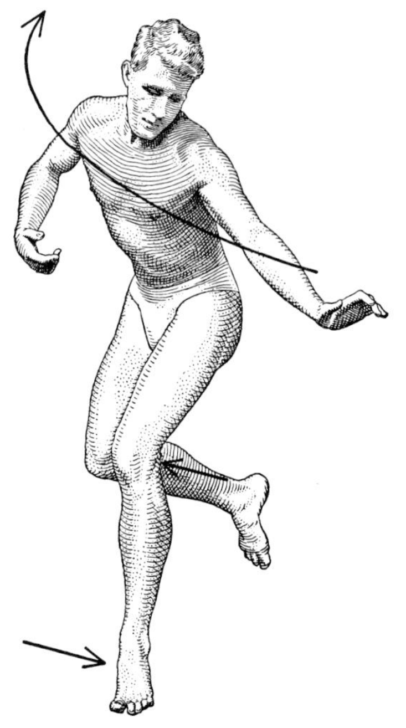

Fig. 13-2. A mechanism of injury commonly resulting in tear of the medial semilunar cartilage. Note flexion of the left knee and violent internal rotation of the femur as the trunk twists to the right while the left tibia and foot are fixed by weight bearing. The anterior portion of the left medial meniscus tends to be displaced posteriorly by the attachment of the tibial collateral ligament and the rotary sweep of the medial femoral condyle.

rapidly. If the leg should be pulled when it is found that he cannot straighten the knee, the athlete may feel something in the knee snap back into place. Instantly he is relieved of the sharp pain and is then usually able to get up and walk. In other instances the history may be that of a coal miner who is working on his knees and who feels something give way in his knee when he turns to shovel coal. Symptoms and signs identical with those of the former case then follow.

If the condition remains untreated, the disability persists for a variable length of time. After the initial accident the feeling of a slipping within the knee, with pain referred to the medial side of the joint, may recur frequently. Such episodes may be precipitated by external rotation and abduction of the foot.

Examination of the affected knee brings to light a number of signs. In acute cases there may be an extensive effusion of blood in the joint. In the presence of this effusion it is difficult to diagnose the exact type of injury. Localized tenderness is present at this stage, either about the anterior tip of the medial semilunar cartilage or about the medial or posterior margins of the joint surfaces. Forced adduction often elicits pain on the medial aspect of the knee. If the meniscus remains displaced, full extension will be impossible. A history of partial locking is present in approximately 70% of the cases in which the definite diagnosis can be made. Hamstring muscle spasm often prevents complete extension of the knee and must be differentiated from true locking. When a posterior tear is accompanied by displacement, it may be impossible to flex the joint fully; this partial locking will persist until the dislocation of the cartilage is reduced. Tears of the posterior half of the cartilage may be demonstrated by the McMurray or "click" test. Holding the tibia in extreme external rotation and abduction, the examiner slowly extends the knee from the fully flexed position—an appreciable clicking and transient pain indicate the presence of a posterior tear.

When the displacement recurs frequently, the patient always feels insecure. Often a little twist or misstep will again throw the cartilage out of its normal position. The patient may learn to reduce the displacement without assistance. Recurrent displacement is usually accompanied by a moderate increase of joint fluid. The swelling will persist as long as any part of the torn cartilage is irritating the synovial membrane. Occasionally in recurrent cases the meniscus is displaced toward the medial side, and a mass can be felt along the medial border of the joint. In such cases the cartilage can usually be pushed in with the fingers. Atrophy of the muscles of the thigh and leg is always present in chronic cases; it is most noticeable in the quadriceps muscle.

Injuries of the lateral semilunar cartilage

The mechanism of injury to the lateral semilunar cartilage is the converse of that causing damage of the medial semilunar cartilage. The foot is usually fixed firmly upon the ground, and the femur rotates outward upon the tibia while the knee is adducted and flexed. The pathologic changes are similar to those in injuries of the medial cartilage.

Clinical picture. The symptoms and signs are essentially the same as those of injury of the medial cartilage, with the exception that they are referred to the lateral side of the knee and are often less severe. After injury of the lateral cartilage, straightening or fully flexing the knee will at times produce a loud crack or snap that can be felt on the lateral side of the joint as the knee jerks into place. This has occasioned the term *trigger knee*. For a posterior tear of the lateral cartilage, the McMurray test is the same as that described for the medial cartilage except that the tibia is placed in extreme internal rotation and adduction instead of external rotation and abduction.

Diagnosis. The diagnosis of a derangement of either meniscus is based on the characteristic history and physical signs. Locking is typical but may occur also in other types of internal derangement. Roentgenograms yield little relevant information but should be taken in order to exclude fracture, loose bodies, exostoses, arthritis, and aberrant calcification. In addition to the usual anteroposterior and lateral roentgenograms, a posteroanterior picture should be taken with the knee flexed 90 degrees; this shows the intercondylar space and is sometimes called the tunnel view. Double-contrast arthrography, in which, in addition to air, a liquid radiopaque contrast medium is injected into the knee joint, is sometimes helpful in the roentgenographic diagnosis of meniscus tears. Direct visualization of the semilunar cartilage can be accomplished with an arthroscope, but as yet this technic is not widely used.

Treatment. For derangement of either cartilage the treatment is essentially the same.

If the joint capsule is greatly distended after a recent injury, the fluid may be aspirated; it is usually bloody but becomes more amber in color as time passes. Application of ice is always helpful in acute knee injuries. If the knee is locked immediately after an initial injury, an attempt should be made to reduce the displacement of the cartilage; this may be done by the application of traction or by resort to manipulation. In the case of the medial cartilage the knee should be flexed, abducted, externally rotated, and then quickly internally rotated and extended. For displacement of the lateral cartilage the knee should be flexed, adducted, internally rotated, and then quickly externally rotated and extended. If the cartilage dislocation can be reduced, the joint should then be immobilized for 3 weeks or more. A dressing of glazed cotton rolls, broad splints, and elastic cloth bandages, or a light plaster cast, may be used. Exercise of the quadriceps muscle should always be employed to aid in maintaining muscle tone and strength. Patella-setting exercises with the knee extended should be started early.

If the knee remains weak and stiff after the original displacement, a program of weight-lifting extension exercises is indicated in order to strengthen the muscles and to restore normal mobility. If pain and limitation of extension persist, the injured meniscus should be removed.

If the displaced cartilage is unquestionably palpable, if displacement has occurred more than once, or if the symptoms recur after immobilization, an exploratory operation is indicated. It is often impossible to make a specific diagnosis before operation. If it is clinically definite that displacement or laceration of the cartilage has occurred, the operation should not be delayed, since arthritic changes may develop in the joint when a torn meniscus is allowed to remain as an irritant to the synovial membrane and hyaline cartilage. When, on exploration of the joint, the meniscus is found to be loosened or torn in any part, it should be removed as completely as possible. In cases of long duration with repeated episodes of knee locking, arthritic changes in the femoral condyle are often found at operation. Such changes are greatest where the articular cartilage of the condyle comes into contact with the torn portion of the meniscus.

Instruction and practice in forceful voluntary contraction of the quadriceps muscle should be begun before operation and resumed on the first postoperative day. Weight bearing on the extended knee may be started as early as a day or two after operation. At the end of a week, more strenuous quadriceps exercises are begun; they are continued until normal mobility and strength have been regained. If the meniscus derangement has been accompanied by injury of the knee ligaments, temporary use of a brace may be advisable. Unrestricted exercise should not be allowed until the quadriceps muscle has acquired good strength, which is usually about 12 weeks after operation. Removal of a meniscus does not in itself impair the strength or function of the joint to any appreciable extent.

Cysts of the semilunar cartilages

Semilunar cartilage cysts are uncommon. They occur six to twelve times as frequently in the lateral meniscus as in the medial meniscus and are seen mostly in young men.

Etiology. The cyst of the semilunar cartilage is thought to represent (1) the end result of a mucoid degenerative process within the cartilage, (2) a congenital defect in the development of the cartilage, or (3) a ganglion-like structure resulting from trauma between the peripheral surface of the cartilage and the synovial membrane.

Pathology. The cysts are more often multiple than single. They contain a soft, gelatinous material and are sometimes lined by cells resembling endothelium.

Clinical picture. The symptoms and signs are similar to those of semilunar cartilage injuries; there is, however, no locking or sudden effusion. Cysts may follow injury of the knee; many years may elapse, however, between the injury and the appearance of the cysts. There is sometimes a continuous, dull ache in the affected joint, most noticeable at night. The discomfort is always worse after activity and relieved by rest. The typical cyst is found on the lateral side of the joint as a tense, slightly tender swelling, which may become as large as a walnut. It is always more prominent with the knee in extension than in flexion.

Treatment. Since the affected cartilage usually contains several small cystic areas in addition to the main cyst, complete excision of the meniscus is indicated. Recurrence of the cyst is common after incomplete excision.

Discoid cartilages

As a result of an uncommon developmental anomaly, the lateral meniscus is sometimes discoid rather than semilunar in form. Only very rarely is the medial meniscus discoid. A discoid cartilage may be present in each knee. It may be associated with a high fibular head. Abnormal attachment of the posterior portion of the meniscus may cause it to be pulled in and out across the joint surface with flexion and extension of the knee.

Clinical picture. The most characteristic clinical feature is a loud click that is felt and heard when the knee joint is flexed or extended; this occurs usually near the limits of knee motion. The joint does not lock. The typical patient is in his teens, has noticed a painless clicking for a variable length of time, and with or without minor trauma of his knee begins to have an aching pain on its lateral aspect and a feeling of weakness in it. Examination confirms the clicking on motion and may disclose tenderness over the lateral meniscus. Because the discoid cartilage is thicker than normal, roentgenograms may show a widening of the cartilage space between the lateral condyles of the femur and the tibia.

Treatment. In the presence of symptoms the treatment is excision of the entire discoid cartilage.

RUPTURE OF THE TIBIAL AND FIBULAR COLLATERAL LIGAMENTS

Etiology. The tibial or medial collateral ligament is ruptured by stresses resulting from forceful abduction of the knee, usually when the knee is slightly flexed and the extensor mechanism relaxed. As pointed out by Slocum and Larson, when abduction stress applied to the flexed knee is associated with forceful external rotation of the tibia on the femur (Fig. 13-2), tears of the deep portion of the medial ligament (the medial capsular ligament) result, causing rotatory instability of the knee. In football, clipping is a frequent cause of rupture of the tibial collateral ligament. The fibular collateral ligament, less frequently injured, may be ruptured by forceful adduction of the internally rotated knee.

Severe ruptures of the collateral ligaments are often accompanied by damage to the cruciate ligaments. Combined injury of the medial collateral ligament and the anterior cruciate ligament is common; with this injury a tear or avulsion of the medial meniscus is often associated (Fig. 13-3).

Clinical picture. There is an increase in the lateral mobility of the knee with tenderness over the injured ligament. When the tibial collateral ligament is ruptured, the tenderness is usually greatest over its inferior attachment. There may be an associated increase of joint fluid, swelling, and ecchymosis. Rarely is an unmistakable defect in the region of the affected ligament palpable. Tenderness may be very persistent. On motion of the knee there are definite weakness, instability, and occasionally a sensation of slight catching within the joint. In old injuries, tenderness may be absent but instability of the knee as tested by abducting the tibia on the femur is present. Rotatory instability is tested with the patient supine, the knee flexed 90 degrees, and the foot supported on the examining table. With the foot and tibia externally rotated 15 degrees the tibia is pulled forward on the femur as in testing for anterior cruciate instability. Abnormal forward movement of the tibia suggests a tear of the medial capsular ligament.

Treatment. If the injury is mild and the ligament is not completely torn, as determined by lack of instability when abduction or adduction stresses are applied to the knee, simple splinting with a soft dressing and an elastic bandage is indicated. The patient is given crutches for 3 or 4 weeks but is permitted partial weight bearing. Quadriceps setting exercises are prescribed during this period and are followed by active resistive exercises after the soreness subsides. In moderate injuries, manifested by a slight degree of knee instability on stress testing, immobiliza-

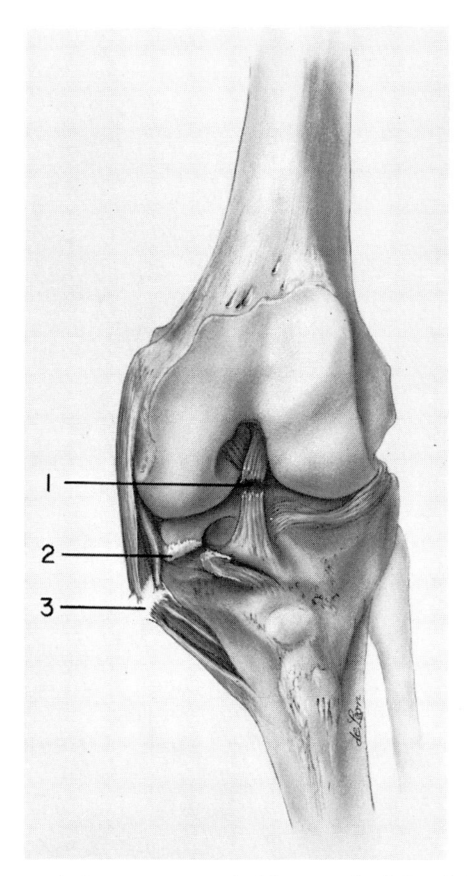

Fig. 13-3. The three internal derangements of this severely injured knee, which exemplifies one type of the unhappy triad of O'Donoghue, are, **1**, rupture of the anterior cruciate ligament, **2**, tear of the medial meniscus, and, **3**, rupture of the superficial and deep layers of the tibial collateral ligament.

tion of the knee in a long cast in a position of slight flexion may be advisable. When the ligamentous rupture is obviously complete and there is gross instability, surgical repair is usually indicated, especially in the young athlete. If repair is done within the first few days after injury, the results are generally better than when surgery is delayed. After suture of the torn ligaments the knee is immobilized in a long cast for 5 to 6 weeks, during which the patient is encouraged to carry out quadriceps setting exercises. After removal of the cast a graduated and intensive exercise program must be followed.

If the complete ligamentous tear is not treated in the acute phase, healing will take place but usually with some laxity of the ligaments and resulting instability of the knee. If this is not excessive and if the patient leads a sedentary existence, he may have no knee symptoms. Such a knee, however, is unsatisfactory for an active athletic life. Late ligamentous repairs or reconstructions, although less effective than those done early, can increase the stability of the knee. For reinforcement or reconstruction of the fibular ligament, a flap of fascia lata and a portion of the biceps tendon may be used. When the tibial collateral ligament is relaxed but not completely ruptured, Mauck's operation of transplanting the distal end of the ligament with its bony attachment downward on the tibia is often successful in improving the stability of the joint. The tibial collateral ligament may be reinforced or reconstructed with the tendons of the gracilis and semitendinosus muscles. The intact semitendinosus tendon may be transplanted into the medial condyle of the femur, as described by Bosworth. Rotatory instability of the knee due to tearing of the deep portion of the tibial collateral ligament may be improved by proximal transplantation of the distal part of the pes anserinus, as described by Slocum and Larson.

RUPTURE OF THE CRUCIATE LIGAMENTS

Etiology. The cruciate ligaments are ruptured only by severe trauma. The anterior cruciate ligament may be damaged by the same type of trauma that causes injury of the medial semilunar cartilage, and the posterior cruciate ligament may be damaged by the same trauma that injures the lateral semilunar cartilage. Forced hyperextension of the knee and internal rotation of the tibia on the femur, with rupture of the tibial collateral ligament, may at the same time produce a rupture of the anterior cruciate ligament. If the ligament itself does not break with this type of violence, avulsion of the medial tubercle of the tibial spine may occur. Any force that displaces the tibia backward on the femur while the knee is flexed may cause rupture of the posterior cruciate ligament. Falling on the flexed knee in such a manner that the force of the impact is received on the upper end of the tibia instead of on the patella may cause rupture of the posterior cruciate ligament.

Clinical picture. Rupture of a cruciate ligament is followed by great swelling and instability of the knee joint. When the anterior cruciate ligament is torn or stretched, the tibia can be displaced forward on the femur. When the posterior cruciate ligament is torn or stretched, the tibia can be displaced backward on the femur when the knee is flexed. Lateral as well as anteroposterior instability may be present. Occasionally both cruciate ligaments are torn; this results in extreme instability and is likely to accompany complete dislocation of the knee.

In acute cruciate ligament injuries, when early and accurate diagnosis is so important, pain and spasm may obscure the degree of instability and make an examination under anesthesia advisable.

Treatment. In athletic young adults, acute rupture of a cruciate ligament is probably best treated by early surgical repair. Especially is operation indicated when there is injury of the anterior cruciate ligament, tibial collateral ligament, and medial meniscus—the unhappy triad of O'Donoghue, in whose hands the results of early operation have been very good.

In older or less vigorous individuals, acute cruciate ligament rupture may be treated by immobilizing the knee in slight flexion, preferably by means of a plaster cast, for a period of at least 2 months. Immobilization may be followed by a snugly fitting knee brace (Fig. 13-4).

Operative reconstructions are occasionally done for old injuries of the cruciate ligaments, the anterior ligament sometimes being reconstructed with fascia lata or

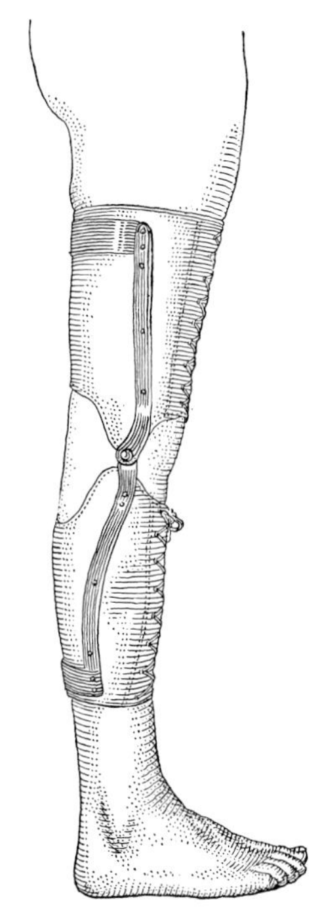

Fig. 13-4. Hinged brace for protection and support of the injured knee. (After Bennett.)

by using the central third of the patellar ligament, as described by Jones. The posterior cruciate ligament may be reconstructed by utilizing the semitendinosus tendon. Unfortunately these procedures cannot be relied upon to restore knee stability of satisfactory degree.

At some stage of treatment all cruciate ligament injuries require a program of vigorous resistive exercises to build up the strength of the quadriceps muscle.

FRACTURE OF THE TIBIAL SPINE

Etiology. Fracture of the tibial spine is usually caused by a mechanism of injury similar to that which results in rupture of the cruciate ligaments. In children this injury is frequently the result of a bicycle accident. Avulsion of the whole tibial spine or of either intercondylar tubercle may occur. With the severe trauma of violent abduction or adduction the whole spine may be broken and one of the intercondylar tubercles depressed. Fracture of the medial tubercle of the spine results from the same type of violence as that which causes injury of the anterior cruciate ligament. The medial tubercle, when detached, usually becomes displaced anteriorly. Fracture of the lateral tubercle of the spine is caused by forcible abduction of the tibia and direct contact with the lateral condyle of the femur.

Clinical picture. Fracture of the tibial spine occurs usually in older children and adolescents. The injury is followed quickly by pain, tenderness, distention of the joint capsule by blood, and inability to extend the knee completely. There is usually a bony block causing locking, which is a characteristic symptom. There may also be considerable anteroposterior instability of the joint.

Treatment. If little or no displacement of the fragment has occurred, immobilization in extension for 4 weeks may be sufficient. Aspiration may be done before immobilization is effected. Resistive quadriceps muscle exercises should be started early and continued. If the fragment is displaced, it may be possible by means of manipulation to force it back into position; immobilization should then be used for about 8 weeks. If the fragment cannot be reduced, however, open operation is indicated. Small fragments that do not weaken the attachment of the cruciate ligaments may be excised. The aftercare is the same as that of arthrotomy for the removal of a semilunar cartilage. If the fragment is large and includes the attachment of one or both cruciate ligaments, it should be replaced in the upper surface of the tibia and fixed by a single screw or by a suture passed through two drill holes from the anterior tibial cortex to the intercondylar area. This procedure should be followed by immobilization for a period of from 8 to 12 weeks.

LOOSE BODIES

Loose bodies are often found in the knee, hip, elbow, and shoulder joints and are frequently referred to as *joint mice.* In one reported series the knee was involved in 90% of the cases. The presence of intra-articular loose bodies is more frequently observed in men than in women.

Etiology. Loose bodies may form in the joint as the result of disease or of trauma. They often consist of a structureless fibrinous material; the so-called rice or melon seed bodies found in association with the chronic synovial reaction of osteoarthritis or tuberculosis are of this type. Occasionally loose bodies arise by the formation and proliferation of cartilage within the synovial villi, forming so-called synovial chondromas.

Timbrell Fisher classified loose bodies as follows: (1) those associated with an underlying pathologic process of recognized nature, such as chronic arthritis, tabes dorsalis, tuberculosis, or pyogenic arthritis; (2) those arising from the cartilage or bone of normal joints, such as the loose bodies of osteochondritis dissecans; detached articular ecchondroses, or detached intra-articular epiphyses; and (3) those bodies, such as the synovial chondromas, which follow an obscure pathologic change within the synovial membrane.

Clinical picture. Loose bodies usually cause a chronic intra-articular inflammation that is attended by an increase of joint fluid. There may be an associated weakness and instability of the joint. After motion there is sometimes a sudden, intense pain, occurring when the loose body becomes wedged between the articular surfaces and causes locking of the joint. In successive episodes of this kind the site of pain may vary widely. Loose bodies that are completely unattached may be found in any part of the joint, and their position may change between successive examinations. This variability of location, commonly determined by roentgenograms, is characteristic. Occasionally the body remains attached by a pedicle, occupies a more constant position in the joint, and can be palpated; such masses may occur in the localized type of villonodular synovitis (p. 365). Loose bodies can cause all

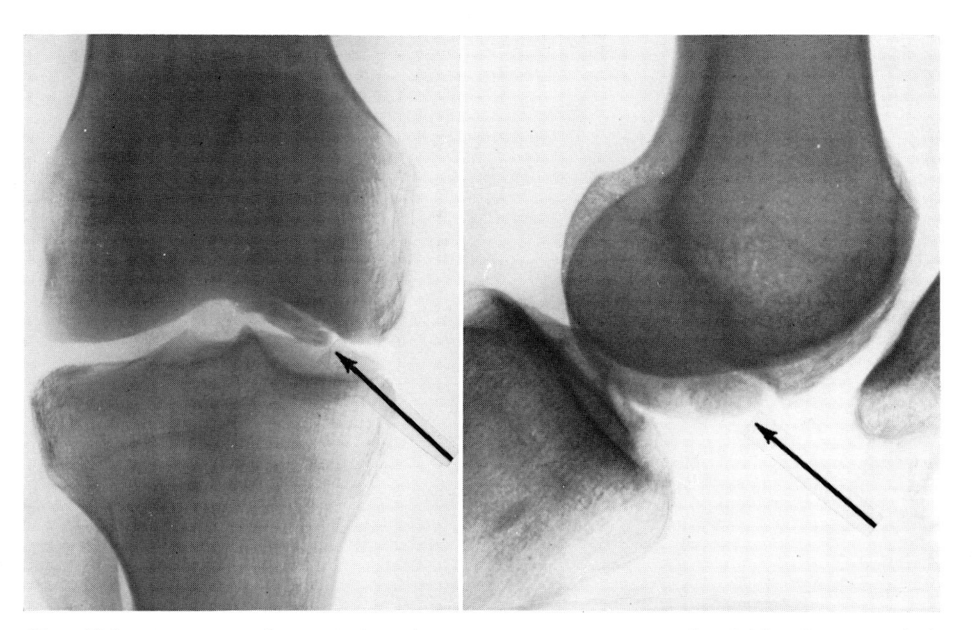

Fig. 13-5. Anteroposterior and lateral roentgenograms of osteochondritis dissecans of the medial condyle of the femur in a man 20 years old. The small fragment of bone and articular cartilage is only partially detached from the condyle.

degrees of joint symptoms and disability, from vague pain to extreme swelling and locking.

Osteochondritis dissecans. Osteochondritis dissecans is a joint affection characterized by partial or complete detachment and necrosis of a fragment of cartilage and underlying bone from the articular surface (Fig. 13-5). In advanced cases the fragment is completely detached, and its area of origin is recognizable as a shallow crater in one of the articular surfaces. The commonest site of osteochondritis dissecans is the lateral portion of the articular surface of the medial condyle of the femur, in the neighborhood of the insertion of the posterior cruciate ligament. Osteochondritis dissecans has been demonstrated also, however, in the ankle, hip, elbow, and shoulder joints. It is seen most commonly in adolescence or early adult life but occurs also in children. Males are affected more frequently than females. The fragmentation is thought by some observers to be the result of trauma, the circulation being impaired and a subsequent disturbance of bone nutrition taking place. It is believed by others to be caused by embolism of minute blood vessels supplying the affected area of bone and cartilage. Another theory of origin is that the fragment represents a separate ossification center that has become detached. The symptoms and signs are in general the same as those occurring with other types of loose bodies and include discomfort, weakness, fatigue, and catching or locking of the joint. However, when the diagnosis is made in children before the fragment has separated, complete rest of the joint from weight bearing for several months may bring about a reversal of the pathologic process, followed by complete disappear-

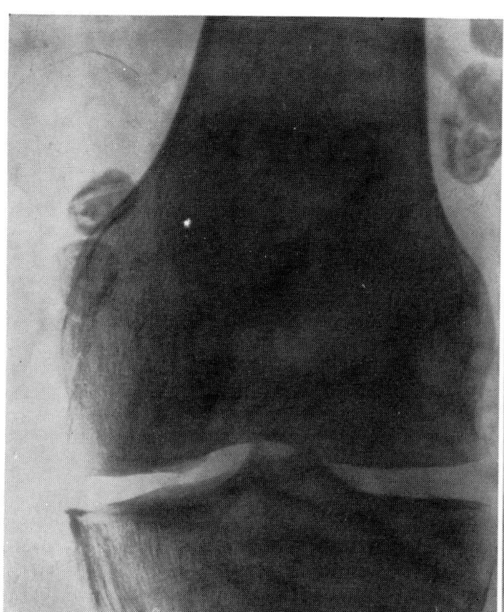
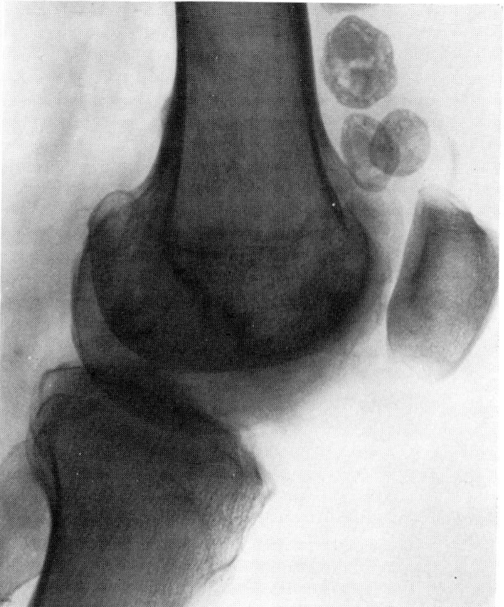

Fig. 13-6. Roentgenograms of synovial chondromatosis (osteochondromatosis) of the knee, showing characteristic large, rounded loose bodies. The patient was a 39-year-old man.

ance of the roentgenographic changes and by clinical cure. In children, anomalies of ossification, which require no treatment, may produce roentgenographic changes similar to those of early osteochondritis dissecans.

Synovial chondromatosis (osteochondromatosis). In synovial chondromatosis, a rare affection, pedunculated and loose osteocartilaginous bodies arise within the synovial membrane. The knee joint is involved most commonly (Fig. 13-6). Cartilage cells develop in the synovial villi, presumably as a result of metaplasia of the connective tissue cells. The cartilaginous bodies may occur singly but are usually numerous. Many of them remain attached to the synovial membrane, which often becomes thickly studded with them. The loose bodies frequently show lamination from the deposition of calcium salts; they may possess a loculated central cavity or a center of bony consistency.

Diagnosis. The diagnosis of intra-articular loose bodies can be made from the history. Roentgenograms demonstrate the presence of loose bodies if they contain calcium or bone.

Treatment. If the loose bodies are producing symptoms of mechanical interference with joint motion, they should be removed. It is often advisable to make a long incision because of the difficulty of finding the loose bodies and the danger of leaving any within the joint. If the bodies contain calcium, roentgenograms made in the operating room may be helpful. The removal of attached bodies is not technically difficult. Joints in which the synovial changes are generalized are best treated by synovectomy. The aftercare consists of a short period of immobilization, followed by physical therapy including vigorous active exercise of the weakened muscles.

HYPERTROPHY AND PINCHING OF THE INFRAPATELLAR FAT PAD AND THE SYNOVIAL MEMBRANE

Infrapatellar fat pad

The large pad of fat behind the patellar ligament may rarely be pinched and caught when the knee is extended. With repeated trauma, hemorrhage occurs into the fatty tissue, and a hard organized swelling appears, which may later become calcified. A similar process may take place on either side of the patellar ligament.

Clinical picture. Pinching of the hypertrophic infrapatellar fat pad is associated with pain behind the patellar ligament. The knee becomes stiff and weak and may show an effusion. Extension may cause a stabbing pain in the anterior aspect of the knee.

Treatment. A trial of nonsurgical therapy is indicated. Raising the heel of the shoe from ½ to 1 inch may afford relief. It may be advisable to apply a hinged knee brace to prevent the last 20 to 30 degrees of extension. If the symptoms persist, surgical exploration and excision of any grossly abnormal part of the fat pad may be helpful.

Synovial membrane

Hypertrophy of the synovial villi is a frequent result of chronic intra-articular inflammation. The villous processes become enlarged and elongated and occasionally form lobulated masses. This may occur in osteoarthritis. Benign proliferation of the synovial membrane is seen also in *pigmented villonodular synovitis.* Although this entity may occur in other joints, as well as in tendon sheaths and bursae, it is seen much more commonly in the knee than in other locations. Pigmented villonodular synovitis may be solitary or diffuse. In the solitary form a single nodule is attached to the synovial lining by a pedicle; such nodules vary in diameter from a few millimeters to several centimeters. In the diffuse form most of the synovial membrane is involved by numerous enlarged villi and nodules. Microscopic sections show hyperplastic synovial tissue containing giant cells, foam cells, and deposits of hemosiderin.

Clinical picture. A solitary lesion often produces symptoms suggesting a loose body, with painful intermittent catching or locking followed by effusion. Diffuse lesions cause chronic swelling and mild discomfort in the knee. Palpation may demonstrate thickening of the synovial membrane, increased joint fluid, and in some instances nodules. Aspiration of bloody fluid from a knee that has not been subjected to trauma should suggest the diagnosis of pigmented villonodular synovitis.

Treatment and prognosis. Solitary lesions should be treated by surgical excision. Diffuse lesions, if causing symptoms, require synovectomy. As a rule the lesions do not recur.

EXOSTOSES

Occasionally a joint may become locked or obstructed by the slipping of a tendon or muscle over a bony projection. This sometimes occurs, for example, at the posterior part of the knee. A common exostosis about the knee arises from the posteromedial surface of the lower end of the femur and is extra-articular. A sense of discomfort and slipping may accompany every movement. More often, however, the symptoms are trivial. The treatment consists of complete surgical removal of the exostosis.

OSGOOD-SCHLATTER DISEASE (PARTIAL SEPARATION OF THE TIBIAL TUBEROSITY, APOPHYSITIS OF THE TIBIAL TUBEROSITY)

Osgood-Schlatter disease is a partial separation of the tonguelike epiphysis of the tibial tuberosity, apparently caused by sudden or continued strain placed upon it by the patellar ligament during exercise (Fig. 13-7). There may be a disturbance of the circulation of the epiphysis since it often shows fragmentation; however, particles of necrotic bone that have been noted on microscopic examination are probably the result of the separation rather than of primary avascular necrosis. Osgood-Schlatter disease occurs usually in active boys between 10 and 15 years of age and is especially common among those who ride bicycles. It is frequently bilateral.

Clinical picture. The disorder is associated with pain over the tibial tuberosity when the patellar ligament is tightened on strong extension or when the patient strongly resists the examiner's attempt to flex the knee. The region of the tuberosity becomes enlarged and often is tender. The symptoms are sometimes acute and the tenderness extreme. Ecchymosis is occasionally observed. There is always aching in the area of the tuberosity on exercise, and particularly on climbing stairs and running.

Roentgenographic picture. The roentgenogram usually shows irregularity and slight separation of the epiphysis of the tibial tuberosity in the early stage and

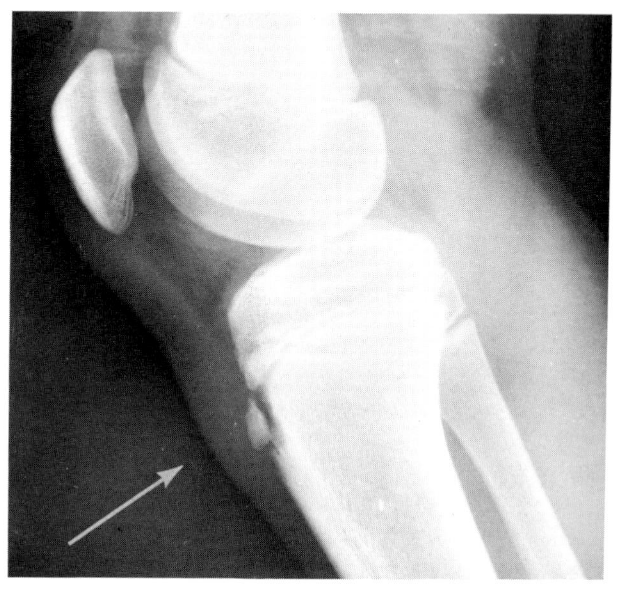

Fig 13-7. Osgood-Schlatter disease, showing irregularity of the epiphysis of the tibial tuberosity and swelling of the overlying soft tissues.

fragmentation of the epiphysis in the later stage. At times, however, little or no change may be visible. A film of the opposite knee may be taken for contrast, but often both knees are affected to the same extent and present an identical roentgenographic appearance.

Prognosis. The outlook for cure is excellent because progression of the affection is self-limited and the symptoms nearly always respond favorably to treatment.

Treatment. Immobilization in extension by means of splints or a plaster cast for a period of at least 5 weeks will usually cause the acute symptoms to subside completely, and no additional treatment may be needed. Weight bearing is not prohibited. After immobilization it is often advisable not to allow full flexion for an additional period of several months. In very mild cases simple restriction of activities that require forceful use of the quadriceps muscle, such as running and bicycle riding, may be sufficient to relieve the symptoms. In the uncommon, persistent case it may be advisable to drill through the tuberosity to the upper end of the tibia in an effort to improve the local circulation. This should not be done unless growth is almost complete, since epiphyseal damage may produce an asymmetric growth disturbance. In untreated cases a fragment of the epiphysis may remain ununited and later require excision.

RECURRENT DISLOCATION OF THE PATELLA (SLIPPING PATELLA)

Recurrent dislocation of the patella occurs more often in females than in males and is usually unilateral. Inherited tendencies resulting in a low lateral condylar ridge on the femur are responsible for some of the cases. Recurrent dislocation of the patella is often associated with genu valgum and a general muscular hypotonia of the lower limbs. Direct trauma or strain is usually the precipitating factor. Recurrent subluxation of the patella in the athlete may be a cause of serious knee disability. The symptoms may suggest other forms of internal derangement.

Clinical picture. The dislocation is almost always a lateral displacement that occurs on sudden contraction of the quadriceps muscle. It may result from a glancing blow on the medial side of the patella. Immediately following the dislocation there is sharp pain, which may cause the patient to fall. The pain is particularly severe when the dislocation is infrequent; in some cases the displacement occurs very often and causes little discomfort. Dislocation of the patella may be followed by a slight or moderate amount of effusion into the joint, especially in the early stages of the disorder. The patient complains of a constant feeling of weakness and insecurity and is timid about taking strenuous exercise.

Treatment. The patella may easily be replaced upon extension of the knee and flexion of the hip. In early cases the leg should be immobilized for a period of from 5 to 6 weeks. This may be followed by raising the inner side of the heel ¼ inch, walking with the toes turned in, and strengthening the quadriceps muscle with resistive exercises.

In repeatedly recurrent cases early operation is indicated and may forestall patellofemoral arthritic changes. Numerous surgical procedures involving the use of bone or soft tissue to hold the patella in place have been devised. One of the most successful operations is transplantation of the tibial tuberosity, together with the

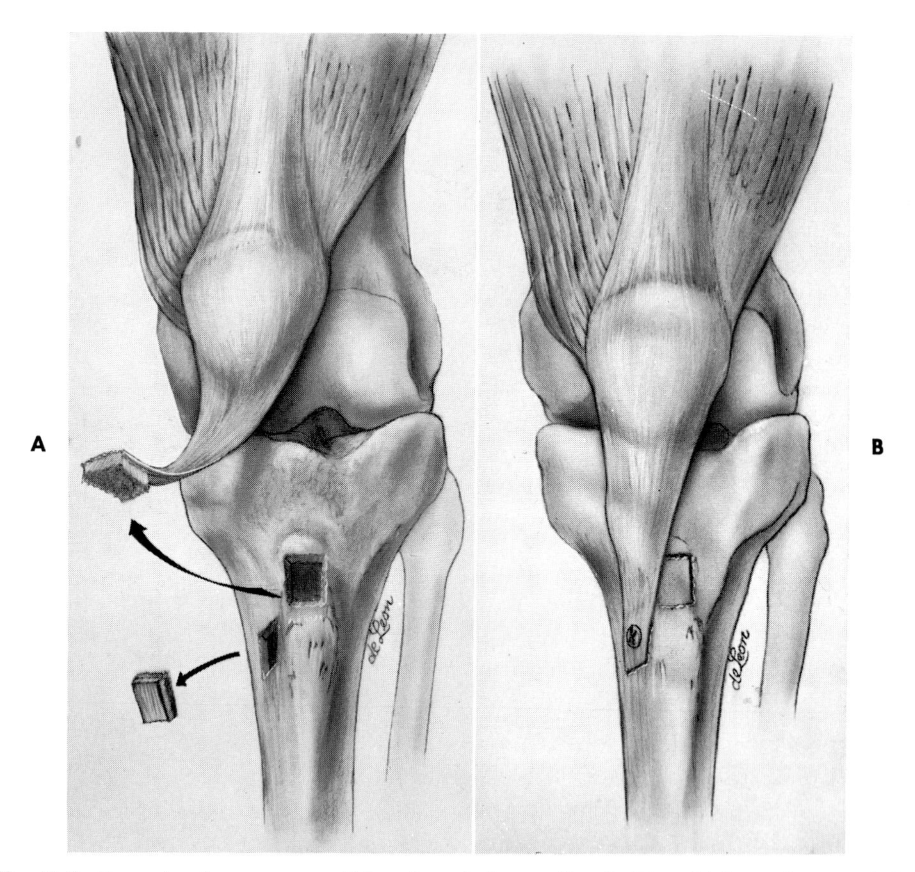

Fig. 13-8. Operation for recurrent dislocation of the patella. **A,** The tibial attachment of the patellar ligament is ready for transplantation medially and distally. **B,** It has been fixed in its new position with a screw. At the same time soft tissue structures on the anterolateral side of the joint may be released and those on the anteromedial side tightened.

patellar ligament, to a more medial and distal site, where it may be fixed in its newly prepared bed with a metal screw (Fig. 13-8). At the same time the joint should be explored for patellar chondromalacia and loose bodies, the anterolateral fascia released, and the anteromedial capsule reefed. Transplantation of the tuberosity should not be done in young children as it endangers growth at the proximal epiphyseal plate of the tibia. Occasionally osteotomy is done to correct an associated genu valgum, and in some instances patellectomy may be indicated.

CHONDROMALACIA OF THE PATELLA

Chondromalacia of the patella is a degenerative process of unknown etiology that involves the cartilage of the articular surface of the patella. The degenerative changes may follow repeated minor or acute severe traumas and often are associated with anomalies of the patella, osteochondritis, and arthritis. The highest incidence of cases is in young adults. When the undersurface of the patella has been examined in arthrotomy of the knee, chondromalacia has been found in approximately 20% of the cases. Articular cartilage changes in the patella have been found in up to 50% of cadavers with supposedly normal knees.

Clinical picture. These patients frequently give a history of having had trauma to the knee, followed by immediate disability and subsequent relief of symptoms. From several months to many years after such an episode there may develop variable pain, catching, instability, locking, weakness, and swelling in the knee, as well as atrophy of the thigh and tenderness of the patella. Descending stairs is often particularly difficult for patients with patellar chondromalacia.

On examination there are usually patellar tenderness, subpatellar crepitation, synovial thickening, and effusion. Routine roentgenograms usually show no diagnostic changes.

Treatment. Mild cases should be treated by rest and heat and protected from further injury by strapping or a brace. Isometric quadriceps-strengthening exercises with the knee in the extended position should be instituted. For the advanced case a partial or total chondrectomy or patellectomy is indicated.

OSSIFICATION OF THE TIBIAL COLLATERAL LIGAMENT (PELLEGRINI-STIEDA DISEASE)

Ossification in the tibial collateral ligament is a fairly common affection, occurring usually in men between the ages of 25 and 40 years. The deposits of new bone usually overlie the medial femoral condyle but have been observed also in the middle portion of the tibial collateral ligament just proximal to the level of the joint space. The ossification may be the result of involvement of the ligament by extension from inflammatory disease in adjacent structures such as bursae, tendons, synovia, and bones. Trauma is most often the exciting cause, however, and may be of a single violent type or may consist of repeated minor injuries of the knee joint.

Clinical picture. After traumatization the medial aspect of the knee becomes sensitive to pressure, especially about the adductor tubercle. Complete extension of the knee is painful, and the joint is usually held in slight flexion. Slight swelling of the knee may be observed. Occasionally an indurated area can be palpated.

Diagnosis. Dislocation of a semilunar cartilage and other internal derangements of the knee must be excluded before the clinical diagnosis can be made. The roentgenographic picture is pathognomonic.

Prognosis and treatment. In many cases the symptoms subside spontaneously without causing prolonged disability. Nonsurgical treatment consists of rest and support during the acute stage. Physical therapy is helpful but should not be started until all of the acute signs have subsided. The surgical treatment consists of excision of the bony plaque and plastic repair of the ligament. Occasionally the ossified mass recurs following resection, especially if it has been removed before reaching a mature stage.

RUPTURE OF THE QUADRICEPS TENDON AND OF THE PATELLAR LIGAMENT

The quadriceps tendon or the patellar ligament may be ruptured by the same type of violence as that which sometimes causes fracture of the patella, that is, sudden and violent contraction of the quadriceps muscle when the knee is flexed. Rupture of the tendon or the ligament is a much less common injury, however, than fracture of the patella.

Clinical picture. Rupture of the quadriceps tendon occurs more frequently than rupture of the patellar ligament and is seen most commonly in elderly persons. A definite tender depression can be seen and felt above the superior margin of the patella. The rupture may be associated with bloody effusion into the knee joint and resultant swelling.

Rupture of the patellar ligament results in upward displacement of the patella. A tender depression below the patella may be palpated but usually cannot be seen because of swelling of the infrapatellar fat pad.

After rupture of either of these structures, active extension of the knee is severely limited. In both conditions active flexion is of normal range but painful, whereas active extension is impossible in complete rupture and weak and painful in partial rupture.

Diagnosis. The diagnosis is made from the history and the physical findings. Tendon rupture is to be differentiated from fracture of the patella by roentgenographic examination.

Treatment. The treatment is surgical repair of the rupture. Suture should be followed by immobilization in extension for from 6 to 8 weeks, after which a program of physical therapy including carefully graded exercise in flexion should be instituted.

SNAPPING KNEE

In adults a snapping noise in the knee may be caused by a displacement of the lateral semilunar cartilage, especially if it is of discoid type, or by a sudden slipping of the biceps tendon or the iliotibial band. In early infancy sudden extension of the knee may cause the tibia to spring forward or rotate outward, producing an audible click; the mechanism is presumably a sudden and violent voluntary contraction of the muscular and tendinous structures about the joint. As a rule the symptoms of snapping knee are of a trivial nature, and there is no disability.

Treatment. An elastic support or plaster splint may be indicated.

INTERMITTENT HYDRARTHROSIS (INTERMITTENT SYNOVITIS)

Intermittent hydrarthrosis is an insidious, painless synovitis that appears most often in young women and is characterized by the accumulation of synovial fluid at regular intervals of from 5 days to a month, each effusion persisting for several days and then regressing spontaneously. It appears most often in the knee but occurs also in the elbow and rarely in other joints. It is not accompanied by increase of local temperature or by pain unless the swelling becomes extremely tense. There is, however, a feeling of stiffness and of slight discomfort about the joint. When the affection has become established, the swelling occurs in regular cycles so that the patient can foretell exactly when the joint will become swollen. Between periods of swelling the joint appears in early cases to be normal, but in older cases it may be somewhat boggy and thickened.

Etiology. Intermittent hydrarthrosis is thought by some authorities to be due to an endocrine disturbance. In young women it may be associated with the menstrual cycle. It has been known to disappear completely during pregnancy and to return after the termination of lactation. The affection has been found to occur in association with the early stages of rheumatoid arthritis. In such cases many of the joints may be involved simultaneously.

Prognosis. The periodic symptoms may continue for years. The outlook for cure is unfavorable except when a causative factor can be found and eliminated.

Treatment. In many cases the treatment is unsatisfactory. The administration of cortisone or iodides has seemed helpful in some cases. Synovectomy offers an excellent chance for relief of the symptoms in a single joint.

BURSITIS

Eighteen or more bursal sacs about the knee have been described, but those of clinical importance (Fig. 13-9) are the prepatellar, deep infrapatellar, superficial pretibial, and popliteal bursae, as well as variable bursae located beneath the tibial collateral ligament.

Prepatellar bursa

The prepatellar bursa lies anterior to the lower half of the patella and the upper half of the patellar ligament; it occurs more constantly than any other bursa about the knee. It often becomes inflamed and thickened from the irritation of repeated or prolonged kneeling. Chronic prepatellar bursitis (Fig. 13-10) has been called housemaid's knee, or nun's knee.

Deep infrapatellar bursa

The deep infrapatellar bursa is situated between the lower portion of the patellar ligament and the tibia. When the bursa becomes swollen, the normal depressions

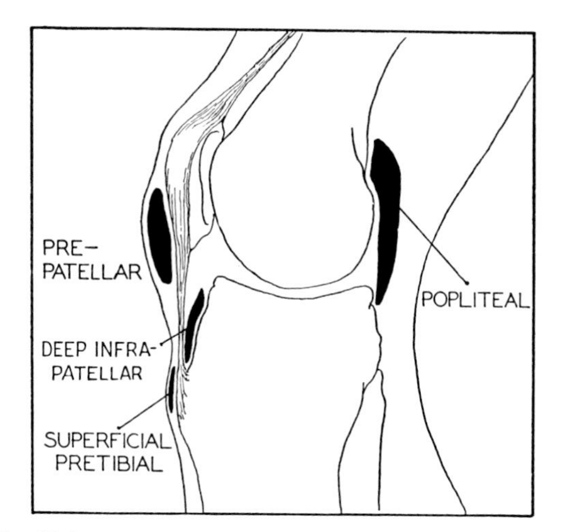

Fig. 13-9. Most commonly affected bursae about the knee.

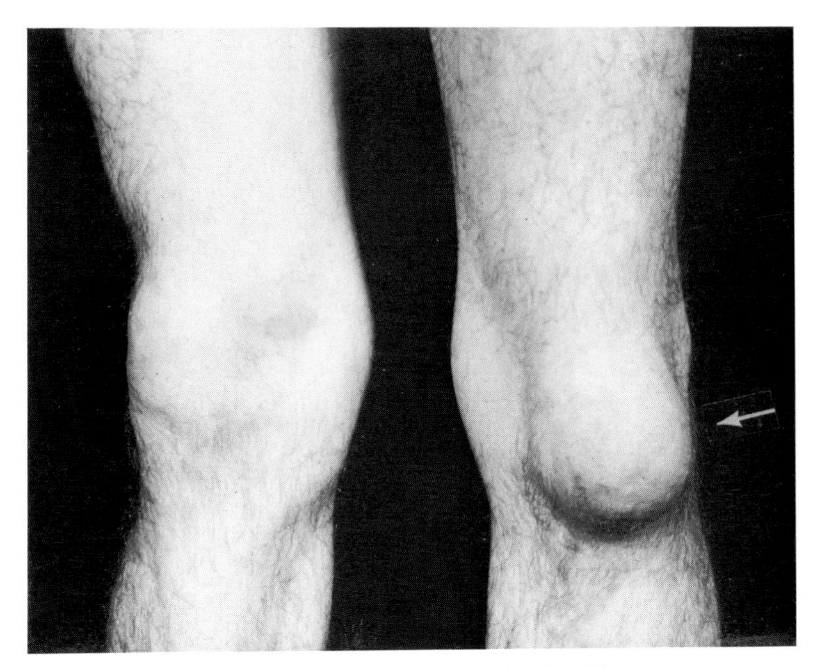

Fig. 13-10. Chronic left prepatellar bursitis.

on either side of the patellar ligament disappear. Active flexion and extension of the knee are painful and limited.

Superficial pretibial bursa

The superficial pretibial bursa overlies the insertion of the patellar ligament into the tibial tuberosity.

Popliteal bursae

The popliteal bursae are numerous and inconstant. Of them, the *gastrocnemio-semimembranosus bursa* is of most clinical importance because its distention by fluid is the usual cause of a *popliteal cyst.* Bursae may develop also about the biceps tendon and between the semitendinosus and gracilis tendons and the tibia (the *anserina bursa*). Occasionally a tubelike extension of the synovial sac of the knee is present beneath the popliteus muscle. The popliteal bursae are often connected with the synovial cavity of the knee. When enlarged, they stand out as firm, hard swellings when the knee is extended but may disappear when it is flexed. Sudden trauma and strain are important factors in causing symptoms.

Bursae beneath the tibial collateral ligament

Variable bursae located beneath the tibial collateral ligament, described by Voshell and Brantigan, may cause localized pain and tenderness, accompanied at times by a visible and palpable swelling.

Clinical picture. Any of these bursae may develop or enlarge without pain or

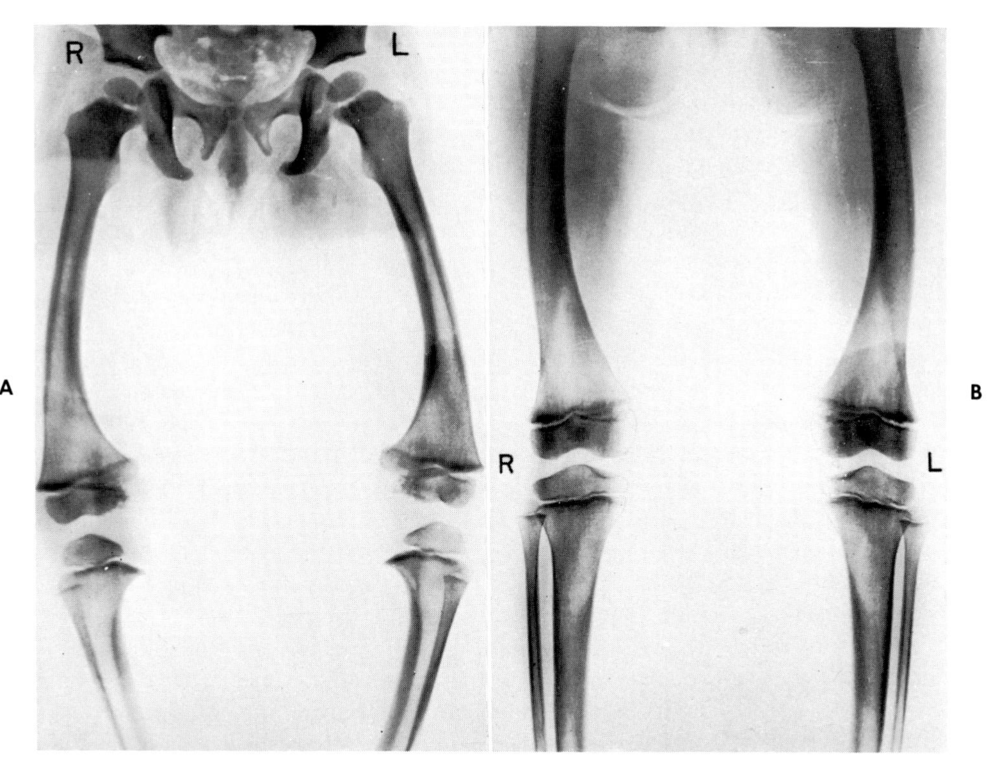

Fig. 13-11. Bilateral genu varum. **A,** Roentgenogram at 2 years of age shows marked bowing of femurs and tibiae with lipping of the medial margins of the metaphyses at the knees. **B,** Two and one-half years later the bowing has improved greatly and clinically the limbs are straight. Surgical treatment was not required.

discomfort. Fatigue and weakness of the knee may gradually appear, especially in affections of the popliteal bursae and of those about the attachment of the patellar ligament. When the bursa becomes distended and tense, the swelling may be accompanied by pain and tenderness of varying degree. The prepatellar bursa is often the seat of an acute pyogenic infection.

Treatment. In most cases the inflammatory symptoms subside with rest of the knee joint and hot applications. If a pyogenic infection is present, aspiration and antibiotic therapy are indicated. The greatly thickened prepatellar bursa resulting from long-standing chronic inflammation or distention is treated by excision. Chronically distended popliteal bursae should also be excised; after the entire sac has been removed the bursa seldom recurs.

GENU VARUM (BOWLEG)

Deforming curvature of the legs occurs so commonly as to merit description as an entity. In *genu varum,* or *bowleg,* the major convexity of the limb is disposed laterally (Fig. 13-11, *A*). Genu varum is usually bilateral but occasionally affects only one leg and may even be associated with the opposite deformity, genu valgum, in the other leg. Curvature in which the major convexity lies anteriorly is occasionally seen and is called *anterior bowleg.*

Etiology and pathology. Mild to moderate bowleg may be normal for most infants and will persist to approximately the twenty-fourth month, after which the legs gradually become straight or even develop knock-knees. An accentuation of this physiologic bowing is frequently seen in the obese child. In rickets, a rare cause of deformity today, bowleg results from the gradual bending of pathologically softened bone under the influence of weight bearing and postural stresses. Bowleg may result from defective growth at the medial side of the upper tibial epiphysis; this is a localized epiphyseal dysplasia sometimes termed *tibia vara,* or *Blount's disease* (p. 376). Infection or injury of the medial side of the tibial or femoral epiphysis may lead to premature growth arrest and unilateral bowleg. Rarely the healed fractures of osteogenesis imperfecta may lead to similar deformities. The *saber shin* of congenital syphilis is differentiated from anterior bowleg by collateral clinical evidence of syphilis, a positive complement-fixation reaction, and the roentgenographic findings. In adults bowleg may occur from the changes of osteitis deformans, osteomalacia, hyperparathyroidism, malunited fracture, or destructive disease of the knee joint such as pyogenic, degenerative, or rheumatoid arthritis affecting predominantly the medial side of the joint.

The curvature may involve the tibia alone, the tibia and femur equally, or occasionally the femur alone; sometimes it is the result of a lateral yielding at the knee joint while the shafts of the femur and tibia remain straight. As the three-cornered column of the tibial shaft bends, torsion necessarily takes place, and the lower end of the tibia becomes internally rotated. Internal tibial torsion and genu varum may be aggravated by certain sitting habits. The infant tends to sleep with his knees drawn up under him and his feet internally rotated; later in childhood he may sit with his feet under his buttocks so twisted that their lateral surfaces rest against the floor. Internal tibial torsion accentuates the apparent deformity in bowleg because as the child attempts to compensate for it he walks with his knees turned outward and slightly flexed. Anterior bowleg may exist independently or in association with the lateral form.

Clinical picture. Bowleg is present if the extended knees are separated when the medial malleoli of the ankles are approximated. Measurements of the space between the knees, with the ankles together, provide a ready clinical estimation of the degree of deformity and of its progress. Because of outward bowing at the knee and inward rotation at the ankle the child tends to walk with his feet widely separated and his toes turned in; this may lead to considerable lateral shift of the body weight with each step and thus to a waddling type of gait.

Treatment. The stage at which the patient is first seen influences the treatment. In the young child, aggravating postural influences in sitting and sleeping should be avoided. If possible, obesity should be corrected. The patient should be reexamined at intervals for a long period since spontaneous correction usually occurs with growth (Fig. 13-11, *B*). Extensive bracing is rarely indicated, but the use of a Denis Browne night splint (Fig. 2-19) may hasten correction of the deformity; it is particularly helpful when severe internal tibial torsion is associated with the bowleg. Persistent bowleg occasionally requires correction by *osteotomy.* At osteotomy any internal

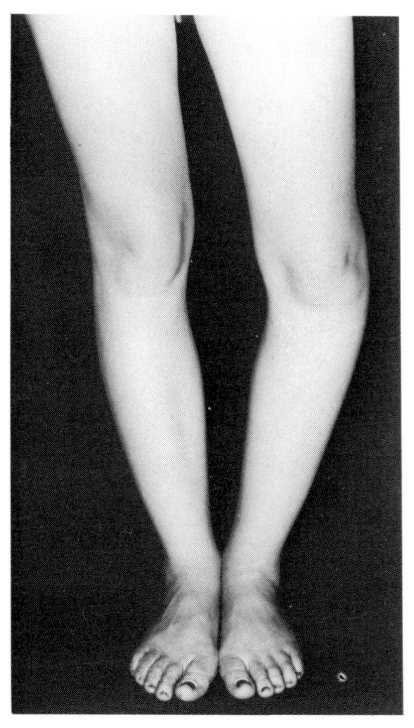

Fig. 13-12. Tibia vara, or Blount's disease, showing a varus deformity of the left leg just below the knee.

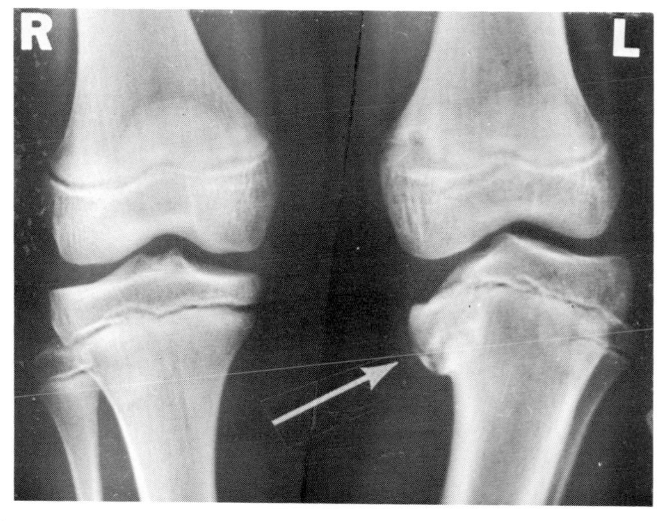

Fig. 13-13. Anteroposterior roentgenograms showing left tibia vara. The medial side of the proximal tibial epiphysis is depressed, and the growth disturbance has resulted in a varus deformity.

rotation can be corrected along with the bowing. In adults, significant bony deformity can be corrected only by osteotomy.

TIBIA VARA (BLOUNT'S DISEASE)

Tibia vara results from retardation of growth at the medial side of the proximal tibial epiphyseal plate while growth on its lateral aspect continues normally, leading to a progressive bowleg deformity (Figs. 13-12 and 13-13). The onset of the disease usually occurs between 1 and 3 years of age; a few less severe cases have begun in adolescence. Bowing is the only clinical finding; it may be unilateral or bilateral, and early may be mistaken for physiologic bowing. The cause of the growth disturbance is not known. Abnormal stresses on the medial side of the epiphyseal plate in the very young have been postulated, and indeed many of these children are overweight.

Treatment. The treatment of progressive cases of Blount's disease is osteotomy of the tibia and fibula. If surgical correction is done early, before the age of 8 years, the correction is often maintained. Surgery done after this age is frequently followed by recurrent deformity, requiring repetition of the osteotomy in adolescence.

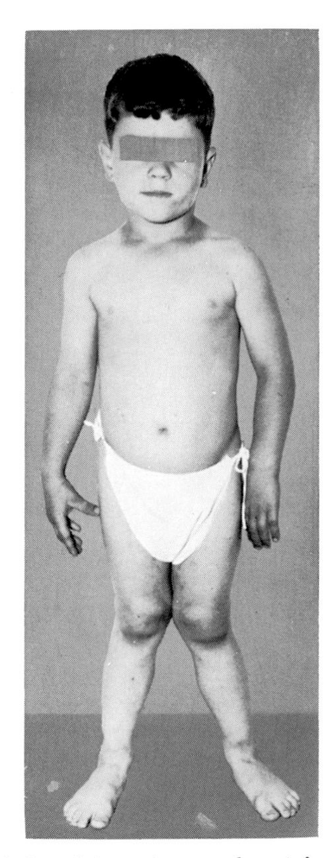

Fig. 13-14. Rachitic knock-knee (genu valgum) in a boy 5 years of age.

GENU VALGUM (KNOCK-KNEE)

In genu valgum, or knock-knee, there is an abnormal curvature or angulation of the lower limb with the apex of the convexity disposed medially at the level of the knee (Fig. 13-14). Although usually bilateral, knock-knee may occur in only one leg. Elements of knock-knee and bowleg may be present at different levels of the same limb.

Etiology and pathology. Normally the medial femoral condyle is slightly longer than the lateral, thus maintaining the horizontal plane of the knee joint despite the inward obliquity of the femur. The axis of weight bearing at the knee, however, passes lateral to the center of the joint so that the lateral condyle bears more weight than the medial condyle. With any relaxation of the tibial collateral ligament, the valgus alignment will increase, resulting in further pressure on the lateral side of the joint. This may produce a slight difference in the growth rates of the medial and lateral sides of the adjacent epiphyseal plates and lead to bony deformity. With increasing knock-knee, external rotation of the tibia takes place from the pull of the lateral hamstring muscles. The foot usually develops marked pronation; in occasional cases, however, compensatory adduction of the anterior part of the foot occurs and leads to a position of varus. Knock-knee is sometimes associated not with obliquity of the joint line but rather with curvature of the tibial or femoral shaft. In most instances, however, knock-knee is not the result of bony changes in the tibia or the femur but is associated with eversion of the foot and exaggerated by relaxation of the tibial collateral ligament. Occasionally genu valgum of slow development may be caused by an impairment of bone growth in the lateral half of the epiphyseal plate as a result of trauma or infection. Knock-knee resulting from vitamin D deficiency is uncommon today, but the deformity may be associated with other abnormalities of calcium metabolism that simulate rickets. Genu valgum may also occur in advanced tuberculosis of the knee, after severe poliomyelitic paralysis, after bumper fractures of the tibia with uncorrected depression of the lateral condyle, and in degenerative or rheumatoid arthritis in which the lateral side of the knee joint is predominantly affected.

Clinical picture. The diagnosis of knock-knee is made on inspection of the extended knee. The patient, as viewed from the front, may stand with his knees overlapping. The gait is altered by an internal rotation of the leg and foot to prevent the knees from striking each other while passing and by an increased lateral sway of the body to carry its weight to a position directly over the foot.

Treatment. The vast majority of nonrachitic knock-knees in children undergo spontaneous correction with growth. In young children a ⅛-inch raise of the medial border of the heel may be helpful. Rarely are braces indicated, but in severe cases the use of a night splint is helpful. When the deformity is the result of a metabolic abnormality, successful treatment of the primary disease will often result in spontaneous correction of the genu valgum.

In the rare severe case that persists into late childhood and shows no tendency toward spontaneous correction, osteotomy may be advisable. The level of section, determined after roentgenographic analysis of the deformity, is usually in the supracondylar region of the femur, but at times tibial and fibular osteotomy is preferable. In older patients the removal of a small wedge of bone may be helpful.

Care must be taken to overcorrect slightly the valgus deformity. The external rotation, estimated from the position of the foot with respect to the patella, can often be corrected by the same osteotomy.

INEQUALITY OF LEG LENGTH

Although inequality of leg length may result from disturbance in any or all parts of the lower limb, its treatment frequently centers about the knee joint. Unequal leg length is a cause of asymmetric posture. When the patient stands, his pelvis tilts downward on the side of the shorter limb, and he has a scoliosis convex to that side (Fig. 10-1). If severe, the inequality may cause a significant limp.

Etiology. The causes of leg length discrepancy are numerous. Until recently the commonest cause was asymmetric paralysis after poliomyelitis. Unequal leg length from shortening of the affected limb is also seen frequently after other types of paralysis in childhood and as a result of congenital defects, malunited fractures, and epiphyseal injuries and infections. A moderate degree of overgrowth of the leg occurs in congenital hemihypertrophy; it is also a frequent late result of fracture or osteomyelitis of the femur or tibia in childhood. Overgrowth may be caused also by vascular lesions such as hemangioma or arteriovenous fistula in childhood.

Treatment. Slight shortening should be treated by a heel raise. Greater disproportion may indicate surgical treatment. Equalization of leg length may be effected either by lengthening the shorter limb or by shortening the longer one. Either the tibia and fibula or the femur may be lengthened, rarely as much as 3 inches, but the end results are often unsatisfactory. The procedure is a gradual stretching of the soft tissues and lengthening of the bone after an osteotomy of special type. After operation the fragments must be maintained in satisfactory apposition and alignment during the lengthening process, and for this purpose many types of apparatus have been devised. Great care and judgment are necessary in the selection of cases suitable for lengthening operations.

In most cases equalization of the length of the legs can be accomplished more satisfactorily by shortening the longer leg, since this is less painful, less hazardous, and less likely to result in inadequate muscular control. The shortening may be accomplished by performing a simple oblique osteotomy of the femur and allowing the fragments to override, or by removing a measured segment of the femur. In either case care must be taken to secure effective fixation of the fragments.

Gross inequality of leg length in children may often be treated satisfactorily by an operation to retard the growth of the longer leg. Local growth in length can be arrested permanently by fusing the epiphysis to the diaphysis. This operation, devised by Phemister and called *epiphyseodesis,* should not be done until the patient is at least 10 years old. In the selection of cases a careful estimate of the relative rate of growth of the two legs and of the duration of further growth must be made. Since two thirds of the growth in length of the leg occurs about the knee, the usual surgical procedure is fusion of the lower femoral epiphysis or the upper tibial and fibular epiphyses, or all three, to the respective metaphyses. The operative technic must be careful and thorough to prevent later deformity of the knee from incomplete, asymmetric arrest of growth. An alternative procedure, developed by Blount, is to insert strong metal staples across the epiphyseal plate into the epiphysis and metaphysis. Later, if resumption of growth is desired, the staples may be removed. The surgical technic of stapling must be meticulous. Many complications have been reported, caused usually by poor judgment or improper technic.

Recently there has been a revival of interest in attempts to stimulate epiphyseal growth in the short leg. As yet, however, no reliable method for producing an adequate acceleration of length growth has been devised.

CHAPTER 14 # Affections of the ankle and foot

The functions of the foot are (1) to serve as a support for the weight of the body and (2) to act as a lever in raising and propelling the body forward in walking and running. The muscles of the leg supply the power, and the heads of the metatarsal bones serve as a fulcrum on which the weight is lifted. The foot contains two main arches formed by bones and supported directly by ligaments and indirectly by tendons and muscles. A normal degree of motion and elasticity in these arches is necessary for proper functioning of the foot.

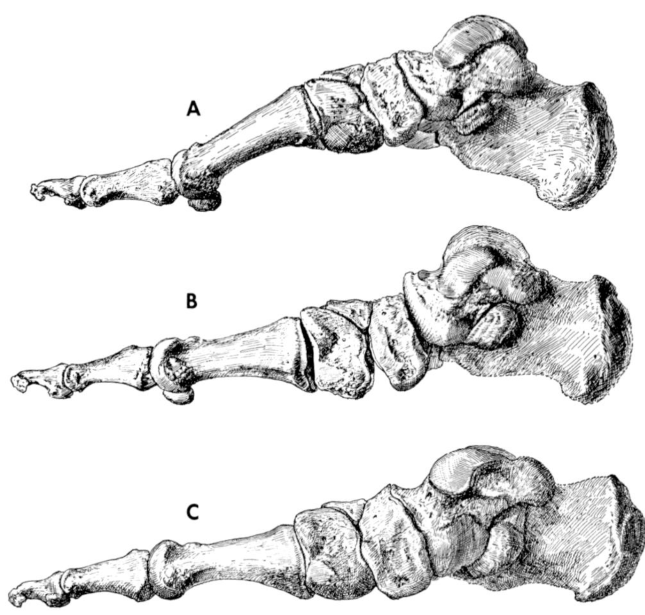

Fig. 14-1. Medial aspect of the bones of the foot. **A,** Normal longitudinal arch. **B,** Loss of arch in flexible flatfoot. **C,** Absence of arch in rigid flatfoot, showing adaptive bone changes.

The longitudinal or long arch is made up of two components, the medial and the lateral. The medial component is the more obvious and comprises the calcaneus, talus, navicular, three cuneiform bones, and the first three metatarsal bones (Fig. 14-1, *A*). This arch rests on the head of the first metatarsal and on the calcaneus. The lateral component of the longitudinal arch consists of the calcaneus, the cuboid, and the fourth and fifth metatarsal bones; it is supported behind by the calcaneus and in front by the heads of the fourth and fifth metatarsals. The anterior, transverse, or metatarsal arch is formed by the heads of the five metatarsal bones. On weight bearing, the anterior arch becomes flattened, but as soon as the weight is removed the metatarsal heads return to their former positions and again form a low arch.

The movements of the foot and ankle are most important in the diagnosis and treatment of disabilities of the foot. The ankle joint permits plantar flexion and dorsiflexion of the foot. The foot is inverted or in the varus position when it is so rotated that its plantar surface faces medially; the foot is everted or in the valgus position when its plantar surface is rotated outwardly (Fig. 14-2). These motions occur in the subtalar joint between talus and calcaneus and in the transverse tarsal or midtarsal joint between talus and navicular and between calcaneus and cuboid. The primary motion of the transverse tarsal joint, however, is adduction or inward swinging and abduction or outward swinging of the anterior portion of the foot. Pronation of the foot is a combination of eversion and abduction of the anterior portion of the foot; supination is a combination of inversion and adduction.

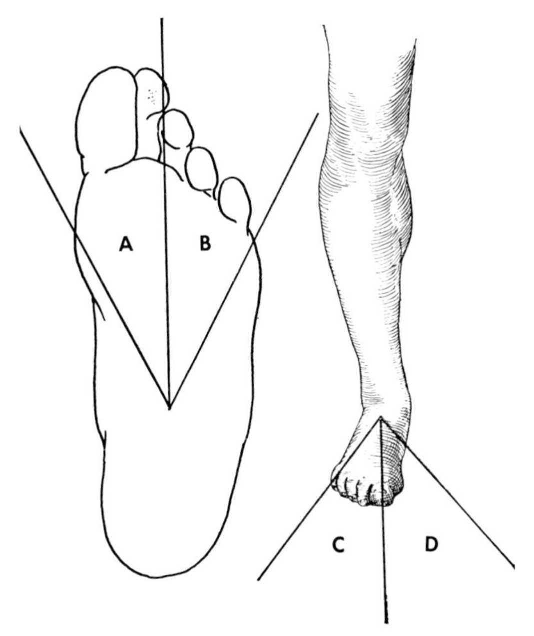

Fig. 14-2. Terminology of motion in the tarsal joints. **A,** Forefoot adduction; **B,** forefoot abduction; **C,** eversion; **D,** inversion. (After Cave and Roberts.)

FOOT STRAIN

Foot strain is encountered in a large group of patients whose disabilities arise in association with weight-bearing stresses on the foot.

Etiology. Factors that may lead to strain of the foot are incorrect shoes, inadequate muscular and ligamentous support, excessive body weight, and excessive exercise.

Incorrect shoes. Stiff leather shoes with pointed ends distort and compress the toes, making it impossible for the muscles to function normally. Muscle atrophy and loss of supporting power result. There can be little doubt that the modern shoe is the most important cause of many disabilities of the foot, especially in women.

Inadequate muscular and ligamentous support. Inadequate support may follow (1) rapid growth when muscle strength does not keep pace with the growth of the bony framework and (2) prolonged illness or severe injury of the leg resulting in muscle atrophy and hypotonicity.

Excessive body weight. Obesity often puts so much stress upon the foot on weight bearing that the strength of the supporting ligaments and muscles is exceeded and the arches become abnormally depressed.

Excessive exercise. Too much standing, walking, or running, especially when the individual is not accustomed to such exercise or wears a flexible rubber-soled shoe, is a frequent cause of foot strain.

When the muscles of the leg and foot fail to contribute their support, the body weight must be borne by the ligaments of the foot. Ligamentous tissues function normally as checkrein structures, and when subjected to prolonged, unrelieved tension, they may become painful. Over a period of time gradual stretching of the ligaments may take place, allowing pronation of the foot and loss of the arch.

Clinical picture. Foot strain may be acute or chronic. Both forms affect adults frequently and are rare in childhood. Acute foot strain is manifested by pain and tenderness in the region of the longitudinal arch to such a degree that the patient is scarcely able to bear weight on the foot.

The patient with chronic foot strain complains of fatigue and aching in his feet. The discomfort is felt chiefly in the region of the longitudinal arch but may extend up into the calves. The symptoms are usually worse toward the end of the day; by evening the feet may feel tight and swollen. Often there is a history of recent weight gain, recent illness, or change to an occupation that involves increased walking or standing.

The most important physical finding is localized tenderness beneath the navicular bone at the apex of the longitudinal arch. In some instances pronation of the foot is obvious when the patient stands. This is not always true because the normal foot and the foot with a high arch are also subject to strain.

Treatment. The immediate problem of foot pain and tenderness responds to local treatment. Hot soaks or contrast baths may relieve the discomfort. Some form of support for the longitudinal arch should be provided. In the acute case this may be accomplished best by adhesive strapping. In chronic strain the use of a felt or sponge rubber arch pad is helpful (Fig. 14-3). A shoe with long counter and firm shank will give increased support. To this may be added a Thomas heel, which is extended forward medially to give support beneath the navicular bone (Fig. 14-4).

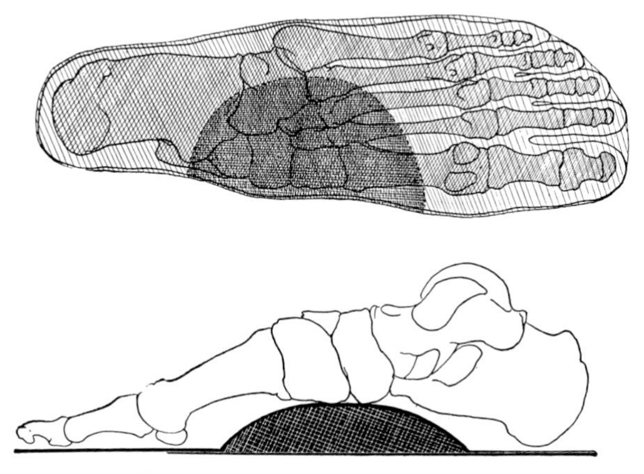

Fig. 14-3. Longitudinal arch support.

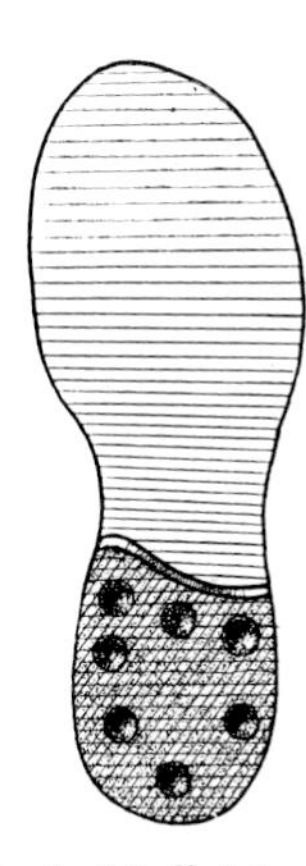

Fig. 14-4. Thomas or orthopaedic heel. Medial border of the heel is extended forward.

In addition the medial border of the heel may be raised 3/16 inch. Exercises of both long and short muscles of the foot aid in restoring strength. The cause of the foot strain should be investigated and, if possible, treated. Weight reduction may be helpful. Modification of the patient's job is occasionally necessary. With proper treatment most cases of foot strain can be satisfactorily relieved by conservative measures.

FLEXIBLE FLATFOOT

Flexible flatfoot is seen frequently in children and in adults.

In children

Some eversion of the foot and relaxation of the longitudinal arch are usual in children in the first few years of life. The degree of eversion varies considerably, and it is impossible to make a clear-cut distinction between what should be considered normal and what abnormal. By the age of 5 or 6 years individuals who are going to acquire a well-developed arch show evidence of it. Many children remain rather flatfooted. Unless the eversion is of severe degree, it is unlikely to lead to discomfort. Rarely does this type of flatfoot cause symptoms in childhood.

The diagnosis of flatfoot in children is usually obvious, but its interpretation may be difficult. When the child stands, absence of the longitudinal arch on the medial side of the foot is apparent. Seen from its posterior aspect, the heel appears everted. Examination of the dorsum shows abduction of the forefoot. The foot is quite mobile and without fixed deformity. There is a normal range of ankle and subtalar motion, and the mobility of the anterior part of the foot may even be increased. Motor power about the foot and ankle is normal. This type of relaxed flatfoot is usually most severe in heavy children; it is often associated with knock-knee or slight back-knee. It may be differentiated from paralytic flatfoot by the presence of normal muscle power, from congenital vertical talus and from peroneal spastic flatfoot by the normal mobility and absence of fixed deformity, and from flatfoot associated with a tight heel cord by the normal range of ankle dorsiflexion.

Treatment. In mild cases no treatment is necessary, other than reassurance of the parents. Children with moderate or severe flatfoot are probably best treated by raising the medial border of the heels of their shoes 1/8 inch. A shoe with long counter and Thomas heel may be helpful. It is permissible and probably advisable to allow the child to go barefoot part of the time during the summer. In children, foot exercises are of doubtful value.

In adults

Asymptomatic flatfoot in the adult requires no treatment. Mild or moderate degrees of pronation do not usually cause symptoms. Severe flatfoot in adults, however, with its disturbed mechanical relationships, is likely to result in degenerative articular changes (Fig. 14-1, *C*), stiffness, and pain on weight bearing. The treatment of symptomatic flatfoot in adults is like that of foot strain, which has been discussed (p. 382).

SPASTIC FLATFOOT

The usual spastic flatfoot is held in marked eversion by spasm of the peroneal muscles (Fig. 14-5). The spasm is not caused by an upper motor neuron lesion but is probably of reflex or protective nature associated with irritative changes in one or more of the joints of the foot. The spasm may be relieved and may disappear with rest, only to return when weight bearing is resumed. Spastic flatfoot may be a sequel of an untreated flexible flatfoot and is seen at times in association with obesity or chronic arthritis.

In the early stages the foot is extremely painful. There is often tenderness over the peroneal tendons, and attempts to invert the foot may cause sharp pain. The foot may be swollen, and the limb may be held in marked external rotation. In acute stages every step is painful and the patient walks with an awkward, shuffling gait. Later the foot may become rigid from the development of secondary bone changes, with proliferation about the joint margins where abnormal pressure and strain are present.

Some rigid, everted flatfeet have a congenital synostosis or synchondrosis be-

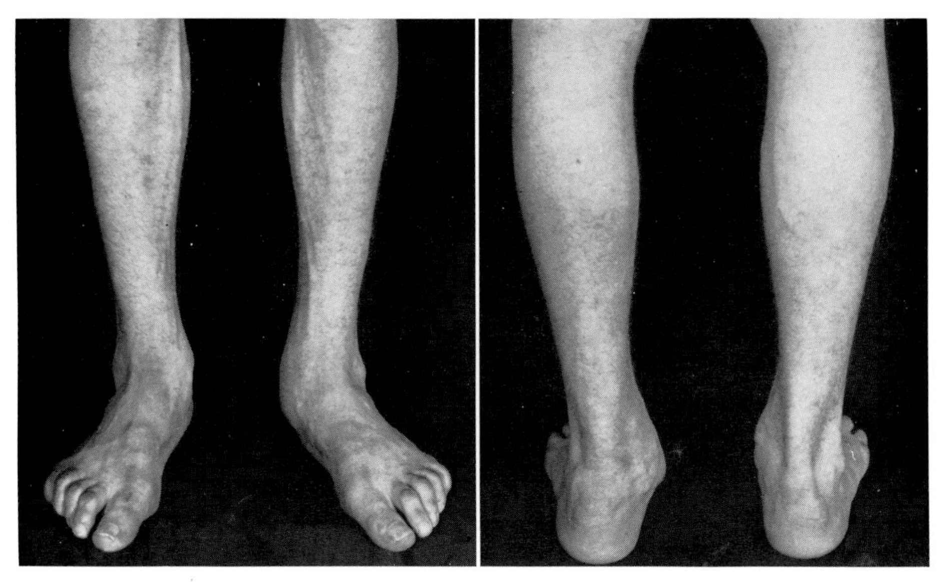

Fig. 14-5. Left spastic flatfoot in a man 22 years of age. Note severe pronation of the left foot, which could not be inverted actively or passively, and atrophy of the left calf. The right foot is slightly pronated but showed no joint stiffness and no tightness of the peroneal muscles.

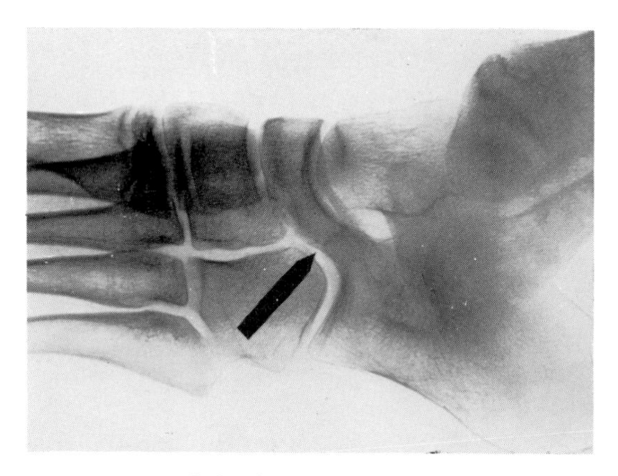

Fig. 14-6. Oblique roentgenogram of the foot of a boy 15 years of age, showing a calcaneonavicular bar. This was bilateral, and a previous diagnosis of spastic flatfeet had been made.

tween the anterior end of the calcaneus and the navicular bone *(calcaneonavicular bar),* which may be seen in oblique roentgenograms (Fig. 14-6). A complete or partial union of the talus and the calcaneus by a bridge of bone situated behind the sustentaculum tali *(talocalcaneal bridge)* may also be found; it can be demonstrated in posterior oblique roentgenograms of the heel. In the presence of either of these anomalies, pain in the partially rigid foot may develop at or shortly before puberty.

Treatment. In the early stages and in the absence of bony abnormalities, correction may be accomplished by the use of casts, which are wedged gradually into inversion. If there is no roentgenographic evidence of bony anomalies, forcible overcorrection under anesthesia, followed by immobilization in plaster, may be helpful. It is often advisable to allow the patient to walk while the cast holds the foot in the inverted position. A support under the longitudinal arch and corrective foot exercises should then be prescribed. If the pain does not respond to such nonoperative measures, arthrodesis of the subtalar and midtarsal joints (triple arthrodesis) is indicated; it is usually followed by complete relief.

In rigid flatfoot, operations for reshaping the longitudinal arch are seldom to be recommended. Relief of pain after resection of a congenital bony bar or bridge may be incomplete, especially in older children with symptoms of long duration and adaptive changes in the tarsal bones. The most effective treatment for such patients is triple arthrodesis.

SHORTENING OF THE ACHILLES TENDON

Shortening of the Achilles tendon, especially in the weak or everted foot, leads to the development of foot strain. In adults the shortening may be due either to a congenital structural change or to reflex muscular spasm from irritation caused by disturbed mechanics of the foot. In the reflex type there is discomfort that may be associated with tenderness and with sharp pain and spasm on attempting to dorsiflex the foot.

Treatment. In cases of the reflex type, raising the heel of the shoe may lead to relief of the symptoms. In the structural type an attempt should be made to stretch the Achilles tendon by wedged plaster casts or by special stretching exercises. Experience has shown that surgical lengthening of the Achilles tendon weakens this type of foot and should be discouraged. In women pain in the feet is sometimes caused by shortening of the Achilles tendon from wearing shoes with very high heels. In such cases the heels should be gradually lowered, and stretching exercises should be carried out. Sudden change from high-heel to low-heel shoes often produces much discomfort in the legs from pull on the gastrocnemius and soleus muscles and should never be advised.

CLAWFOOT

The typical clawfoot has an abnormally high longitudinal arch, a depression of the metatarsal arch, and dorsal contractures of the toes (Fig. 14-7). The plantar fascia is contracted, the anterior half of the foot is drawn downward and sometimes inward, and the Achilles tendon may be shortened. Excessive height of the longitudinal arch is termed *cavus* deformity. It is often seen without other deformities and usually causes few symptoms.

The deformities of clawfoot may be slight or severe. In most instances they show a gradual progession. With the muscular contractures that are associated with clawfoot, there is a loss of the normal elasticity of the arches.

Etiology. In some cases the deformity of clawfoot is the result of an inherited tendency. It may be associated also with the wearing of high heels, which produces a postural equinus, or with excessive use of the leg muscles, as in professional

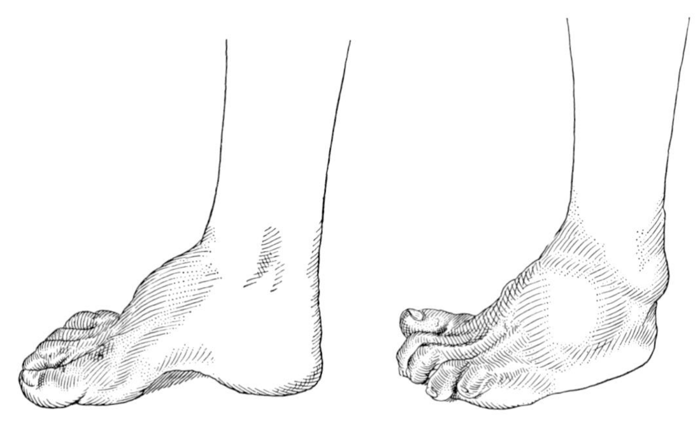

Fig. 14-7. Clawfoot of moderate degree in a young adult. Note high longitudinal arch (cavus), dorsal contracture of the toes, adduction of the forefoot, and slight equinus.

dancers. After poliomyelitis it sometimes develops in the seemingly unparalyzed foot. Imbalance of muscular power in the foot is considered the causative factor in many cases. Clawfoot may occur in the course of progressive lesions of the central nervous system, such as Friedreich's ataxia and peroneal muscular atrophy. Frequently clawfoot has been found in association with spina bifida occulta. It may follow conditions such as cellulitis, rheumatoid arthritis, and ischemic paralysis, and it is sometimes seen after compound fractures of the tibia that have been immobilized in poor position or have received inadequate aftercare. Most commonly, however, clawfoot can be ascribed to no specific etiologic factor; it is then spoken of as idiopathic clawfoot, which is a progressive deformity and may be unilateral or bilateral.

Symptoms. Although considerable deformity may be present without causing symptoms, the patient usually complains of early fatigue with exercise and of tender calluses beneath the metatarsal heads and over the proximal interphalangeal joints.

Treatment and prognosis. In mild cases the treatment may consist of simple stretching of the plantar fascia and Achilles tendon and the wearing of proper shoes. The shoes should be fitted with metatarsal pads or bars to relieve strain on the anterior portion of the foot.

For clawfoot of moderate degree it may be necessary to carry out a plantar fasciotomy and to stretch the foot under anesthesia. Occasionally transplantation of one or more of the extensor tendons of the toes to the distal portion of the metatarsal bones or to the flexor tendons is indicated.

In more severe cases it is sometimes advisable to strip the plantar fascia from its attachment to the calcaneus, section the flexor and extensor tendons of the toes, and lengthen the Achilles tendon.

In still more advanced cases dorsal wedge osteotomy combined with subtalar arthrodesis is the only procedure that will effect satisfactory correction of the deformity (Fig. 14-8). In extreme cases it may be necessary also to excise the metatarsal heads.

In idiopathic clawfoot the deformity may recur after surgical correction, espe-

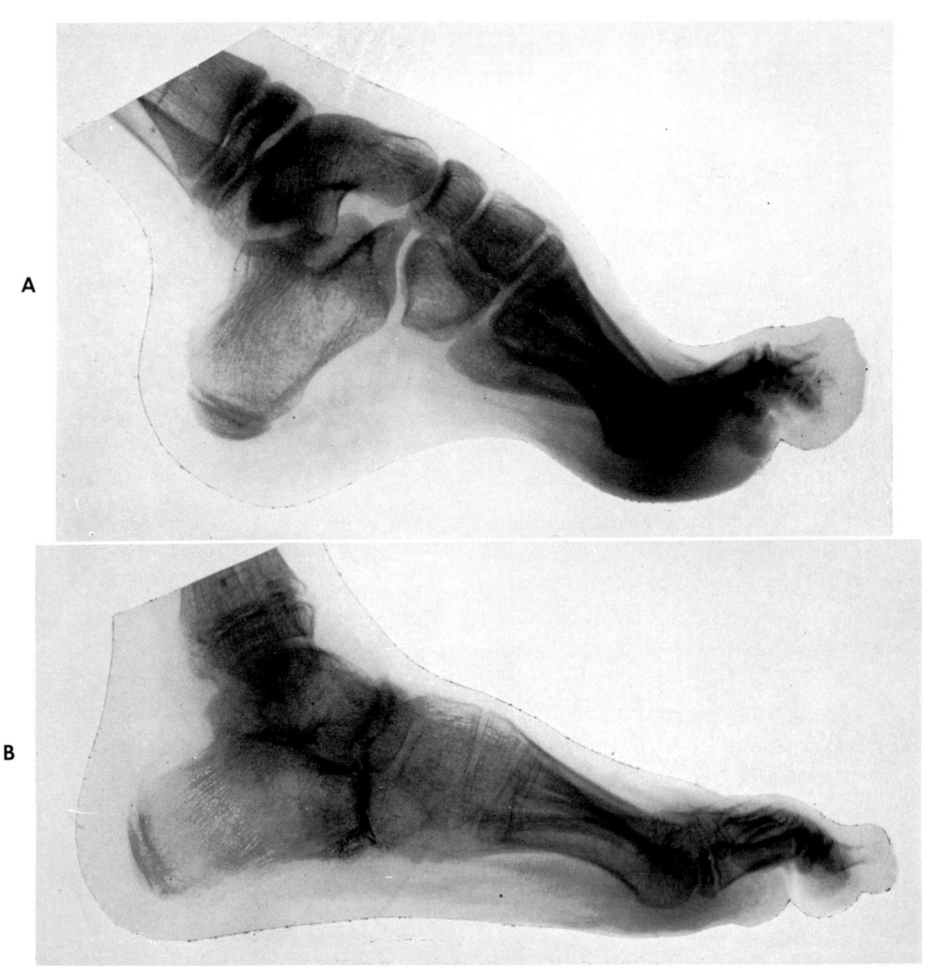

Fig. 14-8. Clawfoot secondary to myelomeningocele in an 11-year-old girl. **A,** Note cavus arch, equinus, and dorsal contracture of toes. **B,** Seven months after operation consisting of plantar fasciotomy and triple arthrodesis. All elements of the deformity are greatly improved.

cially if the operative procedure has involved only tendons and fascia. All of these deformed feet should be fitted with proper shoes and with supports for both the anterior and longitudinal arches. Corrective foot exercises should be started early. The results are often satisfactory when the treatment has been carefully chosen and the follow-up has been adequate.

KÖHLER'S DISEASE (AVASCULAR NECROSIS OR OSTEOCHONDRITIS OF THE NAVICULAR BONE)

Köhler's disease of the tarsal navicular bone (Fig. 14-9) is an uncommon affection that begins insidiously in childhood, about the fifth or sixth year of age. It causes a variable degree of local discomfort and limping and tends toward gradual

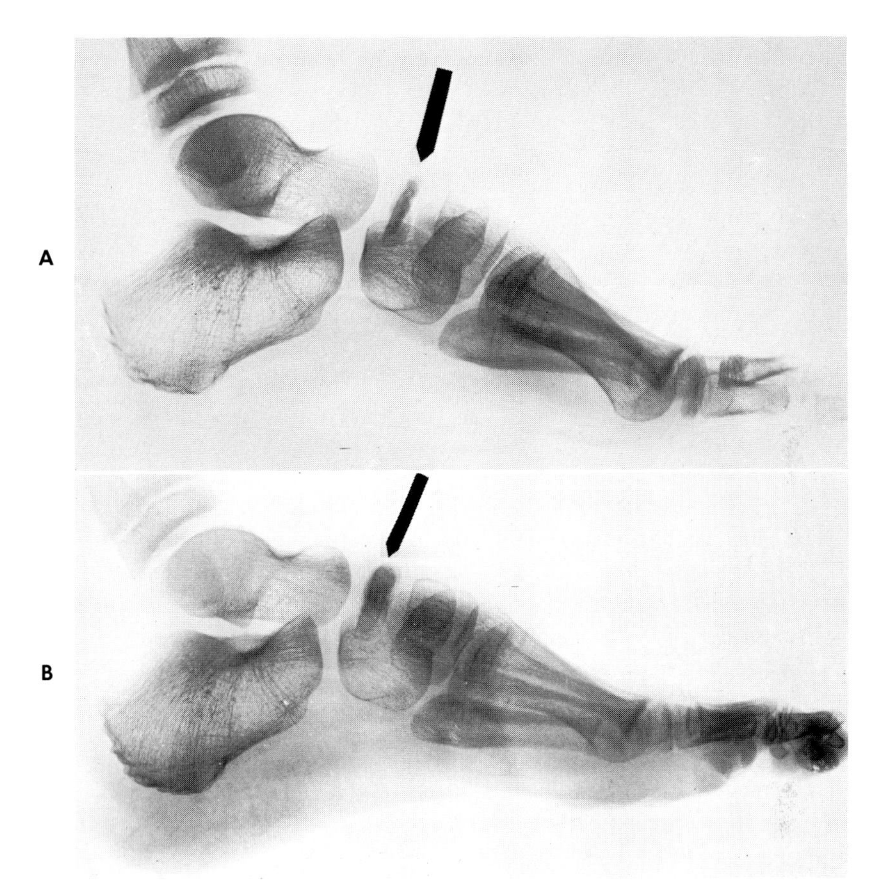

Fig. 14-9. Köhler's disease in a 6-year-old boy. **A,** The navicular bone is abnormally thin and dense. **B,** Five months later. Note improvement in the shape of the navicular bone.

spontaneous recovery. The etiology has not been established; the affection seems to fall into the group of localized and self-limited bone diseases of youth that are ascribed to ischemic degenerative changes. Trauma sometimes appears to be a contributing factor. Clinically there are tenderness and slight thickening over the affected navicular bone. The roentgenographic appearance of the navicular is pathognomonic, the bone being small, dense, and of irregular outline and disordered internal structure as though it had been crushed. Microscopic sections of such bones have shown massive necrosis, which is followed by organization, resorption of dead bone, and the formation of new bone.

Treatment and prognosis. Protection of the diseased bone from excessive traumatization is important. Support of the longitudinal arch and restriction of activity may suffice. If there is much pain on weight bearing, however, it is preferable to immobilize the foot in slight inversion by means of a plaster cast for a period of 6 to 8 weeks. The results are usually satisfactory, there being little or no permanent disability.

ANTERIOR METATARSALGIA

Etiology. Pain beneath the metatarsal heads is common in (1) the everted or abducted foot, (2) the foot with a short Achilles tendon, and (3) the foot with a high longitudinal arch, such as the clawfoot. Disturbances of the metatarsal arch are commonest in individuals over 30 years of age and occur more often in women than in men. The affection may be a result of muscular weakness that allows excessive weight bearing on the middle metatarsal heads. Tight, short, high-heel shoes play an important part in the pathogenesis. A short and tight shoe compresses the anterior part of the foot, elevates the head of the fifth metatarsal bone, and throws weight upon the fourth metatarsal, causing pain about its head.

Clinical picture. Often the first symptom is a burning, cramping pain in the anterior part of the foot, usually under the middle metatarsal heads. Occasionally the pain is preceded by the feeling, experienced on standing or walking, that a bone has slipped out of place. Pain on standing and walking may become severe and disabling. At times the pain radiates to the end of the toe, up into the foot, and even into the leg. The pain is seldom experienced with the shoe off. Tenderness is most often found beneath the fourth metatarsal head, although it may be present also beneath the head of the second or third metatarsal. A tender callus frequently develops under the metatarsal heads.

Examination of the foot often shows depression of the metatarsal arch even when the foot is bearing no weight. The most characteristic features, however, are the tenderness and calluses beneath the metatarsal heads. In severe cases there are usually dorsal contracture of the toes and restriction of their motion. It is impossible to flex the toes fully, and any attempt to do so produces pain in the tender areas. Anterior metatarsalgia may occur in association with hallux valgus or hallux rigidus.

Treatment. The basic objects of treatment are (1) relief of the pain by arch supports and (2) strengthening of the muscles of the foot and ankle by corrective exercises.

The patient should wear a shoe that has a thick sole, adequate width at the toes, a supporting longitudinal arch, and a narrow counter. A small felt or rubber pad, placed immediately behind the metatarsal heads and held in place with circular adhesive strapping, will usually relieve the acute symptoms (Fig. 14-10). Various types of supports, made of leather, felt, rubber, cork, or metal and designed to elevate the metatarsal arch, may be fitted into the shoe. A transverse bar posterior to the metatarsal heads, made of leather and attached to the outside of the shoe (*metatarsal bar,* Fig. 14-11) or placed between the inner and outer soles *(Cook anterior heel),* will often relieve the symptoms of metatarsal strain. The patient should be instructed regarding hot soaks, contrast baths, and corrective foot exercises. The exercises should be continued for several weeks. With gradual improvement of muscle strength, the symptoms may be completely relieved.

Occasionally a sensitive plantar wart may be the source of severe pain. Relieving the pressure upon a tender wart by means of proper pads or supports may lead to its disappearance. Curettement, excision, or roentgentherapy may be indicated in selected cases. In cases of metatarsal pain not responding satisfactorily to conservative measures, pressure on the tender callus can be relieved by an operation to shorten the metatarsal shaft.

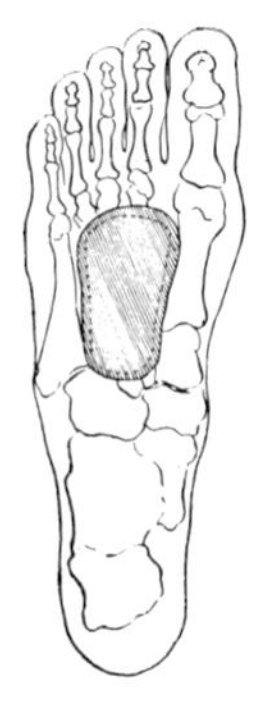

Fig. 14-10

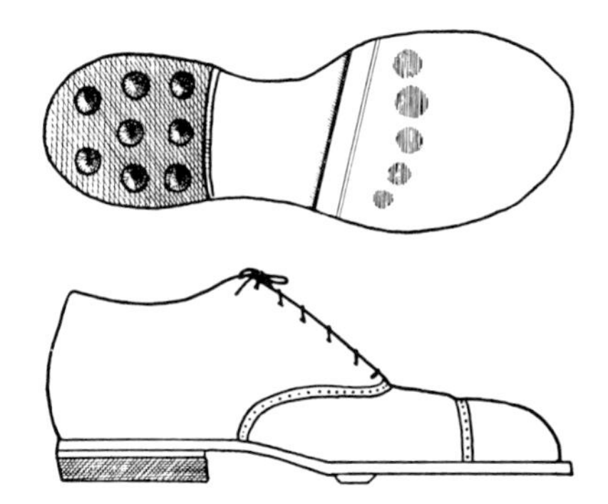

Fig. 14-11

Fig. 14-10. Metatarsal arch support. Note position just behind the metatarsal heads. (After Lewin.)

Fig. 14-11. Shoe with metatarsal bar. Note that the bar is fixed to the sole just behind the metatarsal heads.

MORTON'S TOE (PLANTAR NEUROMA)

In 1876 T. G. Morton described a type of metatarsalgia characterized by sudden attacks of sharp pain usually localized to a single toe. The fourth toe is most commonly involved, and next the third toe. The condition is usually unilateral; women are affected more commonly than men. In the early stages there is a burning sensation in the region of the metatarsal head, which may radiate into the toes and be accompanied by paresthesia and numbness. The characteristic pain, however, is sharp and lancinating; it is often so intense as to demand immediate removal of the shoe and manipulation of the toes. After these sudden attacks tenderness may persist for several days and be followed by numbness of the toes.

As a rule the appearance of the foot is normal. A small area of exquisite tenderness can usually be located by firm palpation of the third web space. In late cases it is sometimes possible to elicit crepitation, together with characteristic pain, by careful palpation with one hand while the other squeezes the metatarsal heads together.

The pain of Morton's toe is associated with a localized thickening of the nerve at its bifurcation in the web space. The enlargement or neuroma is presumably a result of repeated traumatization of the nerve by the metatarsal heads. Microscopically it shows proliferation of neurilemma cells and interstitial fibrosis.

Treatment. The symptoms can sometimes be relieved by a metatarsal arch support. Excision of the enlarged segment of the nerve is a safe and effective procedure, produces little loss of sensation, and is the treatment of choice in most cases. Sometimes exploration discloses a bursa that also should be excised.

STRESS FRACTURE OF A METATARSAL BONE (MARCH OR FATIGUE FRACTURE)

Fracture of a metatarsal shaft, usually the second or third, may result from the repeated stresses associated with an unaccustomed amount of walking. Temporary exhaustion of the muscular and tendinous support of the foot may be a causal factor. A congenital shortening of the first metatarsal bone, *metatarsus atavicus,* which throws an increased leverage on the shaft of the second metatarsal, may also be a predisposing cause. The fracture may be complete or incomplete. Symptoms are often completely absent at the instant of fracture, beginning a week or more later as exuberant callus forms. The pain and swelling that appear at this time may lead to an erroneous diagnosis of malignant tumor. On the other hand, the onset of pain and tenderness may precede the appearance of roentgenographic changes by 1 or 2 weeks, causing the patient to be unjustly suspected of malingering. Stress fractures are especially common in army recruits during their basic training. In a reported series these fractures were found most frequently during the tenth week of training. Stress fractures have been observed in many lower limb bones other than the metatarsals, including with great frequency the calcaneus and occasionally the femur, tibia, or fibula.

Treatment. Extensive experience in the Armed Forces has demonstrated that rest, adhesive strapping, and the use of an anterior arch pad constitute effective treatment for stress fracture of a metatarsal bone. Sometimes, however, a plaster

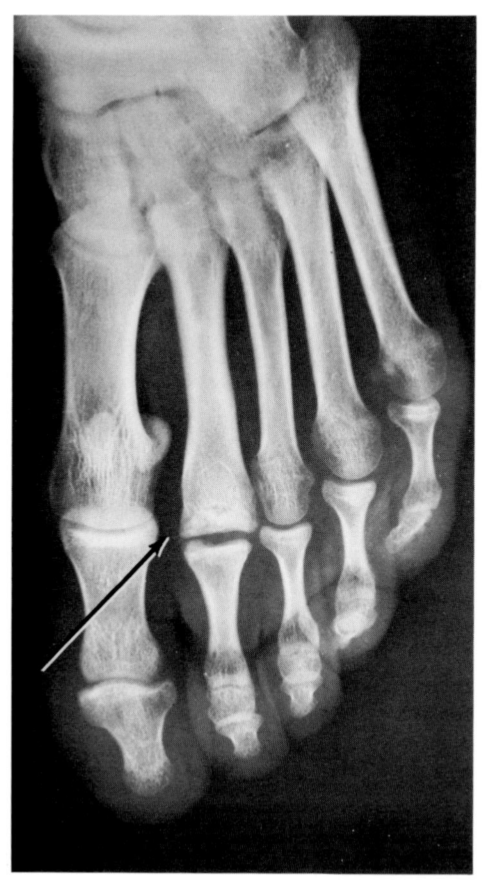

Fig. 14-12. Roentgenogram showing Freiberg's disease. Note deformity and increased density of the head of the second metatarsal bone.

cast is advisable for a period of from 3 to 4 weeks. After the cast has been removed the patient's shoe should be fitted with an anterior arch pad or a metatarsal bar, and he should not engage in excessive walking for several months.

FREIBERG'S DISEASE (AVASCULAR NECROSIS OR OSTEOCHONDRITIS OF A METATARSAL HEAD)

Freiberg's disease is characterized by the gradual development of degenerative changes in the head of the second metatarsal bone (Fig. 14-12) or, rarely, in other metatarsal heads. The affection is uncommon; it occurs usually in adolescent children but may be seen in adults. The etiology is believed to be a disturbance of the circulation that results in a localized ischemic necrosis. Trauma is possibly a contributing factor. The roentgenograms show bone absorption and deformation, with thickening of the distal portion of the affected bone. Clinically there are pain on weight bearing and thickening and tenderness of the affected metatarsal head.

Treatment. In the acute stage of the affection, treatment by means of a plaster boot or anterior arch pad is usually sufficient. Occasionally in late cases it is necessary to resect the excess bone or to excise the deformed metatarsal head in order to relieve the pain and discomfort. After such operations the use of an anterior arch support is advisable.

HALLUX VALGUS

Hallux valgus (Fig. 14-13) is a lateral angulation of the great toe at its metatarsophalangeal joint. There is usually an associated enlargement of the medial side

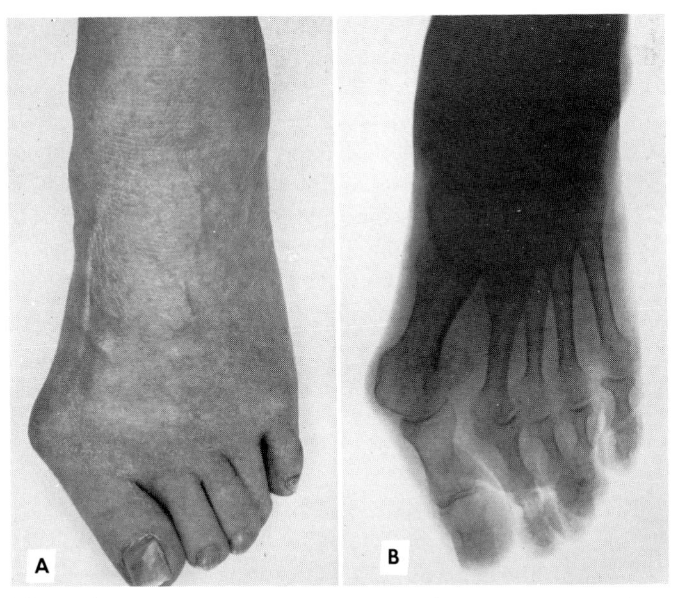

Fig. 14-13. Hallux valgus in a woman 50 years of age. **A,** Photograph showing valgus angulation of first three toes, overlapping of middle toe on fourth toe, and wide, flat metatarsus. **B,** Roentgenogram showing varus of first metatarsal bone, valgus of first three toes, medial enlargement of first metatarsal head, and prominent lateral sesamoid shadow.

of the head of the first metatarsal bone, together with the formation of a bursa and callus over this area. The bony prominence and its overlying bursa constitute a *bunion.*

Etiology and pathology. Hallux valgus is frequently familial. It is much more common in women than in men. Medial deviation of the first metatarsal bone may initiate the deforming changes; narrow, pointed, and short shoes doubtless aggravate them. Medial deviation of the first metatarsal is accompanied by lateral deviation of the great toe; in extreme cases all of the metatarsal bones may incline medially, and all of the toes may incline laterally. The anterior part of the foot is widened and the anterior arch depressed. Contracture of the flexor and extensor hallucis longus muscles is associated with a lateral displacement of the extensor tendons. As the great toe deviates laterally, the second toe may be forced to ride up over it, which often results in a painful callus from friction against the shoe. Occasionally the malalignment and the symptoms are aggravated by degenerative arthritic changes. Atrophy of articular cartilage may be extensive. The pain of hallux valgus may be caused by traumatic arthritis, pressure on digital nerves, and compression and inflammation of the bursa overlying the metatarsal exostosis.

Treatment and prognosis. In mild cases properly fitting shoes and repeated overcorrection by stretching may prevent progression of the deformity and afford relief of the discomfort. The patient should sleep with a pad or other device separating the first and second toes. The depression of the metatarsal arch should be corrected by a supportive pad and exercises. In acute infection of the bursa, rest and hot compresses are indicated.

When considerable deformity is accompanied by disabling pain, surgical treatment is indicated. It is the only available method of correcting the deformity and so relieving the pain. The simplest operation is removal of the exostosis and the bursa. In addition, the ligaments on the medial side may be reefed, the capsule on the opposite side divided, and the extensor hallucis longus tendon sectioned or lengthened *(Silver operation).* Transplantation of the adductor hallucis tendon from phalanx to metatarsal, as described by McBride, is frequently advisable.

Resection of the proximal half of the first phalanx *(Keller* or *Schanz operation)* is often the procedure of choice for the correction of severe deformity. In the absence of arthritic changes a wedge osteotomy of the proximal end of the first metatarsal bone, after removal of the exostosis and bursa, may be the operation of choice; it corrects the varus deformity of the first metatarsal bone, which is an important contributing factor in hallux valgus. Many other surgical procedures for hallux valgus have been described. The choice of procedure depends upon the individual case.

After operations for hallux valgus the great toe should be held in an overcorrected position, by a splint, cast, or soft cushion pad placed between the first and second toes, until it is unlikely that the deformity will recur. Hot soaks, massage, and corrective foot exercises should be employed as long as tenderness and stiffness persist. Some patients recover rapidly from operations for hallux valgus and are permanently relieved of pain, especially those who have had the minimum of surgery performed and who can be relatively inactive in their occupations. Others experience pain for several months after operation. The patient should be observed periodically in an effort to forestall recurrence of the deformity.

HALLUX VARUS

Hallux varus, in contrast to hallux valgus, is a medial angulation of the great toe at the metatarsophalangeal joint. This deformity usually is congenital in origin; however, trauma, infection, muscle imbalance from paralysis of the adductor hallucis, or a bunion operation in which the outer capsule and insertion of the adductor hallucis tendon have been severed, may also be responsible for hallux varus.

Treatment. Mild cases can be corrected surgically by releasing the contracted structures on the medial side of the toe and plicating the lateral part of the joint capsule. Greater degrees of deformity may require additional measures such as osteotomy of the metatarsal head and use of the sectioned extensor hallucis brevis tendon as a tenodesis to maintain the corrected alignment.

HALLUX RIGIDUS

Hallux rigidus is characterized by restriction of motion in the first metatarsophalangeal joint, occasioned usually by trauma, degenerative arthritis, or disuse associated with chronic foot strain. Burning and throbbing pains occur in the affected part of the foot after standing or walking. Since dorsiflexion of the toe is limited, pain is especially great in the latter part of the stance phase of walking, as the patient comes up on the ball of his foot.

Treatment. In mild cases the treatment consists of (1) wearing a thick, inflexible sole on the shoe or (2) having a long steel strip inserted between inner and outer soles of the shoe in order to restrict motion in the affected metatarsophalangeal joint. The use of a metatarsal bar on the shoe is sometimes helpful. Occasionally slight elevation of the medial side of the sole of the shoe will abolish the pain. If the symptoms cannot be relieved in such ways, surgical removal of the proximal half of the first phalanx is helpful. An alternative procedure, often followed by complete relief of pain, is arthrodesis of the first metatarsophalangeal joint.

HAMMER TOE

Hammer toe is characterized by dorsiflexion of the metatarsophalangeal joint and plantar flexion contracture of the proximal interphalangeal joint (Fig. 14-14).

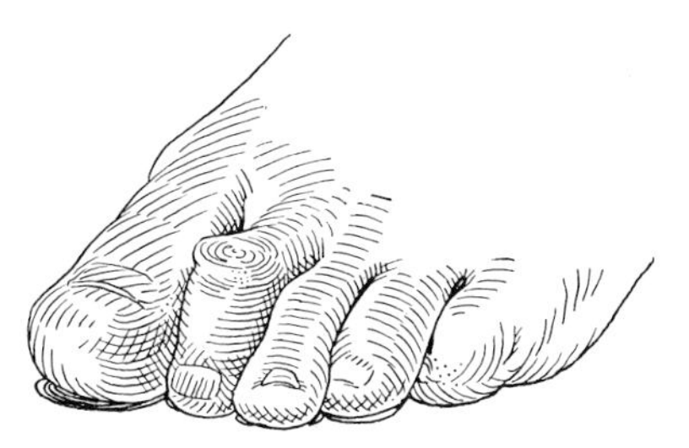

Fig. 14-14. Hammer toe: the second toe shows a flexion contracture of its proximal interphalangeal joint and a dorsal callus.

Although any toe may be so affected, the second is most frequently involved. Tender corns and calluses are usually present on the toe. Hammer toe may be caused by the pressure of a short, narrow shoe upon the end of a long toe. The deformities may begin in early childhood, however, and are often found in association with hallux valgus. Accordingly there may be a familial influence in the etiology.

Treatment. In early cases simple manipulation and splinting of the affected toe may suffice to relieve the moderate discomfort. In older cases arthrodesis of the proximal interphalangeal joint in extension is indicated. Sometimes complete excision of the proximal phalanx yields a very satisfactory result. Hammer-toe deformity of the second toe associated with hallux valgus has sometimes been wrongly treated by amputation of the second toe, which results in an increase of the hallux valgus.

OVERLAPPING OR DORSAL DISPLACEMENT OF THE TOES

Overlapping most often involves the little toe. It may be of congenital origin or may arise from wearing tight shoes. The toe may become very painful because of corns or callus formation.

Treatment. In infants and younger children manipulation and splinting of the toe may correct the deformity. Tenotomy or transplantation of the extensor tendon may be necessary. If the deformity recurs, resection of the proximal phalanx or amputation of the toe may be indicated.

IN-TOEING (PIGEON-TOE)

Habitual turning in of the feet on walking may be encountered in association with flexible flatfoot in small children. It is a physical sign rather than a disease entity and is often found together with hallux varus, metatarsus varus, bowlegs, medial torsion of the tibia, congenital contracture of the internal rotators of the hip, increased anteversion of the femoral neck, or relapsed clubfoot. Some observers believe that medial torsion of the tibia is the commonest cause.

Treatment. Infants and young children should not be allowed to spend too much time in sleeping and sitting positions in which the lower limbs are internally rotated. The deformity is usually corrected spontaneously as the child grows older. Raising the outer border of the soles of the shoes 1/8 to 1/4 inch may hasten the improvement. The child should be instructed to walk with the toes pointed out. If torsion is present in the tibia or the femur, the use of a Denis Browne night splint to hold the lower limbs externally rotated may be very helpful. For older children, roller skating provides excellent corrective exercise.

AFFECTIONS OF THE HEEL

Affections of the heel include the results of inflammation and injury either (1) about the insertion of the Achilles tendon or (2) under the calcaneus on weight bearing.

Inflammation and injury about the insertion of the Achilles tendon

Pain in the back of the heel is the outstanding symptom of several locally disabling affections.

Tenosynovitis

Swelling is usually noticeable in the region of the Achilles tendon, and fine crepitus on motion is often unmistakable. The affection is usually marked by acute local tenderness and considerable disability. The treatment consists of rest, avoidance of pressure, and the application of heat to the tender area. When it is necessary for the patient to walk, a pad should be placed in the shoe to elevate the heel and so lessen the excursion of the tendon. Occasionally the local rest provided by a walking cast for several weeks may be required.

Bursitis

Inflammation of the *retrocalcaneal bursa,* between calcaneus and Achilles tendon, causes pain on motion and local tenderness. The disorder is treated by rest, heat, and elevation of the heel by means of a pad. The irritation of a tight shoe sometimes causes a bursa (the *superficial calcaneal* or *posterior Achilles bursa*) to form between the Achilles tendon and the skin. In inflammation of this bursa, relief of pressure and application of heat are indicated. Chronic pain and tenderness unrelieved by conservative measures may require the excision of either of these bursae and resection of the underlying prominence of the calcaneal tuberosity.

Calcaneal apophysitis or epiphysitis

Sometimes called *Haglund's disease,* calcaneal apophysitis is a somewhat uncommon affection of children, occurring most often in boys between 9 and 14 years of age. It appears to be a low-grade inflammatory reaction, occurring in the posterior calcaneal epiphysis, accompanied by pain and swelling, and ascribed to chronic pressure or strain. Roentgenograms may show the epiphysis to be irregular or segmented with areas of increased density. However, in children roentgenograms of the calcaneus often show such changes in the absence of symptoms. The affection is treated by relieving the local strain and pressure. Raising the heel of the shoe ½ inch for several months usually proves effective. It is occasionally advisable to apply a plaster cast in order to secure complete rest of the affected area.

Rupture of the Achilles tendon

Incomplete rupture of the Achilles tendon may be followed by the formation of small irregular masses of fibrous consistency. Occasionally calcification similar to that of myositis ossificans occurs in these masses; when pain is persistent in such cases, the calcified area should be excised. Complete rupture of the Achilles tendon, usually by indirect violence, is seen in stocky young athletes and in middle-aged, sedentary individuals. The power of voluntary plantar flexion is greatly diminished, and the rupture may be evident on palpation. Recent complete ruptures in physically active persons are best treated by prompt exploration and suture; alternatively, good results can usually be obtained without operation by applying a cast with the ankle in equinus. Rupture of the plantaris tendon, evidenced by a sudden sharp pain like the sting of a whip, may occur, but clinical differentiation from incomplete rupture of the Achilles tendon is often questionable, and most instances of so-called *tennis leg,* formerly thought to be plantaris ruptures, are probably partial tears of the medial belly of the gastrocnemius muscle at its musculotendinous junction. The treatment consists of rest, elastic support, crutches for several days, and a heel elevation for several weeks. Ruptures of other tendons of the ankle and foot are quite rare.

Inflammation and injury under the calcaneus

Localized tenderness beneath the heel, which may cause sharp pain on standing and walking, is frequently found in middle-aged men and less commonly in women. The patient may be severely disabled by heel tenderness, especially if it is bilateral. In some instances the onset of symptoms can be related to an occupation requiring prolonged standing or to an excessive gain in weight. The exact cause, however, and pathologic nature are unknown. The skin and subcutaneous fat of the weight-bearing surface of the heel are specialized structures designed to withstand the tremendous stresses of everyday walking; they may be affected by trauma or inflammation. In the past a low-grade periostitis, fascitis, or bursitis has been postulated. Frequently, ossification at the attachment of plantar fascia to calcaneus appears in the lateral roentgenogram as a calcaneal spur. Whether the spur is related to the pain in most cases is doubtful.

Treatment. In mild cases complete relief can often be obtained by the use of (1) a soft rubber heel pad with a hole cut in its center to relieve pressure on the tender area, (2) a cupped metal plate to eliminate pressure in this area, (3) a special heel that is shaped to bear all the weight at its posterior margin, or (4) a high longitudinal arch pad. Weight reduction and change to sedentary occupation may be helpful. With time the symptoms usually subside spontaneously; rarely is surgical treatment advisable.

EXOSTOSES OF THE BONES OF THE FOOT

At certain points in the foot, pressure from shoes is likely to cause chronic irritation and the resultant formation of exostoses and bursae. The lateral side of the fifth metatarsal head sometimes develops such changes, which are similar to those of a bunion and have been called a bunionette. Other sites are the medial and upper aspects of the navicular bone, the dorsum of the first cuneiform, the plantar aspect of the calcaneus, and the posterior surface of the calcaneus at or above the attachment of the Achilles tendon. Occasionally an exostosis forms beneath a toenail; such localized proliferation is termed a *subungual exostosis.* It is seen most frequently in the great toe and may cause acute tenderness.

Treatment. If removal of the causative pressure cannot be accomplished satisfactorily, the exostosis and bursa should be excised. Subungual exostosis is treated by excision of the nail, when necessary, and of the underlying phalangeal exostosis.

ACCESSORY BONES OF THE FOOT

Twenty-eight accessory bones of the foot have been described. This number includes the *sesamoid bones,* which are embedded in tendons. Two sesamoid bones are found constantly under the base of the great toe. Sometimes these bones are the site of chronic pain and tenderness following local trauma. In such cases the symptoms may be completely relieved by excision of the sesamoid bones. These small bones are sometimes bipartite; this developmental variation must be distinguished from fracture. Painful bursae that sometimes occur in association with the sesamoid bones may require excision.

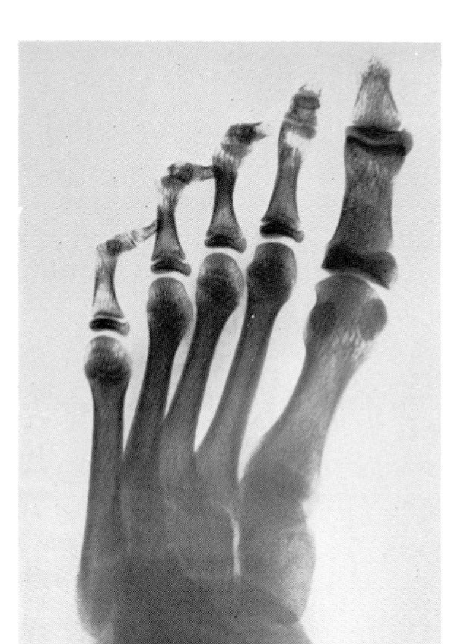

Fig. 14-15. Roentgenogram showing accessory navicular bone (os tibiale externum). Note the well-demarcated, small bone adjacent to the medial border of the navicular bone.

Approximately 25% of the feet of adults and 22% of the feet of children under 16 years of age have roentgenographic evidence of one or more accessory bones in addition to the sesamoid bones. The commonest of the abnormal accessory bones are the following: the *os trigonum,* which is found posterior to the talus; the *os tibiale externum* or *accessory navicular,* located at the medial aspect of the navicular bone (Fig. 14-15); and the *os peroneum,* lying in the peroneus longus tendon on the lateral side of the foot. These bones are often mistaken for fracture fragments and in obscure injuries of the foot should be carefully differentiated. They can usually be recognized without difficulty because of their typical size, shape, position, and smoothly rounded outline. The os trigonum rarely is associated with local pain, requiring symptomatic treatment or even excision. A large accessory navicular bone may be associated with local tenderness from pressure of the shoe and with pain and weakness in the longitudinal arch; in such cases excision of the accessory navicular bone and fixation of the posterior tibial tendon into a groove well beneath the navicular *(Kidner operation)* are frequently indicated. The results of this procedure are usually good.

DISPLACEMENT OF THE PERONEAL TENDONS

Occasionally an abnormal laxity of the retinaculum behind the lateral malleolus or shallowness of the bony groove allows one or both of the peroneal tendons to slip anteriorly over the malleolus. The displacement usually occurs on active dorsiflexion and eversion; it is sometimes accompanied by severe pain. Occasionally this affection is of developmental origin; more commonly it is a sequel of trauma.

Treatment. If the diagnosis is made immediately after the displacement has occurred, manipulative replacement and immobilization by means of a plaster boot may effect a cure. Increasing the height of the heel will help to prevent the displacement. Recurrent cases should be treated surgically. The tendon sheaths may be sutured firmly into their normal position, the depth of the peroneal groove may be increased, a strip of the Achilles tendon may be used to form a stabilizing loop, or a small malleolar ledge may be fashioned to prevent the forward displacement.

CHAPTER 15 Affections of the neck, shoulder, and jaw

Deformities and disabilities of the neck and the shoulder are common; their diagnosis and treatment form an important part of orthopaedic surgery. The differentiation of certain neck and shoulder entities as the cause of upper limb pain in the individual patient often requires experience and judgment. Temporomandibular joint disabilities of orthopaedic interest are seen less commonly; they include derangements of the articular disk, luxations, and ankylosis.

Affections of the neck

TORTICOLLIS (WRYNECK)

Torticollis is a deformity of the neck that, as a rule, includes elements both of rotation and of flexion. In many cases one sternocleidomastoid muscle is shortened. Such shortening results in tilting of the head toward the affected side and rotation of the chin toward the opposite side. Torticollis may be congenital or acquired.

Congenital torticollis (muscular torticollis)

The term *congenital torticollis* includes all cases in which the deformity is present at birth. In most such patients the bony structures of the neck are normal. Since in many cases the deformity is not obvious at birth but develops gradually, some authors prefer the term *muscular torticollis*.

Etiology and pathology. The cause of congenital torticollis is unknown. Many possibilities have been adduced, such as: abnormal position of the head in utero; prenatal injury and interference with the vascular supply to the sternocleidomastoid muscle; a fibroma of prenatal origin in this muscle; and rupture of many of the sternocleidomastoid muscle fibers during birth, with hematoma and scar formation. Congenital torticollis is seen with relative frequency after difficult deliveries, with breech presentations, and in primiparas.

A nontender, cylindric enlargement of the sternocleidomastoid muscle is sometimes observed in the newborn infant or is seen several weeks after birth; it usually regresses slowly in from 3 to 6 months. Incomplete regression may be followed by the development of a permanent contracture of the sternocleidomastoid muscle.

The shortening of the sternocleidomastoid muscle is associated with an increased amount of fibrous tissue in its substance. A secondary contracture develops in the adjacent tissues of the neck.

Clinical picture. Congenital torticollis is seen more commonly in girls than in boys. At first the deformity may be slight. It may not be noticed until the child is able to sit or stand, although a mass may be felt in the muscle in the first few weeks of life. In other cases, however, the deformity is severe, with a characteristic flattening and shortening of the face on the side to which the head is tilted. The facial asymmetry begins to develop within the first 3 months. When correction of the torticollis is accomplished in early years, a gradual decrease of the asymmetry will take place. The chin is rotated away from the side of the shortened muscle, and the head is displaced and tilted toward the side of the shortening (Fig. 15-1). The

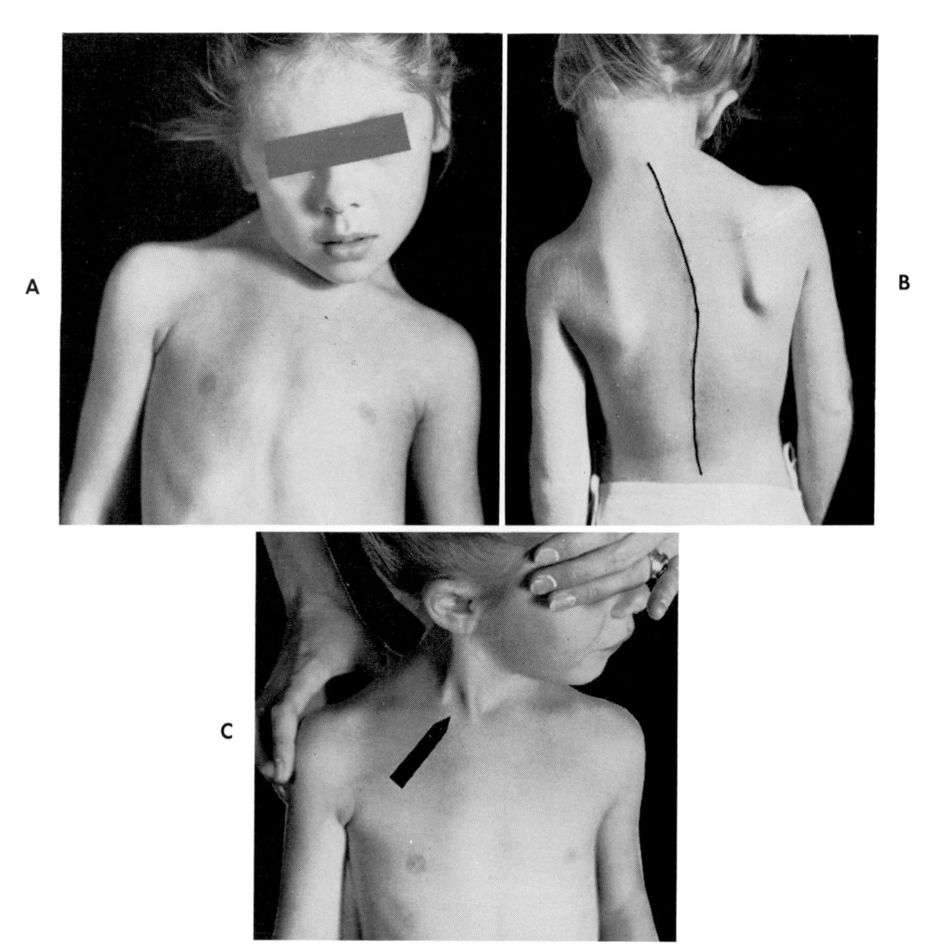

Fig. 15-1. Congenital torticollis in a 5-year-old girl. **A,** Note flexion of the head and neck to the right, pointing of the chin to the left, facial asymmetry, and high right shoulder. **B,** Note elevation of the right shoulder, right thoracic scoliosis, and flexion of the trunk to the left. **C,** Note prominence of the sternal and clavicular heads of the right sternocleidomastoid muscle when it is tightened by correcting the lateral flexion.

shoulder on the affected side may be elevated. Rotation and lateral bending of the neck are restricted, whereas flexion and extension are usually of normal range. Attempts to correct the deformity cause the affected muscle to tighten. There is often an associated cervicodorsal scoliosis. Eyestrain may occur from ocular imbalance secondary to the deformity. In the rare affection of double torticollis the chin is elevated in the midline.

Diagnosis. Tuberculosis of the cervical spine may simulate muscular torticollis. Dislocations and fractures are to be differentiated, as well as rheumatoid arthritis, osteomyelitis, acute myositis, lymphadenitis, and congenital anomalies of the cervical vertebrae. Acute ocular and oral diseases sometimes cause a similar deformity. Of great differential value is the history of painless deformity since birth. The clinical diagnosis should always be supported by negative roentgenograms of the cervical spine.

Prognosis. Proper treatment in infants and children often results in complete cure. Incomplete surgical correction invites recurrence. After correction of the torticollis, the asymmetry of the face gradually becomes less noticeable. In cases corrected before the age of 3 or 4 years the facial deformity may completely disappear.

Treatment. The fundamental principle of treatment is overcorrection of the shortened sternocleidomastoid muscle. The treatment should be started as soon as the deformity is recognized.

Nonsurgical measures. Early cases with slight deformity may often be corrected by passive stretching of the neck into the overcorrected position, positioning of the head during sleep, and, if the child is old enough, active exercises to stretch the short sternocleidomastoid muscle. Continued observation is most important. In the milder cases the results of nonsurgical methods are good.

Surgical measures. Cases of moderate or severe deformity and all late cases require operative treatment. The usual procedure includes: (1) section of the sternocleidomastoid muscle; (2) maintenance of overcorrection of the deformity, which may be accomplished by means of plaster cast, brace, or traction; and, finally (3) a program of exercises for securing muscle balance to maintain the correction permanently. The operation used most commonly is open section of the sternocleidomastoid muscle near its origin from sternum and clavicle. All tight fascial structures, which may include the platysma muscle, should also be sectioned. After operation, corrective exercises should be started and continued until there is no tendency toward recurrence, which may be for as long as 6 months. Many surgeons prefer to use a corrective brace during this period. There is always danger of a recurrence of the deformity; this should be explained to the parents before treatment is started. When the sternocleidomastoid muscle is severely contracted and fibrotic, its complete excision is preferable to myotomy.

Acquired torticollis

Most cases of acquired torticollis occur in the first 10 years of life. Acquired torticollis, unlike congenital torticollis, is often accompanied by pain.

The following types of acquired torticollis are most frequently observed:

1. *Acute,* caused by direct irritation of the muscles from injury or by an inflam-

matory reaction of the muscles (myositis) or of the cervical lymph nodes (lymphadenitis).

2. *Spasmodic,* in which rhythmic convulsive spasms of the muscles take place as a result of an organic disorder of the central nervous system. This will be discussed separately.
3. *Hysterical,* due to psychogenic inability of the patient to control the muscles of the neck.

In addition, torticollis may be associated with arthritis, osteomyelitis, tuberculosis, and injuries of the cervical spine, with contracture of scar tissue in the neck after a burn, with asymmetric paralysis of the neck muscles, with scoliosis, and with meningitis and ocular defects.

Treatment. The treatment of acquired torticollis should be directed toward the removal of local and general causes. Acute traumatic or inflammatory torticollis is sometimes relieved by hot applications, gentle massage, and cervical traction applied in a vertical or horizontal position. If severe contractures of the sternocleidomastoid muscle and the surrounding tissues persist after the causative process has subsided, they should be divided and aftercare as described for congenital torticollis should be given.

Spasmodic torticollis

Etiology. The cause of spasmodic torticollis is thought to be either (1) an organic lesion of the central nervous system or (2) reflex contractions from irritation of the cervical nerve roots by arthritic changes. It seems to occur most often, however, in individuals with a psychoneurotic tendency or a history of stress, overwork, and anxiety.

Clinical picture. The deformity comes on gradually in adult life and is more common in women than in men. The first symptoms are usually stiffness and discomfort of the muscles of one side of the neck. These are followed by a drawing sensation and a momentary twitching or slight contraction that pulls the head toward the affected side. The symptoms slowly become more distinct until convulsive spasms of the muscles develop and draw the head forcibly to one side. The spasms may exhibit either an irregular or a regular rhythm; they are independent of voluntary control and become especially noticeable when the patient grows excited. Spasmodic torticollis may be associated with severe neuralgic pain throughout the head and neck. It is interesting to note that often the convulsive spasms can be inhibited by the light pressure of a finger against the head.

Prognosis. There is little tendency toward spontaneous recovery. The results of surgical treatment are sometimes good.

Treatment. Occasionally conservative methods, consisting of psychotherapy, muscle training, and prolonged mechanical support of the neck with a well-fitting brace or plaster cast, lead to relief. Surgical treatment, however, is usual necessary. It may include section of the spinal accessory nerve and upper cervical rhizotomies.

CERVICAL ROOT SYNDROME

The term *cervical root syndrome* is used to describe the symptoms and signs produced by compression or irritation of cervical nerve roots in or about intervertebral foramina before they divide into anterior and posterior rami.

Etiology. Among the causes of the cervical root syndrome are (1) degenerative disk disease with or without trauma, (2) ruptured intervertebral disks, (3) neck sprains in traffic accidents, (4) neck injuries in contact athletics, and (5) twisting of the neck in sleep.

Hypertrophic spurring in the cervical spine develops over a considerable period of time, allowing ample opportunity for the roots to become adjusted to the new spatial relationships. If disk protrusion or subluxation occurs, it causes compression more readily if spurs are present. Hypertrophic changes are frequently found around the lateral synovial intervertebral joints. Spurring most often occurs at the levels of greatest motion and stress, namely, between the fourth and fifth cervical vertebrae and between the fifth and sixth cervical vertebrae.

Posterior subluxations of the cervical vertebrae produced by hyperextension of the cervical spine may compress the nerve roots by narrowing the vertical diameter of the foramina. Flexing the neck increases the vertical diameter of the foramina and does not result in nerve compression. Cervical disk protrusions are usually lateral in position and compress the nerve root in its foramen. When such protrusions are associated with bony spurs, they are sometimes referred to as lateral hard disks. Much less common but more serious are central disk protrusions, which may be hard or soft and which produce spinal cord rather than root symptoms.

Predisposing causes of cervical root irritation include swelling of the roots after traumatic intervertebral compression, congenital fusion anomalies that alter normal stress patterns, postinflammatory adhesions, and rheumatoid arthritis. Occasionally, a calcified vertebral artery may press on the nerve roots during hyperextension.

Clinical picture. Pain, usually in the neck and arm, is the presenting complaint. The pain is increased by neck motion and often aggravated by coughing and sneezing. It may radiate to the fingertips and is frequently associated with paresthesias in the dermatome of the involved root. Headache, usually occipital, is a common complaint. Other areas of pain radiation include the upper dorsal and scapular regions and occasionally the chest. Weakness of grip or of arm function may be a complaint. Occasionally patients are disturbed by vertigo or blurring of vision, which has been attributed by some observers to spasm or compression of the vertebral artery.

The physical findings include painful limitation of neck motion, tenderness over the cervical spine, and neurologic changes in the upper limb. The patient frequently holds his neck stiff and tilted slightly to one side. The cervical muscles may feel tense. With the patient sitting and his neck slightly hyperextended, pressure on top of the head usually reproduces the pain. Neurologic changes include muscle weakness and atrophy, diminished reflexes, and sensory disturbances corresponding to the level of root irritation.

It will be remembered that in the cervical spine the root emerges above the vertebra of corresponding number. If the sixth root is involved, hypesthesia of the thumb and radial side of the hand, weakness of the biceps muscle, and decrease in the deep tendon reflexes of the biceps and brachioradialis muscles may be noted. Seventh root involvement may be manifested by triceps weakness, diminished triceps muscle reflex, and sensory changes in the middle and ring fingers.

Persistent irritation of the cervical roots may produce reflex stimulation of the cervical sympathetic system, causing symptoms such as blurring of vision, dilatation of the pupils, loss of balance, headache, swelling and stiffness of the fingers, tendinitis, and capsulitis.

Roentgenographic picture. Upright, lateral roentgenograms may be made with the neck in neutral position, in flexion, and in hyperextension. Most patients show loss of the normal lordosis of the cervical spine. Some have a reverse curve in three or four segments. Minor subluxations are often seen; they are considered normal by some observers, pathologic by others. When lateral views show persistent anterior subluxation of a vertebra, a fracture of the neural arch may appear on an oblique flexion view. Open-mouth projections are useful in helping to rule out odontoid fractures and atlantoaxial subluxations.

Myelography may be indicated if thorough conservative therapy has failed and the symptoms and signs strongly suggest a cervical disk lesion. Electromyography and diskography also may be useful in localizing the level of the lesion.

Differential diagnosis. The cervical root syndrome must be differentiated from spinal cord tumor, myelopathy from cervical spondylosis, peripheral nerve irritation, infections and tumors of the cervical vertebrae, reflex pain from somatic or visceral sources, thoracic outlet syndromes, and traumatic subluxations, dislocations, and fractures.

Treatment. It should be explained to the patient that as a rule the outlook for recovery is excellent.

Heat frequently provides relaxation and may relieve root irritation pain; massage may be helpful. Intermittent traction of 15 to 25 pounds with a head halter may relieve the pain. Some patients require bed rest with continuous neck traction, which should be applied with the neck in slight flexion. A Thomas collar for immobilization of the neck may be indicated; if used, it should not place the neck in hyperextension. The collar may be worn until pain on motion of the neck has subsided.

Postural exercises may be indicated. For sleeping, a cylindric pillow that maintains the neck in a neutral position may be helpful. Analgesics and mild sedatives may be needed; habit-forming drugs should be avoided.

In intractable cases, when the symptoms and signs have persisted despite conservative therapy, surgical treatment in the form of foraminotomy, laminectomy, disk excision, or arthrodesis may be advisable.

THORACIC OUTLET SYNDROMES

Although the various forms of the cervical root syndrome just described are the commonest cause of radiating pain in the upper limb, a group of less well-defined disorders associated with pressure on nerves and vessels in the region of the thoracic outlet may produce similar symptoms. They include cervical ribs, the scalenus syndrome, and the costoclavicular syndrome.

Cervical ribs and scalenus syndrome

Cervical ribs are congenital anomalies consisting of supernumerary, independent units of growth similar to the first dorsal ribs. The extra ribs are attached most often to the seventh cervical vertebra but may be attached to the sixth. Cervical ribs are usually bilateral; when this is the case, one rib is always higher and presents a more advanced stage of development than the other.

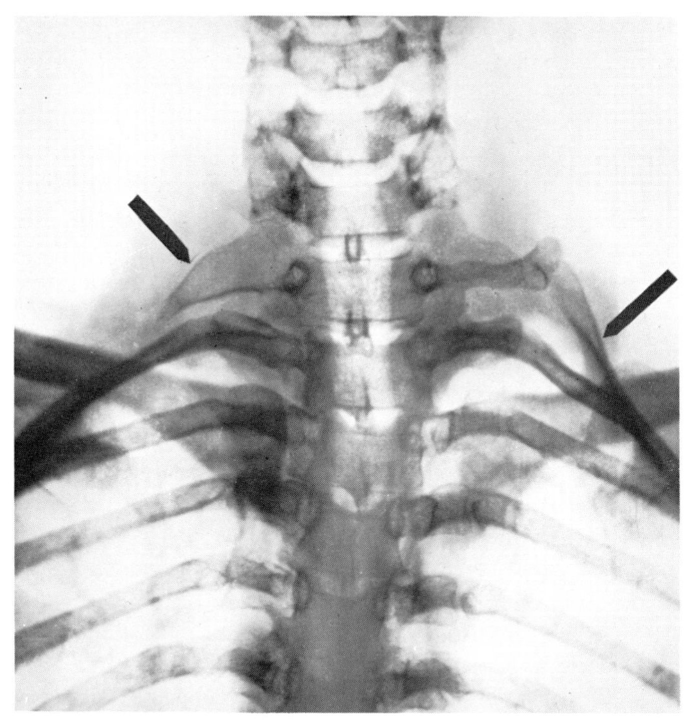

Fig. 15-2. Anteroposterior roentgenogram showing a left cervical rib and overdevelopment of the right transverse process of the seventh cervical vertebra.

The size of the cervical rib varies (Fig. 15-2). There may be only an enlarged transverse process of the seventh cervical vertebra with a fibrous band connecting it with the first rib, an incomplete rib with a fibrous band, or a complete rib articulating or fused with the first rib.

The lower components of the brachial plexus are in close approximation to the anomalous ribs. The seventh cervical nerve, or middle trunk, crosses over the transverse process of the seventh cervical vertebra, and the inferior trunk, made up of the eighth cervical and first thoracic nerve roots, crosses over either the extra rib or the fibrous band. The subclavian vessels also are quite close. The artery lies in front of the rib and often presents at this point a localized dilatation that causes an increased pulsation in the neck. The subclavian vein lies a considerable distance below and in front of the artery and is seldom pressed upon or affected by the cervical rib. The scalenus anterior muscle is attached to the first rib in front of the inferior and middle trunks of the brachial plexus and the subclavian artery.

Clinical picture. Although cervical ribs are often present without causing symptoms, they lead occasionally to a characteristic clinical picture. The symptoms usually appear in adult life, most often at the age of about 30 years. They occur much more commonly in women than in men.

Locally there may be a tumor in the neck, which may be both palpable and visible. It is most frequently found as a firm, rounded, immovable, tender mass 2 to 3 cm. above the middle of the clavicle. Above this there is usually a palpable pulsation caused by the subclavian artery. Auscultation of the supraclavicular area may reveal a loud bruit. About this area lies the brachial plexus. As postural changes result in gradual stretching of the middle and lower portions of the brachial plexus over the rib, friction takes place with use of the shoulder and arm. At first there is no pain, but repeated trauma finally produces a definite brachial neuritis. This may be followed by weakness and atrophy, starting in the intrinsic muscles of the hand and involving finally the entire limb. Occasionally a clawhand deformity develops. Paresthesia of the arm and forearm, especially of the part supplied by the ulnar nerve, is sometimes present and may be accompanied by radiating pain. Tingling and numbness are often the first symptoms, but later, as the pressure becomes more severe, anesthesia and paralysis may develop. At first, relief can be secured by changing the position of the arm. As the condition progresses, however, change of position no longer relieves the pain. Pallor, coldness, and cyanosis may appear in the fingers. The volume of the radial pulse may be diminished, especially while traction is being applied to the arm or while the scalenus anterior muscle is tensed in the *Adson test,* which is carried out by having the patient take a deep breath while his head is turned to the affected side. In addition, the supraclavicular bruit may be accentuated by this maneuver. Some observers believe that these changes result from action of the affected sympathetic nerve fibers in the lower part of the brachial plexus, whereas others believe that they are caused by direct mechanical pressure upon the subclavian artery. In advanced cases thrombosis of the subclavian artery may take place, giving rise to gangrene that starts in the fingers. Gradually a collateral circulation may be established, however, which will compensate for the circulatory impairment. Occasionally the patient presents a scoliosis of the cervical spine with its convexity toward the side of the cervical rib.

Clinical manifestations similar to those of cervical ribs are sometimes encountered in the absence of demonstrable bony abnormality at the base of the neck. In such patients compression of the brachial plexus and the subclavian artery may be caused by spasm of the scalenus anterior muscle, the nerves and artery being pressed against the first rib by the tense muscle. This condition has been called the *scalenus syndrome.* Some observers believe that even in the presence of a cervical rib, spasm of the scalenus anterior muscle may be entirely responsible for the clinical picture by compressing the brachial plexus and the subclavian artery against the extra rib.

Differential diagnosis. Because of the weakness and atrophy of the smaller muscles of the hand, symptoms associated with the presence of a cervical rib are sometimes confused with those of syringomyelia, progressive muscular atrophy, injury of the spinal cord, or ulnar palsy. Subacromial bursitis must occasionally be excluded, as must the shoulder-hand syndrome. Cardiac ischemia or chest lesions such as a Pancoast tumor may be the cause of brachial pain. The several types of cervical root syndrome must be considered. Of them, rupture of an intervertebral disk can often be differentiated by the localization of symptoms and signs to a single nerve root, which is most often the fifth or sixth, whereas thoracic outlet syndromes usually produce irritation of the eighth cervical and first thoracic roots. Any patient whose symptoms fail to respond to conservative orthopaedic measures, however, may warrant neurosurgical consultation concerning the possibility of a cervical disk protrusion.

Treatment. Conservative measures should be tried, especially in cases that seem to be improved by rest. As a psychogenic factor is often present, psychotherapy should be considered. Injection of procaine into the scalenus anterior muscle often produces temporary relief. To relieve the strain on the lower brachial plexus and roots the shoulder should be brought upward and backward with the use of a sling or brace. Hot baths, massage, and active exercises for the shoulder and arm may be given. Emphasis should be placed on exercises for strengthening the trapezius and levator scapulae muscles. In some cases change to an occupation that does not put strains on the shoulder or arm will result in complete relief of the symptoms.

When the response to conservative treatment is unsatisfactory, operation may be indicated. Relief can often be obtained by myotomy of the scalenus anterior. Because of its technical ease and safety, this is the operation of choice and should be done before excising a cervical rib. At the time of such myotomy all other structures compressing the neurovascular bundle should be resected. Occasionally in severe obliterative vascular conditions, a portion of the first rib or of the clavicle may have to be removed to provide relief.

Costoclavicular syndrome

Neurovascular compression may occur in the space between the clavicle and the first rib, producing symptoms quite similar to those caused by cervical ribs. The commonest type of compression in this area is probably postural, resulting from gradual sagging of the shoulder girdle in middle life. Holding the shoulder back in a military position may also narrow the costoclavicular space and cause symptoms. Narrowing of the space may be produced by tumors of the clavicle or first rib or by excessive callus about an ununited or malunited fracture of the clavicle.

Diagnosis. Helpful in distinguishing the costoclavicular syndrome from the scalenus syndrome is the obliteration of the radial pulse when the shoulder is abducted. In this position a bruit may be heard in the infraclavicular area. The scalenus anterior tests are negative. Angiography may be most useful in establishing the diagnosis.

Treatment. For mild cases the treatment consists of postural exercises designed to elevate the shoulder girdle by strengthening the trapezius and the levator scapulae muscles. If conservative treatment fails or if signs of serious neurovascular impairment appear, resection of the first rib or the clavicle may be advisable.

Affections of the shoulder

The shoulder possesses less stability and less mechanical protection than any other large joint of the body. The glenoid fossa of the scapula is shallow, presenting only a slight concavity for the large globe of the humeral head. The concavity is disposed vertically, moreover, and gravity, instead of increasing joint stability as in the case of hip or knee, exerts through the entire weight of the arm a constant force tending to produce a shearing strain. Not only does the bony configuration of the shoulder joint afford little stability but the capsular ligaments, necessarily long and loose to permit the wide range of shoulder motion, oppose little strength to any force tending to cause subluxation. It is therefore to be expected that the

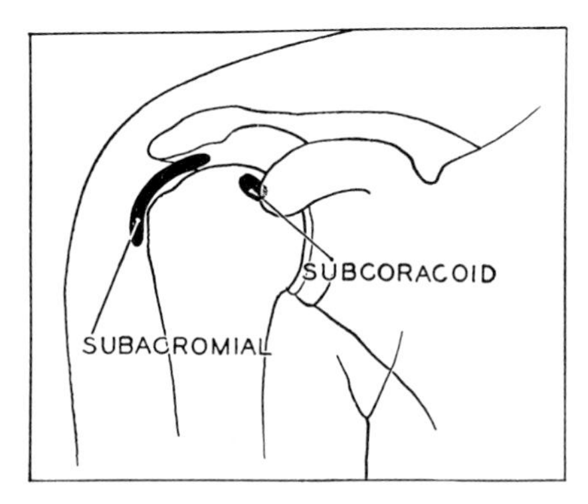

Fig. 15-3. Most commonly affected bursae about the shoulder.

muscles and tendons about the shoulder play a relatively important part in maintaining joint stability. They form, in fact, a flexible but strong protective covering closely investing the entire joint, and in part the tendons are actually fused with the capsule. To provide for motion with a minimum of friction the shoulder bursae (Fig. 15-3) are important; they are in close relationship with the tendons and with the capsule.

These anatomic considerations are of significance in the interpretation of common clinical findings in affections of the shoulder. The intimate association of synovial membrane, ligaments, tendons, and bursae makes multiple involvement likely when the shoulder is injured, renders specific diagnostic analysis difficult, and leads to a confused terminology of shoulder affections. That any acute injury of the shoulder is likely to clear up slowly and incompletely, leading to symptoms of chronic nature, is a common clinical observation. Increasing the likelihood of a chronic disability are the constant drag of the unsupported arm on the shoulder joint and also the fact that in adduction, the position in which the injured joint is most conveniently held, contractures and atrophy quickly develop and decrease the range of painless motion.

MINOR INJURIES: TRAUMATIC SYNOVITIS, SPRAIN, AND STRAIN

Clinical picture. After most shoulder injuries pain is severe; it is sometimes felt not only about the joint itself but throughout the arm, particularly on its lateral aspect about the insertion of the deltoid muscle. Tenderness also may be generalized. There is usually pain on motion, especially on abduction or external rotation, which results in the arm's being held closely against the side of the thorax. If the arm is allowed to remain in this position, a vicious circle is set up: adhesions, contractures, and atrophy develop and further increase the disability.

Differentiation of the various types of injury at the shoulder is important to treatment. *Acute traumatic synovitis* undoubtedly is of common occurrence but is relatively of less importance than in the weight-bearing joints; it is associated usually with injury of one or more of the surrounding structures. Shoulder *sprain,* or stretching or slight tearing of the capsular ligaments, accompanied by extravasation of blood, is a common athletic injury characterized clinically by tenderness over the joint capsule, pain on motion, and slow recovery. Muscular *strain* about the shoulder is suspected when particular pain is elicited by contraction of a specific muscle against resistance and when localized tenderness is present over the muscle or its tendon.

Treatment. The first essential in the treatment of any acute injury of the shoulder is rest. This may be facilitated by the use of a sling, often in conjunction with a shoulder cap of adhesive plaster. In the more severe cases it may be preferable to support the joint in abduction, which is the position of physiologic rest and the one in which adhesions and contractures are least likely to occur. If the shoulder is seen within the first few hours after injury, application of cold by means of an ice cap may retard extravasation of blood. Afterward heat is applied constantly to relieve pain and to hasten absorption of the extravasated blood. However, rest must not be continued for too long a time. Massage, followed by gradually increasing active exercise, should be started as soon as the acute symptoms have begun definitely to subside.

SUBACROMIAL BURSITIS

The subacromial or subdeltoid bursa is a large, flat, thin-walled sac covering laterally the upper end of the humerus and the shoulder joint (Fig. 15-3). It consists of an upper or subacromial portion, which overlies the tendon insertions into the shoulder capsule and the greater tubercle of the humerus and which is in turn covered by the acromion, and of a lower or subdeltoid portion, which overlies the greater tubercle and is covered by the deltoid muscle. The two portions are occasionally separated by a thin membrane; normally they do not communicate with the cavity of the shoulder joint. In opening the normal shoulder the bursa is scarcely recognizable as a definite structure, but in common inflammatory states it may be so distended and thickened as to be palpable even through the deltoid muscle.

Subacromial bursitis is commonly accepted as a disease entity, although clinically it often cannot be differentiated with certainty from lesions of other closely related structures about the shoulder joint. Its usual cause is presumably degenerative or minor traumatic changes in the musculotendinous cuff, to which the bursal floor is closely adherent. Pyogenic infection of the subacromial bursa is a relatively rare entity. Occasionally pain and tenderness localized to the region of the coracoid process result from inflammation of the subcoracoid bursa.

Clinical picture. Subacromial bursitis is seen most often in patients between 40 and 50 years of age, is more common in women than in men, and occurs with relatively high frequency in sedentary individuals. The patient with *acute bursitis* experiences severe pain in the shoulder, sometimes referred down the entire limb as well. This pain may permit the arm to be swung in the sagittal plane but sharply limits any rotation or abduction. Often when abduction is carried out passively, an angle can be identified at which pressure of the inflamed sac beneath the acromion causes a sudden increase in the intensity of the pain. Tenderness may be generalized over the shoulder joint but is often greatest over the central portion of the bursa just beneath the acromion. In more chronic cases the signs are much less definite, and since in such cases there must almost surely be associated pathologic changes

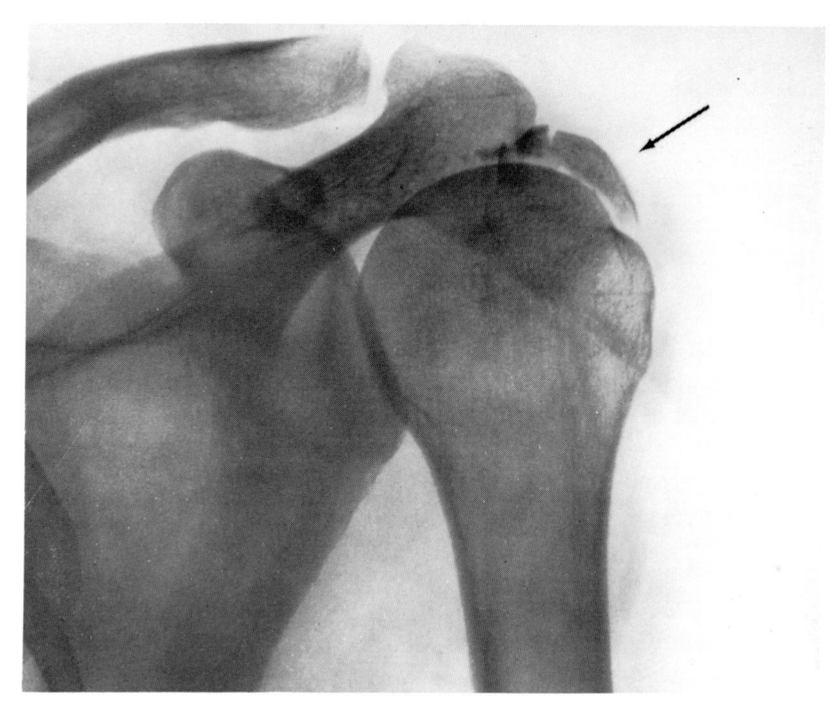

Fig. 15-4. Roentgenogram of the shoulder showing extensive calcification in the subacromial bursa.

in other structures about the joint and contiguous to the bursa, the term *periarthritis* is often applied to them (p. 414). In *chronic bursitis* an old tear of the musculotendinous cuff of the shoulder joint or degenerative changes in the tendon of the long head of the biceps muscle may be present. Long-standing cases frequently present disabling adduction contracture, limitation of rotation, and muscle weakness and atrophy. Roentgenograms are indicated to investigate the possible presence of calcification in the bursa or the rotator cuff (Fig. 15-4) and of arthritic changes in the glenohumeral or the acromioclavicular joint.

Treatment. The first essential in the treatment of acute bursitis is rest and support, preferably in bed and with the arm in as much abduction as can be secured. Pillows or splints may be used. A simple Velpeau dressing may give sufficient immobilization to relieve the pain, but the position of adduction must not be long maintained. Frequently, light skin traction gives the best combination of immobilization, support, and gentle corrective force. Heat is indicated and may afford full relief of the pain. It is best applied by the use of compresses or electric pad. Diathermy treatments are sometimes effective in relieving the acute symptoms. Ultrasonic therapy has also been considered helpful. In some cases, particularly those in which heat seems to aggravate the pain, the application of ice packs is very effective. Procaine block of the suprascapular nerve, which carries a large part of the sensory supply of the shoulder joint, is often most helpful. Injection of a weak solution of procaine into the inflamed bursal sac sometimes leads to instant relief of the pain. If this form of therapy is employed, the bursa should be punctured by numerous needle holes from different directions. Sometimes it is possible to irrigate the bursa with physiologic saline by placing two needles from different angles into the sac. Part or all of a calcific deposit may at times be removed in this manner. Injection of hydrocortisone into the bursa may decrease the inflammatory reaction. Phenylbutazone has been reported to give effective relief in 24 to 48 hours. If the severe pain of an acute bursitis fails to respond satisfactorily to these measures, excision of the bursa and of any calcific material in the rotator cuff may be indicated.

As the discomfort and tenderness become less acute, massage and exercise of the shoulder should be started. Full range-of-motion exercises should be done briefly each day as soon as the decreasing pain will allow; they should be continued for at least 6 weeks in order to regain and preserve full mobility of the shoulder.

In chronic or recurrent cases irrigation of the subacromial bursa sometimes relieves the pain. Excision of the bursa often affords complete relief. Excision of the small calcified bodies or soft calcareous materials, which in the roentgenogram are often visible near the floor of the bursa (Fig. 15-4), may be carried out when their presence is associated with obstinate pain and tenderness. Otherwise their removal is not indicated, since they have frequently been observed to undergo resorption.

BICIPITAL TENOSYNOVITIS

Bicipital tenosynovitis is a frequently observed affection causing pain and stiffness in the shoulder joint; it is an inflammation of the tendon-tendon sheath gliding mechanism of the long head of the biceps muscle, which results in adhesions that bind the tendon to the bicipital groove and inner surface of the rotator cuff. It may be bilateral. It occurs most commonly in women in their early forties. In older individuals bicipital tenosynovitis tends to persist longer than in young adults and to lead to partial stiffness of the shoulder.

Etiology. Anomalies such as inadequate depth of the bicipital groove, or abnormal ridges that develop in association with certain occupations requiring excessive use of the arms, may produce trauma to the tendon and an inflammatory reaction. These factors in addition to normal physiologic wear and tear are responsible for degenerative changes, which increase in severity with each successive decade. Such changes include fraying, shredding, and fasciculation of the tendon and roughening of the bicipital groove. Adhesions may be formed in the tendon sheath. Tenosynovitis is a common sequel of fractures and fracture-dislocations involving the bicipital groove.

Clinical picture. The pain may have an insidious onset or be precipitated by strenuous activity, such as shoveling snow or playing tennis. The pain is located first over the anterior and medial region of the shoulder and then may radiate to the belly of the biceps muscle and flexor surface of the forearm. Pain also may occur at the insertion of the deltoid muscle. Pain may first be noticed on putting the arm back of the body or head. On palpation, exquisite tenderness may be elicited over the intertubercular sulcus or on rolling the biceps tendon with the examining fingers.

Treatment. The early treatment of bicipital tenosynovitis is conservative, consisting of rest, moist heat, and voluntary elimination of the painful arcs of motion.

Procaine and hydrocortisone injection of the tender area, done with strict aseptic precautions, may give prompt relief. When conservative management does not relieve the pain and disability and the recurrences are frequent, surgical treatment is indicated. The tendon of the long head of the biceps may be sectioned and reattached to the humerus below the bicipital groove, or to the coracoid process.

ADHESIVE CAPSULITIS (FROZEN SHOULDER, PERIARTHRITIS)

The term *frozen shoulder* has been used to designate a shoulder showing severe limitation of motion as a result of degenerative changes involving to a varying degree the musculotendinous cuff, synovial membrane, bicipital tendon and tendon sheath, and surrounding structures.

These changes consist of edema, fibrosis, and round cell infiltration indicating a low-grade inflammatory process. This results in loss of elasticity of the periarticular tissues, which become shortened and fibrotic, thereby firmly fixing the humeral head in the glenoid cavity. Muscle atrophy becomes pronounced. The contracted coracohumeral ligament and subscapularis tendon prevent external rotation of the head of the humerus.

This condition, occurring most frequently after middle life, may be initiated by muscular inactivity in the presence of a degenerative alteration such as bicipital tenosynovitis or changes in the musculotendinous cuff. It is sometimes a part of the shoulder-hand syndrome (p. 416).

Clinical picture. The condition may have an insidious onset, may follow direct or indirect local trauma, or may be a sequel to injuries of the distal part of the limb. It also may follow cerebrovascular accidents or come about as the result of referred shoulder pain from cardiac or cervical nerve root affections. It frequently follows subacromial bursitis, calcific tendinitis, and tenosynovitis of the long head of the biceps muscle. Because degenerative changes in the rotator cuff and long head of the biceps are present in most individuals beyond the age of 40 years, any painful affection of the upper limb that causes a person of this age to keep his arm in an adducted, internally rotated position may trigger pathologic changes leading to a frozen shoulder. The pain is accentuated by attempts at scapulohumeral joint motion, particularly abduction, external rotation, and extension. The pain may radiate, as in bicipital tenosynovitis, to the anterolateral aspect of the shoulder region, biceps muscle belly, flexor surface of the forearm, and inferior angle of the scapula. Tenderness may be elicited over the intertubercular sulcus and the tendon of the biceps muscle. The pain is often greatest at night. The main physical finding in frozen shoulder is decreased active and passive mobility in the scapulohumeral joint. This restriction first involves the movements of rotation and abduction; the stiffness may progress to almost complete loss of scapulohumeral motion.

The clinical course of the frozen shoulder is variable. After periods of pain and dysfunction the inflammatory process may subside, with resolution of the adhesions, disappearance of pain, and restoration of muscle activity. This process of recovery may take more than a year. In a minority of cases the pathologic changes remain static for very long periods, with persistent pain and dysfunction. Muscle activity, which is essential to normal tissue metabolism, is necessary to abort or reverse the pathologic processes responsible for frozen shoulder; however, the extent of normal muscle action is limited by the intensity of the pain.

Treatment. Initial treatment of the frozen shoulder syndrome is conservative, consisting of the use of moist heat, gravity-free exercises within painless arcs of motion, and sedation. Later, antigravity exercises such as fingertip wall-climbing should be instituted. Procaine blocks of the sympathetic ganglia may relieve the pain. Adhesive traction with the shoulder abducted may be most helpful. Manipulation of the shoulder is contraindicated if any active inflammation remains or if the process is of long duration. It may be possible to help some of these patients in the late stage by surgical measures such as exploration of the subacromial bursa and biceps tendon or *acromioplasty* (partial resection of the acromion).

RUPTURE OF THE SUPRASPINATUS TENDON AND TEARS OF THE MUSCULOTENDINOUS CUFF

Rupture of the supraspinatus tendon is a common and important cause of shoulder disability. In severe cases the tear may be extensive, involving not only the supraspinatus tendon but also nearly the entire musculotendinous or rotator cuff of the shoulder.

Pathology. The flattened tendon of the supraspinatus muscle, after blending intimately with the adjacent infraspinatus tendon and the capsule of the shoulder joint, inserts into the highest facet of the greater tubercle of the humerus. The supraspinatus tendon thus forms both the roof of the shoulder joint and the floor of the subacromial bursa.

Rupture of the supraspinatus tendon usually occurs near its insertion (Fig. 15-5). Retraction of the muscle after full-thickness tears leaves a direct opening between the subacromial bursa and the shoulder joint. This communication may be demonstrated by arthrography, or roentgenograms taken after the injection of a radiopaque dye into the shoulder joint. Minor lacerations of the cuff are probably common in later adult years and have been found in a large percentage

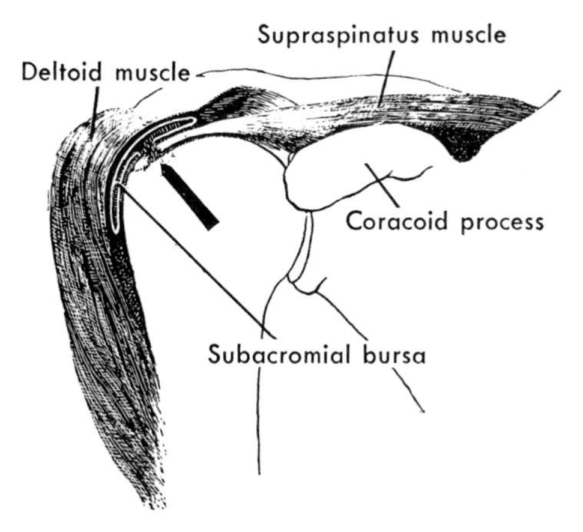

Fig. 15-5. Rupture of the supraspinatus tendon. Arrow points to the rupture. Note that the shoulder joint communicates with the subacromial bursa through the rupture.

of routine postmortem examinations. At times a small fragment of the greater tubercle of the humerus is torn away, together with avulsion of the tendon near its insertion.

With time the pathologic changes accompanying complete rupture become more pronounced. Progressive retraction of the muscle widens the rent in the joint capsule. The distal stub of the tendon, at first a sharply outlined mass, becomes atrophic and may disappear, whereas the edge of the tubercle gradually becomes rounded and smooth.

Etiology. There is almost always a history of trauma, and at times the rupture accompanies an anterior dislocation of the shoulder. The mechanism is presumably an indirect violence associated with sudden powerful elevation of the arm in an attempt to regain balance or to cushion a fall. The presence of a heavy object in the hand increases the likelihood of rupture. Degenerative changes in the cuff predispose to and generally precede rupture.

Clinical picture. Rupture of the supraspinatus tendon is commonest in laborers past the age of 40 years. The rupture is accompanied by a transient sharp pain in the shoulder, and a few hours later there begins a steady ache that may last for several days.

Examination with the humerus adducted discloses a tender point below the acromion, and a sulcus between acromion and tendon insertion may be palpable. As a rule, the ability to initiate and maintain abduction of the shoulder is not lost when the supraspinatus tendon alone is ruptured but is absent in complete or massive avulsions of the cuff.

When the arm is carried into abduction, there are transient pain and slight crepitus as the torn insertion of the tendon passes under the edge of the acromion; the same pain is felt at the corresponding angle as the arm is again lowered.

Subsequently the patient may experience severe shoulder pain for months. This is often most severe at night and is so aggravated after exertion during the day that any type of active work becomes impossible. In late cases atrophy of the supraspinatus and infraspinatus muscles is a constant sign.

Treatment. Small, incomplete tears should be treated by physical therapy. Splinting in abduction may be helpful. For complete rupture of the supraspinatus tendon and extensive tears of the musculotendinous cuff, surgical repair is indicated. When the diagnosis is in doubt, arthrography may help in establishing the diagnosis. For the repair a wide exposure is necessary and may be gained by extending an incision posteriorly with osteotomy of the acromion, exposing the whole roof of the shoulder joint. It is often advisable to divide the coracoacromial ligament. Temporary use of an abduction brace after operation is sometimes helpful. Chief reliance, however, is placed on active exercises of the shoulder. These are started 2 weeks after operation, increased gradually, and continued until maximum function has been regained.

SHOULDER-HAND SYNDROME

The shoulder-hand syndrome is a painful shoulder disability associated with swelling and pain in the homolateral hand. The shoulder changes are identical with those seen in some forms of frozen shoulder.

Etiology. This syndrome, which is thought to be a manifestation of reflex sympathetic dystrophy, may follow any painful shoulder lesion. It may be a sequel to myocardial infarction, pleurisy, other painful intrathoracic lesions, cerebrovascular accidents, trauma, rupture of a cervical disk, or cervical arthritis. In all of these affections the patient tends to keep the painful shoulder immobile. The shoulder-hand syndrome may also arise without known predisposing cause.

Clinical picture. Pain and stiffness of the shoulder with diffuse, nonpitting edema and stiffness and pain in the hand develop acutely or over several months. Symptoms in shoulder and hand may occur together, or either the shoulder or the hand may be affected first. After 3 to 6 months there is gradual relief, but atrophy and stiffness of the hand, with flexion deformity of the fingers and extension contractures of the metacarpophalangeal joints, become more pronounced as the swelling recedes. Recovery occurs by slow stages. Shoulder symptoms and hand swelling require about a year to resolve; trophic changes may disappear in 2 to 4 years. Residual stiffness of the metacarpophalangeal joints may be permanent.

Roentgenographic picture. The bones become osteoporotic; the joints show little roentgenographic change.

Differential diagnosis. The differential diagnosis should include rheumatoid arthritis, the thoracic outlet syndromes, scleroderma, postinfarction sclerodactyly, causalgia, and Sudeck's atrophy limited to the hand. Diagnosis depends on the following findings: a painful, stiff shoulder, uniform swelling of the hand, and a restriction of the motion of all finger joints. There is no true joint swelling, and the sedimentation rate is normal.

Treatment. The treatment is conservative. Daily range-of-motion exercises improve shoulder function. Active and passive exercises of the hands should be frequent and prolonged. Heat or cold may give relief. Procaine blocks of the brachial plexus or stellate ganglion are of value early in the course of the disease, as is orally given cortisone to a lesser extent. Sympathetic blocking drugs may also be of help.

Prognosis. The prognosis is usually good, but recovery is slow and some patients will have permanent stiffness of the shoulder or the fingers. Psychologic factors influence the prognosis in that the neurasthenic patient may refuse to exercise the shoulder and fingers in the presence of pain.

RUPTURE OF THE BICEPS BRACHII MUSCLE

Rupture of the biceps brachii is of infrequent occurrence. The long head may rupture at or near its origin from the glenoid tubercle, at some point within the bicipital groove, or at the musculotendinous junction. Rupture may also occur within the muscle belly of one or both heads, or at their insertion into the tuberosity of the radius.

Etiology. Rupture of the biceps muscle is ordinarily a result of sudden indirect violence, usually without direct injury of the overlying tissues. Rupture of the long head as it traverses the bicipital groove is said to be associated always with a previous weakening of the tendon by degenerative changes. In older individuals such ruptures may follow violence of trivial degree.

Clinical picture. At the moment of rupture there is usually sharp pain and occasionally an audible snap. The most characteristic physical finding is a sharply

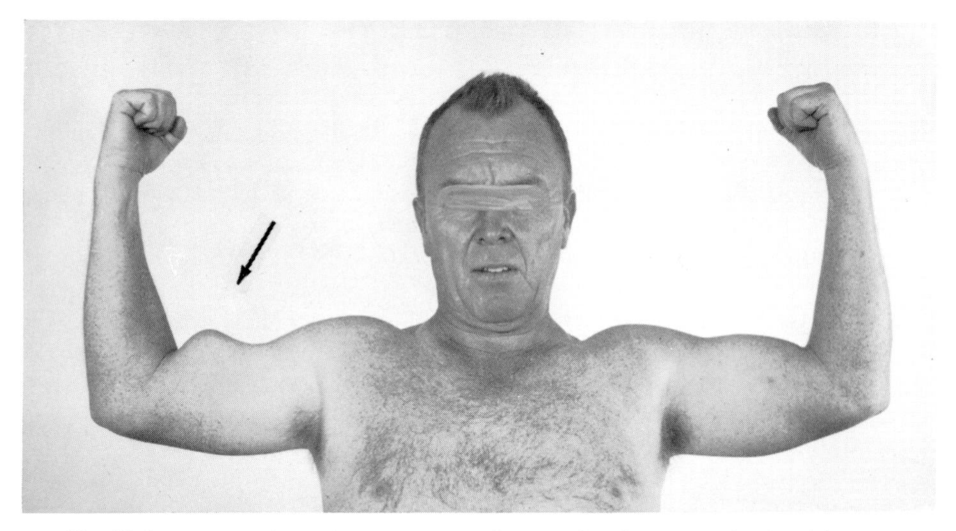

Fig. 15-6. Rupture of the right biceps muscle. Note localized bulge in the right arm.

convex bulge near the middle of the arm (Fig. 15-6). There may be slight local tenderness and weakness of flexion of the elbow and supination of the forearm as compared with these motions in the normal arm. The roentgenogram sometimes shows a small avulsion fracture of the glenoid rim. Rupture of the long head is sometimes difficult to differentiate clinically from dislocation of this tendon out of the bicipital groove.

Treatment. Early cases in active individuals should be treated by exploration and suture. When separation from the glenoid tubercle or rupture of the tendon of the long head has occurred, the proximal portion of the tendon may be excised and the distal portion fixed to the coracoid process or into the floor of the bicipital groove. Adequate postoperative immobilization should be followed by a program of graded exercise. In older or inactive individuals rupture of the long head, in contrast to rupture near the radial tuberosity, causes little disability and may not require operation.

SNAPPING SHOULDER

A shoulder in which an audible click or snap can be elicited by appropriate muscular contractions is occasionally observed. This condition sometimes becomes habitual or involuntary, especially after the patient learns how to cause the snap and elicits it repeatedly. The sound is produced by an incomplete luxation of the joint or by the slipping of a taut tendon over a bony prominence. Rarely a structural change, such as the presence of an anomalous group of muscle fibers, has been demonstrated at exploratory operation. As a rule the symptoms are trivial and no treatment is required.

RECURRENT DISLOCATION OF THE SHOULDER

Traumatic anterior dislocation of the shoulder joint, even when treated according to accepted principles, is sometimes followed by repeated dislocations.

In two reported series the percentages of recurrence were 50% and 57%, respectively, by far the highest percentage being in persons under 20 years of age. The incidence of recurrence is somewhat less in patients whose shoulder has been immobilized for at least 3 weeks after reduction of the first dislocation. Cases of recurrent dislocation are common enough to form an important clinical entity, especially in the military services. Recurrent posterior dislocation is rarely encountered.

Etiology and pathology. Factors that may facilitate repeated dislocation are incompletely healed tears or relaxation of the capsular ligaments, weakness of the surrounding musculature, and congenital or acquired changes in the contour of the humeral head or of the glenoid fossa. In most cases, however, avulsion of the glenoid labrum from the anterior rim of the glenoid cavity seems to be of chief importance, together with erosion of the glenoid rim and a grooved defect of the posterolateral aspect of the head of the humerus.

Clinical picture. The majority of patients are young adults. The condition is seen with especial frequency in athletes and in persons subject to the trauma of epileptic seizures. At first the dislocation occurs only after severe trauma, but successive recurrences require less and less force, and finally dislocation may follow any movement, however trivial, which involves abduction of the shoulder. With successive recurrences, reduction of the displacement becomes correspondingly easier. The pain attending dislocation similarly decreases, but the disability remains extreme, and the patient may feel a constant dread of impending displacement. Atrophy of the muscles of the shoulder girdle sometimes occurs, and rarely signs of involvement of the brachial plexus develop.

Treatment. The nonsurgical treatment of recurrent dislocation consists of (1) preventing the displacement by limiting abduction of the arm and (2) carrying out resistive exercises to strengthen the muscles that internally rotate the shoulder. The displacement may be prevented by strapping the arm to the thorax, by pinning the sleeve to the coat, or by applying a chest belt and an arm band connected by a short flexible strap. Such devices are inconvenient and of only temporary value.

The treatment of choice is surgical. Many procedures have been devised. Bankart's suture of the capsule to the glenoid rim, a difficult but highly effective procedure, is the operation of choice of many surgeons. The Putti-Platt operation, the principle of which is to limit external rotation of the shoulder, is used in many clinics. In this operation the subscapularis tendon and the capsule are divided 1 inch from their humeral attachment, the cut end of the lateral portion is sutured to the anterior rim of the glenoid fossa, and the cut end of the medial portion is attached to the tendinous cuff over the greater tubercle of the humerus; thus, the medial portion of the tendon overlaps the lateral portion. There have been many modifications of these procedures.

After all operations for recurrent dislocation of the shoulder, carefully graded exercises should be carried out for many months, with precautions against strenuous exercise requiring abduction of the shoulder.

OLD DISLOCATION OF THE SHOULDER

Cases of unreduced anterior or posterior dislocation of the shoulder of long duration are occasionally seen and present a difficult therapeutic problem. The patients may show an obvious deformity and on attempting to use the affected shoulder may have considerable pain and disability. The nerves of the brachial plexus are sometimes affected (p. 201). The diagnosis of dislocation is confirmed by anteroposterior and axillary roentgenograms.

Pathology. When the humeral head remains displaced from the glenoid cavity, a series of pathologic changes makes reduction more and more difficult as time goes on. The glenoid cavity fills with granulation tissue, and the torn capsule contracts. The head of the humerus becomes bound down by scar tissue, and the muscles about the joint become shortened and fibrotic. Atrophy of the humerus develops rapidly and may constitute a formidable obstacle to manipulation.

Treatment. If the dislocation has been present for not more than 8 weeks, it can occasionally be reduced by closed manipulation. In order to provide adequate relaxation, the anesthetic agent should be supplemented with a muscle-relaxant drug given intravenously. Care must be taken not to employ sudden or excessive force that might injure the brachial plexus, rupture the axillary vessels, or fracture the humerus at its surgical neck. After reduction the arm may be bandaged to the side or may be immobilized in plaster in slight abduction and flexion. Two weeks later a program of daily physical therapy, including active exercise, is started.

For older dislocations and those that have failed to be reduced by closed manipulation, operation may be indicated. A wide exposure is necessary. After reduction the arm is immobilized at the side by a plaster or adhesive dressing. Again the aftertreatment consists of the prolonged use of daily physical therapy.

Occasionally it is advisable to excise the head of the humerus. The functional result is sometimes satisfactory. In some cases, because of the patient's age or poor general condition, the long duration of the dislocation, or the absence of significant pain or limitation of motion, strenuous efforts to effect replacement are not to be advised.

OLD ACROMIOCLAVICULAR DISLOCATION

Dislocation of the acromioclavicular joint is a common injury. It occurs frequently in football players as a consequence of falls or blows on the point of the shoulder, and the resulting disability when the arm is elevated makes it impossible to pass the ball effectively. The diagnosis is usually evident on inspection and palpation but should be supported by roentgenographic examination to exclude the possibility of fracture of the acromial end of the clavicle. The roentgenogram should always be taken with the patient standing and the arm unsupported or carrying a weight, since often in the supine position the dislocation becomes spontaneously reduced and the instability will not be evident. Unless carefully treated, acromioclavicular dislocations are likely to become chronic since the weight of the dependent limb tends to maintain the dislocation.

Pathology. The usual displacement is a dropping of the acromion downward and forward because of the weight of the arm; the acromial end of the clavicle thereby acquires undue upward prominence and mobility. More important than the tearing of the articular capsule and rupture of the acromioclavicular ligaments is rupture of the powerful coracoclavicular ligaments that normally prevent upward displacement of the lateral end of the clavicle.

Treatment and prognosis. In persons who engage in no athletics and do no heavy work, old acromioclavicular dislocations may cause little disability and require no treatment. In more active individuals, disabling symptoms can be satisfactorily relieved by operation. The acromioclavicular ligaments should be repaired, and new ligaments should be constructed of fascia lata to tie the clavicle down to the coracoid process. After operation the position of reduction may be maintained for 8 weeks by means of temporary internal fixation with wires or pins through the acromion into the clavicle. Lifting of heavy objects should be avoided for an additional 8 weeks.

An alternative procedure, consisting of excision of the lateral end of the clavicle, usually produces a satisfactory functional result. Fusion of the acromioclavicular joint, which limits mobility of the shoulder, is inadvisable.

OLD STERNOCLAVICULAR DISLOCATION

Dislocation of the sternal end of the clavicle occurs much less frequently than acromioclavicular dislocation. Sternoclavicular dislocation may result from a violent fall or blow upon the shoulder, rupturing the sternoclavicular ligaments and usually leaving the intra-articular fibrocartilage attached to the clavicle. The displacement tends to persist unless corrected by treatment. In debilitated individuals chronic or recurrent dislocation is sometimes found without history of acute trauma. The dislocation also may develop gradually when there is a paralysis of the pectoral muscles. In sternoclavicular dislocation the end of the clavicle is ordinarily displaced anteriorly, and the deformity is usually evident upon inspection and palpation.

Treatment. Rarely is surgical treatment indicated, and only when the deformity is accompanied by pain that cannot be relieved by other measures. The end of the clavicle may be sutured to the sternum or the first rib with fascia lata. After operation the corrected position must be maintained for 6 to 8 weeks by means of a pressure pad and a strong adhesive dressing. Lifting of heavy objects should be avoided for an additional period of several weeks.

Affections of the jaw

Disability of the temporomandibular joint, although forming a relatively uncommon condition in orthopaedic surgery, may nevertheless represent to the individual patient an affliction of peculiarly distressing nature. Two clinical entities of fairly frequent occurrence, snapping jaw and ankylosis of the jaw, can be efficiently relieved by resort to joint surgery.

SNAPPING JAW

The term *snapping jaw* is applied to a common clinical condition characterized by a clicking sensation in the temporomandibular joint on opening and closing the mouth. Usually the symptoms are trivial, but a few patients experience severe pain or locking of the jaw and require careful investigation and treatment.

Pathology. Except for cases attributable to malocclusion of the teeth, intractable snapping of the jaw is caused by one of two pathologic entities: (1) mechanical derangement of the articular disk or (2) recurrent dislocation of the temporomandibular joint, with irregularity of the joint surfaces. In the etiology both developmental and traumatic factors are presumably concerned. Derangements of the articular disk are in some respects analogous to disorders of the semilunar cartilages of the knee. In dislocation of the jaw the condyle of the mandible, which slips forward when the mouth is opened, reaches a position anterior to the articular tubercle of the temporal bone.

Clinical picture. The condition of snapping jaw is usually of gradual onset during adolescence or early adult life, occurring most commonly in young women. The slipping may at first take place only when the mouth is opened widely, as in yawning, but often becomes increasingly frequent and troublesome and in some cases leads to intense discomfort. The diagnosis may be evident from the history and physical findings but should be investigated further with roentgenograms of the temporomandibular joints taken with the mouth in open and closed positions.

Treatment. Early cases may be treated with local heat, a soft diet, and avoidance of any unnecessary motion of the jaw in talking. Dental consultation concerning the occlusion may be advisable. Partial rest secured by the use of a Barton bandage or a Thomas collar may be helpful. Complete rest of the jaws, obtained by wiring the teeth for several weeks, is often curative. In some cases, however, operation becomes necessary, the type of procedure depending upon the pathologic changes demonstrated at exploration. For lesions of the fibrocartilage, simple excision of the disk has given satisfactory results. Reefing of the joint capsule is sometimes indicated. Recurrent dislocation may be treated effectively by the creation of a bone block, as described by Mayer.

ANKYLOSIS OF THE JAW

Loss of motion in the temporomandibular joint constitutes a deforming and extremely disabling affection that can be relieved by surgical measures.

Etiology. Ankylosis of the jaw is most frequently seen as the late result of a local traumatic lesion. It is at times a sequel of osteomyelitis of the temporal bone following infection of the middle ear. Ankylosis may also result from rheumatoid arthritis of the temporomandibular joint.

Treatment. The operation of choice is an arthroplastic procedure in which the head of the mandible and the upper portion of the ramus are resected. A layer of fascia or fat may be placed around the end of the ramus but is not essential to a good clinical result.

CHAPTER 16 Affections of the elbow, wrist, and hand

In many instances affections of the elbow, wrist, and hand may lead to serious functional and economic loss. Accordingly their prompt recognition and careful treatment are of the greatest importance.

Affections of the elbow

The elbow is a hinge joint with a range of motion of from 0 to 140 degrees of flexion. Pronation and supination are made possible by rotation of the head of the radius on the capitulum of the humerus. Despite the wide extent of elbow motion, stability is well maintained by the fitting of the trochlea of the humerus deeply into the trochlear notch of the ulna, by powerful muscular and tendinous supports, and by strong collateral ligaments. Stability of the radiohumeral joint is not due to bony configuration since the head of the radius presents only a shallow concavity for articulation with the capitulum, but rather is due largely to the strong annular ligament that holds the head of the radius close to the radial notch of the ulna.

The wide range of motion in the elbow region renders its structures peculiarly liable to strains and sprains. Because of its tendinous investment and its exposed position, the elbow may suffer attacks of bursitis, and because of the proximity of important structures on its anterior aspect, injuries of the elbow occasionally lead to serious circulatory and neurologic complications.

STRAINS AND SPRAINS

Muscular strains about the elbow are common results of isolated episodes of slight trauma or of unaccustomed repeated trauma of minor degree.

Diagnosis. An effort should always be made to allocate the lesion to one muscle or muscle group; this is best done by careful localization of the maximum pain and tenderness and by analysis of the type of resisted active motion that causes most discomfort. In differential diagnosis ligamentous sprain, bursitis, and minor fractures must be kept in mind. In youth epiphyseal separations in the elbow region are particularly important because of the large number of epiphyses and the wide range in the time of their consolidation; the first of these epiphyses, the capitular, makes its roentgenographic appearance in early infancy, and the last union is not complete until the age of 20 years. It is advisable to make roentgenograms of the injured elbow, and in children views of the normal elbow are valuable for detailed comparison with those of the traumatized area.

Treatment. Strain is treated by temporary immobilization of the affected muscle in a position of relaxation by means of removable plaster splints, adhesive plaster, or sling. Physical therapy consisting of heat and massage is started at once and is followed by graded active exercise.

Ligamentous sprain, although resembling muscle strain in causation, symptoms, and treatment, may be slower in subsiding. Sprain of the radial collateral ligament occurs fairly commonly among baseball, tennis, and golf players.

OLECRANON BURSITIS

About the elbow joint as many as ten bursae have been described; most of them are inconstant and of little clinical significance. The olecranon bursa (Fig. 16-1), situated between the tip of the olecranon and the skin, is frequently involved by inflammatory changes. A single episode of local injury may produce olecranon bursitis, but continued traumatization of slight degree is a more common cause. The diagnosis is usually evident from the history of injury, from the localization of pain, swelling, and tenderness posteriorly only, and from the slight restriction of flexion by pain as the inflamed structure is placed under tension.

Treatment. The type of therapy varies with the acuteness of the individual case. Purulent bursitis is uncommon; its treatment includes rest, heat, aspirations, antibiotics, and sometimes incision and drainage. The more frequent nonspecific inflammation may usually be cured by aspiration of the fluid, application of a compression bandage and local heat, and care to prevent repetition of the causative trauma. In chronic cases the bursal sac may become greatly thickened; this con-

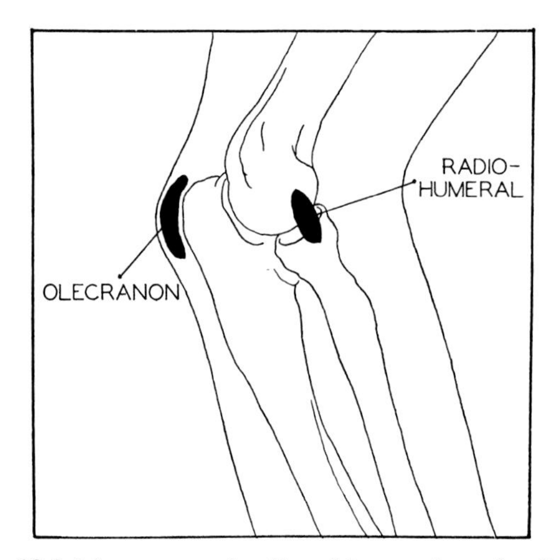

Fig. 16-1. Most commonly affected bursae about the elbow.

dition is sometimes called *miner's elbow.* The sac sometimes contains many small loose bodies. Such chronic changes, when causing symptoms, indicate excision of the bursa. It is occasionally necessary to remove, in addition to the bursa, a small bony spur that may lie at the tip of the olecranon and may extend for a short distance into the triceps tendon.

TENNIS ELBOW (LATERAL EPICONDYLITIS, RADIOHUMERAL BURSITIS)

The common disorder called tennis elbow exhibits typical clinical characteristics; its cause, however, has not been established. In some instances irritation of bursal tissue overlying the radiohumeral joint capsule (Fig. 16-1) seems to be responsible. However, at operation a definite bursal sac is seldom found. Many observers believe that the usual lesion is a partial rupture of the origin of the extensor muscles with a secondary traumatic periostitis at the lateral epicondyle. Other observers believe that the lesion is a radiohumeral synovitis due to pinching of synovial tissue between radial head and capitulum.

Clinical picture. The usual onset is a gradually increasing discomfort after continued slight traumatization in the region of the radiohumeral joint. Tennis playing, which requires repeated pronation and supination of the forearm with extension of the elbow, is a common cause. The condition may begin after forceful supination, such as is required in using a screwdriver, particularly in persons unaccustomed to such activity. The disorder is frequent among athletes, butchers, carpenters, and the like. Pain is experienced in the lateral aspect of the elbow, particularly when the patient reaches forward to pick up an object or to turn a doorknob, and the discomfort may spread down the entire forearm; it may be very persistent and annoying. Examination shows a small area of tenderness over the lateral epicondyle of the humerus and the radiohumeral joint, and sometimes extending distally 1 or 2 cm. over the extensor muscles. The grip may be weak because forced grasp causes pain over the lateral epicondyle. Passive motion of the elbow is unaffected, but active extension and especially supination against resistance may cause intense discomfort. Roentgenograms are usually negative.

Treatment. Temporary immobilization of the elbow with a sling, adhesive dressing, or plaster cast and the use of heat will usually produce relief. The application of a dorsiflexion splint to the wrist, which relaxes the extensor tendons, is often helpful. The tender area may be injected with procaine or hydrocortisone. Occasionally manipulation of the elbow has been helpful. In cases that fail to respond to these measures, surgical treatment may be indicated. The common extensor origin may be released from the epicondyle; any bursal tissue found between the common extensor tendon and the joint capsule should be excised.

RADIOHUMERAL SUBLUXATION IN CHILDREN

A common occurrence in children, usually between 2 and 5 years of age, is subluxation of the proximal end of the radius on the capitulum of the humerus. This condition has been termed *nursemaid's elbow* or *pull syndrome* because of the mechanism by which it is produced: a sudden, direct pull on the elevated limb with the elbow extended and the forearm pronated. This may occur when a child falls while his hand is being held by an adult. The diagnosis is made on the history, together with characteristic physical findings. The child refuses to use the arm, and the elbow is held slightly flexed. Because of failure to use the arm the condition may be mistaken for a paralysis, such as that caused by injury of the brachial plexus. All movements are of essentially normal range except supination. The forearm is held in neutral position as regards rotation, and attempts to supinate cause pain and a sensation of mechanical blocking.

Treatment. A very brief manipulation without anesthesia is indicated. With the child's elbow flexed to a right angle, the forearm is supinated quickly while pressure is exerted on the radial head by the operator's thumb. As the point of obstruction is passed, a definite click is usually felt; after this the child almost immediately resumes use of the arm. If for any reason prompt reduction is not carried out, the subluxation will usually undergo spontaneous reduction within a few days.

OLD DISLOCATION OF THE ELBOW

Old unreduced posterior dislocation of the radius and ulna on the humerus is occasionally encountered and presents a characteristic clinical picture. The deformity is usually obvious to inspection and palpation, the olecranon and the triceps tendon appearing abnormally prominent posteriorly. There may be severe pain on resisted motion, and the mobility, particularly flexion, is much restricted.

Treatment. When the dislocation has been present for as long as 2 or 3 weeks, an open reduction is usually necessary, since forced manipulation is likely to damage the soft tissues or to fracture the humeral condyles. A wide operative exposure is essential and is often gained by the posterior route with division of the triceps tendon. Care must be exercised to avoid injury of the ulnar nerve. After reduction, suture of the triceps tendon, and closure, the elbow is immobilized in flexion. In the aftertreatment physical therapy is of value but must be carefully supervised and should not include passive stretching.

VOLKMANN'S ISCHEMIC CONTRACTURE

The very disabling contracture of fingers and wrist first described in 1875 by von Volkmann is an occasional complication of injuries of the upper limb and particularly of supracondylar fractures of the humerus. An analogous condition is sometimes observed in the lower limb, where it has been known to follow the use of Bryant's overhead traction for fractures of the femur. The anterior, posterior, and peroneal compartment syndromes that occur in the leg are similar entities. In Volkmann's contracture the muscles of the forearm become indurated and rigid, the joints present unsightly contracture deformities, and the hand may become almost completely useless. Except in the mildest cases, one or more of the major nerve trunks may be involved as well as the muscles. The condition is seen much more often in children (Fig. 16-2) than in adults.

Etiology and pathology. The cause of the contracture is severe ischemia resulting from spasm of the brachial artery just above its bifurcation, associated with contusion, thrombosis, or laceration of the artery and with reflex spasm of the collateral vessels. Early exploration at the site of injury may show a severe segmental spasm of the brachial artery extending down into the radial and ulnar

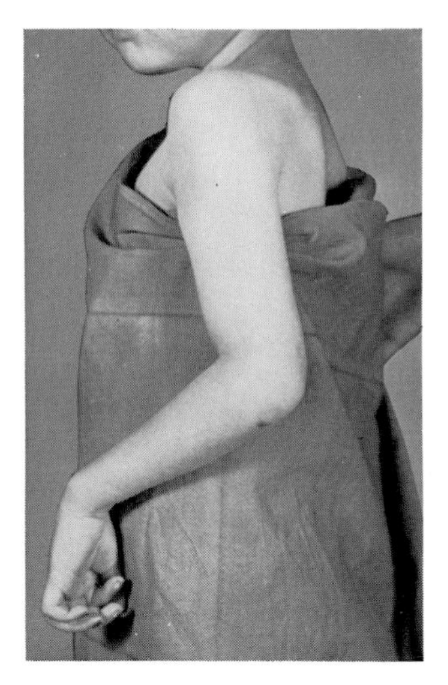

Fig. 16-2. Volkmann's ischemic contracture in a 6-year-old boy after supracondylar fracture of the left humerus. Note deforming contractures of the wrist and fingers with atrophy of the forearm.

branches, narrowing them to the size of a string, and obliterating their lumina. A visible injury or thrombosis of the vessel may or may not be present.

The muscle bellies, which require a great deal of blood, are rapidly and profoundly affected by the ischemia. They usually develop an ellipsoidal area of infarction that follows the course of the anterior interosseous artery with greatest damage in the midforearm. The flexor digitorum profundus and flexor pollicis longus muscles are most severely affected; the flexor digitorum sublimis and wrist flexors are also involved. The median nerve, which passes near the center of the infarct, is more extensively damaged than the ulnar nerve. Microscopically, evidence of widespread degeneration and necrosis of the muscle fibers is quickly followed by round cell infiltration and by extensive formation of fibrous tissue that later undergoes progressive contraction. Nerves to the forearm and hand may be damaged by the original injury as well as by the subsequent ischemia.

Clinical picture. Ischemic contracture occurs most commonly after fracture of the lower end of the humerus. Severe pain should at once suggest the onset of this complication, but pain is not always present. The most important clinical sign is absence of the radial pulse. The hand may quickly become cyanotic, slightly swollen, and cold, and the fingers may be insensitive and powerless. Ischemia of 48 hours' duration usually results ultimately in a severe degree of contracture. As fibrosis proceeds, the forearm becomes hardened and shrunken, and the hand develops severe claw deformity (Fig. 16-2) and extreme disability. Flexion of the interphalangeal joints in the clawhand increases with attempts to extend the wrist. In severe cases there may be fixed flexion of the elbow and pronation of the forearm. Examination may show also a paralysis that has resulted from damage of one or more major nerves, particularly the median nerve.

Treatment. The early treatment of Volkmann's contracture is exceedingly important and should be instituted as soon as possible. Prophylactic treatment should begin immediately after the original injury because ischemia lasting 6 hours will cause some contracture. When elbow swelling is minimal and circulation to the forearm and hand is satisfactory, closed reduction of the fracture may be carried out at once or may be started by elevation in traction. If reduction by manipulation is elected, several precautions are to be observed. Flexion should never be forced against resistance and must never be carried to the point of decreasing the strength of the radial pulse. After gentle manipulation, immobilization is secured by a posterior splint of plaster applied with care to avoid constriction. Observation during the next few days must be frequent and critical, and if severe pain or the signs of circulatory obstruction appear, immediate loosening or removal of the bandages is obligatory.

If at manipulation sufficient flexion cannot be obtained to ensure satisfactory alignment, the fracture should be treated in traction. Either skin traction or skeletal traction by means of a Kirschner wire through the olecranon may be used.

In borderline cases in which ischemia appears possible, the safest treatment is to regard the fracture as of decidedly secondary importance and to concentrate all efforts on improving the circulation. Elevation of the arm by means of traction, with the elbow in partial flexion, and the application of cold are indicated. If immediate improvement in the circulation is not apparent, surgical exploration of the antecubital area should be carried out at once.

If ischemia is manifest when the patient is first seen, the fracture should be reduced and if prompt return of the circulation does not occur, immediate surgical intervention is indicated. Arteriography may be helpful but should not be allowed to delay the operation. The hematoma should be evacuated through an anterior incision, and distended fascial compartments of the forearm should be opened. The artery must be exposed at the level of injury, care being taken to avoid damage of any of its branches. If warm saline applications, gentle massage of the vessels, and the application of a papaverine solution fail to restore the radial pulse, or if a lesion of the arterial wall is manifest, the constricted segment should be resected between ligatures. Collateral vessels, thus freed of reflex spasm, will in most cases maintain an adequate circulation. Vessel grafting is usually unnecessary. The fracture may be reduced gently before the wound is closed. In the next few days repeated procaine block of the brachial plexus or the stellate and cervical sympathetic ganglia may be indicated. A sympatholytic drug may be used. The wrist and fingers should be splinted in functional position to forestall contractures.

Treatment of the late case is frequently unsatisfactory. Moderate improvement of the deformity and recovery of power often follow gradual stretching of the contracted muscles. This may be accomplished by means of a splint fitted with elastic finger traction and hinged for progressive extension of the wrist, or by the gradual wedging of a well-padded, well-molded plaster cast. Vigorous active exercises should be carried out. If contractures and paralyses remain, surgical treatment is indicated. Good results have been reported after extensive resection of fibrotic muscle and scar tissue, accompanied by muscle and tendon lengthening and transplantation if indicated. A muscle-slide operation in which the origins of all flexor muscles are completely detached, together with neurolysis of the median and ulnar nerves, probably gives the best results. The ultimate prognosis is relatively favorable if the operation is done within 6 months of the injury. Occasionally a nerve graft may be indicated. Contracture of the intrinsic muscles of the hand (p. 439) may require surgical release.

TRAUMATIC MYOSITIS OSSIFICANS

Another important complication of injuries about the elbow, seen also in other parts of the body, is the formation of excessive amounts of new bone that may restrict the range of joint motion. Such new bone may be the result of any of several related pathologic processes. Excessive bone formation resulting simply from an active healing process and limited in area to the actual site of bony or periosteal injury is called *exuberant callus*. Bone produced by the organization and ossification of a hematoma, a process directly analogous to callus formation, may be found beneath the periosteum or extending out into the muscle or other soft tissues and is called *ossifying hematoma*. Intramuscular ossification following local injury is known as traumatic myositis ossificans. It is not to be confused with progressive myositis ossificans, a rare systemic disease described on p. 65.

Traumatic myositis ossificans may result either from repeated slight injuries or from a single episode of more severe trauma. Examples of the chronic type are the *rider's bone,* which forms within the thigh adductors of horsemen, and the ossification within the pectoralis major, which in soldiers sometimes follows the habitual firing of a rifle. Intramuscular ossification following a single injury may occur in various regions of the body but is most frequent in the thigh, in the upper part of the arm, and about the elbow joint. Ossification in and about the elbow (Fig. 16-3) is of particular importance because it may cause a serious limitation of joint motion.

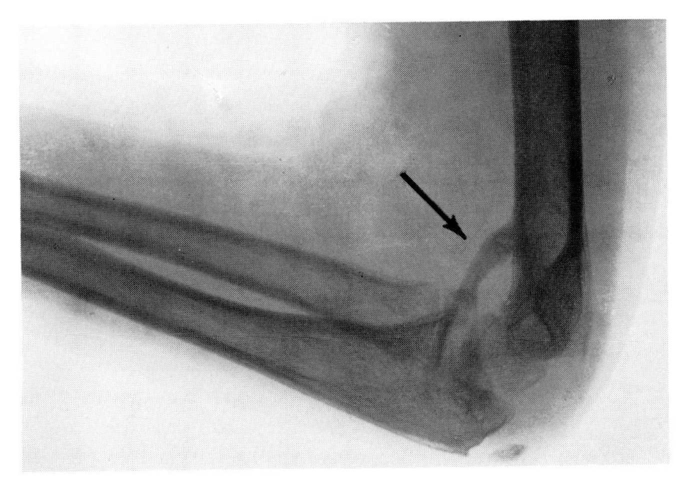

Fig. 16-3. Roentgenogram of traumatic myositis ossificans in the elbow of an 11-year-old boy. Note bony projection anterior to the distal end of the humerus. Eleven weeks previously the medial epicondyle of the humerus had been fractured.

Etiology and pathology. The exact genesis of the new bone is uncertain. Interstitial hemorrhage may lead to the proliferation of collagenoblasts. Osteoblasts from nearby injured periosteum may play a part, whereas in many cases a metaplasia of local connective tissue cells into bone-forming cells seems to take place. Microscopically this issue may simulate osteogenic sarcoma. Some observers have attributed importance to a hypothetical systemic factor predisposing to ossification. Of interest in this connection is the fact that periarticular myositis ossificans is quite common in patients with traumatic paraplegia or old, severe burns.

In traumatic myositis ossificans about the elbow the new bone forms commonly in the lower portion of the belly of the brachialis muscle, anterior to the elbow joint. It may at first consist of minute osteocartilaginous spicules that lie parallel to the muscle fibers; these later coalesce, forming an irregular bony mass that gradually becomes more rounded, homogeneous, and dense. Rarely a complete bridge of bone may develop, forming an extra-articular ankylosis. Usually the bone formation ceases within 3 to 6 months. The bony mass may be connected to the shaft by an osseous pedicle or a fibrous septum or may lie entirely separate from it. Spontaneous regression of the mass occurs frequently and may result in its complete disappearance.

Clinical picture. Traumatic myositis ossificans is seen most frequently in patients between the ages of 15 and 30 years. In the case of the elbow joint the antecedent injury is most often a posterior dislocation or a fracture of the lower end of the humerus, but the condition may follow an uncomplicated strain or contusion. The diagnosis is to be suspected when tenderness, swelling, limitation of elbow motion, or pain on motion fails to clear up within the usual period, or when after an initial improvement these signs begin to become worse. The area of ossification is usually palpable and is demonstrable in the roentgenogram after 3 or 4 weeks. Early osteogenic sarcoma must be considered in the differential diagnosis but can as a rule be definitely excluded by the roentgenogram.

Treatment. The stage at which the process is encountered determines the treatment. Since vigorous physical therapy is believed to stimulate the ossification, massage and strenuous exercise are to be avoided after elbow injuries; this is not likely to lead to contractures, since in the elbow—and particularly in young individuals—motion may be readily regained after immobilization. When a diagnosis of myositis ossificans in the region of the elbow has been made, prolonged rest of the elbow is indicated. Although guarded active motion may do no harm, the most reliable method of treatment is immobilization. This is best obtained by means of a light plaster cast, extending from the upper arm to the metacarpal heads with the elbow at a right angle and the forearm in neutral rotation. Immobilization should be continued until arrest of the process of ossification is definitely demonstrable in the roentgenograms. Active motion may then be gradually restored, but massage and passive stretching should be avoided.

Operation is contraindicated in the early stages of the ossification because additional trauma will usually be followed by an increased formation of bone. When, despite an adequate period of waiting, which should be at least 6 to 12 months, the bony tumor has failed to be absorbed and is causing either limitation of joint motion or serious discomfort, surgical intervention is indicated. Complete excision of the bony mass should be carried out with care to avoid unnecessary trauma and to control carefully all hemorrhage. Immobilization for a period of 2 weeks after operation is advisable.

Traumatic myositis ossificans in other parts of the body presents essentially the same clinical picture and is treated in the same way.

Affections of the wrist and hand

The importance of the wrist and hand as compared with other parts of the upper limb cannot be overemphasized. The more proximal portions of the limb subserve primarily the purpose of putting the hand in positions where it can best carry out its functions.

Disabilities of the wrist and hand are extremely common. They fall largely into two major groups: (1) traumatic lesions and (2) infections. Many of the traumatic disorders are important not so much because of their immediate symptomatology as because of their disabling sequelae. Acute infections and injuries of the hand require surgical treatment based on detailed anatomic and clinical study; for reference the standard texts on hand surgery are listed in the bibliography. In addition to frank traumatic and infectious lesions the wrist and hand are subject to a number of affections of chronic type, the causes of which are less well known. With the numerous and well-planned technics for treating disabilities of the hand that have been devised during and since World War II, hand surgery has become highly developed. This has resulted in better care of affections of the hand and improved functional results.

Anatomy. The wrist joint is of double-hinge type, permitting approximately 180 degrees of motion in the anteroposterior plane and approximately 80 degrees in the lateral plane. In volar flexion there is more motion in the midcarpal joints than in the radiocarpal joints; in dorsiflexion the reverse is true. The carpometacarpal joints contribute little to wrist motion. Stability of the wrist is maintained not by bony configuration but by ligaments and the numerous tough, fibrous tendon sheaths that closely invest them.

On their volar surface the wrist bones form a concave arch that is converted into a tunnel by the transverse carpal ligament. Through this carpal tunnel pass the tendons of the flexor digitorum sublimis and profundus, the median nerve, and the flexor pollicis longus tendon. In this snug compartment, swelling from any of several causes may compress the median nerve and produce the *carpal tunnel syndrome* (p. 205). Where long flexor tendons must pull at an angle, as in the wrist, distal palm, and digits, they are invested in a double-layered synovial sac or tendon sheath. This delicate membrane facilitates the smooth gliding of the tendon. To prevent the flexor tendons from bowstringing in the distal palm and fingers, the tendons and their synovial sheaths are encased in ligamentous tunnels. In these narrow passages, the "no man's land" of Bunnell, fibrous adhesions of the tendon, sheath, and tunnel wall may immobilize the tendon and destroy its function. Within the digital tunnel the sublimis tendon, which proximally lies superficial to the profundus tendon, splits and passes dorsal to the profundus to insert near the base of the middle phalanx. The profundus tendon continues distally to insert near the base of the distal phalanx.

The long extensor tendons on the dorsum of the hand do not traverse such tight compartments but rather fan out flatly over the dorsum of each metacarpophalangeal joint to form complex dorsal hoods. The central slip of a hood inserts on the base of the middle phalanx, whereas the lateral bands to either side of the central slip continue distally to insert on the dorsum of the base of the distal phalanx. The long extensor muscles are primarily extensors of the metacarpophalangeal, or proximal, joints. The dorsal hoods are joined, just distal to the metacarpophalangeal joints, by the lumbricales and interossei. These intrinsic muscles act as flexors of the proximal joints of the fingers and extensors of their middle and distal joints.

Physical examination. Examination of the hand is essentially the clinical application of a good working knowledge of its anatomy. The origin, insertion, innervation, and function of each of the muscles that control the hand must be known before the clinician can appreciate the effects of lesions involving these structures.

AFFECTIONS OF TENDONS AND TENDON SHEATHS

Tendon lacerations

Tendon lacerations are common, serious problems. They frequently cause substantial disability of the digit or of the entire hand and, as industrial injuries, result in the loss of many man-hours of work. In most instances early diagnosis and proper care of tendon injuries will yield good or excellent results. Failure to make the diagnosis or inadequate treatment may lead to permanent disability.

Tendon lacerations may be caused by any sharp instrument that penetrates the skin over the course of a tendon. They frequently result from accidents such as grasping a sharp blade in the hand or falling on pieces of glass, or from the careless use of a power tool. The skin opening may be small and clean-cut or an extensive, jagged wound; in either case there may be associated injury of neurovascular and bony structures. Tendon injury should be suspected in the presence of every laceration of wrist, hand, or digit. The diagnosis is strengthened by demonstration of the patient's inability to move the joint served by the tendon in question. This test may be difficult in the case of the severely injured hand, but with careful, gentle examination the diagnosis can usually be made. In the case of a volar laceration near the proximal finger crease, only the sublimis tendon may be severed and the diagnosis may not be obvious, since the profundus will flex both the middle and distal joints. The test of isolated sublimis function is carried out with the adjacent fingers held in extension, which prevents flexion by the profundus tendon. The diagnosis of tendon laceration is confirmed at operation when the wound is treated.

Treatment. The treatment is based on knowledge of the detailed anatomy of the hand and of the physiology of tendon healing. In the case of fresh, clean lacerations, primary repair of the tendon may be carried out, provided that it has not been severed in the region of the tight flexor tunnels, that is, between the distal palmar crease and the middle phalanx ("no man's land"). For lacerations in this area, simple closure of the skin wound followed by a delayed tendon graft is advisable.

Tendon ruptures

Rupture may occur in almost any tendon of the hand. Ruptures are more common in extensor than in flexor tendons. As a rule, ruptures of flexor tendons occur at their site of insertion. Ruptures within the tendon are usually the result of local changes from disease or injury, rheumatoid arthritis and tenosynovitis (p. 155) being common offenders.

Rupture of the extensor pollicis longus tendon

The tendon of the extensor pollicis longus muscle is occasionally ruptured as a late effect of (1) rheumatoid arthritis, (2) occupational trauma, as seen in kettledrummers, or (3) fracture of the lower end of the radius. The rupture is preceded by attenuation from chronic inflammation or from friction about Lister's tubercle or the fragments of an imperfectly reduced fracture.

Clinical picture. Unlike the rupture of a normal tendon, the separation may occur without pain or sensation of snapping. After fracture of the radius the rupture often takes place on exertion from 1 to 3 months after the original trauma. The diagnostic signs are inability to extend the distal phalanx of the thumb against resistance and absence of the subcutaneous bowstring formed by the normal tendon when the thumb is actively extended.

Treatment. Exploration, followed by tendon suture, graft, or transference, is indicated. Cautious early mobilization is to be employed postoperatively.

Rupture of the central extensor slip

Rupture of the central extensor slip near its insertion into the base of the middle phalanx causes a flexion deformity of the proximal interphalangeal joint. The lateral bands to either side of the central slip gradually subluxate to the sides of the joint. They continue to exert force on the distal joint, which becomes hyperextended. This tendon lesion and finger attitude are known as a boutonniere (or buttonhole) deformity. The treatment consists of suturing the central slip, repositioning the lateral bands, and releasing them distally.

Mallet finger (baseball or dropped finger)

Sudden forcible flexion of a distal phalanx may cause an avulsion of the extensor tendon at its insertion. A small portion of the posterior lip of the phalangeal base is often torn away with the tendon. The injury is common among athletes and has been called mallet, baseball, or dropped finger.

Clinical picture. The history of injury together with inability to extend actively the distal phalanx is diagnostic. In early cases swelling and tenderness may obscure the loss of power in the finger. Roentgenographic examination should always be made.

Treatment. The distal interphalangeal joint should be held constantly in hyperextension (Fig. 16-4) for from 4 to 6 weeks while the proximal interphalangeal joint is held flexed. This can be done with a plaster splint or a small metal splint bent to fit the finger. The insertion of a small wire or pin through the distal phalanx, across the hyperextended distal phalangeal joint, and into the middle phalanx has proved at times to be an effective method. The results of adequate treatment

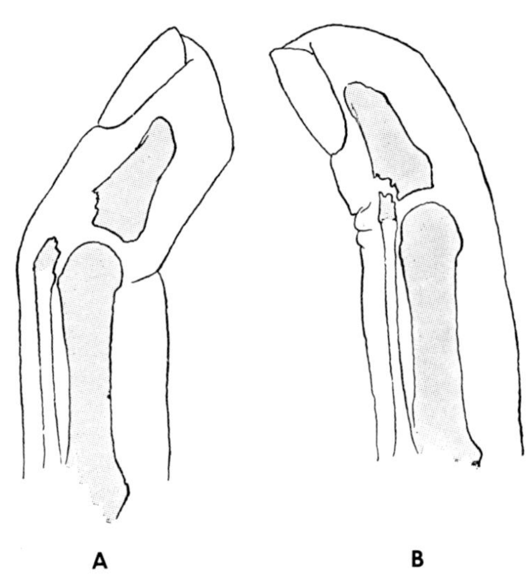

Fig. 16-4. Mallet finger. **A,** Typical deformity due to avulsion of the tendon together with a small fragment of bone. **B,** Position of reduction. Whereas in treatment the distal joint of the finger is immobilized in hyperextension, the middle and proximal joints should be held in the better functional position of flexion.

are usually good. In late cases one may resort to suture of the tendon or of the bone fragment to the phalanx, but the results are often unsatisfactory. The disability from dropped finger, however, is usually very slight.

Tenosynovitis

Tenosynovitis is an inflammation of the synovial sheaths covering the tendons. It is seen frequently as a result of trauma, rheumatoid arthritis, or infection.

Traumatic tenosynovitis

Traumatic tenosynovitis is a frequent clinical entity following minor occupational injury of the wrist region. It also occurs frequently in the sheaths of the anterior and posterior tibial and the Achilles tendons.

Etiology and pathology. Traumatic tenosynovitis is usually a result of strenuous, oft-repeated, or unaccustomed use of the adjacent joint. Serous or fibrinous fluid may accumulate in the affected tendon sheath, sometimes resulting later in chronic sclerosing changes and stenosis.

Clinical picture. Pain on motion of the affected tendon is the presenting symptom. Swelling usually occurs but may not be conspicuous, and redness is usually absent. Tenderness on pressure over the tendon sheath is a constant finding. On motion of the affected tendon unmistakable crepitation can usually be elicited. Considerable disability may develop rapidly.

Diagnosis. In differential diagnosis other lesions resulting from minor trauma in the affected region must be excluded. *Muscular strain* and *ligamentous sprain* are usually preceded by a single traumatic episode and are accompanied by pain and tenderness limited to the structure involved. At the wrist, fractures with little displacement are common, and when the history is at all suggestive, fracture must be ruled out by roentgenographic examination, including oblique projections to visualize the scaphoid bone.

Treatment. Acute traumatic tenosynovitis often responds quickly to complete immobilization. In the wrist and forearm an elastic bandage, adhesive strapping, or a leather cuff may be used, but in most cases it is preferable to begin the treatment with the more effective support of well-fitting anterior and posterior plaster splints. In the ankle and lower leg, adhesive strapping or plaster immobilization is usually indicated. The application of heat is often a helpful adjunct. Immobilization should be continued until all tenderness has subsided, when carefully graded activity is to be resumed. Hydrocortisone injected into the tendon sheath space is sometimes effective in relieving the pain.

Stenosing tenosynovitis (de Quervain's tenosynovitis)

First described by de Quervain in 1895, stenosing tenosynovitis commonly involves the abductor pollicis longus and extensor pollicis brevis tendons, producing a stenosis of their sheaths in their common osteofibrous canal. The condition is most common in middle-aged and elderly women. The patient complains of disabling pain that is most severe near the styloid process of the radius. There may be local swelling. Pain is present on moving the wrist and thumb, especially with ulnar deviation and opposition of the thumb to the little finger. The pain may shoot into the thumb or up into the wrist and forearm. The thickened tendon sheath can sometimes be palpated as a hard, tender nodule over the styloid process of the radius. The involvement is frequently bilateral. Splinting of the wrist and thumb by means of a light plaster cast and injection of hydrocortisone into the tendon sheath may relieve the symptoms. Recurrence after such therapy is common. Release of the constriction by longitudinal incision or by partial resection of the sheath is usually curative. The operation should include search for anomalous tendons, which are common in this region.

Snapping finger or thumb (trigger finger)

Occasionally a finger or thumb is seen that, at a single constant angle, shows a partial obstruction of the movement of flexion or of extension. Motion past the angle of obstruction, which often must be accomplished passively, may be accompanied by a definite sensation of snapping. This affection, called snapping finger or trigger finger, is usually of gradual and painless development. Its cause is unknown; however, trauma may be a contributing factor. The pathologic changes consist of a localized stenosis of the flexor tendon sheath, located usually near the metacarpophalangeal joint, and a nodular thickening of the tendon. Usually a small, tender mass is palpable clinically. When this condition is found in young children, it is thought to be congenital in origin. Several members of the same family may be affected. In some instances the finger or thumb remains flexed and cannot be extended.

Treatment. A trial of nonsurgical treatment, consisting of immobilizing the digit by means of a plaster cast for several weeks, may be made. If conservative treatment fails, operation is usually indicated; the skin incision should be transverse. Sometimes simple longitudinal section of the fibrous sheath at the point of constriction will afford complete relief, but resection of a small portion of the sheath is advisable. After operation active motion should be started at once.

Acute suppurative tenosynovitis

Acute suppurative tenosynovitis results from an infection by staphylococci or streptococci commonly introduced into the tendon sheath through a puncture wound. The puncture is usually on the volar surface, the most vulnerable areas being the volar digital creases where the sheaths are unprotected by subcutaneous tissue. Occasionally an infection of the pulp space of the fingertip, or *felon* (possibly secondary to a phalangeal osteomyelitis, or *bone felon*), may extend into the synovial sheath. The rapid accumulation of pus under pressure distends the sheath; if allowed to continue, it ultimately cuts off the blood supply to the tendon, causing necrosis and sloughing. In the middle and ring fingers the infection may break into the midpalmar space and thence extend through the lumbrical canals to the dorsum of the hand. The tendon sheath of the thumb is continuous with the radial bursa at the wrist, and that of the small finger is usually continuous with the ulnar bursa. Therefore involvement of these digits is especially serious, since rapid proximal spread of infection may take place.

Clinical picture. When the infection is limited to a single tendon sheath, the patient presents a painful, swollen, tense finger held in a partially flexed position. There is acute tenderness over the course of the tendon sheath, and any attempt to extend the finger is extremely painful. The patient usually has a leukocytosis and moderate fever. He may experience some relief of pain as the infection breaks out of the sheath to spread to a palmar space or to the forearm. If untreated, an extensive infection may result in an almost useless hand.

Treatment. The treatment must be started early if satisfactory results are to be obtained. The most important measure is immediate and adequate drainage of the involved tendon sheath and other secondarily infected spaces. Antibiotic treatment is important but unless instituted in the earliest phase of the disease cannot alone be expected to abort or control the infection. The meager blood supply of the tendon, which may be completely compromised by the infection, limits the effectiveness of antibiotics. Careful splinting in functional position is important. If the hand infection progresses to the point of necrosis and sloughing of the tendon, the prognosis for future function is unfavorable.

Tuberculous tenosynovitis

Tuberculous tenosynovitis is encountered as a diffuse granulomatous, and often purulent, involvement of the tendon sheath or as a cystic expansion of the sheath containing particles of fibrin known as *rice bodies;* its treatment varies with the individual case but in general consists of excision of the pathologic tissue and temporary immobilization of the wrist. The postoperative result is usually satisfactory. Streptomycin and isoniazid are important adjuvants to surgery, and in some cases excellent results have been reported from their use alone. The streptomycin should be accompanied by para-aminosalicylic acid to decrease the likelihood that the organisms will become streptomycin resistant.

Acute calcific tendinitis (peritendinitis calcarea)

Calcium deposits may occur in or about the tendons of the wrist and hand and present a problem similar to that seen at the shoulder. The commonest site is in the flexor carpi ulnaris tendon near the pisiform bone. The onset of symptoms is usually sudden and the pain intense. Occasionally there is a history of acute or chronic trauma. Swelling and localized tenderness, as well as pain on motion of the involved tendon, are usual findings. Roentgenograms show a minute deposit of calcium in the affected area. The condition is usually self-limited; the duration of the symptoms may be shortened by injecting a small amount of local anesthetic or hydrocortisone into the tender area.

Ganglion

The wrist is the commonest site of the ganglion of joint or tendon sheath, a frequently seen but little understood lesion of benign character.

Pathology and etiology. The typical ganglion is a small, smooth, cystic structure containing a thick, clear, mucinous fluid. It is usually connected to the capsule of an adjacent joint or to a tendon sheath by a narrow pedicle without a lumen; sometimes no pedicle is demonstrable. Ganglia may reach considerable dimensions and may be multilocular or multiple.

The etiology is obscure. It was formerly thought that these cysts result from herniation of the lining membrane of a joint or tendon sheath. Recent observations have led to the belief that they are produced by colloid degeneration, which occurs locally in the connective tissue and which may be related to ischemic changes. It has also been suggested that the cyst formation is possibly analogous to the developmental process that leads, in tissue of similar origin, to the formation of normal joint cavities.

Clinical picture. Ganglia occur most commonly in patients between the ages of 15 and 35 years. They are most frequent on the radial side of the back of the wrist (Fig. 16-5) but are found also on the volar aspect. Small ganglia near the joints of the fingers are common. About the knee, ganglia may appear in the popliteal space, laterally in front of the tendon of the biceps or anteriorly beneath the patellar ligament. The front of the ankle and the dorsum of the foot are other common sites.

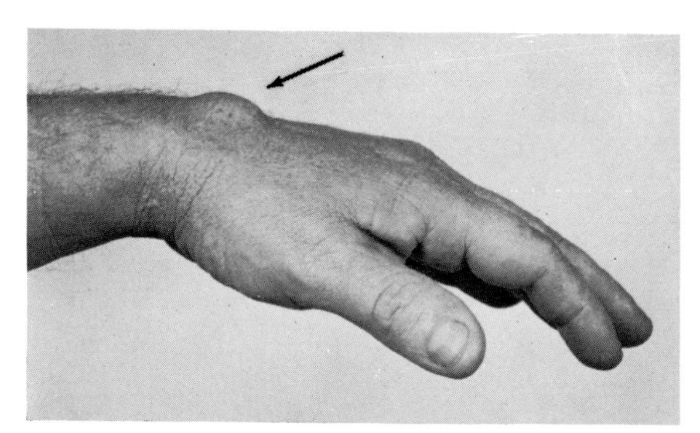

Fig. 16-5. Ganglion of the wrist. Note rounded, well-demarcated swelling on the dorsum of the wrist.

The swelling is of slow growth and causes few symptoms. Occasionally there is discomfort in the adjacent joint or tendons, particularly after overuse or strain. The swelling is tense or fluctuant, rounded, nontender, and not fixed to the skin. It often varies in prominence with motion of the adjacent joint. Ganglia frequently regress or disappear spontaneously. In differential diagnosis ganglia must be distinguished from benign soft tissue tumors such as fibromas and lipomas, from bursitis, and from tuberculous tenosynovitis.

Treatment. The old treatment of rupturing the cyst by trauma or pressure is painful and likely to be followed by recurrence. The combination of aspiration, chemical cauterization, and application of a pressure bandage is sometimes employed. The most successful results, however, are obtained from surgical excision, although recurrences sometimes take place. The operation must be planned with the realization that the ganglion may be found to arise not from a superficial structure but from the articular capsule.

CONTRACTURES OF THE WRIST AND HAND

Contractures are the result of shortening of the soft tissues. The shortening may involve skin, muscle, or ligament and is usually the result of replacement of a part of these tissues by scar. Contractures in the hand limit its mobility and function. The principal causes of hand contractures are trauma, infection, nonspecific inflammatory processes, nerve lesions, ischemia, and congenital changes.

Skin contractures

The commonest causes of skin contractures are scars secondary to lacerations, poorly placed surgical incisions, keloids, and burns. A laceration that crosses perpendicularly a skin crease in the wrist, hand, or finger will result in a scar that gradually hypertrophies, contracts, and pulls the underlying joint into deformity. Surgical incisions should avoid skin folds.

Treatment. The treatment of skin contractures involves revision of the scar by technics such as Z-plasty. When the scar is extensive, its replacement by a free or pedicle skin graft is often necessary.

Metacarpophalangeal and interphalangeal contractures

Metacarpophalangeal joint contractures are frequent complications of injuries to the upper limb and injudicious immobilization. Extension contracture of these joints is the result of shortening of their collateral ligaments. These ligaments are eccentrically attached in such a way that they are relaxed when the joint is extended and are tight in flexion. Thus abduction and adduction are possible in the extended position but quite limited when the metacarpophalangeal joints are flexed. A cast or splint extending beyond the distal palmar crease may immobilize these joints in extension. If the collateral ligaments are allowed to shorten in this position, flexion will no longer be possible. Only in rare instances should the metacarpophalangeal or other finger joints be immobilized in extension, and then only for the shortest possible time. The use of the straight tongue depressor as a splint for injured fingers is to be condemned because it has frequently led to a permanently straight, stiff finger.

Treatment. Since a primary cause of extension contracture of the finger joints is immobilization in the extended position, the treatment is first prophylactic. Careful attention to the details of splinting in functional position, and early, active flexion exercises are important. Established contractures are difficult to eradicate. Active flexion exercises and the use of knuckle-bender elastic splints are usually indicated. Surgical release is occasionally helpful.

Intrinsic contracture of the hand

Intrinsic contracture of the hand is the result of a shortening of the small muscles of the hand, usually secondary to trauma, ischemia, or rheumatoid arthritis. Contracture of the lumbrical and interosseous muscles results in a flexion deformity of the metacarpophalangeal joints and hyperextension of the proximal interphalangeal joints. The distal joints may become flexed because of the pull of the profundus tendons. Involvement of the thumb muscles results in an adduction deformity with flexion of the proximal joint and extension of the distal joint. This combination of deformities is referred to as an *intrinsic-plus hand* (Fig. 5-11). It is the opposite of the clawhand seen after combined median and ulnar nerve injuries, the *intrinsic-minus hand* (Fig. 6-16), in which the metacarpophalangeal joints are hyperextended and the interphalangeal joints flexed. The clinical test for intrinsic contracture is carried out by first flexing fully the metacarpophalangeal joints to relax the tight lumbricales and interossei. This will permit flexion of the interphalangeal joints. Then the metacarpophalangeal joints are extended, which tightens the intrinsic muscles, and if they are contracted, it is impossible to flex the interphalangeal joints.

Treatment. The treatment of intrinsic contracture consists of gradual stretching with elastic splints, supplemented by surgical release of the extensor components of the intrinsic muscle mechanisms.

Dupuytren's contracture

Dupuytren's contracture is a common affection, first analyzed and described by Dupuytren in 1832. It is a slowly progressive contracture of the palmar fascia,

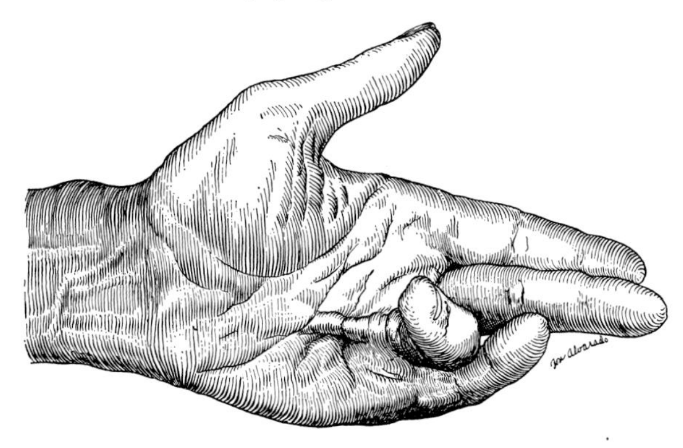

Fig. 16-6. Dupuytren's contracture of the ring finger. Note flexion deformity of the ring finger and prominent band caused by localized contracture of the palmar fascia. (After Colonna.)

occurring most often in men between 55 and 75 years of age. When it occurs in women, it is nearly always in a younger group. As a rule, the changes involve the ring finger, little finger, or both (Fig. 16-6). Dupuytren's contracture is usually bilateral; when unilateral, it is found more frequently in the right hand. It occurs in 2% to 3% of the older population and is uncommon in Negroes.

Pathology. The essential morbid process is a chronic inflammation of the palmar fascia, with progressive fibrosis and contracture, principally of the pretendinous bands of the palmar aponeurosis to the fingers. In advanced cases the skin is secondarily involved. Changes in flexor tendons and digital joints are slight and occur only in late and extreme cases.

Etiology. The cause is unknown. The possible etiologic role of chronic trauma has not been established by analysis of statistics. In some cases there is a hereditary factor.

Clinical picture. The contracture is first evidenced by the appearance of a small nodular, painless thickening in the palmar fascia overlying a flexor tendon in the region of a metacarpophalangeal joint, often associated with a dimpling of the skin at this point. A thickened longitudinal band is gradually formed, and flexion contracture of the finger progressively increases. The metacarpophalangeal and adjacent interphalangeal joints become flexed, whereas the distal interphalangeal joint, controlled by no prolongations of the palmar fascia, retains a normal range of extension. In extreme cases the tip of the finger may be drawn into constant contact with the palm. As a rule, there is little pain or tenderness. Frequently both hands are involved; usually the affection appears much earlier, however, in one hand than in the other. In some cases the plantar fascia shows small nodular thickenings similar to those in the palmar fascia.

Differential diagnosis. Flexion contractures secondary to congenital malforma-

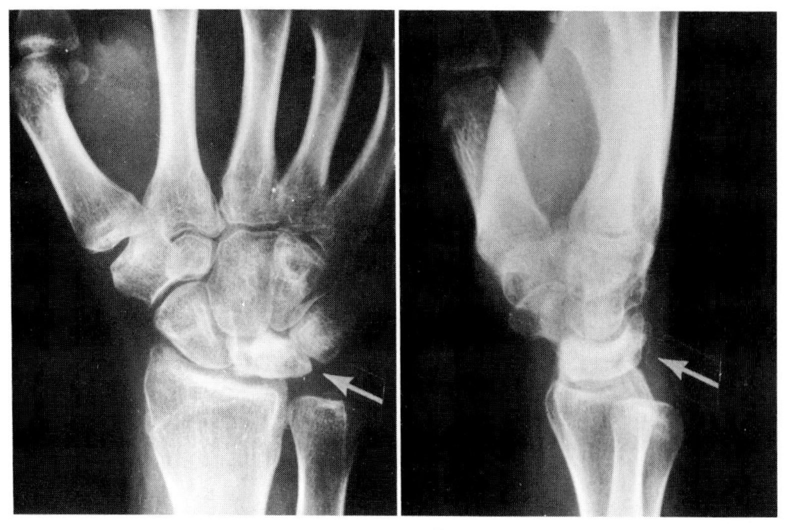

Fig. 16-7. Kienböck's disease. Anteroposterior and lateral roentgenograms of the wrist showing increased density of the lunate bone.

tion, spasticity, trauma, and infection must be excluded. The insidious and painless onset, the presence of flexion at the metacarpophalangeal level, and the inability to extend the proximal interphalangeal joint even when the wrist is flexed are characteristic of Dupuytren's contracture. It should be remembered that in some instances the microscopic appearance of the lesion in its early stage may simulate that of fibrosarcoma.

Treatment. In the earliest cases frequent forced extension may be of some value. For all other cases the only effective treatment is surgical. Frequently the procedure of choice is a careful and thorough excision of the abnormal palmar fascia. Postoperatively a splint is worn for a few days, but active and passive motion must be begun as soon as healing of the skin permits. Excision of the fascia obviates, as a rule, the danger of recurrence. However, it is a major procedure that may be followed by necrosis of palmar skin. The simple operation of multiple subcutaneous divisions of contracted fascial bands with a tenotomy knife is sometimes preferred, particularly for cases in which the contracture is well localized. For satisfactory results all of the constricting bands must be divided, and the postoperative treatment, including the use of an extension splint at night in addition to exercises, must be prolonged in an effort to prevent recurrence of the contracture.

AFFECTIONS OF THE BONES OF THE WRIST AND HAND

Avascular necrosis of the carpal bones

Avascular necrosis of one or more of the bones of the carpus, a rare and little understood affection, is characterized by the gradual development of pain and disability in the wrist, associated with typical roentgenographic changes. A history of antecedent injury is often obtainable, and the condition is sometimes termed *traumatic osteitis* or *osteoporosis*. The etiologic mechanism is usually considered to be a local nutritive disturbance dependent upon circulatory changes induced by trauma. The pathologic changes are those of avascular necrosis. Involvement of the lunate bone is often known as *Kienböck's disease* (Fig. 16-7); involvement of the scaphoid bone is often called *Preiser's disease*.

Clinical picture. The majority of patients are men between the ages of 20 and 40 years. The initial trauma may be trivial or severe and may cause pain and swelling that persist for several days or weeks. A period of months or even years then elapses in which no symptoms are present. Thereafter a gradual development of aching pain takes place, at first felt only on jarring or on exertion but later persistent and severe. Local swelling and tenderness appear, dorsiflexion becomes limited, and the disability may become extreme. When the lunate bone is severely affected, the head of the third metacarpal may lose its normal dorsal prominence because of a proximal displacement of the entire bone, and longitudinal thrusts upon the metacarpal in a proximal direction will cause pain. The range of wrist motion often becomes restricted.

Roentgenographic picture. In early films there may be no definite changes, but as the condition advances characteristic alterations of the affected bone appear. It becomes flattened and may be abnormally dense. In some cases irregular areas of rarefaction are conspicuous, suggesting fragmentation. In adjacent bones the changes of an early traumatic arthritis often develop.

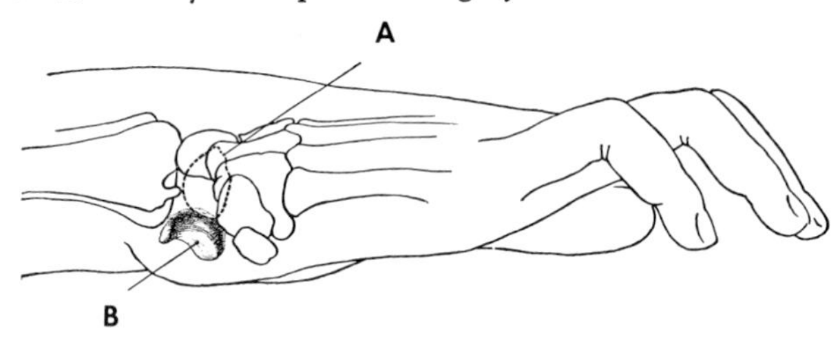

Fig. 16-8. Anterior dislocation of the lunate bone. **A,** Normal position of the lunate bone. **B,** Position after dislocation. (Drawing from roentgenogram.)

Treatment. Nonsurgical treatment is usually indicated. The wrist is immobilized in a short dorsiflexion splint, which may be removed daily for gentle underwater exercises. If after several months the improvement is found to be unsatisfactory, excision of the affected bone or bones may be advisable; postoperatively the wrist should be immobilized for 1 month, after which exercise is gradually resumed. In late cases with pain and advanced arthritic changes, arthrodesis of the wrist is sometimes indicated.

Old dislocation of the lunate bone

Two lesions of the carpus are frequently seen after episodes of moderate or severe trauma: (1) fracture of the scaphoid bone and (2) anterior dislocation of the lunate bone. Cases of neglected scaphoid injuries are discussed on p. 269. Unreduced dislocation of the lunate bone is an important entity since it results in severe disability of the hand.

Pathology. When, by a fall upon the hyperextended hand, the lunate bone has been squeezed from its position between the radius and the capitate, it lies anteriorly, with its concave distal surface rotated forward (Fig. 16-8). In this position the lunate bone is located immediately beneath the structures crossing the volar aspect of the wrist and is likely to cause involvement of the flexor tendons by friction and of the median nerve by pressure.

Clinical picture. Enlargement of the anterior aspect of the wrist is usually obvious, and in some cases the displaced bone is palpable. There may be noticeable atrophy of the muscles of the hand. The wrist and fingers are held slightly flexed, extension is limited and painful, and the grip is greatly weakened. In a large proportion of the cases the characteristic anesthesia and paralysis that result from involvement of the median nerve will be found.

Treatment. In late cases the displaced lunate bone should be excised, and after its removal a plaster splint or cast should be applied for immobilization. Physical therapy is begun after a week and should usually be continued for an extended period.

Accessory bones of the wrist and hand

According to the findings of comparative anatomy and embryology, a large variety of accessory or supernumerary bones may be present in the carpal region. Clinically, however, accessory bones of the hand are much less frequent and less important than those of the foot.

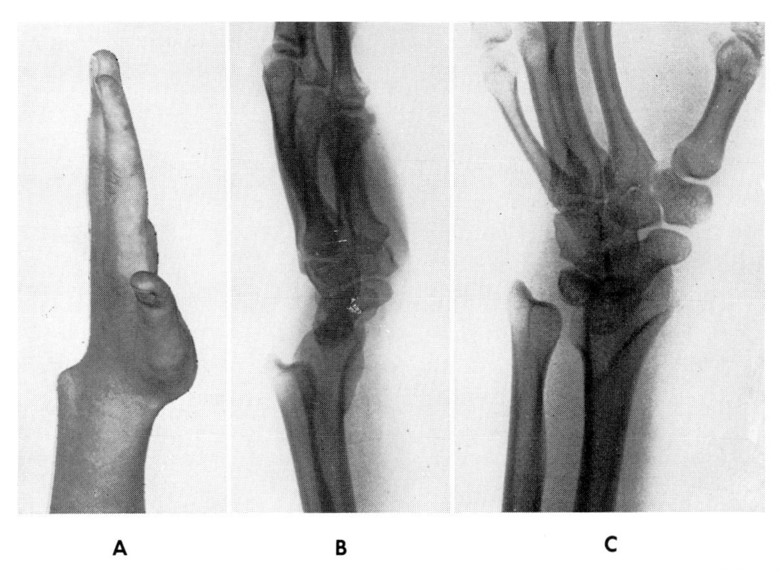

Fig. 16-9. Madelung's deformity. **A,** In this photograph note dorsal prominence of the distal end of the ulna. **B** and **C,** In these roentgenograms note distortion of the distal end of the radius and changes in the radiocarpal relationships.

Of chief diagnostic significance is the rare *divided scaphoid,* resulting from failure of radial and ulnar anlagen of the scaphoid bone to fuse. Roentgenographically this condition resembles an ununited fracture, but most authorities agree that it is of purely developmental origin. It is possible that imperfect fusion is a predisposing factor in traumatic fractures of the scaphoid bone. The hamulus of the hamate bone may develop separately, and the ulnar styloid process may fail to unite with the distal end of the ulna. In differential diagnosis it must be remembered, however, that these developmental variations are extremely uncommon.

Developmental fusion of one or more adjacent carpal bones sometimes occurs. An accessory epiphysis is rarely seen at the proximal end of the second or fifth metacarpal bone.

Madelung's deformity

Madelung's deformity is an uncommon affection of the wrist, characterized by dorsal prominence of the lower end of the ulna, instability of the distal radioulnar articulation, and local changes in the conformation of the radius and the ulna (Fig. 16-9).

Etiology. The cause has not been definitely established but is apparently a local nutritional or growth disturbance resulting from changes secondary to congenital abnormality. Growth at the ulnar side of the distal radial epiphyseal plate is retarded. In some instances there is a hereditary factor.

Clinical picture. The condition is usually seen in adolescents and is not infrequently bilateral. The patient complains of deformity of the wrist and of a feeling of weakness and insecurity. The wrist may appear enlarged, and on palpation there may be demonstrable instability of the radioulnar joint. The end of the radius is displaced anteriorly, carrying with it the carpus and the hand. There are a resulting dorsal prominence of the lower end of the ulna and an abnormal limitation of wrist dorsiflexion. In severe cases the range of supination and pronation may be decreased.

Treatment. In all except the most advanced cases a trial of nonsurgical treatment is indicated. The wrist is immobilized by a dressing that incorporates a short dorsiflexion splint and a pressure pad over the prominent distal end of the ulna. Graded exercises may be given to strengthen the local musculature. Excision of the distal end of the ulna is sometimes helpful. In advanced cases it may be advisable to reinforce surgically the ligaments at the distal radioulnar joint. In severe late lesions osteotomy of the radius or the ulna may be necessary in order to correct the deformity.

Bibliography

General textbooks

Adams, J. C. Outline of orthopedics, ed. 6. Baltimore, 1967, The Williams & Wilkins Co.

Aegerter, E. E., and Kirkpatrick, J. A., Jr. Orthopedic diseases: physiology, pathology, radiology, ed. 3. Philadelphia, 1968, W. B. Saunders Co.

American Academy of Orthopaedic Surgeons. Instructional course lectures, vols. 1-15. Ann Arbor, 1943-1958, J. W. Edwards

Instructional course lectures, vols. 16-19. St. Louis, 1959-1961, 1970, The C. V. Mosby Co.

Selective bibliography of orthopaedic surgery, including a basic science supplement on the musculoskeletal system, ed. 2. St. Louis, 1970, The C. V. Mosby Co.

American College of Surgeons Committee on Pre- and Postoperative Care. Manual of preoperative and postoperative care. Philadelphia, 1967, W. B. Saunders Co.

Apley, A. G. A system of orthopaedics and fractures, ed. 3. New York, 1968, Appleton-Century-Crofts

Banks, S. W., and Laufman, H. An atlas of surgical exposures of the extremities. Philadelphia, 1953, W. B. Saunders Co.

Bechtol, C. O., Ferguson, A. B., Jr., and Laing, P. G. Metals and engineering in bone and joint surgery. Baltimore, 1959, The Williams & Wilkins Co.

Bleck, E. E., Duckworth, N., and Hunter, N. Atlas of plaster cast techniques. Chicago, 1956, Year Book Medical Publishers, Inc.

Caffey, J. Pediatric x-ray diagnosis, ed. 5. Chicago, 1967, Year Book Medical Publishers, Inc.

Collins, D. H. Pathology of bone. Washington, D. C., 1966, Butterworth & Co.

Colonna, P. C. Principles of orthopaedic surgery. Boston, 1960, Little, Brown & Co.

Cozen, L. An atlas of orthopedic surgery. Philadelphia, 1966, Lea & Febiger

Crenshaw, A. H., editor. Campbell's operative orthopaedics, ed. 5. St. Louis, 1971, The C. V. Mosby Co.

Curran, W. J. Tracy's the doctor as a witness, ed. 2. Philadelphia, 1965, W. B. Saunders Co.

General textbooks—cont'd

English, M.	Plaster of Paris technique	Edinburgh, 1957, E. & S. Livingstone, Ltd.
Ferguson, A. B., Jr.	Orthopedic surgery in infancy and childhood, ed. 3	Baltimore, 1968, The Williams & Wilkins Co.
Frankel, V. H., and Burstein, A. H.	Orthopaedic biomechanics	Philadelphia, 1970, Lea & Febiger
Gartland, J. J.	Fundamentals of orthopaedics	Philadelphia, 1965, W. B. Saunders Co.
Greenfield, G. B.	Radiology of bone diseases	Philadelphia, 1969, J. B. Lippincott Co.
Greulich, W. W., and Pyle, S. I.	Radiographic atlas of skeletal development of the hand and wrist, ed. 2	Stanford, Calif., 1959, Stanford University Press
Henry, A. K.	Extensile exposure, ed. 2	Baltimore, 1957, The Williams & Wilkins Co.
Howorth, M. B.	A textbook of orthopaedics	Philadelphia, 1952, W. B. Saunders Co.
Kaplan, E. B.	Surgical approaches to the neck, cervical spine and upper extremity	Philadelphia, 1966, W. B. Saunders Co.
Köhler, A. (translated by J. T. Case)	Borderlands of the normal and early pathologic in skeletal roentgenology, ed. 11	New York, 1966, Grune & Stratton, Inc.
Krusen, F. H., Kottke, F. J., and Ellwood, P. M., Jr.	Handbook of physical medicine and rehabilitation	Philadelphia, 1965, W. B. Saunders Co.
Licht, S., editor	Therapeutic exercise, ed. 2 (revised)	New Haven, 1965, Elizabeth Licht, Publisher
Lichtenstein, L.	Diseases of bone and joints	St. Louis, 1970, The C. V. Mosby Co.
Luck, J. V.	Bone and joint diseases: pathology correlated with roentgenological and clinical features	Springfield, Ill., 1950, Charles C Thomas, Publisher
MacAusland, W. R., Jr., and Mayo, R. A.	Textbook of orthopedics	Boston, 1965, Little, Brown & Co.
McBride, E. D.	Disability evaluation: principles of treatment of compensable injuries, ed. 6 (revised)	Philadelphia, 1963, J. B. Lippincott Co.
Mercer, Sir W., and Duthie, R. B.	Orthopaedic surgery, ed. 6	Baltimore, 1964, The Williams & Wilkins Co.
Moore, F. D.	Metabolic care of the surgical patient	Philadelphia, 1959, W. B. Saunders Co.
Moseley, H. F.	An atlas of musculoskeletal exposures	Philadelphia, 1955, J. B. Lippincott Co.
Nicola, T.	Atlas of orthopaedic exposures	Baltimore, 1966, The Williams & Wilkins Co.
O'Donoghue, D. H.	Treatment of injuries to athletes, ed. 2	Philadelphia, 1970, W. B. Saunders Co.
Rang, M. C.	Anthology of orthopaedics	Baltimore, 1966, The Williams & Wilkins Co.
Rosse, C., and Clawson, D. K.	Introduction to the musculoskeletal system	New York, 1971, Harper & Row, Publishers

General textbooks—cont'd

Salter, R. B.	Textbook of disorders and injuries of the musculoskeletal system	Baltimore, 1970, The Williams & Wilkins Co.
Schmeisser, G., Jr.	A clinical manual of orthopedic traction techniques	Philadelphia, 1963, W. B. Saunders Co.
Scuderi, C. S.	Atlas of orthopedic traction procedures	St. Louis, 1954, The C. V. Mosby Co.
Steindler, A.	Kinesiology of the human body	Springfield, Ill., 1955, Charles C Thomas, Publisher
Turek, S. L.	Orthopaedics: principles and their application, ed. 2	Philadelphia, 1967, J. B. Lippincott Co.
Wiles, P., and Sweetnam, R.	Essentials of orthopaedics, ed. 4	Boston, 1965, Little, Brown & Co.
Willard, H. S., and Spackman, C. S.	Occupational therapy, ed. 3	Philadelphia, 1963, J. B. Lippincott Co.
Young, H. H., editor	The year book of orthopedics and traumatic surgery	Chicago, 1971, Year Book Medical Publishers, Inc.

Chapter 1
Introduction to orthopaedics

American Academy of Orthopaedic Surgeons	Joint motion: method of measuring and recording	Chicago, 1965, American Academy of Orthopaedic Surgeons
Andry, N.	Orthopaedia: or the art of correcting and preventing deformities in children	Paris, 1741; London, 1743; Philadelphia, 1961, reproduced by J. B. Lippincott Co.
Bassett, C. A. L.	Current concepts of bone formation	American Academy of Orthopaedic Surgeons Instructional Courses, J. Bone Joint Surg. **44-A:**1217-1244, 1962
Bick, E. M.	Source book of orthopaedics, ed. 2 (facsimile reprint of 1948 edition)	New York, 1968, Hafner Publishing Co., Inc.
Bourne, G. H.	The biochemistry and physiology of bone	New York, 1956, Academic Press, Inc.
Brower, T. D., and Hsu, W-Y.	Normal articular cartilage	Clin. Orthop. **64:**9-17, 1969
Cohen, J., and Harris, W. H.	The three-dimensional anatomy of haversian systems	J. Bone Joint Surg. **40-A:**419-434, 1958
Curtiss, P. H., Jr.	Changes produced in the synovial membrane and synovial fluid by disease	American Academy of Orthopaedic Surgeons Instructional Courses, J. Bone Joint Surg. **46-A:**873-888, 1964
Daniels, L., Williams, M., and Worthingham, C.	Muscle testing: techniques of manual examination, ed. 2	Philadelphia, 1956, W. B Saunders Co.
Duchenne, G. B. (translated and edited by E. B. Kaplan)	Physiology of motion	Philadelphia, 1949, J. B. Lippincott Co.
Evans, F. G.	Biomechanical studies of the musculoskeletal system	Springfield, Ill., 1961, Charles C Thomas, Publisher

Chapter 1: Introduction to orthopaedics—cont'd

Fox, T. A., editor	Manual of orthopaedic surgery, ed. 5	Chicago, 1966, American Orthopaedic Association
Frost, H. M.	The laws of bone structure	Springfield, Ill., 1964, Charles C Thomas, Publisher
Furey, J. G., Clark, W. S., and Brine, K. L.	The practical importance of synovial-fluid analysis	J Bone Joint Surg. **41-A:**167-174, 1959.
Gardner, E. D.	The development and growth of bones and joints	American Academy of Orthopaedic Surgeons Instructional Courses, J. Bone Joint Surg. **45-A:**856-862, 1963
Hall, M. C.	The locomotor system: functional histology	Springfield, Ill., 1965, Charles C Thomas, Publisher
Hamerman, D., Rosenberg, L. C., and Schubert, M.	Review article: Diarthrodial joints revisited	J. Bone Joint Surg. **52-A:**725-774, 1970
Howell, D. S.	Review article: Current concepts of calcification	J. Bone Joint Surg. **53-A:**250-258, 1971
Howorth, B.	Examination and diagnosis of the spine and extremities	Springfield, Ill., 1962, Charles C Thomas, Publisher
Keith, Sir A.	Menders of the maimed	London, 1919, Oxford University Press; Philadelphia, 1951, J. B. Lippincott Co.
Kendall, H. O., and Kendall, F. P.	Muscles: testing and function	Baltimore, 1949, The Williams & Wilkins Co.
LaCroix, P. (translated by S. Gilder)	The organization of bones	London, 1951, J. & A. Churchill, Ltd.
Langman, J.	Medical embryology	Baltimore, 1963, The Williams & Wilkins Co.
MacConaill, M. A.	The movements of bones and joints. 1. Fundamental principles with particular reference to rotation movement	J. Bone Joint Surg. **30-B:**322-326, 1948
	The movements of bones and joints. 2. Function of the musculature	J. Bone Joint Surg. **31-B:**100-104, 1949
	The movements of bones and joints. 3. The synovial fluid and its assistants	J. Bone Joint Surg. **32-B:**244-252, 1950
McLean, F. C., and Budy, A. M.	Radiation, isotopes, and bone	New York, 1964, Academic Press, Inc.
McLean, F. C., and Urist, M. R.	Bone: an introduction to the physiology of skeletal tissue, ed. 3	Chicago, 1968, University of Chicago Press
Mankin, H. J.	The articular cartilages: a review	American Academy of Orthopaedic Surgeons Instructional Course Lectures, St. Louis, 1970, The C. V. Mosby Co., vol. 19, pp. 204-224
Neuman, W. F., and Neuman, M. W.	The chemical dynamics of bone mineral	Chicago, 1958, University of Chicago Press
Osgood, R. B.	The evolution of orthopaedic surgery	St. Louis, 1925, The C. V. Mosby Co.
Raney, R. B.	Andry and the orthopaedia	J. Bone Joint Surg. **31-A:**675-682, 1949

Chapter 1: Introduction to orthopaedics—cont'd

Robinson, R. A., and Cameron, D. A.	The organic matrix of bone and epiphyseal cartilage	Clin. Orthop. **9:**16-29, 1957
Rusk, H. A.	Rehabilitation medicine, ed. 3	St. Louis, 1971, The C. V. Mosby Co.
Shands, A. R., Jr.	The early orthopaedic surgeons of America	St. Louis, 1970, The C. V. Mosby Co.
Simon, G.	Principles of bone x-ray diagnosis, ed. 2	Washington, D. C., 1965, Butterworth & Co.
Sognnaes, R. F., editor	Calcification in biological systems	Washington, D. C., 1960., American Association for Advancement of Science
Sognnaes, R. F., editor	Mechanisms of hard tissue destruction	Washington, D. C., 1963, American Association for Advancement of Science
Weinmann, J. P., and Sicher, H.	Bone and bones: fundamentals of bone biology, ed. 2	St. Louis, 1955, The C. V. Mosby Co.
White, R. K.	The rheology of synovial fluid	American Academy of Orthopaedic Surgeons Instructional Courses, J. Bone Joint Surg. **45-A:**1084-1090, 1963

Chapter 2

Congenital deformities

Congenital dysplasia of the hip

Badgley, C. E.	Etiology of congenital dislocation of the hip	J. Bone Joint Surg. **31-A:**341-356, 1949
Bosworth, D. M., Fielding, J. W., Ishizuka, T., and Ege, R.	Hip-shelf operation in adults	J. Bone Joint Surg. **43-A:**93-106, 1961
Caffey, J., Ames, R., Silverman, W. A., Ryder, C. T., and Hough, G.	Contradiction of the congenital dysplasia—predislocation hypothesis of congenital dislocation of the hip	Pediatrics **17:**632-641, 1956
Carter, C. O., and Wilkinson, J. A.	Genetic and environmental factors in the etiology of congenital dislocation of the hip	Clin. Orthop. **33:**119-128, 1964
Coleman, S. S.	Congenital dysplasia of the hip in the Navajo infant	Clin. Orthop. **56:**179-193, 1968
Colonna, P. C.	Capsular arthroplasty for congenital dislocation of the hip: indications and technique	J. Bone Joint Surg. **47-A:**437-449, 1965
Dunlap, K., Shands, A. R., Jr., Hollister, L. C., Jr., Gaul, J. S., Jr., and Streit, H. A.	A new method for determination of torsion of the femur	J. Bone Joint Surg. **35-A:**289-311, 1953
Hart, V. L.	Congenital dysplasia of the hip joint and sequelae	Springfield, Ill., 1952, Charles C Thomas, Publisher
Hass, J.	Congenital dislocation of the hip	Springfield, Ill., 1951, Charles C Thomas, Publisher
Heubelin, G. W., Greene, G. S., and Conforti, V. P.	Hip joint arthrography	Amer. J. Roentgen. **68:**736-748, 1952

Chapter 2: Congenital deformities—cont'd
Congenital dysplasia of the hip—cont'd

Heyman, C. H.	Long-term results following a bone-shelf operation for congenital and some other dislocations of the hip in children	J. Bone Joint Surg. **45-A:**1113-1146, 1963
Hirsch, C., and Scheller, S.	Result of treatment from birth of unstable hips	Clin. Orthop. **62:**162-166, 1969
Howorth, B.	The etiology of congenital dislocation of the hip	Clin. Orthop. **29:**164-179, 1963
Langenskiöld, A., and Laurent, L. E.	Development of the concepts of pathogenesis and treatment of congenital dislocation of the hip	Clin. Orthop. **44:**41-49, 1966
Larson, R. L., Neumann, R. F., and Meredith, D. C.	Congenital subluxation and dislocation of the hip	J.A.M.A. **178:**14-18, 1961
MacKenzie, I. G., Seddon, H. J., and Trevor, D.	Congenital dislocation of the hip	J. Bone Joint Surg. **42-B:**689-705, 1960
Massie, W. K.	Vascular epiphyseal changes in congenital dislocation of the hip	J. Bone Joint Surg. **33-A:**284-306, 1951
McCarroll, H. R.	Congenital dislocation of the hip after the age of infancy	American Academy of Orthopaedic Surgeons Instructional Course Lectures, Ann Arbor, 1955, J. W. Edwards, vol. 12, pp. 69-89
McCarroll, H. R., Coleman, S. S., Schottstaedt, E. R., and Petrie, J. G.	Symposium: Congenital subluxation and dislocation of the hip in infancy	American Academy of Orthopaedic Surgeons Instructional Courses, J. Bone Joint Surg. **47-A:**589-618, 1965
McFarland, B.	Some observations on congenital dislocation of the hip	J. Bone Joint Surg. **38-B:**54-69, 1956
Mëdbo, I.	Follow-up study of hip joint dysplasia treated from the newborn stage	Acta Orthop. Scand. **35:**338-347, 1965
Palmén, K.	Preluxation of the hip joint	Acta Paediat. **50,** supp. 129: 1-71, 1961
Pemberton, P. A.	Pericapsular osteotomy of the ilium for treatment of congenital subluxation and dislocation of the hip	J. Bone Joint Surg. **47-A:**65-86, 1965
Platou, E.	Luxatio coxae congenita, a follow-up study of 406 cases of closed reduction	J. Bone Joint Surg. **35-A:**843-866, 1953
Platt, Sir H.	Congenital dislocation of the hip, its early recognition and treatment.	Brit. J. Surg. **45:**438-442, 1958
Ponseti, I. V.	Non-surgical treatment of congenital dislocation of the hip	American Academy of Orthopaedic Surgeons Instructional Courses, J. Bone Joint Surg. **48-A:**1392-1403, 1966

Chapter 2: Congenital deformities—cont'd

Congenital dysplasia of the hip—cont'd

Ponseti, I. V., and Frigerio, E. R.	Results of treatment of congenital dislocation of the hip	J. Bone Joint Surg. **41-A:**823-846, 1959
Putti, V.	Early treatment of congenital dislocation of the hip	J. Bone Joint Surg. **15:**16-21, 1933
Ring, P. A.	The treatment of unreduced congenital dislocation of the hip in adults	J. Bone Joint Surg. **41-B:**299-313, 1959
Salter, R. B.	Innominate osteotomy in the treatment of congenital dislocation and subluxation of the hip	J. Bone Joint Surg. **43-B:**518-539, 1961
	Role of innominate osteotomy in the treatment of congenital dislocation and subluxation of the hip in the older child	American Academy of Orthopaedic Surgeons Instructional Courses, J. Bone Joint Surg. **48-A:**1413-1439, 1966
	Etiology, pathogenesis and possible prevention of congenital dislocation of the hip	Canad. Med. Ass. J. **98:**933-945, 1968
Salter, R. B., Kostuik, J., and Dallas, S.	Avascular necrosis of the femoral head as a complication of treatment for congenital dislocation of the hip in young children: a clinical and experimental investigation	Canad. J. Surg. **12:**44-61, 1969
Severin, E.	Congenital dislocation of the hip, development of the joint after closed reduction	J. Bone Joint Surg. **32-A:**507-518, 1950
Shands, A. R., Jr., and Steele, M. K.	Torsion of the femur	J. Bone Joint Surg. **40-A:**803-816, 1958
Somerville, E. W., and Scott, J. C.	The direct approach to congenital dislocation of the hip	J. Bone Joint Surg. **39-B:**623-640, 1957
Stanisavljevic, S., and Mitchell, C. L.	Congenital dysplasia, subluxation, and dislocation of the hip in stillborn and newborn infants	J. Bone Joint Surg. **45-A:**1147-1158, 1963
von Rosen, S.	Diagnosis and treatment of congenital dislocation of the hip joint in the new-born	J. Bone Joint Surg. **44-B:**284-291, 1962

Congenital dislocation of other joints

Almquist, E. E., Gordon, L. H., and Blue, A. I.	Congenital dislocation of the head of the radius	J. Bone Joint Surg. **51-A:**1118-1127, 1969
Cozen, L.	Congenital dislocation of the shoulder and other anomalies	Arch. Surg. **35:**956-966, 1937
Curtis, B. H., and Fisher, R. L.	Congenital hyperextension with anterior subluxation of the knee	J. Bone Joint Surg. **51-A:**255-269, 1969
Niebauer, J. J., and King, D. E.	Congenital dislocation of the knee	J. Bone Joint Surg. **42-A:**207-225, 1960

Chapter 2: Congenital deformities—cont'd

Congenital talipes equinovarus (congenital clubfoot)

Abrams, R. C.	Relapsed club foot, the early results of an evaluation of Dillwyn Evans' operation	J. Bone Joint Surg. **51-A:**270-282, 1969
Bell, J. F., and Grice, D. S.	Treatment of congenital talipes equinovarus with the modified Denis Browne splint	J. Bone Joint Surg. **26:**799-811, 1944
Browne, D.	Talipes equino-varus	Lancet **2:**969-974, 1934
Duthie, R. B., and Townes, P. L.	The genetics of orthopaedic conditions	J. Bone Joint Surg. **49-B:**229-248, 1967
Dwyer, F. C.	The treatment of relapsed club foot by the insertion of a wedge into the calcaneum	J. Bone Joint Surg. **45-B:**67-75, 1963
Evans, D.	Relapsed club foot	J. Bone Joint Surg. **43-B:**722-733, 1961
Garceau, G. J.	Congenital talipes equinovarus	American Academy of Orthopaedic Surgeons Instructional Course Lectures, St. Louis, 1961, The C. V. Mosby Co., vol. 18, pp. 178-183
Garceau, G. J., and Palmer, R. M.	Transfer of the anterior tibial tendon for recurrent club foot, a long-term follow-up	J. Bone Joint Surg. **49-A:**207-231, 332, 1967
Hersh, A.	The role of surgery in the treatment of club feet	American Academy of Orthopaedic Surgeons Instructional Courses, J. Bone Joint Surg. **49-A:**1684-1696, 1967
Irani, R. N., and Sherman, M. S.	The pathological anatomy of club foot	J. Bone Joint Surg. **45-A:**45-52, 1963
Kite, J. H.	Principles involved in the treatment of congenital clubfoot	J. Bone Joint Surg. **21:**595-606, 1939
	The clubfoot	New York, 1964, Grune & Stratton, Inc.
Kuhlmann, R. F., and Bell, J. F.	A clinical evaluation of operative procedures for congenital talipes equinovarus	J. Bone Joint Surg. **39-A:**265-283, 1957
Lovell, W. W., and Hancock, C. I.	Treatment of congenital talipes equinovarus	Clin. Orthop. **70:**79-86, 1970
MacEwen, G. D., Scott, D. J., Jr., and Shands, A. R., Jr.	Follow-up survey of clubfoot	J.A.M.A. **175:**427-430, 1961
McCauley, J. D., Jr.	Treatment of clubfoot	American Academy of Orthopaedic Surgeons Instructional Course Lectures, St. Louis, 1959, The C. V. Mosby Co., vol. 16, pp. 93-99
Palmer, R. M.	The genetics of talipes equinovarus	J. Bone Joint Surg. **46-A:**542-556, 1964
Ponseti, I. V., and Smoley, E. N.	Congenital clubfoot, the results of treatment	J. Bone Joint Surg. **45-A:**261-275, 1963
Settle, G. W.	The anatomy of congenital talipes equinovarus: sixteen dissected specimens	J. Bone Joint Surg. **45-A:**1341-1354, 1963

Chapter 2: Congenital deformities—cont'd

Congenital talipes equinovarus (congenital clubfoot)—cont'd

Swann, M., Lloyd-Roberts, G. C., and Catterall, A.	The anatomy of uncorrected club feet	J. Bone Joint Surg. **51-B**:263-269, 1969
Turco, V. J.	Surgical correction of resistant clubfoot: 1-stage posteromedial release with internal fixation: a preliminary report	J. Bone Joint Surg. **53-A**:477-497, 1971

Other congenital deformities

Angle, C. R.	Congenital bowing and angulation of long bones	Pediatrics **13**:257-268, 1954
Arnold, W. D.	Congenital absence of the fibula	Clin. Orthop. **14**:20-29, 1959.
Badgley, C. E., O'Connor, S. J., and Kudner, D. F.	Congenital kyphoscoliotic tibia	J. Bone Joint Surg. **34-A**:349-371, 1952
Baker, C. J., and Rudolph, A. J.	Congenital ring constrictions and intrauterine amputations	Amer. J. Dis. Child. **121**:393-400, 1971
Banta, J. V., and Nichols, O.	Sacral agenesis	J. Bone Joint Surg. **51-A**:693-703, 1969
Barenberg, L. H., and Greenberg, B.	Intrauterine amputations and constriction bands	Amer. J. Dis. Child. **64**:87-92, 1942
Barsky, A. J.	Congenital anomalies of the hand	J. Bone Joint Surg. **33-A**: 35-64, 1951
	Cleft hand: classification, incidence, and treatment	J. Bone Joint Surg. **46-A**:1707-1720, 1964
Beals, R. K., and Eckhardt, A. L.	Hereditary onycho-osteodysplasia (nail-patella syndrome)	J. Bone Joint Surg. **51-A**:505-516, 1969
Bevan-Thomas, W. H., and Millar, E. A.	A review of proximal focal femoral deficiencies	J. Bone Joint Surg. **49-A**:1376-1388, 1967
Blackfield, H. M., and Hause, D. P.	Syndactylism	Plast. Reconstr. Surg. **16**:37-46, 1955
Bonola, A.	Surgical treatment of the Klippel-Feil syndrome	J. Bone Joint Surg. **38-B**:440-449, 1956
Böök, J. A., and Hesselvik, L.	Acrocephalosyndactyly (Apert's syndrome)	Acta Paediat. **42**:359-364, 1953
Browne, D.	Congenital deformities of mechanical origin	Arch. Dis. Child. **30**:37-41, 1955
Bryan, R. S., Lipscomb, P. R., and Chatterton, C. C.	Orthopedic aspects of congenital hypertrophy	Amer. J. Surg. **96**:654-659, 1958
Coleman, S. S., Stelling, F. H., III, and Jarrett, J.	Pathomechanics and treatment of congenital vertical talus	Clin. Orthop. **70**:62-72, 1970
Coventry, M. B.	Some skeletal changes in the Ehlers-Danlos syndrome	J. Bone Joint Surg. **43-A**:855-860, 1961
Coventry, M. B., and Johnson, E. W., Jr.	Congenital absence of the fibula	J. Bone Joint Surg. **34-A**:941-956, 1952
Delorme, T. L.	Treatment of congenital absence of the radius by trans-epiphyseal fixation	J. Bone Joint Surg. **51-A**:117-129, 1969
Duraiswami, P. K.	Experimental causation of congenital skeletal defects and its significance in orthopaedic surgery	J. Bone Joint Surg. **34-B**:646-698, 1952

Chapter 2: Congenital deformities—cont'd

Other congenital deformities—cont'd

Fahlstrom, S.	Radio-ulnar synostosis	J. Bone Joint Surg. **14:**395, 403, 1932
Farmer, A. W., and Laurin, C. A.	Congenital absence of the fibula	J. Bone Joint Surg. **42-A:** 1-12, 1960
Fishbein, M., editor	First International Conference on Congenital Malformations	Philadelphia, 1960, J. B. Lippincott Co.
Forland, M.	Cleidocranial dysostosis	Amer. J. Med. **33:**792-799, 1962
Frantz, C. H., and O'Rahilly, R.	Congenital skeletal limb deficiencies	J. Bone Joint Surg. **43-A:**1202-1224, 1961
Fraser, F. C.	Causes of congenital malformations in human beings	J. Chronic Dis. **10:**97-100, 1959
Freund, E.	Congenital defects of femur, fibula and tibia	Arch. Surg. **33:**349-391, 1936
Friedlander, H. L., Westin, G. W., and Wood, W. L., Jr.	Arthrogryposis multiplex congenita, a review of forty-five cases	J. Bone Joint Surg. **50-A:**89-112, 1968
Gibson, D. A., and Urs, N. D. K.	Arthrogryposis multiplex congenita	J. Bone Joint Surg. **52-B:**483-493, 1970
Gordon, G. C.	Congenital deformities	Baltimore, 1961, The Williams & Wilkins Co.
Gray, S. W., Romaine, C. B., and Skandalakis, J. E.	Congenital fusion of the cervical vertebrae	Surg. Gynec. Obstet. **118:**373-385, 1964
Harris, L. E., and Steinberg, A. G.	Abnormalities observed during the first six days of life in 8,716 live-born infants	Pediatrics **14:**314-326, 1954
Heyman, C. H.	The diagnosis and treatment of congenital convex pes valgus or vertical talus	American Academy of Orthopaedic Surgeons Instructional Course Lectures, St. Louis, 1959, The C. V. Mosby Co., vol. 16, pp. 117-126
Heyman, C. H., Herndon, C. H., and Heiple, K. G.	Congenital posterior angulation of the tibia with talipes calcaneus	J. Bone Joint Surg. **41-A:**476-488, 1959
Jeannopoulos, C. L.	Congenital elevation of the scapula	J. Bone Joint Surg. **34-A:**883-892, 1952
Kahn, A. J., Jr., and Fulmer, J.	Acrocephalosyndactylism (Apert's syndrome)	New Eng. J. Med. **252:**379-382, 1955
Keck, C.	Congenital vertical talus	Southern Med. J. **61:**628-633, 1968
Kelikian, H., and Doumanian, A.	Congenital anomalies of the hand. Parts I and II	J. Bone Joint Surg. **39-A:**1002-1019, 1249-1266, 1957
Kite, J. H.	Torsion of the lower extremities in small children	J. Bone Joint Surg. **36-A:**511-520, 1954
	Arthrogryposis multiplex congenita	Southern Med. J. **48:**1141-1146, 1955
	Congenital metatarsus varus	American Academy of Orthopaedic Surgeons Instructional Courses, J. Bone Joint Surg. **49-A:**388-397, 1967
Lloyd-Roberts, G. C., and Lettin, A. W. F.	Arthrogryposis multiplex congenita	J. Bone Joint Surg. **52-B:**494-508, 1970

Chapter 2: Congenital deformities—cont'd

Other congenital deformities—cont'd

Lloyd-Roberts, G. C., and Spence, A. J.	Congenital vertical talus	J. Bone Joint Surg. **40-B**:33-41, 1958
Lyons, C. G., and Sawyer, J. G.	Cleidocranial dysostosis	Amer. J. Roentgen. **51**:215-219, 1944
MacCollum, D. W.	Webbed fingers	Surg. Gynec. Obstet. **71**:782-789, 1940
Madsen, E. T.	Congenital angulations and fractures of the extremities	Acta Orthop. Scand. **25**:242-280, 1956
McCormick, D. W., and Blount, W. P.	Metatarsus adductovarus, "skewfoot"	J.A.M.A. **141**:449-453, 1949
McIntosh, R., Merritt, K. K., Richards, M. R., Samuels, M. H., and Bellows, M. T.	The incidence of congenital malformations: a study of 5,964 pregnancies	Pediatrics **14**:505-522, 1954
Mead, N. G., Lithgow, W. C., and Sweeney, H. J.	Arthrogryposis multiplex congenita	J. Bone Joint Surg. **40-A**:1285-1309, 1958
Milch, R. A., and McKusick, V. A., editors	Symposium: Genetics of congenital deformity	Clin. Orthop. **33**:9-179, 1964
Miles, P. W.	Cleidocranial dysostosis: a survey of six new cases and 126 from the literature	J. Kansas Med. Soc. **41**:462-468, 1940
Pardini, A. G., Jr.	Radial dysplasia	Clin. Orthop. **57**:153-177, 1968
Peabody, C. W.	Hemihypertrophy and hemiatrophy; congenital total unilateral somatic asymmetry	J. Bone Joint Surg. **18**:466-474, 1936
Perry, J., Bonnett, C. A., and Hoffer, M. M.	Vertebral pelvic fusions in the rehabilitation of patients with sacral agenesis	J. Bone Joint Surg. **52-A**:288-294, 1970
Ponseti, I. V., and Becker, J. R.	Congenital metatarsus adductus: the results of treatment	J. Bone Joint Surg. **48-A**:702-711, 1966
Pygott, F.	Arachnodactyly (Marfan's syndrome) with a report of two cases	Brit. J. Radiol. **28**:26-29, 1955
Ring, P. A.	Congenital short femur	J. Bone Joint Surg. **41-B**:73-79, 1959
Ringrose, R. E., Jabbour, J. T., and Keele, D. K.	Hemihypertrophy	Pediatrics **36**:434-448, 1965
Ross, L. J.	Arachnodactyly	Amer. J. Dis. Child. **78**:417-436, 1949
Russell, H. E., and Aitken, G. T.	Congenital absence of the sacrum and lumbar vertebrae with prosthetic management	J. Bone Joint Surg. **45-A**:501-508, 1963
Skerik, S. K., and Flatt, A. E.	The anatomy of congenital radial dysplasia	Clin. Orthop. **66**:125-143, 1969
Smith, R. J., and Kaplan, E. B.	Camptodactyly and similar atraumatic flexion deformities of the proximal interphalangeal joints of the fingers	J. Bone Joint Surg. **50-A**:1187-1203, 1249, 1968
Stevenson, S. S., Worcester, J., and Rice, R. G.	677 congenitally malformed infants and associated gestational characteristics	Pediatrics **6**:37-50, 208-222, 1950

Chapter 2: Congenital deformities—cont'd
Other congenital deformities—cont'd

Straub, L. R.	Congenital absence of the ulna	Amer. J. Surg. **109:**300-305, 1965
Sulamaa, M., and Ryoppy, S.	Congenital absence of the tibia	Acta Orthop. Scand. **34:**337-348, 1964
Summer, G. K.	The Ehlers-Danlos syndrome	Amer. J. Dis. Child. **91:**419-428, 1956
Thompson, T. C., Straub, L. R., and Arnold, W. D.	Congenital absence of the fibula	J. Bone Joint Surg. **39-A:**1229-1237, 1957
Wallace, H. M., Baumgartner, L., and Rich, H.	Congenital malformations and birth injuries in New York City	Pediatrics **12:**525-535, 1953
Warkany, J.	Production of congenital malformations by dietary measures	J.A.M.A. **168:**2020-2023, 1958
Yamazaki, J. N., Wright, S. W., and Wright, P. M.	Outcome of pregnancy in women exposed to the atomic bomb in Nagasaki	Amer. J. Dis. Child. **87:**448-463, 1954

Chapter 3
General affections of bone

Albright, F., Butler, A. M., Hampton, A. O., and Smith, P.	Syndrome characterized by osteitis fibrosa disseminata, areas of pigmentation and endocrine dysfunction, with precocious puberty in females	New Eng. J. Med. **216:**727-746, 1937
Amstutz, H. C., and Carey, E. J.	Skeletal manifestations and treatment of Gaucher's disease: review of twenty cases	J. Bone Joint Surg. **48-A:**670-701, 1966
Amuso, S. J.	Diastrophic dwarfism	J. Bone Joint Surg. **50-A:**113-122, 1968
Bailey, J. A., II	Orthopaedic aspects of achondroplasia	J. Bone Joint Surg. **52-A:**1285-1301, 1970
Baldursson, H., Evans, E. B., Dodge, W. F., and Jackson, W. T.	Tumoral calcinosis with hypophosphatemia: a report of a family with incidence in four siblings	J. Bone Joint Surg. **51-A:**913-925, 1969
Barnicot, N. A., and Datta, S. P.	Vitamin A and bone	In Bourne, G. H., editor: The biochemistry and physiology of bone, New York, 1956, Academic Press, Inc., pp. 507-538
Barry, H. C.	Paget's disease of bone	Baltimore, 1970, The Williams & Wilkins Co.
	Fractures of the femur in Paget's disease of bone in Australia	J. Bone Joint Surg. **49-A:**1359-1370, 1967
Bassan, J., Frame, B., and Frost, H.	Osteoporosis: a review of pathogenesis and treatment	Ann. Intern. Med. **58:**539-550, 1963
Bickel, W. H., Ghormley, R. K., and Camp, J. D.	Osteogenesis imperfecta	Radiology **40:**145-154, 1943

Chapter 3: General affections of bone—cont'd

Bondy, P. K., editor	Duncan's diseases of metabolism: endocrinology and nutrition, ed. 6	Philadelphia, 1969, W. B. Saunders Co.
	Duncan's diseases of metabolism: genetics and metabolism, ed. 6	Philadelphia, 1969, W. B. Saunders Co.
Bourne, G. H.	Vitamin C and bone	In Bourne, G. H., editor: The biochemistry and physiology of bone, New York, 1956, Academic Press, Inc., pp. 539-580
Burkhart, J. M., Burke, E. C., and Kelly, P. J.	The chondrodystrophies	Mayo Clin. Proc. **40**:481-499, 1965
Caffey, J., and Silverman, W. A.	Infantile cortical hyperostoses	Amer. J. Roentgen. **54**:1-16, 1945
Campbell, C. J., Papademetriou, T., and Bonfiglio, M.	Melorheostosis	J. Bone Joint Surg. **50-A**:1281-1304, 1968
Campbell, C. S.	Melorheostosis of the upper limb	J. Bone Joint Surg. **37-B**:471-473, 1955
Chalmers, J.	Osteomalacia: review of 93 cases	J. Roy. Coll. Surg. Edinb. **13**:255-275, 1968
Chalmers, J., Conacher, W. D. H., Gardner, D. L., and Scott, P. J.	Osteomalacia—a common disease in elderly women	J. Bone Joint Surg. **49-B**:403-423, 1967
Christensen, W. R., Lin, R. K., and Berghout, J.	Dysplasia epiphysialis multiplex	Amer. J. Roentgen. **74**:1059-1067, 1955
Coventry, M. B.	Some skeletal changes in the Ehlers-Danlos syndrome	J. Bone Joint Surg. **43-A**:855-860, 1961
Crocker, A. C., and Farber, S.	Niemann-Pick disease: a review of eighteen patients	Medicine **37**:1-95, 1958
Daves, M. L., and Yardley, J. H.	Fibrous dysplasia of bone	Amer. J. Med. Sci. **234**:590-606, 1957
Dent, C. E., Friedman, M., and Watson, L.	Hereditary pseudo-vitamin D deficiency rickets	J. Bone Joint Surg. **50-B**:708-719, 1968
Dickson, W., and Horrocks, R. H.	Hypophosphatasia	J. Bone Joint Surg. **40-B**:64-74, 1958
Dixon, T. F., Mulligan, L., Nassim, R., and Stevenson, F. H.	Myositis ossificans progressiva	J. Bone Joint Surg. **36-B**:445-449, 1954
Drucker, W. R., Hubay, C. A., Holden, W. D., and Bukovnic, J. A.	Pathogenesis of post-traumatic sympathetic dystrophy	Amer. J. Surg. **97**:454-465, 1959
Elmore, S. M.	Pycnodysostosis: a review	J. Bone Joint Surg. **49-A**:153-162, 1967
Evans, J. A.	Reflex sympathetic dystrophy: report on 57 cases	Ann. Intern. Med. **26**:417-426, 1947
Fairbank, Sir T.	An atlas of general affections of the skeleton	Edinburgh, 1951, E. & S. Livingstone, Ltd.
Fisher, R. H.	Multiple lesions of bone in Letterer-Siwe disease	J. Bone Joint Surg. **35-A**:445-464, 1953
Fourman, P., and Royer, P.	Calcium metabolism and the bone, ed. 2	Oxford, 1968, Blackwell Scientific Publications

Chapter 3: General affections of bone—cont'd

Fowles, J. V., and Bobechko, W. P.	Solitary eosinophilic granuloma in bone	J. Bone Joint Surg. **52-B**:238-243, 1970
Gardner, L. I., editor	Endocrine and genetic diseases of childhood	Philadelphia, 1969, W. B. Saunders Co.
Goidanich, I. F., and Lenzi, L.	Morquio-Ullrich disease: a new mucopolysaccharidosis	J. Bone Joint Surg. **46-A**:734-746, 1964
Goldenberg, R. R.	The skull in Paget's disease	J. Bone Joint Surg. **33-A**:911-922, 1951
Grollman, A.	Clinical endocrinology and its physiologic basis	Philadelphia, 1964, J. B. Lippincott Co.
Hammarsten, J. F., and O'Leary, J.	The features and significance of hypertrophic osteoarthropathy	Arch. Intern. Med. **99**:431-441, 1957
Harris, L. J.	Vitamin D and bone	In Bourne, G. H., editor: The biochemistry and physiology of bone, New York, 1956, Academic Press, Inc., pp. 581-622
Harris, W. H., Dudley, H. R., Jr., and Barry, R. J.	The natural history of fibrous dysplasia	J. Bone Joint Surg. **44-A**:207-233, 1962
Harrison, H. E.	The Fanconi syndrome	J. Chronic Dis. **7**:346-355, 1958
Harrison, H. E., and Harrison, H. C.	Hereditary metabolic bone diseases	Clin. Orthop. **33**:147-163, 1964
Hasenhuttl, K.	Osteopetrosis	J. Bone Joint Surg. **44-A**:359-370, 1962
Hess, W. E., and Street, D. M.	Melorheostosis	J. Bone Joint Surg. **32-A**:422-427, 1950
Hinkel, C. L.	Developmental affections of the skeleton characterized by osteosclerosis	Clin. Orthop. **9**:85-106, 1957
Holling, H. E., and Brodey, R. S.	Pulmonary hypertrophic osteoarthropathy	J.A.M.A. **178**:977-982, 1961
Hsia, D. Y.-Y.	Inborn errors of metabolism. Part 1. Clinical aspects, ed. 2	Chicago, 1966, Year Book Medical Publishers, Inc.
Jackson, W. P. U.	Calcium metabolism and bone disease	London, 1967, Edward Arnold, Ltd.
Karlen, A. G., and Cameron, J. A. P.	Dysplasia epiphysialis punctata	J. Bone Joint Surg. **39-B**:293-301, 1957
Kettelkamp, D. B., Campbell, C. J., and Bonfiglio, M.	Dysplasia epiphysealis hemimelica: a report of 15 cases and a review of the literature	J. Bone Joint Surg. **48-A**:746-766, 1966
Key, J. A.	Brittle bones and blue sclera	Arch. Surg. **13**:523-567, 1926
King, J. D., and Bobechko, W. P.	Osteogenesis imperfecta: an orthopaedic description and surgical review	J. Bone Joint Surg. **53-B**:72-89, 1971
Lennon, E. A., Schechter, M. M., and Hornabrook, R. W.	Engelmann's disease	J. Bone Joint Surg. **43-B**:273-284, 1961
Leroy, J. G., and Crocker, A. C.	Clinical definition of the Hurler-Hunter phenotypes: a review of 50 patients	Amer. J. Dis. Child. **112**:518-530, 1966
Levin, E. J.	Osteogenesis imperfecta in the adult	Amer. J. Roentgen. **91**:973-978, 1964

Chapter 3: General affections of bone—cont'd

Lichtenstein, L.	Histiocytosis X (eosinophilic granuloma of bone, Letterer-Siwe disease, and Schüller-Christian disease)	J. Bone Joint Surg. **46-A:**76-90, 1964
Lieberman, P. H., Jones, C. R., Dargeon, H. W. K., and Begg, C. F.	A reappraisal of eosinophilic granuloma of bone, Hand-Schüller-Christian syndrome, and Letterer-Siwe syndrome	Medicine **48:**375-410, 1969
Lucaya, J.	Histiocytosis X	Amer. J. Dis. Child. **121:**289-295, 1971
McGavran, M. H., and Spady, H. A.	Eosinophilic granuloma of bone	J. Bone Joint Surg. **42-A:**979-992, 1960
McKenna, R. J., Schwinn, C. P., Soong, K. Y., and Higinbotham, N. L.	Osteogenic sarcoma arising in Paget's disease	Cancer **17:**42-66, 1964
McKusick, V. A.	Heritable disorders of connective tissue, ed. 3	St. Louis, 1966, The C. V. Mosby Co.
McKusick, V. A., Kaplan, D., Wise, D., et al.	The genetic mucopolysaccharidoses	Medicine **44:**445-483, 1965
McLean, F. C., and Urist, M. R.	Bone: fundamentals of the physiology of skeletal tissue, ed. 3	Chicago, 1968, University of Chicago Press
Margolis, J.	Ollier's disease	Arch. Intern. Med. **103:**279-284, 1959
Mercer, W., and Duthie, R. B.	Some observations on osteitis deformans of Paget	J. Roy. Coll. Surg. Edinb. **1:** 58-74, 1955
Morris, J. M., Samilson, R. L., and Corley, C. L.	Melorheostosis	J. Bone Joint Surg. **45-A:**1191-1206, 1963
Nicholas, J. A., and Killoran, P.	Fracture of the femur in patients with Paget's disease	J. Bone Joint Surg. **47-A:**450-461, 1965
Nicholas, J. A., and Wilson, P. D.	Diagnosis and treatment of osteoporosis	J.A.M.A. **171:**2279-2284, 1959
Nicholas, J. A., Saville, P. D., and Bronner, F.	Osteoporosis, osteomalacia, and the skeletal system	American Academy of Orthopaedic Surgeons Instructional Courses, J. Bone Joint Surg. **45-A:**391-405, 1963
Nordin, B. E. C.	Osteomalacia, osteoporosis and calcium deficiency	Clin. Orthop. **17:**235-257, 1960
	International patterns of osteoporosis	Clin. Orthop. **45:**17-30, 1966
Paget, Sir J.	On a form of chronic inflammation of bones (osteitis deformans)	Med.-Chir. Trans. **60:**37-64, 1877
Pease, C. N.	Focal retardation and arrestment of growth of bones due to vitamin A intoxication	J.A.M.A. **182:**980-985, 1962
Pierce, D. S., Wallace, W. M., and Herndon, C. H.	Long-term treatment of vitamin-D resistant rickets	J. Bone Joint Surg. **46-A:**978-997, 1964
Plewes, L. W.	Sudeck's atrophy in the hand	J. Bone Joint Surg. **38-B:**195-203, 1956

Chapter 3: General affections of bone—cont'd

Ponseti, I. V.	Bone lesions in eosinophilic granuloma, Hand-Schüller-Christian disease, and Letterer-Siwe disease	J. Bone Joint Surg. **30-A:**811-833, 1948
	Skeletal growth in achondroplasia	J. Bone Joint Surg. **52-A:**701-716, 1970
Pygott, F.	Arachnodactyly (Marfan's syndrome) with a report of two cases	Brit. J. Radiol. **28:**26-29, 1955
Rees, T. D., Wood-Smith, D., and Converse, J. M.	The Ehlers-Danlos syndrome with a report of 3 cases	Plast. Reconstr. Surg. **32:**39-44, 1963
Reed, R. J.	Fibrous dysplasia of bone: a review of 25 cases	Arch. Path. **75:**480-495, 1963
Riley, H. D., Jr., and Christie, A.	Myositis ossificans progressiva	Pediatrics **8:**753-767, 1951
Rosenkrantz, J. A., Wolf, J., and Kaicher, J. J.	Paget's disease (osteitis deformans): review of 111 cases	Arch. Intern. Med. **90:**610-633, 1952
Ross, L. J.	Arachnodactyly	Amer. J. Dis. Child. **78:**417-436, 1949
Rubin, P.	Dynamic classification of bone dysplasias	Chicago, 1964, Year Book Medical Publishers, Inc.
Shorbe, H. B.	Infantile scurvy	Clin. Orthop. **1:**49-55, 1953
Silverman, F. N.	The roentgen manifestations of unrecognized skeletal trauma in infants	Amer. J. Roentgen. **69:**413-427, 1953
Singleton, E. B., Thomas, J. R., Worthington, W. W., and Hild, J. R.	Progressive diaphyseal dysplasia (Engelmann's disease)	Radiology **67:**233-241, 1956
Solomon, L.	Hereditary multiple exostosis	J. Bone Joint Surg. **45-B:**292-304, 1963
Staheli, L. T., Church, C. C., and Ward, B. H.	Infantile cortical hyperostosis (Caffey's disease): 16 cases with a late follow-up of 8	J.A.M.A. **203:**384-388, 1968
Stanbury, J. B., Wyngaarden, J. B., and Fredrickson, D. S., editors	The metabolic basis of inherited disease, ed. 2	New York, 1966, Blakiston Division, McGraw-Hill Book Co.
Stewart, M. J., Gilmer, W. S., Jr., and Edmonson, A. S.	Fibrous dysplasia of bone	J. Bone Joint Surg. **44-B:**302-318, 1962
Tampas, J. P., Van Buskirk, F. W., Peterson, O. S., Jr., and Soule, A. B.	Infantile cortical hyperostosis	J.A.M.A. **175:**491-493, 1961
Winters, R. W., Graham, J. B., Williams, T. F., McFalls, V. W., and Burnett, C. H.	A genetic study of familial hypophosphatemia and vitamin D resistant rickets with a review of the literature	Medicine **37:**97-142, 1958

Chapter 4
Infections of bones and joints

Pyogenic osteomyelitis and pyogenic arthritis

Almquist, E. E.	The changing epidemiology of septic arthritis in children	Clin. Orthop. **68:**96-99, 1970

Chapter 4: Infections of bones and joints—cont'd

Pyogenic osteomyelitis and pyogenic arthritis—cont'd

Blanche, D. W.	Osteomyelitis in infants	J. Bone Joint Surg. **34-A:**71-85, 1952
Blockey, N. J., and Watson, J. T.	Acute osteomyelitis in children	J. Bone Joint Surg. **52-B:**77-87, 1970
Brailsford, J. F.	Brodie's abscess and its differential diagnosis	Brit. Med. J. **2:**119-123, 1938
Browne, H. J.	Malignancy in osteomyelitic sinuses	Surg. Gynec. Obstet. **123:**1252-1254, 1966
Buchman, J.	Osteomyelitis	American Academy of Orthopaedic Surgeons Instructional Course Lectures, St. Louis, 1959, The C. V. Mosby Co., vol. 16, pp. 232-245
Bulmer, J. H.	Septic arthritis of the hip in adults	J. Bone Joint Surg. **48-B:**289-298, 1966
Chacha, P. B.	Suppurative arthritis of the hip joint in infancy: a persistent diagnostic problem and possible complication of femoral venipuncture	J. Bone Joint Surg. **53-A:**538-544, 1971
Chartier, Y., Martin, W. J., and Kelly, P. J.	Bacterial arthritis: experiences in the treatment of 77 patients	Ann. Intern. Med. **50:**1462-1474, 1959
Clawson, D. K., and Dunn, A. W.	Management of common bacterial infections of bones and joints	J. Bone Joint Surg. **49-A:**164-182, 1967
Coventry, M. B., and Mitchell, W. C.	Osteitis pubis, observations based on a study of 45 patients	J.A.M.A. **178:**898-905, 1961
Curtiss, P. H., Jr., and Klein, L.	Destruction of articular cartilage in septic arthritis. II. In vivo studies	J. Bone Joint Surg. **47-A:**1595-1604, 1965
David, J. R., and Black, R. L.	Salmonella arthritis	Medicine **39:**385-403, 1960
Dennison, W. M.	Unilateral limb lengthening associated with haematogenous osteitis in childhood	Arch. Dis. Child. **27:**54-59, 1952
	Haematogenous osteitis in the newborn	Lancet **2:**474-476, 1955
Diggs, L. W.	Bone and joint lesions in sickle-cell disease	Clin. Orthop. **52:**119-143, 1967
Downey, J. W., and Simon, H. E.	Brodie's abscess, two case reports	Amer. J. Surg. **70:**86-94, 1945
Eyre-Brook, A. L.	Septic arthritis of the hip and osteomyelitis of the upper end of the femur in infants	J. Bone Joint Surg. **42-B:**11-20, 1960
Fox, M. J., and Gilbert, J.	Meningococcus infections with articular complications	Amer. J. Med. Sci. **208:**63-69, 1944
Garcia, A., Jr., and Grantham, S. A.	Hematogenous pyogenic vertebral osteomyelitis	J. Bone Joint Surg. **42-A:**429-436, 1960
Garrod, L. P., and Scowen, E. F.	The principles of therapeutic use of antibiotics	Brit. Med. Bull. **16:**23-28, 1960
Gilmour, W. N.	Acute haematogenous osteomyelitis	J. Bone Joint Surg. **44-B:**841-853, 1962
Green, M., Nyhan, W. L., Jr., and Fousek, M. D.	Acute hematogenous osteomyelitis	Pediatrics **17:**368-382, 1956

Chapter 4: Infections of bones and joints—cont'd

Pyogenic osteomyelitis and pyogenic arthritis—cont'd

Hagemann, P. O., Hendin, A., Lurie, H. H., and Stein, M.	Therapy of gonococcal arthritis	Ann. Intern. Med. **36**:77-89, 1952
Hall, J. E., and Silverstein, E. A.	Acute hematogenous osteomyelitis	Pediatrics **31**:1033-1038, 1963
Harris, N. H., and Kirkaldy-Willis, W. H.	Primary subacute pyogenic osteomyelitis	J. Bone Joint Surg. **47-B**:526-532, 1965
Henson, S. W., and Coventry, M. B.	Osteomyelitis of the vertebrae as the result of infection of the urinary tract	Surg. Gynec. Obstet. **102**:207-214, 1956
Hess, E. V., Hunter, D. K., and Ziff, M.	Gonococcal antibodies in acute arthritis	J.A.M.A. **191**:531-534, 1965
Hook, E. W., Campbell, C. G., Weens, H. S., and Cooper, G. R.	Salmonella osteomyelitis in patients with sickle-cell anemia	New Eng. J. Med. **257**:403-407, 1957
Horwitz, T., and Lambert, R. G.	Chronic osteomyelitis complicating war compound fractures	Surg. Gynec. Obstet. **82**:573-578, 1946
Kagan, B. M.	Antimicrobial therapy	Philadelphia, 1970, W. B. Saunders Co.
Kelly, P. J., Martin, W. J., and Coventry, M. B.	Bacterial (suppurative) arthritis in the adult	J. Bone Joint Surg. **52-A**:1595-1602, 1970
	Chronic osteomyelitis. II. Treatment with closed irrigation and suction	J.A.M.A. **213**:1843-1848, 1970
Kelly, P. J., Martin, W. J., Schirger, A., and Weed, L. A.	Brucellosis of the bones and joints	J.A.M.A. **174**:347-353, 1960
Kernwein, G. A., and Capps, R. B.	Typhoid osteomyelitis, case report	Amer. J. Surg. **60**:433-437, 1943
Knight, M. P., and Wood, G. O.	Surgical obliteration of bone cavities following traumatic osteomyelitis	J. Bone Joint Surg. **27**:547-556, 1945
Kulowski, J.	Management of hematogenous pyogenic osteomyelitis	Surgery **40**:1094-1104, 1956
Leventen, E. O.	Closed irrigation-suction of spine infections	J.A.M.A. **196**:761-764, 1966
Meyer, T. L., Kieger, A. B., and Smith, W. S.	Antibiotic management of staphylococcal osteomyelitis, with particular reference to antibiotic-resistant infections	J. Bone Joint Surg. **47-A**:285-292, 1965
Moe, J. H.	Prevention and treatment of infections in bone	J. Lancet **79**:2-7, 1959
Neligan, G. A., and Elderkin, F. M.	Treatment of acute haematogenous osteitis in children assessed in a consecutive series of selected cases	Brit. Med. J. **1**:1349-1352, 1965
Obletz, B. E.	Acute suppurative arthritis of the hip in the neonatal period	J. Bone Joint Surg. **42-A**:23-30, 1960
Orr, H. W.	Osteomyelitis and compound fractures and other infected wounds	St. Louis, 1929, The C. V. Mosby Co.

Chapter 4: Infections of bones and joints—cont'd

Pyogenic osteomyelitis and pyogenic arthritis—cont'd

Ortiz, A. C., and Miller, W. E.	Treatment of a septic joint	Southern Med. J. **54**:594-599, 1961
Reynolds, F. C., and Zaepfel, F.	Management of chronic osteomyelitis secondary to compound fractures	J. Bone Joint Surg. **30-A**:331-338, 1948
Rowling, D. E.	The positive approach to chronic osteomyelitis	J. Bone Joint Surg. **41-B**:681-688, 1959
Sedlin, E. D., and Fleming, J. L.	Epidermoid carcinoma arising in chronic osteomyelitic foci	J. Bone Joint Surg. **45-A**:827-838, 1963
Silver, H. K., Simon, J. L., and Clement, D. H.	Salmonella osteomyelitis and abnormal hemoglobin disease	Pediatrics **20**:439-447, 1957
Spitzer, N., and Steinbrocker, O.	The treatment of gonorrheal arthritis with penicillin	Amer. J. Med. Sci. **218**:138-144, 1949
Trueta, J.	The three types of acute haematogenous osteomyelitis	J. Bone Joint Surg. **41-B**:671-680, 1959
Trueta, J., and Morgan, J. D.	Late results in the treatment of 100 cases of acute haematogenous osteomyelitis	Brit. J. Surg. **41**:449-457, 1954
Waldvogel, F. A., Medoff, G., and Schwartz, M. N.	Osteomyelitis: a review of clinical features, therapeutic considerations, and unusual aspects	New Eng. J. Med. **282**:198-206, 260-266, 316-322, 1970
Watkins, M. B., Samilson, R. L., and Winters, D. M.	Acute suppurative arthritis	J. Bone Joint Surg. **38-A**:1313-1320, 1956
West, W. F., Kelly, P. J., and Martin, W. J.	Chronic osteomyelitis. I. Factors affecting the results of treatment in 186 patients	J.A.M.A. **213**:1837-1842, 1970
Wiltse, L. L., and Frantz, C. H.	Non-suppurative osteitis pubis in the female	J. Bone Joint Surg. **38-A**:500-516, 1956
Winters, J. L., and Cahen, I.	Acute hematogenous osteomyelitis	J. Bone Joint Surg. **42-A**:691-704, 1960
Zammit, F.	Undulant fever spondylitis	Brit. J. Radiol. **31**:683-690, 1958

Syphilis of bones and joints

Borella, L., Goobar, J. E., and Clark, G. M.	Synovitis of the knee joints in late congenital syphilis, Clutton's joints	J.A.M.A. **180**:190-192, 1962
Ehrenpreis, B.	Syphilis of bone in the adult	New York J. Med. **57**:3486-3488, 1957
Johns, D.	Syphilitic disorders of the spine: report of two cases	J. Bone Joint Surg. **52-B**:724-731, 1970
Morton, J. J.	Syphilis of bone	Urol. Cutan. Rev. **43**:72-78, 1939
Thomason, H. A., and Mayoral, A.	Syphilitic osteomyelitis	J. Bone Joint Surg. **22**:203-206, 1940

Tuberculosis of bones and joints

Ahern, R. T.	Tuberculosis of the femoral neck and greater trochanter	J. Bone Joint Surg. **40-B**:406-419, 1958

Chapter 4: Infections of bones and joints—cont'd

Tuberculosis of bones and joints—cont'd

Allen, A. R., and Stevenson, A. W.	The results of combined drug therapy and early fusion in bone tuberculosis	J. Bone Joint Surg. **39-A:**32-42, 1957
Bosworth, D. M.	Femoro-ischial transplantation	J. Bone Joint Surg. **24:**38-46, 1942
	Treatment of bone and joint tuberculosis in children	J. Bone Joint Surg. **41-A:**1255-1266, 1959
	Treatment of tuberculosis of bone and joint	Bull. N. Y. Acad. Med. **35:**167-177, 1959
Bosworth, D. M., Wright, H. A., and Fielding, J. W.	The treatment of bone and joint tuberculosis	J. Bone Joint Surg. **34-A:**761-771, 1952
Butler, R. W.	Paraplegia in Pott's disease, with special reference to the pathology and etiology	Brit. J. Surg. **22:**738-768, 1935
Carrell, W. B., and Childress, H. M.	Tuberculosis of the large long bones of the extremities	J. Bone Joint Surg. **22:**569-588, 1940
Chan, K. P., and Shin, J. S.	Brittain ischiofemoral arthrodesis for tuberculosis of the hip	J. Bone Joint Surg. **50-A:**1341-1352, 1968
Charnley, J., and Baker, S. L.	Compression arthrodesis of the knee	J. Bone Joint Surg. **34-B:**187-199, 1952
Cleveland, M., and Bosworth, D. M.	The pathology of tuberculosis of the spine	J. Bone Joint Surg. **24:**527-546, 1942
Cleveland, M., Bosworth, D. M., Fielding, J. W., and Smyrnis, P.	Fusion of the spine for tuberculosis in children	J. Bone Joint Surg. **40-A:**91-106, 1958
Cremin, B. J., Fisher, R. M., and Levinsohn, M. W.	Multiple bone tuberculosis in the young	Brit. J. Radiol. **43:**638-645, 1970
Crofton, J.	The chemotherapy of tuberculosis	Brit. Med. Bull. **16:**55-60, 1960
Evans, D.	The diagnosis and treatment of skeletal tuberculosis	Brit. J. Clin. Pract. **12:**811-820, 1958
Friedman, B.	Chemotherapy of tuberculosis of the spine	J. Bone Joint Surg. **48-A:**451-474, 1956
Hartung, E. F.	Tuberculous arthritis	J.A.M.A. **158:**818-821, 1955
Hibbs, R. A., and Risser, J. C.	Treatment of vertebral tuberculosis by the spine fusion operation: a report of 286 cases	J. Bone Joint Surg. **10:**805-815, 1928
Hodgson, A. R., and Stock, F. E.	Anterior spine fusion for the treatment of tuberculosis of the spine	J. Bone Joint Surg. **42-A:**295-310, 1960
Hodgson, A. R., Skinsnes, O., and Leong, C. Y.	The pathogenesis of Pott's paraplegia	J. Bone Joint Surg. **49-A:**1147-1156, 1967
Hodgson, A. R., Yau, A., Kwon, J. S., and Kim, D.	A clinical study of 100 consecutive cases of Pott's paraplegia	Clin. Orthop. **36:**128-150, 1964
Johnson, R. W., Jr., Hillman, J. W., and Southwick, W. O.	The importance of direct surgical attack upon lesions of the vertebral bodies, particularly in Pott's disease	J. Bone Joint Surg. **35-A:**17-25, 1953
Karlén, A.	Early drainage of paraspinal tuberculous abscesses in children	J. Bone Joint Surg. **41-B:**491-498, 1959

Chapter 4: Infections of bones and joints—cont'd

Tuberculosis of bones and joints—cont'd

Katayama, R., Itami, Y., and Marumo, E.	Treatment of hip and knee-joint tuberculosis: an attempt to retain motion	J. Bone Joint Surg. **44-A:**897-917, 1962
Kessler, A. D., Scott, R. B., Kelley, C. H., and Steinman, R.	Cystic tuberculosis of the bones in children: report of two cases	Amer. J. Dis. Child. **88:**201-209, 1954
Lipscomb, P. R., and McCaslin, F. E., Jr.	Arthrodesis of the hip	J. Bone Joint Surg. **43-A:**923-938, 1961
Martin, N. S.	Tuberculosis of the spine: a study of the results of treatment during the last twenty-five years	J. Bone Joint Surg. **52-B:**613-628, 1970
Mercer, W.	The management of the tuberculous hip joint	J. Bone Joint Surg. **36-A:**1123-1128, 1954
O'Connor, B. T., Steel, W. M., and Sanders, R.	Disseminated bone tuberculosis	J. Bone Joint Surg. **52-A:**537-542, 1970
Pimm, L. H., and Waugh, W.	Tuberculous tenosynovitis	J. Bone Joint Surg. **39-B:**91-101, 1957
Poppel, M. H., Lawrence, L. R., Jacobson, H. G., and Stein, J.	Skeletal tuberculosis, a roentgenographic survey with reconsideration of diagnostic criteria	Amer. J. Roentgen. **70:**936-963, 1953
Seddon, H. J.	Pott's paraplegia	In Platt, H., editor: Modern trends in orthopedics (second series), New York, 1956, Paul B. Hoeber, Inc.
Smith, A. DeF.	Tuberculosis of the spine: results in 70 cases treated at the New York Orthopaedic Hospital from 1945 to 1960	Clin. Orthop. **58:**171-176, 1968
Soholt, S. T.	Tuberculosis of the sacro-iliac joint	J. Bone Joint Surg. **33-A:**119-130, 1951
Somerville, E. W., and Wilkinson, M. C., editors	Girdlestone's tuberculosis of bone and joint, ed. 3	New York, 1965, Oxford University Press, Inc.
Stevenson, F. H.	Chemotherapy and tuberculosis of bone and joint	Tubercle **38:**355-364, 1957
Stevenson, F. H., Cholmeley, J. A., and Jory, H. I.	Tuberculosis of the hip in children: seven years use of chemotherapy	Tubercle **38:**164-174, 1957
	Tuberculosis of the knee: results with chemotherapy between 1948 and 1956	Tubercle **39:**1-6, 1958
Strange, F. C. St. C.	The prognosis in sacroiliac tuberculosis	Brit. J. Surg. **50:**561-571, 1963
Wilkinson, M. C.	The treatment of bone and joint tuberculosis	Ann. Roy. Coll. Surg. Eng. **37:**19-39, 1965
	Tuberculosis of the spine treated by chemotherapy and operative débridement: a long-term follow-up study	J. Bone Joint Surg. **51-A:**1331-1342, 1969

Chapter 4: Infections of bones and joints—cont'd

Tuberculosis of bones and joints—cont'd

Wilkinson, M. C.—cont'd	Tuberculosis of the hip and knee treated by chemotherapy, synovectomy, and débridement: a follow-up study	J. Bone Joint Surg. **51-A:**1343-1359, 1969.
Wilson, J. N.	Tuberculosis of the elbow	J. Bone Joint Surg. **35-B:**551-560, 1953

Fungus infections of bones and joints

Aidem, H. P.	Intra-articular amphotericin B in the treatment of coccidioidal synovitis of the knee	J. Bone Joint Surg. **50-A:**1663-1668, 1968
Allen, J. H., Jr.	Bone involvement with disseminated histoplasmosis	Amer. J. Roentgen. **82:**250-254, 1959
Altner, P. C., and Turner, R. R.	Sporotrichosis of bones and joints: review of the literature and report of six cases	Clin. Orthop. **68:**138-148, 1970
Carnesale, P. L., and Stegman, K. F.	Blastomycosis of bone: report of 4 cases	Ann. Surg. **144:**252-257, 1956
Colonna, P. C., and Gucker, T., III	Blastomycosis of the skeletal system	J. Bone Joint Surg. **26:**322-328, 1944
Conaty, J. P., Biddle, M., and McKeever, F. M.	Osseous coccidioidal granuloma	J. Bone Joint Surg. **41-A:**1109-1122, 1959
Cope, V. Z.	Actinomycosis of bone with special reference to infection of the vertebral column	J. Bone Joint Surg. **33-B:**205-214, 1951
Cullen, C. H., and Sharp, M. E.	Infection of wounds with actinomyces	J. Bone Joint Surg. **33-B:**221-227, 1951
Duran, R. J., Coventry, M. B., Weed, L. A., and Kierland, R. R.	Sporotrichosis	J. Bone Joint Surg. **39-A:**1330-1342, 1957
Dykes, J., Segesman, J. K., and Birsner, J. W.	Coccidioidomycosis of bone in children	Amer. J. Dis. Child. **85:**34-42, 1953
Hildick-Smith, G., Blank, H., and Sarkany, I.	Fungus diseases and their treatment	Boston, 1964, Little, Brown & Co.
Oyston, J. K.	Madura foot	J. Bone Joint Surg. **43-B:**259-267, 1961
Pollock, S. F., Morris, J. M., and Murray, W.	Coccidioidal synovitis of the knee	J. Bone Joint Surg. **49-A:**1397-1407, 1967
Reeves, R. J., and Pedersen, R.	Fungous infection of bone	Radiology **62:**55-60, 1954
Rhangos, W. C., and Chick, E. W.	Mycotic infections of bone	Southern Med. J. **57:**664-674, 1964
Toone, E. C., Jr., and Kelly, J.	Joint and bone disease due to mycotic infection	Amer. J. Med. Sci. **231:**263-273, 1956
Wolfe, J. N., and Jacobson, G.	Roentgen manifestations of torulosis (cryptococcosis)	Amer. J. Roentgen. **79:**216-227, 1958

Chapter 5

Chronic arthritis

Ahlberg, Å.	Treatment and prophylaxis of arthropathy in severe hemophilia	Clin. Orthop. **53:**135-146, 1967

Chapter 5: Chronic arthritis—cont'd

American Rheumatism Association Committee	Nineteenth rheumatism review	Arthritis Rheum. **13**:457-711, 1970
Bauer, G. C. H., Insall, J., and Koshino, T.	Tibial osteotomy in gonarthrosis (osteo-arthritis of the knee)	J. Bone Joint Surg. **51-A**:1545-1563, 1969
Bick, E. M.	Vertebral osteophytosis, pathologic basis of its roentgenology	Amer. J. Roentgen. **73**:979-983, 1955
Bland, J. H., Davis, P. H., London, M. G., Van Buskirk, F. W., and Duarte, C. G.	Rheumatoid arthritis of cervical spine	Arch. Intern. Med. **112**:892-898, 1963
Blount, W. P.	Osteotomy in the treatment of osteo-arthritis of the hip	American Academy of Orthopaedic Surgeons Instructional Courses, J. Bone Joint Surg. **46-A**:1297-1325, 1964
Blumberg, B., and Ragan, C.	The natural history of rheumatoid spondylitis	Medicine **35**:1-31, 1956
Brewer, E. J., Jr.	Juvenile rheumatoid arthritis	Philadelphia, 1970, W. B. Saunders Co. (Vol. VI in series Major problems in clinical pediatrics)
Bundens, W. D., Jr., Brighton, C. T., and Weitzman, G.	Primary articular-cartilage calcification with arthritis (pseudogout syndrome)	J. Bone Joint Surg. **47-A**:111-122, 1965
Calabro, J. J.	An appraisal of the medical and surgical management of ankylosing spondylitis	Clin. Orthop. **60**:125-148, 1968
Calabro, J. J., Katz, R. M., and Maltz, B. A.	A critical reappraisal of juvenile rheumatoid arthritis	Clin. Orthop. **74**:101-119, 1971
Carter, M. E., editor and translator	Symposium on radiological aspects of rheumatoid arthritis	Internat. Congress Series No. 61, Amsterdam, 1963, Excerpta Medica Foundation
Clayton, M. L.	Surgery of the thumb in rheumatoid arthritis	J. Bone Joint Surg. **44-A**:1376-1386, 1962
	Surgery of the lower extremity in rheumatoid arthritis	American Academy of Orthopaedic Surgeons Instructional Courses, J. Bone Joint Surg. **45-A**:1517-1536, 1963
	Surgery of the rheumatoid hand	Clin. Orthop. **36**:47-65, 1964
Copeman, W. S. C., editor	Textbook of the rheumatic diseases, ed. 4	Edinburgh, 1969, E. & S. Livingstone, Ltd.
Coventry, M. B.	Osteotomy of the upper portion of the tibia for degenerative arthritis of the knee	J. Bone Joint Surg. **47-A**:984-990, 1965
Curtiss, P. H., Jr.	Changes produced in the synovial membrane and synovial fluid by disease	American Academy of Orthopaedic Surgeons Instructional Courses, J. Bone Joint Surg. **46-A**:873-888, 1964
DePalma, A. F.	Hemophilic arthropathy	Clin. Orthop. **52**:145-165, 1967
Duthie, J. J. R., Brown, P. E., Knox, J. D. E., and Thompson, M.	Course and prognosis in rheumatoid arthritis	Ann. Rheum. Dis. **16**:411-424, 1957

Chapter 5: Chronic arthritis—cont'd

Eggers, G. W. N., Evans, E. B., Blumel, J., Nowlin, D. H., and Butler, J. K.	Cystic change in the iliac acetabulum	J. Bone Joint Surg. **45-A:**669-686, 1963
Empire Rheumatism Council	Gold therapy in rheumatoid arthritis	Ann. Rheum. Dis. **19:**95-119, 1960
Ferguson, A. B., Jr.	The pathological changes in degenerative arthritis of the hip and treatment by rotational osteotomy	American Academy of Orthopaedic Surgeons Instructional Courses, J. Bone Joint Surg. **46-A:**1337-1352, 1964
Fernandez de Valderrama, J. A., and Matthews, J. M.	The haemophilic pseudotumour or haemophilic subperiosteal haematoma	J. Bone Joint Surg. **47-B:**256-265, 1965
Flatt, A. E.	The care of the rheumatoid hand, ed. 2	St. Louis, 1968, The C. V. Mosby Co.
Forestier, J., Jacqueline, F., and Rotes-Querol, J. (translated by A. U. Desjardins)	Ankylosing spondylitis	Springfield, Ill., 1956, Charles C Thomas, Publisher
France, W. G., and Wolf, P.	Treatment and prevention of chronic haemorrhagic arthropathy and contractures in haemophilia	J. Bone Joint Surg. **47-B:**247-255, 1965
Gardner, D. L.	Pathology of the connective tissue diseases	Baltimore, 1966, The Williams & Wilkins Co.
Geckeler, E. O., and Quaranta, A. V.	Patellectomy for degenerative arthritis of the knee, late results	J. Bone Joint Surg. **44-A:**1109-1114, 1962
Geens, S.	Synovectomy and debridement of the knee in rheumatoid arthritis. I. Historical review	J. Bone Joint Surg. **51-A:**617-625, 1969
Geens, S., Clayton, M., Leidholt, J. D., Smyth, C. J., and Bartholomew, B. A.	Synovectomy and debridement of the knee in rheumatoid arthritis. II. Clinical and roentgenographic study of thirty-one cases	J. Bone Joint Surg. **51-A:**626-642, 1969
Harris, W. H.	Role of intertrochanteric osteotomy in treatment of arthritis of the hip	Surg. Clin. N. Amer. **49:**775-778, 1969
Harris, W. R., and Kostuik, J. P.	High tibial osteotomy for osteoarthritis of the knee	J. Bone Joint Surg. **52-A:**330-336, 1970
Hart, F. D.	Corticosteroids in treatment of rheumatic disorders	Brit. Med. J. **1:**493-496, 1960
Hench, P. S., and Rosenberg, E. F.	Palindromic rheumatism	Arch. Intern. Med. **73:**293-321, 1944
Henderson, E. D., and Lipscomb, P. R.	Surgical treatment of rheumatoid hand	J.A.M.A. **175:**431-436, 1961
Hollander, J. L.	Intra-articular hydrocortisone in arthritis and allied conditions	J. Bone Joint Surg. **35-A:**983-990, 1953
Hollander, J. L., editor	Arthritis and allied conditions, ed. 7	Philadelphia, 1966, Lea & Febiger
Holman, H. R.	The L. E. cell phenomenon	Ann. Rev. Med. **11:**231-242, 1960

Chapter 5: Chronic arthritis—cont'd

Houpt, J. B.	The effect of allopurinol (HPP) in the treatment of gout	Arthritis Rheum. **8:**899-904, 1965
Ishmael, W. K., editor	Symposium: Gout	Clin. Orthop. **71:**2-98, 1970
Ivins, J. C., Benson, W. F., Bickel, W. H., and Nelson, J. W.	Arthroplasty of the hip for idiopathic degenerative joint disease	Surg. Gynec. Obstet. **125:**1281-1284, 1967
Johnson, L. C.	Kinetics of osteoarthritis	Lab. Invest. **8:**1223-1241, 1959
Johnson, N. J., and Dodd, K.	Juvenile rheumatoid arthritis	Med. Clin. N. Amer. **39:**459-487, 1955
Jones, J. B., editor	Symposium: Orthopaedic surgery in rheumatoid arthritis	Clin. Orthop. **36:**5-108, 1964
Jordan, H. H.	Hemophilic arthropathies	Springfield, Ill., 1958, Charles C Thomas, Publisher
Kuhns, J. G., Jr., and Potter, T. A.	Problems of the knee joint in rheumatoid arthritis of the knee	Curr. Pract. Orthop. Surg. **1:**65-110, 1963
Larmon, W. A., and Kurtz, J. F.	The surgical management of chronic tophaceous gout	J. Bone Joint Surg. **40-A:**743-772, 1958
Laskar, F. H., and Sargison, K. D.	Ochronotic arthropathy: a review with four case reports	J. Bone Joint Surg. **52-B:**653-666, 1970
Law, W. A.	Osteotomy of the spine	Clin. Orthop. **66:**70-76, 1969
Lipscomb, P. R.	Synovectomy of the wrist for rheumatoid arthritis	J.A.M.A. **194:**655-659, 1965
Lipscomb, P. R., moderator	Symposium on surgery for rheumatoid arthritis	American Academy of Orthopaedic Surgeons Instructional Courses, J. Bone Joint Surg. **50-A:**575-617, 1968
Lockie, L. M.	Steroid therapy in rheumatoid diseases	J.A.M.A. **170:**1063-1066, 1959
Lowman, E. W., Lee, P. R., and Rusk, H. A.	Total rehabilitation of the rheumatoid arthritic cripple	J.A.M.A. **158:**1335-1344, 1955
McDuffie, F. C.	Nature and possible significance of rheumatoid factor	Southern Med. J. **58:**1126-1130, 1965
McEwen, C.	Early synovectomy in treatment of rheumatoid arthritis	New Eng. J. Med. **279:**420-422, 1968
McEwen, C., Ziff, M., Carmel, P., DiTata, D., and Tanner, M.	The relationship to rheumatoid arthritis of its so-called variants	Arthritis Rheum. **1:**481-496, 1958
McFarland, G., Jr., and Sherman, M. S.	The synovial reactions of rheumatoid arthritis	Clin. Orthop. **36:**10-21, 1964
McKusick, V. A.	Genetic factors in diseases of connective tissue	Amer. J. Med. **26:**283-302, 1959
McMurray, T. P.	Osteo-arthritis of the hip joint	J. Bone Joint Surg. **21:**1-11, 1939
Mellors, R. C., Nowoslawski, A., Korngold, L., and Sengson, B. L.	Rheumatoid factor and the pathogenesis of rheumatoid arthritis	J. Exp. Med. **113:**475-484, 1961
Milch, R. A.	Surgery of arthritis	Baltimore, 1964, The Williams & Wilkins Co.
Mongan, E. S., Boger, W. M., Gilliland, B. C., and Meyerowitz, S.	Synovectomy in rheumatoid arthritis	Arthritis Rheum. **13:**761-768, 1970
Montgomery, M. M., Pilz, C. G., and Aronson, A. R.	Early diagnosis of arthritis and allied disorders	Med. Clin. N. Amer. **44:**29-48, 1960

Chapter 5: Chronic arthritis—cont'd

Mueller, M. N., Sorensen, L. B., Strandjord, N., and Kappas, A.	Alkaptonuria and ochronotic arthropathy	Med. Clin. N. Amer. **49:**101-115, 1965
Nalebuff, E. A., and Potter, T. A.	Rheumatoid involvement of tendon and tendon sheaths in the hand	Clin. Orthop. **59:**147-159, 1968
Nordby, E. J., and Sachtjen, K. M.	Femoral head prosthesis for hypertrophic arthritis of the hip	Clin. Orthop. **57:**191-202, 1968
Polley, H. F., and Slocumb, C. H.	Rheumatoid spondylitis: a study of 1,035 cases	Ann. Intern. Med. **26:**240-249, 1947
Portis, R. B.	Pathology of chronic arthritis of children (Still's disease)	Amer. J. Dis. Child. **55:**1000-1017, 1938
Potter, T. A., and Nalebuff, E. A., editors	Surgical management of rheumatoid arthritis	Surg. Clin. N. Amer. **49:**731-950, 1969
Preston, R. L.	The surgical management of rheumatoid arthritis	Philadelphia, 1968, W. B. Saunders Co.
Rhinelander, F. W.	The effectiveness of splinting and bracing on rheumatoid arthritis	Arthritis Rheum. **2:**270-277, 1959
Ropes, M. W., and Bauer, W.	Synovial fluid changes in joint disease	Cambridge, Mass., 1953, Harvard University Press
Rose, G. K.	The surgical management of ankylosing spondylitis	Rheumatism **17:**63-69, 1961
Shands, A. R., Jr., and Muehe, C. C.	Rheumatoid arthritis in childhood, involvement of the hip	Delaware Med. J. **32:**1-12, 1960
Sherman, M. S.	Psoriatic arthritis	J. Bone Joint Surg. **34-A:**831-852, 1955
Smith, C. F., Pugh, D. G., and Polley, H. F.	Physiologic vertebral ligamentous calcification: an aging process	Amer. J. Roentgen. **74:**1049-1058, 1955
Sokoloff, L.	The biology of degenerative joint disease	Chicago, 1969, University of Chicago Press
Steel, W. M., Duthie, R. B., and O'Connor, B. T.	Haemophilic cysts	J. Bone Joint Surg. **51-B:**614-626, 1969
Stevens, J.	Osteoarthritis of the hip: a review with special consideration of the problem of bilateral malum coxae senilis	Clin. Orthop. **71:**152-181, 1970
Straub, L. R.	The rheumatoid hand	Clin. Orthop. **15:**127-139, 1959
Straub, L. R., and Ranawat, C. S.	The wrist in rheumatoid arthritis, surgical treatment and results	J. Bone Joint Surg. **51-A:**1-20, 1969
Talbott, J. H., and Seegmiller, J. E.	Gout, ed. 3	New York, 1967, Grune & Stratton, Inc.
Tarnay, T. J.	Surgery in the hemophiliac	Springfield, Ill., 1968, Charles C Thomas, Publisher
Traut, E. F.	Degenerative arthritis: its causes, recognition and management	Med. Clin. N. Amer. **40:**63-78, 1956
Trueta, J.	Studies on the etiopathology of osteoarthritis of the hip	Clin. Orthop. **31:**7-19, 1963

Chapter 5: Chronic arthritis—cont'd

Watson-Jones, R., and Robinson, W. C.	Arthrodesis of the osteoarthritic hip joint	J. Bone Joint Surg. **38-B:**353-377, 1956
Wilson, P. D., and Levine, D. B.	Compensatory pelvic osteotomy for ankylosing spondylitis	J. Bone Joint Surg. **51-A:**142-148, 1969
Wood, K., Omer, A., and Shaw, M. T.	Haemophilic arthropathy, a combined radiological and clinical study	Brit. J. Radiol. **42:**498-505, 1969

Ankylosis and arthroplasty

Aufranc, O. E.	Constructive surgery of the hip	St. Louis, 1962, The C. V. Mosby Co.
Barr, J. S., Donovan, J. F., and Florence, D. W.	Arthroplasty of the hip	J. Bone Joint Surg. **46-A:**249-266, 1964
Charnley, J., editor	Symposium: Total hip replacement	Clin. Orthop. **72:**2-241, 1970
Clouse, M. E., Aufranc, O. E., and Weber, A. L.	Roentgenologic and clinical evaluation of Vitallium mold arthroplasty of the hip	Surg. Gynec. Obstet. **127:**1042-1050, 1968
Hagen, R.	Contracture of the quadriceps muscle in children: a report of 12 cases	Acta Orthop. Scand. **39:**565-578, 1968
Hurri, L., Pulkki, T., and Vainio, K.	Arthroplasty of the elbow in rheumatoid arthritis	Acta Chir. Scand. **127:**459-465, 1964
Johnston, R. C., and Larson, C. B.	Results of treatment of hip disorders with cup arthroplasty	J. Bone Joint Surg. **51-A:**1461-1479, 1969
Jones, G. B.	Arthroplasty of the knee by the Walldius prosthesis	J. Bone Joint Surg. **50-B:**505-510, 1968
Kelikian, H.	A method of mobilizing the temporomandibular joint	J. Bone Joint Surg. **32-A:**113-131, 1950
Knight, R. A., and Van Zandt, I. L.	Arthroplasty of the elbow, an end-result study	J. Bone Joint Surg. **34-A:**610-618, 1952
Law, W. A.	Late results in Vitallium-mold arthroplasty of the hip	J. Bone Joint Surg. **44-A:**1497-1517, 1962
Lipscomb, P. R.	Reconstructive surgery for bilateral hip-joint disease in the adult	J. Bone Joint Surg. **47-A:**1-30, 1965
Lloyd-Roberts, G. C., and Thomas, T. G.	The etiology of quadriceps contracture in children	J. Bone Joint Surg. **46-B:**498-502, 1964
Moore, A. T.	The self-locking metal hip prosthesis	J. Bone Joint Surg. **39-A:**811-827, 1957
Ring, P. A.	Complete replacement arthroplasty of the hip by the Ring prosthesis	J. Bone Joint Surg. **50-B:**720-731, 1968
Smith-Petersen, M. N.	Evolution of mould arthroplasty of the hip joint	J. Bone Joint Surg. **30-B:**59-75, 1948
Stinchfield, F. E., and Chamberlin, A. C.	Arthroplasty of the hip	American Academy of Orthopaedic Surgeons Instructional Courses, J. Bone Joint Surg. **48-A:**564-581, 1966
Swanson, A. B.	Silicone rubber implants for replacement of arthritic or destroyed joints in the hand	Surg. Clin. N. Amer. **48:**1113-1127, 1968

Chapter 5: Chronic arthritis—cont'd

Ankylosis and arthroplasty—cont'd

Swanson, A. B.—cont'd	Arthroplasty in traumatic arthritis of the joints of the hand	Orthop. Clin. N. Amer. **1**:285-298, 1970
Thompson, T. C.	Quadricepsplasty	Ann. Surg. **121**:751-755, 1945

Chapter 6

Neuromuscular affections

Cerebral palsy

Anderson, G. W., Hicks, S. P., Palmer, M. F., Hughes, J. G., Arey, J. B., Yakovlev, P. I., Unna, K. R., and Byers, R. K.	Symposium on cerebral palsy	J. Pediat. **40**:340-375, 489-524, 606-633, 1952
Baker, L. D.	A rational approach to the surgical needs of the cerebral palsy patient	J. Bone Joint Surg. **38-A**:313-323, 1956
Baker, L. D., and Hill, L. M.	Foot alignment in the cerebral palsy patient	J. Bone Joint Surg. **46-A**:1-15, 1964
Banks, H. H., and Green, W. T.	The correction of equinus deformity in cerebral palsy	J. Bone Joint Surg. **40-A**:1359-1379, 1958
	Adductor myotomy and obturator neurectomy for the correction of adduction contracture of the hip in cerebral palsy	J. Bone Joint Surg. **42-A**:111-126, 1960
Cooper, W.	The diagnosis and treatment of cerebral palsy	American Academy of Orthopaedic Surgeons Instructional Course Lectures, Ann Arbor, 1957, J. W. Edwards, vol. 14, pp. 293-299
Crothers, B., and Paine, R. S.	The natural history of cerebral palsy	Cambridge, Mass., 1959, Harvard University Press
Deaver, G. G.	Cerebral palsy: methods of treating the neuromuscular disabilities	Arch. Phys. Med. **37**:363-367, 1956
Denhoff, E., Holden, R. H., and Silver, M. L.	Prognostic studies in children with cerebral palsy	J.A.M.A. **161**:781-784, 1956
Eggers, G. W. N.	Transplantation of hamstring tendons to femoral condyles in order to improve hip extension and to decrease knee flexion in cerebral spastic paralysis	J. Bone Joint Surg. **34-A**:827-830, 1952
Eggers, G. W. N., and Evans, E. B.	Surgery in cerebral palsy	American Academy of Orthopaedic Surgeons Instructional Courses, J. Bone Joint Surg. **45-A**:1275-1305, 1963

Chapter 6: Neuromuscular affections—cont'd

Cerebral palsy—cont'd

Goldner, J. L.	Reconstructive surgery of the hand in cerebral palsy and spastic paralysis resulting from injury to the spinal cord	J. Bone Joint Surg. **37-A:**1141-1154, 1955
Green, W. T., and Banks, H. H.	Flexor carpi ulnaris transplant and its use in cerebral palsy	J. Bone Joint Surg. **44-A:**1343-1352, 1962
Hohman, L. B.	Intelligence levels in cerebral palsied children	Amer. J. Phys. Med. **32:**282-290, 1953
Keats, S.	Cerebral palsy	Springfield, Ill., 1965, Charles C Thomas, Publisher
	Surgical treatment of the hand in cerebral palsy: correction of thumb-in-palm and other deformities	J. Bone Joint Surg. **47-A:**274-284, 1965
	Operative orthopaedics in cerebral palsy	Springfield, Ill., 1970, Charles C Thomas, Publisher
Kenney, W. E., editor	Symposium: Cerebral palsy: upper extremity	Clin. Orthop. **46:**3-108, 1966
Kenney, W. E., and Denhoff, E., editors	Symposium: Cerebral palsy: lower extremity	Clin. Orthop. **47:**3-139, 1966
Kirman, B. H.	Epilepsy and cerebral palsy	Arch. Dis. Child. **31:**1-7, 1956
Koven, L. J., and Lamm, S. S.	The athetoid syndrome in cerebral palsy, clinical aspects	Pediatrics **14:**181-192, 1954
Levitt, S.	Physiotherapy in cerebral palsy	Springfield, Ill., 1962, Charles C Thomas, Publisher
Miller, E., and Rosenfeld, G. B.	The psychologic evaluation of children with cerebral palsy and its implications in treatment	J. Pediat. **41:**613-621, 1952
Minear, W. L.	A classification of cerebral palsy	Pediatrics **18:**841-852, 1956
Norfleet, G. M.	Motor concepts in muscle training	New York, 1969, Vantage Press Inc.
Perlstein, M. A.	Cerebral palsy	Arch. Pediat. **79:**289-298, 1962
Perlstein, M. A., and Hood, P. N.	Etiology of postneonatally acquired cerebral palsy	J.A.M.A. **188:**850-854, 1964
Phelps, W. M.	Description and differentiation of types of cerebral palsy	Nerv. Child **8:**107-127, 1949
	Classification of athetosis with special reference to the motor classification	Amer. J. Phys. Med. **35:**24-31, 1956
	Long-term results of orthopaedic surgery in cerebral palsy	J. Bone Joint Surg. **39-A:**53-59, 1957
Pollock, G. A.	Surgical treatment of cerebral palsy	J. Bone Joint Surg. **44-B:**68-81, 1962
Pollock, G. A., and English, T. A.	Transplantation of the hamstring muscles in cerebral palsy	J. Bone Joint Surg. **49-B:**80-86, 1967
Samilson, R. L., and Morris, J. M.	Surgical improvement of the cerebral-palsied upper limb	J. Bone Joint Surg. **46-A:**1203-1216, 1964
Saturen, P., and Tobis, J. S.	Evaluation and management of motor disturbance in brain-damaged children	J.A.M.A. **175:**588-591, 1961

Chapter 6: Neuromuscular affections—cont'd

Cerebral palsy—cont'd

Schwartz, P.	Birth injuries of the newborn: morphology, pathogenesis, clinical pathology and prevention	New York, 1961, Hafner Publishing Co., Inc.
Silver, C. M., and Simon, S. D.	Gastrocnemius-muscle recession (Silfverskiold operation) for spastic equinus deformity in cerebral palsy	J. Bone Joint Surg. **41-A:**1021-1028, 1959
Stamp, W. G.	Bracing in cerebral palsy	American Academy of Orthopaedic Surgeons Instructional Courses, J. Bone Joint Surg. **44-A:**1457-1476, 1962
Strayer, L. M., Jr.	Gastrocnemius recession	J. Bone Joint Surg. **40-A:**1019-1030, 1958
Sutherland, D. H., Schottstaedt, E. R., Larsen, L. J., Ashley, R. K., Callander, J. N., and James, P. M.	Clinical and electromyographic study of seven spastic children with internal rotation gait	J. Bone Joint Surg. **51-A:**1070-1082, 1969
Swanson, A. B.	Surgery of the hand in cerebral palsy and muscle origin release procedures	Surg. Clin. N. Amer. **48:**1129-1138, 1968
Towbin, A.	The pathology of cerebral palsy	Springfield, Ill., 1961, Charles C Thomas, Publisher

Poliomyelitis

Baker, A. B., and Cornwell, S.	Poliomyelitis: the spinal cord	Arch. Path. **61:**185-206, 1956
Clippinger, F. W., Jr., and Irwin, C. E.	The opponens transfer, analysis of end results	Southern Med. J. **55:**33-36, 1962
Grice, D. S.	Further experience with extra-articular arthrodesis of the subtalar joint	J. Bone Joint Surg. **37-A:**246-259, 1955
Grulee, C. G., Jr.	Differential diagnosis of poliomyelitis	J.A.M.A. **152:**1587-1590, 1953
Herndon, C. H.	Tendon transplantation at the knee and foot	American Academy of Orthopaedic Surgeons Instructional Course Lectures, St. Louis, 1961, The C. V. Mosby Co., vol. 18, pp. 145-168
Herndon, C. H., Strong, J. M., and Heyman, C. H.	Transposition of the tibialis anterior in the treatment of paralytic talipes calcaneus	J. Bone Joint Surg. **38-A:**751-760, 1956
Hoke, M.	An operation for stabilizing paralytic feet	J. Orthop. Surg. **3:**494-507, 1921
Hunt, J. C., and Brooks, A. L.	Subtalar extra-articular arthrodesis for correction of paralytic valgus deformity of the foot	J. Bone Joint Surg. **47-A:**1310-1314, 1965
Ingram, A. J., and Hundley, J. M.	Posterior bone block of the ankle for paralytic equinus, an end-result study	J. Bone Joint Surg. **33-A:**679-691, 1951

Chapter 6: Neuromuscular affections—cont'd

Poliomyelitis—cont'd

Irwin, C. E.	Iliotibial band, its role in producing deformity in poliomyelitis	J. Bone Joint Surg. **31-A**:141-146, 1949
	The calcaneus foot	Southern Med. J. **44**:191-197, 1951
Irwin, C. E., and Eyler, D. L.	Surgical rehabilitation of the hand and forearm disabled by poliomyelitis	J. Bone Joint Surg. **33-A**:825-835, 1951
Kendall, H. O., and Kendall, F. P.	Muscles, testing and function	Baltimore, 1949, The Williams & Wilkins Co.
Kettelkamp, D. B., and Larson, C. B.	Evaluation of the Steindler flexorplasty	J. Bone Joint Surg. **45-A**:513-518, 1963
Kuhlmann, R. F., and Bell, J. F.	A clinical evaluation of tendon transplantations for poliomyelitis affecting the lower extremities	J. Bone Joint Surg. **34-A**:915-926, 1952
Lowman, C. L.	Fascial transplants in paralysis of abdominal and shoulder girdle muscles	American Academy of Orthopaedic Surgeons Instructional Course Lectures, Ann Arbor, 1957, J. W. Edwards, vol. 14, pp. 300-304
Mayer, L.	The physiological method of tendon transplants reviewed after forty years	American Academy of Orthopaedic Surgeons Instructional Course Lectures, Ann Arbor, 1956, J. W. Edwards, vol. 13, pp. 116-120
Mayer, L., and Green, W.	Experiences with the Steindler flexorplasty at the elbow	J. Bone Joint Surg. **36-A**:775-789, 1954
Mortens, J., and Pilcher, M. F.	Tendon transplantation in the prevention of foot deformities after poliomyelitis in children	J. Bone Joint Surg. **38-B**:633-639, 1956
Mustard, W. T.	A follow-up study of iliopsoas transfer for hip instability	J. Bone Joint Surg. **41-B**:289-298, 1959
Patterson, R. L., Jr., Parrish, F. F., and Hathaway, E. N.	Stabilizing operations on the foot, a study of the indications, techniques used, and end results	J. Bone Joint Surg. **32-A**:1-26, 1950
Pollock, J. H., and Carrell, B.	Subtalar extra-articular arthrodesis in the treatment of paralytic valgus deformities, a review of 112 procedures in 100 patients	J. Bone Joint Surg. **46-A**:533-541, 1964
Reidy, J. A., Broderick, T. F., Jr., and Barr, J. S.	Tendon transplantations in lower extremity: a review of end results in poliomyelitis. I. Tendon transplantations about the foot and ankle. II. Tendon transplantations at the knee	J. Bone Joint Surg. **34-A**:900-914, 1952

Chapter 6: Neuromuscular affections—cont'd

Poliomyelitis—cont'd

Sabin, A. B., Michaels, R. H., Spigland, I., Pelon, W., Rhim, J. S., and Wehr, R. E.	Community-wide use of oral poliovirus vaccine	Amer. J. Dis. Child. **101**:546-567, 1961
Salk, J. E.	Studies in human subjects on active immunization against poliomyelitis	J.A.M.A. **151**:1081-1098, 1953
Schottstaedt, E. R., Larsen, L. J., and Bost, F. C.	The surgical reconstruction of the upper extremity paralyzed by poliomyelitis	J. Bone Joint Surg. **40-A**:633-643, 1958
Schwartzmann, J. R., and Crego, C. H., Jr.	Hamstring-tendon transplantation for the relief of quadriceps femoris paralysis in residual poliomyelitis, a follow-up study of 134 cases	J. Bone Joint Surg. **30-A**:541-549, 1948
Segal, A., Seddon, H. J., and Brooks, D. M.	Treatment of paralysis of the flexors of the elbow	J. Bone Joint Surg. **41-B**:44-50, 1959
Sharrard, W. J. W.	Muscle recovery in poliomyelitis	J. Bone Joint Surg. **37-B**:63-79, 1955
Steindler, A.	The pathomechanics of the paralytic gait	American Academy of Orthopaedic Surgeons Instructional Course Lectures, Ann Arbor, 1955, J. W. Edwards, vol. 12, pp. 189-200
Straub, L. R., Harvey, J. P., Jr., and Fuerst, C. E.	A clinical evaluation of tendon transplantation in the paralytic foot	J. Bone Joint Surg. **39-A**:1-16, 1957
Thompson, T. C.	A modified operation for opponens paralysis	J. Bone Joint Surg. **24**:632-640, 1942
Watkins, M. B., Jones, J. B., Ryder, C. T., Jr., and Brown, T. H., Jr.	Transplantation of the posterior tibial tendon	J. Bone Joint Surg. **36-A**:1181-1189, 1954
Waugh, R. R., Wagner, J., and Stinchfield, F. E.	An evaluation of pantalar arthrodesis: a follow-up study of one hundred and sixteen operations	J. Bone Joint Surg. **47-A**:1315-1322, 1965
Westin, G. W.	Tendon transfers about the foot, ankle, and hip in the paralyzed lower extremity	American Academy of Orthopaedic Surgeons Instructional Courses, J. Bone Joint Surg. **47-A**:1430-1443, 1965
Westin, G. W., and Hall, C. B.	Subtalar extra-articular arthrodesis	J. Bone Joint Surg. **39-A**:501-512, 1957
Wilson, F. C., Jr., Fay, G. F., Lamotte, P., and Williams, J. C.	Triple arthrodesis: a study of the factors affecting fusion after 301 procedures	J. Bone Joint Surg. **47-A**:340-348, 1965

Other neuromuscular disabilities from involvement of the brain and spinal cord

Amick, L. D.	Bracing the hemiplegic: criteria of short-leg braces	Southern Med. J. **57**:312-316, 1964
Beetham, W. P., Jr., Kaye, R. L., and Polley, H. F.	Charcot's joints	Ann. Intern. Med. **58**:1002-1012, 1963

Chapter 6: Neuromuscular affections—cont'd

Other neuromuscular disabilities from involvement of brain and spinal cord—cont'd

Bluestone, S. S., and Deaver, G. G.	Habilitation of the child with spina bifida and myelomeningocele	J.A.M.A. **161**:1248-1251, 1956
Bosch, A., Stauffer, S., and Nickel, V. L.	Incomplete traumatic quadriplegia: a ten-year review	J.A.M.A. **216**:473-478, 1971
Burdick, W. F., Whipple, D. V., and Freeman, W.	Amyotonia congenita (Oppenheim): report of 5 cases with necropsy	Amer. J. Dis. Child. **69**:295-307, 1945
Cameron, A. H.	The spinal cord lesion in spina bifida cystica	Lancet **2**:171-174, 1956
Covalt, D. A., Cooper, I. S., Hoen, T. I., and Rusk, H. A.	Early management of patients with spinal cord injury	J.A.M.A. **151**:89-94, 1953
Delano, P. J.	The pathogenesis of Charcot's joint	Amer. Roentgen. **56**:189-200, 1946
Eichenholtz, S. N.	Charcot joints	Springfield, Ill., 1966, Charles C Thomas, Publisher
Ford, F. R.	Diseases of the nervous system in infancy, childhood and adolescence, ed. 5	Springfield, Ill., 1966, Charles C Thomas, Publisher
Garrett, A. L., Perry, J., and Nickel, V. L.	Traumatic quadriplegia	J.A.M.A. **187**:7-11, 1964
Gilroy, J., and Meyer, J. S.	Medical neurology	Toronto, 1969, The Macmillan Co. of Canada Ltd.
Gordon, E. E.	Physiological approach to ambulation in paraplegia	J.A.M.A. **161**:686-688, 1956
Haslam, E. T., and Wickstrom, J.	Orthopedic reconstructive procedures in the lower extremity in adults with acquired spastic paralysis	Southern Med. J. **57**:1322-1325, 1964
Hayes, J. T., Gross, H. P., and Dow, S.	Surgery for paralytic defects secondary to myelomeningocele and myelodysplasia	American Academy of Orthopaedic Surgeons Instructional Courses, J. Bone Joint Surg. **46-A**:1577-1597, 1964
Johnson, J. T. H.	Neuropathic fractures and joint injuries	J. Bone Joint Surg. **49-A**:1-30, 1967
Kessler, H.	Traumatic paraplegia—rationale of therapy	Ann. Intern. Med. **40**:905-923, 1954
Kilfoyle, R. M., Foley, J. J., and Norton, P. L.	Spine and pelvic deformity in childhood and adolescent paraplegia, a study of 104 cases	J. Bone Joint Surg. **47-A**:659-682, 1965
Lamm, S. S.	Pediatric neurology	New York, 1959, Landsberger Medical Books, Inc.
Lawyer, T., Jr., and Netsky, M. G.	Amyotrophic lateral sclerosis: a clinicoanatomic study of fifty-three cases	Arch. Neurol. Psychiat. **69**:171-192, 1953
Makin, M.	The surgical management of Friedreich's ataxia	Ann. Roy. Coll. Surg. Eng. **22**:1-10, 1958
Menelaus, M. B.	Dislocation and deformity of the hip in children with spina bifida cystica	J. Bone Joint Surg. **51-B**:238-251, 1969

Chapter 6: Neuromuscular affections—cont'd

Other neuromuscular disabilities from involvement of brain and spinal cord—cont'd

Merritt, H. H.	A textbook of neurology, ed. 4	Philadelphia, 1967, Lea & Febiger
Michaelis, L. S.	Orthopaedic surgery of the limbs in paraplegia	Berlin, 1964, Springer-Verlag
Nickel, V. L., editor	Symposium: Orthopaedic management of stroke	Clin. Orthop. **63**:3-161, 1969
Perret, G.	Diagnosis and treatment of diastematomyelia	Surg. Gynec. Obstet. **105**:69-83, 1957
Perry, J.	Lower-extremity bracing in hemiplegia	Clin. Orthop. **63**:32-38, 1969
Samilson, R. L., Sankaran, B., Bersani, F. A., and Smith, A. D.	Orthopedic management of neuropathic joints	Arch. Surg. **78**:115-121, 1959
Shands, A. R., Jr.	Neuropathies of the bones and joints: report of a case of an arthropathy of the ankle due to a peripheral nerve lesion	Arch. Surg. **20**:614-636, 1930
Sharrard, W. J. W.	Posterior iliopsoas transplantation in the treatment of paralytic dislocation of the hip	J. Bone Joint Surg. **46-B**:426-444, 1964
Sharrard, W. J. W., and Grosfield, I.	The management of deformity and paralysis of the foot in myelomeningocele	J. Bone Joint Surg. **50-B**:456-465, 1968
Soto-Hall, R., and Haldeman, K. O.	The diagnosis of neuropathic joint disease (Charcot joint): an analysis of forty cases	J.A.M.A. **114**:2076-2078, 1940
Tachdjian, M. O., and Matson, D. D.	Orthopaedic aspects of intraspinal tumors in infants and children	J. Bone Joint Surg. **47-A**:223-248, 1965
Talbot, H. S.	Adjunctive care of spinal cord injury	Surg. Clin. N. Amer. **48**:737-757, 1968
Taylor, R. G., and Gleave, J. R. W.	Incomplete spinal cord injuries, with Brown-Séquard phenomena	J. Bone Joint Surg. **39-B**:438-450, 1957
Walton, J. N.	Amyotonia congenita, a follow-up study	Lancet **1**:1023-1027, 1956

Peripheral nerve injuries

Barnes, R.	Traction injuries of the brachial plexus in adults	J. Bone Joint Surg. **31-B**:10-16, 1949
	Peripheral nerve injuries	In Platt, Sir H., editor: Modern trends in orthopaedics, second series, New York, 1956, Paul B. Hoeber, Inc.
Bateman, J. E.	Peripheral nerve injuries	American Academy of Orthopaedic Surgeons Instructional Course Lectures, Ann Arbor, 1956, J. W. Edwards, vol. 13, pp. 85-100
	Trauma to nerves in limbs	Philadelphia, 1962, W. B. Saunders Co.
Beasley, R. W.	Tendon transfers for radial nerve palsy	Orthop. Clin. N. Amer. **1**:439-445, 1970

Chapter 6: Neuromuscular affections—cont'd

Peripheral nerve injuries—cont'd

Bonney, G.	Prognosis in traction lesions of the brachial plexus	J. Bone Joint Surg. **41-B**:4-35, 1959
Brand, P. W.	Tendon transfers for median and ulnar nerve paralysis	Orthop. Clin. N. Amer. **1**:447-454, 1970
Childress, H. M.	Recurrent ulnar-nerve dislocation at the elbow	J. Bone Joint Surg. **38-A**:978-984, 1956
Clawson, D. K., and Seddon, H. J.	The results of repair of the sciatic nerve	J. Bone Joint Surg. **42-B**:205-212, 1960
Combes, M. A., Clark, W. K., Gregory, C. F., and James, J. A.	Sciatic nerve injury in infants, recognition and prevention of impairment resulting from intragluteal injections	J.A.M.A. **173**:1336-1339, 1960
Cullen, C. H.	Causalgia, diagnosis and treatment	J. Bone Joint Surg. **30-B**:467-477, 1948
Garland, H., Sumner, D., and Clark, J. M. P.	Carpal-tunnel syndrome, with particular reference to surgical treatment	Brit. Med. J. **1**:581-584, 1963
Gay, J. R., and Love, J. G.	Diagnosis and treatment of tardy paralysis of the ulnar nerve	J. Bone Joint Surg. **29**:1087-1097, 1947
Goldner, J. L., and Kelley, J. M.	Radial nerve injuries	Southern Med. J. **51**:873-883, 1958
Haymaker, W.	The pathology of peripheral nerve injuries	Milit. Surg. **102**:448-459, 1948
Haymaker, W., and Woodhall, B.	Peripheral nerve injuries, principles of diagnosis, ed. 2	Philadelphia, 1953, W. B. Saunders Co.
Hendry, A. M.	The treatment of residual paralysis after brachial plexus injuries	J. Bone Joint Surg. **31-B**:42-49, 1949
Howard, F. M.	Ulnar-nerve palsy in wrist fractures	J. Bone Joint Surg. **43-A**:1197-1201, 1961
Johnson, J. T. H., and Kendall, H. O.	Isolated paralysis of the serratus anterior muscle	J. Bone Joint Surg. **37-A**:567-574, 1955
King, T., and Morgan, F. P.	Late results of removing the medial humeral epicondyle for traumatic ulnar neuritis	J. Bone Joint Surg. **41-B**:51-55, 1959
Kopell, H. P., and Thompson, W. A. L.	Peripheral entrapment neuropathies	Baltimore, 1963, The Williams & Wilkins Co.
Larsen, R. D., and Posch, J. L.	Nerve injuries in the upper extremity	Arch. Surg. **77**:469-482, 1958
Leffert, R. D., and Seddon, Sir H.	Infraclavicular brachial plexus injuries	J. Bone Joint Surg. **47-B**:9-22, 1965
Littler, J. W.	Tendon transfers and arthrodeses in combined median and ulnar nerve paralysis	J. Bone Joint Surg. **31-A**:225-234, 1949
McGowan, A. J.	The results of transposition of the ulnar nerve for traumatic ulnar neuritis	J. Bone Joint Surg. **32-B**:293-301, 1950
Nicolle, F. V., and Woolhause, F. M.	Nerve compression syndromes of the upper limb	J. Trauma **5**:313-318, 1965
Parry, C. B. W.	Electrodiagnosis	J. Bone Joint Surg. **43-B**:222-236, 1961

Chapter 6: Neuromuscular affections—cont'd

Peripheral nerve injuries—cont'd

Phalen, G. S. The carpal-tunnel syndrome, a 17 years' experience in diagnosis and treatment of 654 hands. J. Bone Joint Surg. **48-A**:211-228, 1966

Riordan, D. C. Surgery of the paralytic hand. American Academy of Orthopaedic Surgeons Instructional Course Lectures, St. Louis, 1959, The C. V. Mosby Co., vol. 16, pp. 79-90

Rosenthal, A. M. Electrodiagnostic testing in neuromuscular disease. J.A.M.A. **177**:829-833, 1961

Sakellarides, H. A follow-up study of 172 peripheral nerve injuries in the upper extremity in civilians. J. Bone Joint Surg. **44-A**:140-148, 1962

Semple, J. C., and Cargill, A. O. Carpal-tunnel syndrome: results of surgical decompression. Lancet **1**:918-919, 1969

Shea, J. D., and McClain, E. J. Ulnar-nerve compression syndromes at and below the wrist. J. Bone Joint Surg. **51-A**:1095-1103, 1969

Stern, W. E. Recognition and treatment of peripheral nerve injuries in civilian practice. J.A.M.A. **178**:462-467, 1961

Sunderland, S. Nerves and nerve injury. Baltimore, 1968, The Williams & Wilkins Co.

Tanzer, R. C. The carpal-tunnel syndrome, a clinical and anatomical study. J. Bone Joint Surg. **41-A**:626-634, 1959

Tracy, J. F., and Brannon, E. W. Management of brachial-plexus injuries (traction type). J. Bone Joint Surg. **40-A**:1031-1042, 1958

Versaci, A. D. Tendon transfers in ulnar-nerve injuries. New Eng. J. Med. **262**:801-804, 1960

Woodhall, B. Common injuries of peripheral nerves. American Academy of Orthopaedic Surgeons Instructional Course Lectures, Ann Arbor, 1954, J. W. Edwards, vol. 11, pp. 269-273

Yeoman, P. M., and Seddon, H. J. Brachial plexus injuries: treatment of the flail arm. J. Bone Joint Surg. **43-B**:493-500, 1961

Obstetric paralysis

Adler, J. B., and Patterson, R. L., Jr. Erb's palsy, long-term results of treatment in eighty-eight cases. J. Bone Joint Surg. **49-A**:1052-1064, 1967

Granberry, W. M., and Lipscomb, P. R. Tendon transfers to the hand in brachial palsy. Amer. J. Surg. **108**:840-844, 1964

Sever, J. W. Obstetrical paralysis. Surg. Gynec. Obstet. **44**:547-549, 1927

Wickstrom, J., Haslam, E. T., and Hutchinson, R. H. The surgical management of residual deformities of the shoulder following birth injuries of the brachial plexus. J. Bone Joint Surg. **37-A**:27-36, 1955

Neuritis

Bardenwerper, H. W. Serum neuritis from tetanus antitoxin. J.A.M.A. **179**:763-766, 1962

Chapter 6: Neuromuscular affections—cont'd

Neuritis—cont'd

Haymaker, W., and Kernohan, J. W.	The Landry-Guillain-Barré syndrome, a clinicopathologic report of 50 fatal cases and a critique of the literature	Medicine **28:**59-141, 1949
Magee, K. R., and De-Jong, R. N.	Paralytic brachial neuritis, discussion of clinical features with review of 23 cases	J.A.M.A. **174:**1258-1262, 1960
Peterman, A. F., Daly, D. D., Dion, F. R., and Keith, H. M.	Infectious neuronitis (Guillain-Barré syndrome) in children	Neurology **9:**533-539, 1959
Von Hagen, K. O., and Baker, R. N.	Infectious neuronitis, present concepts of etiology and treatment	J.A.M.A. **151:**1465-1472, 1953

Involvement of muscles

Adams, R. D., Denny-Brown, D., and Pearson, C. M.	Diseases of muscle, ed. 2	New York, 1962, Paul B. Hoeber, Inc.
Curtis, B. H.	Orthopaedic management of muscular dystrophy and related disorders	American Academy of Orthopaedic Surgeons Instructional Course Lectures, St. Louis, 1970, The C. V. Mosby Co., vol. 19, pp. 78-89
Dawson, C. W., and Roberts, J. B.	Charcot-Marie-Tooth disease	J.A.M.A. **188:**659-661, 1964
Engel, W. K., editor	Symposium: Review of current concepts of myopathies	Clin. Orthop. **39:**3-125, 1965
Jacobs, J. E., and Carr, C. R.	Progressive muscular atrophy of the peroneal type (Charcot-Marie-Tooth disease), orthopaedic management and end-result study	J. Bone Joint Surg. **32-A:**27-38, 1950
Kenrick, M. M.	Certain aspects of managing patients with muscular dystrophy	Southern Med. J. **58:**996-1000, 1965
Miller, J.	Management of muscular dystrophy	American Academy of Orthopaedic Surgeons Instructional Courses, J. Bone Joint Surg. **49-A:**1205-1211, 1967
Ramsey, R. H., and Mc-Carroll, H. R.	Problem of muscular dystrophies	J.A.M.A. **150:**659-662, 1952
Ross, A. T.	The myopathies and the atrophies	American Academy of Orthopaedic Surgeons Instructional Course Lectures, Ann Arbor, 1956, J. W. Edwards, vol. 13, pp. 71-78
Spencer, G. E., Jr.	Orthopaedic care of progressive muscular dystrophy	American Academy of Orthopaedic Surgeons Instructional Courses, J. Bone Joint Surg. **49-A:**1201-1204, 1967
Vignos, P. J., Jr.	Diagnosis of progressive muscular dystrophy	American Academy of Orthopaedic Surgeons Instructional Courses, J. Bone Joint Surg. **49-A:**1212-1220, 1967

Chapter 6: Neuromuscular affections—cont'd

Involvement of muscles—cont'd

Vignos, P. J., Jr., Spencer, G. E., Jr., and Archibald, K. C.	Management of progressive muscular dystrophy of childhood	J.A.M.A. **184**:89-96, 1963
Walton, J. N., editor	Disorders of voluntary muscle	Boston, 1964, Little, Brown & Co.
Zundel, W. S., and Tyler, F. H.	The muscular dystrophies	New Eng. J. Med. **273**:537-543, 596-601, 1965

Chapter 7

Tumors

Tumors and tumorlike affections of bone

Ackerman, L. V., and Spjut, H. J.	Tumors of bone and cartilage	Washington, D. C., 1962, Armed Forces Institute of Pathology
Barnes, R., and Catto, M.	Chondrosarcoma of bone	J. Bone Joint Surg. **48-B**:729-764, 1966
Bhansali, S. K., and Desai, P. B.	Ewing's sarcoma, observation on 107 cases	J. Bone Joint Surg. **45-A**:541-553, 1963
Biesecker, J. L., Marcove, R. C., Huvos, A. G., and Miké, V.	Aneurysmal bone cysts, a clinicopathologic study of 66 cases	Cancer **26**:615-625, 1970
Boseker, E. H., Bickel, W. H., and Dahlin, D. C.	Clinicopathologic study of simple unicameral bone cysts	Surg. Gynec. Obstet. **127**:550-560, 1968
Campbell, C. J., and Harkess, J.	Fibrous metaphyseal defect of bone	Surg. Gynec. Obstet. **104**:329-336, 1957
Changus, G. W., Speed, J. S., and Stewart, F. W.	Malignant angioblastoma of bone, a reappraisal of adamantinoma of long bone	Cancer **10**:540-559, 1957
Coley, B. L.	Neoplasms of bone and related conditions, ed. 2	New York, 1960, Paul B. Hoeber, Inc.
Compere, C. L., and Coleman, S. S.	Nonosteogenic fibroma of bone	Surg. Gynec. Obstet. **105**:588-598, 1957
Coran, A. G., Banks, H. H., Aliapoulios, M. A., and Wilson, R. E.	The management of pathologic fractures in patients with metastatic carcinoma of the breast	Surg. Gynec. Obstet. **127**:1225-1230, 1968
Dahlin, D. C.	Bone tumors, general aspects and data on 3,987 cases, ed. 2	Springfield, Ill., 1967, Charles C Thomas, Publisher
Dahlin, D. C., and Coventry, M. B.	Osteogenic sarcoma	J. Bone Joint Surg. **49-A**:101-110, 1967
Dahlin, D. C., Coventry, M. B., and Scanlon, P. W.	Ewing's sarcoma	J. Bone Joint Surg. **43-A**:185-192, 1961
Donaldson, W. F., Jr.	Aneurysmal bone cyst	J. Bone Joint Surg. **44-A**:25-40, 1962
Dwinnell, L. A., Dahlin, D. C., and Ghormley, R. K.	Parosteal (juxtacortical) osteogenic sarcoma	J. Bone Joint Surg. **36-A**:732-744, 1954

Chapter 7: Tumors—cont'd

Tumors and tumorlike affections of bone—cont'd

Eyre-Brook, A. L., and Price, C. H. G.	Fibrosarcoma of bone	J. Bone Joint Surg. **51-B:**20-37, 1969
Falk, S., and Alpert, M.	The clinical and roentgen aspects of Ewing's sarcoma	Amer. J. Med. Sci. **250:**492-508, 1965
Feldman, F., Hecht, H. L., and Johnston, A. D.	Chondromyxoid fibroma of bone	Radiology **94:**249-260, 1970
Francis, K. C.	Prophylactic internal fixation of metastatic osseous lesions	Cancer **13:**75-76, 1960
Freiberger, R. H., Loitman, B. S., Helpern, M., and Thompson, T. C.	Osteoid osteoma, a report on 80 cases	Amer. J. Roentgen. **82:**194-205, 1959
Garceau, G. J., and Gregory, C. F.	Solitary unicameral bone cyst	J. Bone Joint Surg. **36-A:**267-280, 1954
Ghormley, R. K., and Adson, A. W.	Hemangioma of vertebrae	J. Bone Joint Surg. **23:**887-895, 1941
Gilmer, W. S., Jr., Higley, G. B., Jr., and Kilgore, W. E.	Atlas of bone tumors, including tumorlike lesions	St. Louis, 1963, The C. V. Mosby Co.
Goldenberg, R. R., Campbell, C. J., and Bonfiglio, M.	Giant-cell tumor of bone: an analysis of two hundred and eighteen cases	J. Bone Joint Surg. **52-A:**619-664, 1970
Gorji, J., and Francis, K. C.	Multiple myeloma	Clin. Orthop. **38:**106-119, 1965
Greenwald, C. M., Meaney, T. F., and Hughes, C. R.	Chordoma—uncommon destructive lesion of cerebrospinal axis	J.A.M.A. **163:**1240-1244, 1957
Haggart, G. E., Johnston, D. O., and Creeden, F.	Supervoltage radiation for sarcomata of the pelvis and lower extremities	J. Bone Joint Surg. **40-A:**870-876, 1958
Harkess, J. W.	Parosteal osteosarcoma	Amer. Surg. **30:**730-736, 1964
Henderson, E. D., and Dahlin, D. C.	Chondrosarcoma of bone—a study of 288 cases	J. Bone Joint Surg. **45-A:**1450-1458, 1963
Hustu, H. O., and Pinkel, D.	Lymphosarcoma, Hodgkin's disease and leukemia in bone	Clin. Orthop. **52:**83-93, 1967
Jaffe, H. L.	Tumors and tumorous conditions of the bones and joints	Philadelphia, 1958, Lea & Febiger
Jaffe, H. L., and Lichtenstein, L.	Osteoid-osteoma	J. Bone Joint Surg. **22:**645-682, 1940
	Solitary benign enchondroma of bone	Arch. Surg. **46:**480-493, 1943
Johnson, E. W., Jr., and Dahlin, D. C.	Treatment of giant-cell tumor of bone	J. Bone Joint Surg. **41-A:**895-904, 1959
Jones, J. B.	Symposium: Metastatic bone disease	Clin. Orthop. **73:**2-108, 1970
Kendrick, J. I., and Evarts, C. M.	Osteoid-osteoma: a critical analysis of 40 tumors	Clin. Orthop. **54:**51-59, 1967
Knutson, C. O., and Spratt, J. S., Jr.	The natural history and management of mammary cancer metastatic to the femur	Cancer **26:**1199-1203, 1970

Chapter 7: Tumors—cont'd

Tumors and tumorlike affections of bone—cont'd

Korst, D. R., Clifford, G. O., Fowler, W. M., Louis, J., Will, J., and Wilson, H. E.	Multiple myeloma. II. Analysis of cyclophosphamide therapy in 165 patients	J.A.M.A. **189:**758-762, 1964
Kunkel, M. G., Dahlin, D. C., and Young, H. H.	Benign chondroblastoma	J. Bone Joint Surg. **38-A:**817-826, 1956
Kyle, R. A., Maldonado, J. E., and Bayard, E. D.	Treatment of multiple myeloma	Mayo Clin. Proc. **43:**730-737, 1968
Lichtenstein, L.	Bone tumors, ed. 3	St. Louis, 1965, The C. V. Mosby Co.
Lichtenstein, L., and Jaffe, H. L.	Ewing's sarcoma of bone	Amer. J. Path. **23:**43-78, 1947
Lichtenstein, L., and Sawyer, W. R.	Benign osteoblastoma, further observations and report of 20 additional cases	J. Bone Joint Surg. **46-A:**755-765, 1964
Lumb, G., and Mackenzie, D. H.	Round-cell tumours of bone	Brit. J. Surg. **43:**380-389, 1956
Magnus, H. A., and Wood, H. L.-C.	Primary reticulo-sarcoma of bone	J. Bone Joint Surg. **38-B:**258-278, 1956
McCarroll, H. R.	Practical considerations in the management of malignant bone tumors	J.A.M.A. **152:**297-300, 1953
McKenna, R. J., Schwinn, C. T., Soong, K. Y., and Higinbotham, N. L.	Sarcomata of the osteogenic series (osteosarcoma, fibrosarcoma, chondrosarcoma, parosteal osteogenic sarcoma, and sarcomata arising in abnormal bone), an analysis of 552 cases	J. Bone Joint Surg. **48-A:**1-26, 1966
Marcove, R. C., Miké, V., Hajek, J. V., Levin, A. G., and Hutter, R. V. P.	Osteogenic sarcoma under the age of twenty-one: a review of one hundred and forty-five operative cases	J. Bone Joint Surg. **52-A:**411-423, 1970
Mnaymneh, W. A., Dudley, H. R., and Mnaymneh, L. G.	Giant-cell tumor of bone	J. Bone Joint Surg. **46-A:**63-75, 1964
Moersch, R. N., Bickel, W. H., and Clagett, O. T.	Surgical resection of pulmonary metastatic lesions secondary to tumors of the head, trunk, or extremities	J. Bone Joint Surg. **45-A:**1030-1042, 1963
Moon, N. F.	Adamantinoma of the appendicular skeleton	Clin. Orthop. **43:**189-213, 1965
Morton, J. J., and Mider, G. B.	Chondrosarcoma	Ann. Surg. **126:**895-931, 1947
Neer, C. S., II, Francis, K. C., Marcove, R. C., Terz, J., and Carbonara, P. N.	Treatment of unicameral bone cyst: a follow-up study of 175 cases	J. Bone Joint Surg. **48-A:**731-745, 1966

Chapter 7: Tumors—cont'd

Tumors and tumorlike affections of bone—cont'd

Parrish, F. F., and Murray, J. A.	Surgical treatment for secondary neoplastic fractures: a retrospective study of ninety-six patients	J. Bone Joint Surg. **52-A:**665-686, 1970
Phalen, G. S.	Hodgkin's disease of bone	Clin. Orthop. **13:**234-244, 1959
Price, C. H. G., and Goldie, W.	Paget's sarcoma of bone	J. Bone Joint Surg. **51-B:**205-224, 1969
Ralph, L. L.	Chondromyxoid fibroma of bone	J. Bone Joint Surg. **44-B:**7-24, 1962
Reed, R. J., and Rothenberg, M.	Lesions of bone that may be confused with aneurysmal bone cyst	Clin. Orthop. **35:**150-162, 1964
Scaglietti, O., and Calandriello, B.	Ossifying parosteal sarcoma	J. Bone Joint Surg. **44-A:**635-647, 1962
Schajowicz, F.	Giant-cell tumors of bone (osteoclastoma)	J. Bone Joint Surg. **43-A:**1-29, 1961
Schajowicz, F., and Gallardo, H.	Epiphysial chondroblastoma of bone	J. Bone Joint Surg. **52-B:**205-226, 1970
	Chondromyxoid fibroma (fibromyxoid chondroma) of bone: a clinical-pathological study of 32 cases	J. Bone Joint Surg. **53-B:**198-216, 1971
Schajowicz, F., and Lemos, C.	Osteoid osteoma and osteoblastoma	Acta Orthop. Scand. **41:**272-291, 1970
Selby, S.	Metaphyseal cortical defects in the tubular bones of growing children	J. Bone Joint Surg. **43-A:**395-400, 1961
Sherman, M. S.	Osteoid osteoma, review of the literature and report of thirty cases	J. Bone Joint Surg. **29:**918-930 1947
Sklaroff, D. M., and Charkes, N. D.	Diagnosis of bone metastasis by photoscanning with strontium 85	J.A.M.A. **188:**1-4, 1964
Slowick, F. A., Campbell, C. J., and Kettelkamp, D. B.	Aneurysmal bone cyst	J. Bone Joint Surg. **50-A:**1142-1152, 1968
Sweetnam, R.	Osteosarcoma	Ann. Roy. Coll. Surg. Eng. **44:**38-58, 1969
Sweetnam, R., and Ross, K.	Surgical treatment of pulmonary metastases from primary tumours of bone	J. Bone Joint Surg. **49-B:**74-79, 1967
Troup, J. B., and Bickel, W. H.	Malignant disease of the extremities treated by exarticulation, analysis of 264 consecutive cases with survival rates	J. Bone Joint Surg. **42-A:**1041-1050, 1960
Van Der Heul, R. O., and Von Ronnen, J. R.	Juxtacortical osteosarcoma	J. Bone Joint Surg. **49-A:**415-439, 1967
Wang, C. C., and Fleischli, D. J.	Primary reticulum cell sarcoma of bone, with emphasis on radiation therapy	Cancer **22:**994-998, 1968

Chapter 7: Tumors—cont'd

Tumors and tumorlike affections of bone—cont'd

Watt, J.	The use of cytotoxic drugs in the surgery of malignant disease	J. Bone Joint Surg. **50-B:**511-523, 1968
Weinfeld, M. S., and Dudley, H. R., Jr.	Osteogenic sarcoma	J. Bone Joint Surg. **44-A:**269-276, 1962
Windeyer, B. W.	Chordoma	Proc. Roy. Soc. Med. **52:**1088-1100, 1959

Tumors of soft tissues

Bennett, G. A.	Malignant neoplasms originating in synovial tissues (synoviomata)	J. Bone Joint Surg. **29:**259-291, 1947
Bulmer, J. H.	Smooth muscle tumours of the limbs	J. Bone Joint Surg. **49-B:**52-58, 1967
Codman, N. L., Soule, E. H., and Kelly, P. J.	Synovial sarcoma, an analysis of 134 tumors	Cancer **18:**613-627, 1965
Enzinger, F. M., and Shiraki, M.	Alveolar rhabdomyosarcoma, an analysis of 110 cases	Cancer **24:**18-31, 1969
Galloway, J. D. B., Broders, A. C., and Ghormley, R. K.	Xanthoma of tendon sheaths and synovial membranes, a clinical and pathologic study	Arch. Surg. **40:**485-538, 1940
Ghormley, R. K.	Soft-tissue tumors of the extremities exclusive of skin and lymphatics	Arch. Surg. **72:**817-823, 1956
Hampole, M. K., and Jackson, B. A.	Analysis of 25 cases of malignant synovioma	Canad. Med. Ass. J. **99:**1025-1029, 1968
Hunt, J. C., and Pugh, D. G.	Skeletal lesions in neurofibromatosis	Radiology **76:**1-20, 1961
Ivins, J. C.	Fibrosarcoma of the soft tissues of the extremities	American Academy of Orthopaedic Surgeons Instructional Course Lectures, Ann Arbor, 1954, J. W. Edwards, vol. 11, pp. 18-22
Jenkins, S. A.	Solitary tumours of peripheral nerve trunks	J. Bone Joint Surg. **34-B:**401-411, 1952
Johnson, E. W., Jr., Ghormley, R. K., and Dockerty, M. B.	Hemangiomas of the extremities	Surg. Gynec. Obstet. **102:**531-538, 1956
Lichtenstein, L.	Tumors of synovial joints, bursae and tendon sheaths	Cancer **8:**816-830, 1955
Linscheid, R. L., Soule, E. H., and Henderson, E. D.	Pleomorphic rhabdomyosarcomata of the extremities and limb girdles	J. Bone Joint Surg. **47-A:**715-726, 1965
McCarroll, H. R.	Soft-tissue neoplasms associated with congenital neurofibromatosis	J. Bone Joint Surg. **38-A:**717-731, 1956
Pack, G. T.	End results in the treatment of sarcomata of the soft somatic tissues	J. Bone Joint Surg. **36-A:**241-263, 1954
Pack, G. T., and Ariel, I. M., editors	Treatment of cancer and allied diseases, ed. 2	New York, 1964, Paul B. Hoeber Inc., Medical Division, Harper & Row, Publishers, vol. 8

Chapter 7: Tumors—cont'd

Tumors of soft tissues—cont'd

Pinkel, D., and Pickren, J.	Rhabdomyosarcoma in children	J.A.M.A. **175:**293-298, 1961
Raben, M., Calabrese, A., Higinbotham, N. L., and Phillips, R.	Malignant synovioma	Amer. J. Roentgen. **93:**145-153, 1965
Rabhan, W. N., and Rosai, J.	Desmoplastic fibroma	J. Bone Joint Surg. **50-A:**487-502, 1968
Shallow, T. A., Eger, S. A., and Wagner, F. B., Jr.	Primary hemangiomatous tumors of skeletal muscle	Ann. Surg. **119:**700-740, 1944
Sharpe, J. C., and Young, R. H.	Recklinghausen's neurofibromatosis, clinical manifestations in 31 cases	Arch. Intern. Med. **59:**299-328, 1937
Shugart, R. R., Soule, E. H., and Johnson, E. W., Jr.	Glomus tumor	Surg. Gynec. Obstet. **117:**334-340, 1963
Soule, E. H.	Lipomatous tumors: classification, pathology, and diagnosis	American Academy of Orthopaedic Surgeons Instructional Course Lectures, Ann Arbor, 1957, J. W. Edwards, vol. 14, pp. 311-320
	Myomatous tumors of the extremities: classification and pathology	American Academy of Orthopaedic Surgeons Instructional Course Lectures, Ann Arbor, 1957, J. W. Edwards, vol. 14, pp. 321-328
Soule, E. H., Mahour, G. H., Mills, S. D., and Lynn, H. B.	Soft-tissue sarcomas of infants and children: a clinicopathologic study of 135 cases	Mayo Clin. Proc. **43:**313-326, 1968
Stout, A. P.	Tumors of the peripheral nervous system	Washington, D. C., 1949, Armed Forces Institute of Pathology
	Tumors of the soft tissues	Washington, D. C., 1953, Armed Forces Institute of Pathology
Sullivan, C. R., Dahlin, D. C., and Bryan, R. S.	Lipoma of the tendon sheath	J. Bone Joint Surg. **38-A:**1275-1280, 1956
White, N. B.	Neurilemomas of the extremities	J. Bone Joint Surg. **49-A:**1605-1610, 1967

Chapter 8

Fracture principles, fracture healing

Fracture principles

Adams, J. C.	Outline of fractures, including joint injuries, ed. 4	Baltimore, 1964, The Williams & Wilkins Co.
Altemeier, W. A., Culbertson, W. R., Vetto, M., and Cole, W.	Problems in the diagnosis and treatment of gas gangrene	Arch. Surg. **74:**839-845, 1957
Anderson, L. D.	Compression plate fixation and the effect of different types of internal fixation on fracture healing	American Academy of Orthopaedic Surgeons Instructional Courses, J. Bone Joint Surg. **47-A:**191-208, 1965

Chapter 8: Fracture principles, fracture healing—cont'd

Fracture principles—cont'd

Aufranc, O. E.	Care of the patient with multiple injuries	J.A.M.A. **168**:2091-2094, 1958
Barker, W. F.	Treatment of shock: practical aspects	American Academy of Orthopaedic Surgeons Instructional Courses, J. Bone Joint Surg. **44-A**:767-776, 1962
Bechtol, C. O., Ferguson, A. B., Jr., and Laing, P. G.	Metals and engineering in bone and joint surgery	Baltimore, 1959, The Williams & Wilkins Co.
Blount, W. P.	Fractures in children	Baltimore, 1954, The Williams & Wilkins Co.
Böhler, L.	The treatment of fractures, ed. 5	New York, 1956 and 1966, Grune & Stratton, Inc., vols. 1-3 and 1966 supplementary volume
Bowers, W. F.	Surgery of trauma	Philadelphia, 1953, J. B. Lippincott Co.
Bradford, D. S., Foster, R. R., and Nossel, H. L.	Coagulation alterations, hypoxemia, and fat embolism in fracture patients	J. Trauma **10**:307-321, 1970
Brav, E. A.	The management of open fractures of the extremities	American Academy of Orthopaedic Surgeons Instructional Course Lectures, Ann Arbor, 1956, J. W. Edwards, vol. 13, pp. 227-233
Cave, E. F., editor	Fractures and other injuries	Chicago, 1958, Year Book Medical Publishers, Inc.
Charnley, J.	The closed treatment of common fractures, ed. 3	Baltimore, 1958, The Williams & Wilkins Co.
Clowes, G. H. A., Jr.	Metabolic responses to injury	J. Trauma **3**:149-175, 1963
Cobb, C. A., Jr., and Hillman, J. W.	Fat embolism	American Academy of Orthopaedic Surgeons Instructional Course Lectures, St. Louis, 1961, The C. V. Mosby Co., vol. 18, pp. 122-129
Collins, D. H.	Structural changes around nails and screws in human bones	J. Path. Bact. **65**:109-121, 1953
Committee on Trauma, American College of Surgeons	The management of fractures and soft tissue injuries	Philadelphia, 1965, W. B. Saunders Co.
Compere, E. L., Banks, S. W., and Compere, C. L.	Pictorial handbook of fracture treatment, ed. 5	Chicago, 1963, Year Book Medical Publishers, Inc.
Conwell, H. E., and Reynolds, F. C.	Key and Conwell's management of fractures, dislocations, and sprains, ed. 7	St. Louis, 1961, The C. V. Mosby Co.
Eggers, G. W. N.	Internal contact splint	J. Bone Joint Surg. **30-A**:40-52, 1948
	The internal fixation of fractures of the shafts of long bones	In Carter, B. N., et al., editors: Monographs on surgery, Baltimore, 1952, The Williams & Wilkins Co.

Chapter 8: Fracture principles, fracture healing—cont'd

Fracture principles—cont'd

Estes, W. L., Jr.	The surgeon's responsibility in rehabilitation of the injured	Surg. Gynec. Obstet. **100:**619-621, 1955
Fine, J., Palmerio, C., and Rutenburg, S.	New developments in therapy of refractory traumatic shock	Arch. Surg. **96:**163-175, 1968
Flatt, A. E.	The care of minor hand injuries, ed. 2	St. Louis, 1963, The C. V. Mosby Co.
Frankel, C. J., Bateman, J. E., Eaton, G. O., Kessler, H. H., and McBride, E. D.	Symposium on disability evaluation	American Academy of Orthopaedic Surgeons Instructional Course Lectures, St. Louis, 1960, The C. V. Mosby Co., vol. 17, pp. 331-350
Godfrey, J. D.	Major and extensive soft-tissue injuries complicating skeletal fractures	American Academy of Orthopaedic Surgeons Instructional Courses, J. Bone Joint Surg. **44-A:**753-766, 1962
Gurd, F. N., editor	Symposium on shock	J. Trauma **2:**355-423, 1962
Hamit, H. F.	Current trends of therapy and research in shock	Surg. Gynec. Obstet. **120:**835-854, 1965
Hardaway, R. M., James, P. M., Anderson, R. W., Bredenberg, C. E., and West, R. L.	Intensive study and treatment of shock in man	J.A.M.A. **199:**779-790, 1967
Harris, W. R.	Epiphyseal injuries	American Academy of Orthopaedic Surgeons Instructional Course Lectures, Ann Arbor, 1958, J. W. Edwards, vol. 15, pp. 206-214
Higinbotham, N. L., and Marcove, R. C.	The management of pathological fractures	J. Trauma **5:**792-798, 1965
Hoover, N. W., and Ivins, J. C.	Wound debridement	Arch. Surg. **79:**701-710, 1959
Howard, J. M.	Fluid replacement in shock and hemorrhage	J.A.M.A. **173:**516-518, 1960
Hughes, C. W., and Bowers, W. F.	Traumatic lesions of peripheral vessels	Springfield, Ill., 1961, Charles C Thomas, Publisher
Klingensmith, W., Oles, P., and Martinez, H.	Fractures with associated blood vessel injury	Amer. J. Surg. **110:**849-852, 1965
Liebenson, H. A.	The doctor in personal injury cases	Chicago, 1956, Year Book Medical Publishers, Inc.
Lowery, B. D., Cloutier, C. T., and Cary, L. C.	Blood gas determinations in the severely wounded in hemorrhagic shock	Arch. Surg. **99:**330-338, 1969
McCarroll, H. R.	Orthopedic management of the severely injured patient	J.A.M.A. **165:**1913-1916, 1957
McLaughlin, H. L.	Internal fixation of fractures	Surgery **39:**892-899, 1956
	Trauma	Philadelphia, 1959, W. B. Saunders Co.
Mazet, R., Jr.	A manual of closed reduction of closed fractures and dislocations	Springfield, Ill., 1967, Charles C Thomas, Publisher
Moore, J. R.	The closed fracture of the long bones	J. Bone Joint Surg. **42-A:**869-874, 1960

Chapter 8: Fracture principles, fracture healing—cont'd

Fracture principles—cont'd

Moyer, C. A.	Fluid balance and trauma	American Academy of Orthopaedic Surgeons Instructional Course Lectures, Ann Arbor, 1954, J. W. Edwards, vol. 11, pp. 275-278
Müller, M. E., Allgöwer, M., and Willenegger, H.	Technique of internal fixation of fractures	New York, 1965, Springer-Verlag
Owens, J. C.	The management of arterial trauma	Surg. Clin. N. Amer. **43**:371-385, 1963
Parrish, F. F., and Murray, J. A.	Surgical treatment for secondary neoplastic fractures	J. Bone Joint Surg. **52-A**:665-686, 1970
Patman, R. D., Poulos, E., and Shires, G. T.	The management of civilian arterial injuries	Surg. Gynec. Obstet. **118**:725-738, 1964
Peltier, L. F.	Fat embolism, a current concept	Clin. Orthop. **66**:241-253, 1969
Peterson, L. T.	Principles of internal fixation with plates and screws	Arch. Surg. **64**:345-354, 1952
Ralston, E. L.	Handbook of fractures	St. Louis, 1967, The C. V. Mosby Co.
Rhoads, J. E., and Howard, J. M.	The chemistry of trauma	Springfield, Ill., 1963, Charles C Thomas, Publisher
Rifkind, D.	The diagnosis and treatment of gas gangrene	Surg. Clin. N. Amer. **43**:511-517, 1963
Schmeisser, G., Jr.	A clinical manual of orthopedic traction techniques	Philadelphia, 1963, W. B. Saunders Co.
Schumer, W., and Sperling, R.	Shock and its effect on the cell	J.A.M.A. **205**:215-219, 1968
Scuderi, C.	Atlas of orthopedic traction procedures	St. Louis, 1954, The C. V. Mosby Co.
Seeley, S. F.	Emergency care of wounds	J.A.M.A. **173**:518-521, 1960
Skudder, P. A., and McCarroll, J. R.	Current status of tetanus control	J.A.M.A. **188**:625-627, 1964
Steel, H. H.	Surgical infections—orthopaedic considerations	American Academy of Orthopaedic Surgeons Instructional Course Lectures, St. Louis, 1961, The C. V. Mosby Co., vol. 18, pp. 288-293
Stein, A. H., Jr.	Diagnosis of arterial injury in the extremities	American Academy of Orthopaedic Surgeons Instructional Course Lectures, St. Louis, 1960, The C. V. Mosby Co., vol. 17, pp. 47-60
Stetler, C. J., and Moritz, A. R.	Doctor and patient and the law, ed. 4	St. Louis, 1962, The C. V. Mosby Co.
Stokes, J. M.	Surgical management of vascular injuries associated with long bone fracture	American Academy of Orthopaedic Surgeons Instructional Course Lectures, St. Louis, 1960, The C. V. Mosby Co., vol. 17, pp. 61-67
Street, D. M.	Medullary nailing of the femur, comparative study of skeletal traction, dual plating and medullary nailing	J.A.M.A. **143**:709-714, 1950

Chapter 8: Fracture principles, fracture healing—cont'd

Fracture principles—cont'd

Strode, J. E.	Postoperative wound infections	Arch. Surg. **79**:141-143, 1959
Teschan, P. E.	Management of patients with posttraumatic renal insufficiency	J. Trauma **3**:181-188, 1963
Wade, P. A.	Fractures in children	Amer. J. Surg. **107**:531-536, 1964
Watson-Jones, Sir R.	Fractures and joint injuries, ed. 4	Baltimore, 1952-1955, The Williams & Wilkins Co.
Wiles, P.	Fractures, dislocations and sprains	Boston, 1960, Little, Brown & Co.
Wilson, J. N.	Rational approach to management of clinical shock	Arch. Surg. **91**:92-120, 1965

Bone formation and repair

Bassett, C. A. L.	Current concepts of bone formation	American Academy of Orthopaedic Surgeons Instructional Courses, J. Bone Joint Surg. **44-A**:1217-1244, 1962
Bourne, G. H.	The biochemistry and physiology of bone	New York, 1956, Academic Press, Inc.
Clark, J. M. P., editor	Modern trends in orthopaedics, science of fractures, fourth series	Washington, D. C., 1964, Butterworth, Inc.
Eggers, G. W. N., Shindler, T. O., and Pomerat, C. M.	Influence of contact-compression factor on osteogenesis in surgical fractures	J. Bone Joint Surg. **31-A**:693-716, 1949
Ham, A. W.	Some histophysiological problems peculiar to calcified tissues	J. Bone Joint Surg. **34-A**:701-728, 1952
Howard, J. E.	Some current concepts on the mechanism of calcification	J. Bone Joint Surg. **33-A**:801-806, 1951
McLean, F. C., and Budy, A. M.	Radioisotopes in the study of bone	Clin. Orthop. **24**:178-197, 1962
Phemister, D. B.	Repair of bone in the presence of aseptic necrosis resulting from fractures, transplantations, and vascular obstruction	J. Bone Joint Surg. **12**:769-787, 1930
Rhinelander, F. W.	The normal microcirculation of diaphyseal cortex and its response to fracture	American Academy of Orthopaedic Surgeons Instructional Course Lectures, J. Bone Joint Surg. **50-A**:784-800, 1968
Robinson, R. A.	An electron-microscopic study of the crystalline inorganic component of bone and its relationship to the organic matrix	J. Bone Joint Surg. **34-A**:389-435, 1952

Delayed union, nonunion, and malunion

Abbott, L. C., Schottstaedt, E. R., Saunders, J. B. De C. M., and Bost, F. C.	The evaluation of cortical and cancellous bone as grafting material	J. Bone Joint Surg. **29**:381-414, 1947

Chapter 8: Fracture principles, fracture healing—cont'd

Delayed union, nonunion, and malunion—cont'd

Barr, J. S., et al.	Fracture of the carpal navicular (scaphoid) bone, an end-result study in military personnel	J. Bone Joint Surg. **35-A:**609-625, 1953
Blount, W. P.	Proximal osteotomies of the femur	American Academy of Orthopaedic Surgeons Instructional Course Lectures, Ann Arbor, 1952, J. W. Edwards, vol. 9, pp. 1-29
Bonfiglio, M., and Voke, E. M.	Aseptic necrosis of the femoral head and nonunion of the femoral neck	J. Bone Joint Surg. **50-A:**48-66, 1968
Boyd, H. B.	Avascular necrosis of the head of the femur	American Academy of Orthopaedic Surgeons Instructional Course Lectures, Ann Arbor, 1957, J. W. Edwards, vol. 14, pp. 196-204
Boyd, H. B., and Brindley, H. H.	Nonunion of the neck of the femur, study of 347 cases	Arch. Surg. **65:**169-180, 1952
Boyd, H. B., Lipinski, S. W., and Wiley, J. H.	Observations on non-union of the shafts of the long bones, with a statistical analysis of 842 patients	J. Bone Joint Surg. **43-A:**159-168, 1961
Boyd, H. B., and Sage, F. P.	Congenital pseudarthrosis of the tibia	J. Bone Joint Surg. **40-A:**1245-1270, 1958
Boyd, H. B., Wray, J. B., Brashear, H. R., and Hohl, M.	Symposium: Treatment of ununited fractures of the long bones	American Academy of Orthopaedic Surgeons Instructional Courses, J. Bone Joint Surg. **47-A:**167-190, 1965
Brav, E. A., and Blair, J. D.	Reconstruction procedures for forearm bone defects	Amer. J. Surg. **86:**139-144, 1953
Campbell, W. C.	Malunited Colles' fractures	J.A.M.A. **109:**1105-1108, 1937
	Onlay bone graft for ununited fractures	Arch. Surg. **38:**313-327, 1939
Capener, N.	Reconstructive surgery of the hip joint	In Platt, Sir H., editor: Modern trends in orthopaedics, second series, New York, 1956, Paul B. Hoeber, Inc., Medical Division, Harper & Row, Publishers
Carr, C. R., and Hyatt, G. W.	Clinical evaluation of freeze-dried bone grafts	J. Bone Joint Surg. **37-A:**549-566, 1955
Cave, E. F.	The healing of fractures and nonunion of bone	Surg. Clin. N. Amer. **43:**337-349, 1963
Charnley, J.	Congenital pseudarthrosis of the tibia treated by the intramedullary nail	J. Bone Joint Surg. **38-A:**283-290, 1956
Cleveland, M., and Fielding, J. W.	A continuing end-result study of intracapsular fracture of the neck of the femur	J. Bone Joint Surg. **36-A:**1020-1030, 1954

Chapter 8: Fracture principles, fracture healing—cont'd

Delayed union, nonunion, and malunion—cont'd

Colonna, P. C.	The trochanteric reconstruction operation for ununited fractures of the upper end of the femur	J. Bone Joint Surg. **42-B**:5-10, 1960
Connelly, J. R.	Plastic surgery in bone problems	Plast. Reconstr. Surg. **17**:129-167, 1956
Coventry, M. B.	The Phemister bone graft in ununited fractures of the long bones	Clin. Orthop. **2**:194-202, 1953
Dickson, J. A.	The high geometric osteotomy, with rotation and bone graft, for ununited fractures of the neck of the femur	J. Bone Joint Surg. **29**:1005-1018, 1947
	The "unsolved" fracture	J. Bone Joint Surg. **35-A**:805-822, 1953
Dineen, J. R., and Gresham, R. B.	Rib osteoperiosteal grafts	J. Bone Joint Surg. **44-A**:1653-1658, 1962
Flanagan, J. J., and Burem, H. S.	Reconstruction of defects of the tibia and femur with apposing massive grafts from the affected bone	J. Bone Joint Surg. **29**:587-597, 1947
Garber, C. Z., and Bush, L. F.	The bone bank and the use of homogenous bone	In Carter, B. N., editor: Monographs on surgery for 1950, New York, 1950, Thomas Nelson & Sons
Gibson, A., and Loadman, B.	The bridging of bone defects	J. Bone Joint Surg. **30-A**:381-396, 1948
Green, W. T., and Rudo, N.	Pseudarthrosis and neurofibromatosis	Arch. Surg. **46**:639-651, 1943
Hallock, H.	Arthrodesis of the ankle joint for old painful fractures	J. Bone Joint Surg. **27**:49-58, 1945
Herndon, C. H.	Principles of bone graft surgery—different methods of operative procedure and indications for each	American Academy of Orthopaedic Surgeons Instructional Course Lectures, St. Louis, 1960, The C. V. Mosby Co., vol. 17, pp. 149-164
Horwitz, T., and Lambert, R. G.	Massive iliac bone grafts in the treatment of ununited fractures and large defects of long bones: the combined bone graft–metallic plate technique	Surg. Gynec. Obstet. **84**:435-450, 1947
Jackson, R. W., and Macnab, I.	Fractures of the shaft of the tibia, a clinical and experimental study	Amer. J. Surg. **97**:543-557, 1959
Johnson, E. W., Jr., and Collins, H. R.	Nonunion of the clavicle	Arch. Surg. **87**:963-966, 1963
King, D., and Secor, C.	Bow elbow (cubitus varus)	J. Bone Joint Surg. **33-A**:572-576, 1951
Lowe, H. G.	Radio-ulnar fusion for defects in the forearm bones	J. Bone Joint Surg. **45-B**:351-359, 1963

Chapter 8: Fracture principles, fracture healing—cont'd

Delayed union, nonunion, and malunion—cont'd

MacAusland, W. R.	Total patellectomy	Amer. J. Surg. **87**:221-226, 1954
Mazet, R., Jr., and Hohl, M.	Fractures of the carpal navicular	J. Bone Joint Surg. **45-A**:82-112, 1963
McBride, E. D.	Disability evaluation: principles of treatment of compensable injuries, ed. 6 (revised)	Philadelphia, 1963, J. B. Lippincott Co.
McCarroll, H. R.	The surgical management of ununited fractures of the tibia	J.A.M.A. **175**:578-583, 1961
McFarland, B.	Pseudarthrosis of the tibia in childhood	J. Bone Joint Surg. **33-B**:36-46, 1951
McKeever, F. M.	Fractures of tarsal and metatarsal bones	Surg. Gynec. Obstet. **90**:735-745, 1950
Milch, H.	Fractures of the external humeral condyle	J.A.M.A. **160**:641-646, 1956
Moore, J. R.	Congenital pseudarthrosis of the tibia	American Academy of Orthopaedic Surgeons Instructional Course Lectures, Ann Arbor, 1957, J. W. Edwards, vol. 14, pp. 222-237
Phemister, D. B.	Treatment of ununited fractures by onlay bone grafts without screw or tie fixation and without breaking down of the fibrous union	J. Bone Joint Surg. **29**:946-960, 1947
Rankin, E. A., and Metz, C. W., Jr.	Management of delayed union in early weight-bearing treatment of the fractured tibia	J. Trauma **10**:751-759, 1970
Ray, R. D., Sankaran, B., and Fetrow, K. O.	Delayed union and non-union of fractures	American Academy of Orthopaedic Surgeons Instructional Courses, J. Bone Joint Surg. **46-A**:627-643, 1964
Reich, R. S.	Treatment of ununited fractures of the neck of the femur	American Academy of Orthopaedic Surgeons Instructional Course Lectures, St. Louis, 1960, The C. V. Mosby Co., vol. 17, pp. 68-77
Sakellarides, H. T., Freeman, P. A., and Grant, B. D.	Delayed union and non-union of tibial-shaft fractures: a review of 100 cases	J. Bone Joint Surg. **46-A**:557-569, 1964
Shands, A. R., Jr.	Malunited fractures of lower end of humerus	Amer. J. Surg. **36**:679-693, 1937
Sherman, M. S., and Phemister, D. B.	The pathology of ununited fractures of the neck of the femur	J. Bone Joint Surg. **29**:19-40, 1947
Smith-Petersen, M. N.	Treatment of fractures of the neck of the femur by internal fixation	Surg. Gynec. Obstet. **64**:287-295, 1937
Stewart, M. J.	Fractures of the carpal navicular (scaphoid): a report of 436 cases	J. Bone Joint Surg. **36-A**:998-1006, 1954

Chapter 8: Fracture principles, fracture healing—cont'd

Delayed union, nonunion, and malunion—cont'd

Stewart, M. J., and Wells, R. E.	Osteotomy and osteotomy combined with bone-grafting for non-union following fracture of the femoral neck	J. Bone Joint Surg. **38-A**:33-49, 1956
Taylor, L. W.	Principles of treatment of fractures and non-union of the shaft of the femur	American Academy of Orthopaedic Surgeons Instructional Courses, J. Bone Joint Surg. **45-A**:191-198, 1963
Urist, M. R.	Bone: transplants, implants, derivatives, and substitutes—a survey of research of the past decade	American Academy of Orthopaedic Surgeons Instructional Course Lectures, St. Louis, 1960, The C. V. Mosby Co., vol. 17, pp. 184-195
Urist, M. R., Mazet, R., Jr., and McLean, F. C.	The pathogenesis and treatment of delayed union and non-union: a survey of 85 ununited fractures of the shaft of the tibia and 100 control cases with similar injuries	J. Bone Joint Surg. **36-A**:931-968, 1954
Van Nes, C. P.	Congenital pseudarthrosis of the leg	J. Bone Joint Surg. **48-A**:1467-1483, 1966
Watson-Jones, R., and Coltart, W. D.	Slow union of fractures, with a study of 804 fractures of the shafts of the tibia and femur	Brit. J. Surg. **30**:260-276, 1943
Wilson, P. D.	Fracture of the lateral condyle of the humerus in childhood	J. Bone Joint Surg. **18**:301-318, 1936
ZumBrunnen, J. L., and Brindley, H. H.	Nonunion of the shafts of the long bones	J.A.M.A. **203**:637-640, 1968

Chapter 9

Amputations, prostheses, braces

Amputations

Aitken, G. T.	Surgical amputation in children	American Academy of Orthopaedic Surgeons Instructional Courses, J. Bone Joint Surg. **45-A**:1735-1741, 1963
Aitken, G. T., and Frantz, C. H.	Management of the child amputee	American Academy of Orthopaedic Surgeons Instructional Course Lectures, St. Louis, 1960, The C. V. Mosby Co., vol. 17, pp. 246-295
Alldredge, R. H.	Amputations and artificial limbs	In Christopher, F., editor: Textbook of surgery, ed. 6, Philadelphia, 1956, W. B. Saunders Co.
Bailey, R. W., and Stevens, D. B.	Radical exarticulation of the extremities for the curative and palliative treatment of malignant neoplasms	J. Bone Joint Surg. **43-A**:845-854, 1961

Chapter 9: Amputations, prostheses, braces—cont'd

Amputations—cont'd

Brav, E. A., et al.	Cineplasty	J. Bone Joint Surg. **39-A**:59-76, 1957
Burgess, E. M., Romano, R. L., and Zettl, J. H.	The management of lower extremity amputees	Washington, D. C., 1967, Government Printing Office
Burgess, E. M., Traub, J. E., and Wilson, A. B.	Immediate post-surgical prostheses	Washington, D. C., 1967, Government Printing Office
Compere, C. L., and Thompson, R. G.	Amputations and modern prosthetics	Surg. Clin. N. Amer. **37**:103-118, 1957
Dale, R. H.	Electrical accidents, a discussion with illustrative cases	Brit. J. Plast. Surg. **7**:44-66, 1954
Elftman, H.	Biomechanics of muscle, with particular application to studies of gait	American Academy of Orthopaedic Surgeons Instructional Courses, J. Bone Joint Surg. **48-A**:363-377, 1966
Falconer, M. A.	Surgical treatment of intractable phantom-limb pain	Brit. Med. J. **1**:299-304, 1953
Gillis, L.	Amputations	London, 1954, William Heinemann, Ltd.
Glattly, H. W.	A statistical study of 12,000 new amputees	Southern Med. J. **57**:1373-1378, 1964
Hall, C. B., Brooks, M. B., and Dennis, J. F.	Congenital skeletal deficiencies of the extremities, classification and fundamentals of treatment	J.A.M.A. **181**:590-599, 1962
Harris, R. I.	Syme's amputation	J. Bone Joint Surg. **38-B**:614-632, 1956
Kelly, P. J., and Janes, J. M.	Criteria for determining the proper level of amputation in occlusive vascular disease	J. Bone Joint Surg. **39-A**:883-891, 1957
Kihn, R. B., Golbranson, F. L., Hutchinson, R. H., et al.	The immediate post-operative prosthesis in lower extremity amputations	Arch. Surg. **101**:40-44, 1970
Lambert, C. N.	Amputations of the lower extremity	Surg. Clin. N. Amer. **45**:147-156, 1965
Larmon, W. A.	Amputations and limb substitution	In Davis, L., editor: Christopher's textbook of surgery, ed. 8, Philadelphia, 1964, W. B. Saunders Co.
Marmor, L., editor	Symposium: Amputations and prostheses	Clin. Orthop. **37**:7-112, 1964
Mazet, R., Jr.	Cineplasty	J. Bone Joint Surg. **40-A**:1389-1400, 1958
	Syme's amputation: a follow-up study of fifty-one adults and thirty-two children	J. Bone Joint Surg. **50-A**:1549-1563, 1968
Mazet, R., Jr., Taylor, C. L., and Bechtol, C. O.	Upper-extremity amputation surgery and prosthetic prescription	J. Bone Joint Surg. **38-A**:1185-1198, 1956
Mercer, W.	Syme's amputation	Artif. Limbs **6**:1-3, 1961
Morris, H. D.	Amputations for congenital anomalies of the lower extremities	American Academy of Orthopaedic Surgeons Instructional Course Lectures, Ann Arbor,

Chapter 9: Amputations, prostheses, braces—cont'd

Amputations—cont'd

Morris, H. D.—cont'd		1958, J. W. Edwards, vol. 15, pp. 255-261
Moseley, H. F.	The forequarter amputation	Philadelphia, 1957, J. B. Lippincott Co.
Pack, G. T.	Major exarticulations for malignant neoplasms of the extremities	J. Bone Joint Surg. **38-A:**249-262, 1956
Pedersen, H. E.	Lower extremity amputations for gangrene	American Academy of Orthopaedic Surgeons Instructional Course Lectures, Ann Arbor, 1958, J. W. Edwards, vol. 15, pp. 262-281
Russell, W. R., and Spalding, J. M. K.	Treatment of painful amputation stumps	Brit. Med. J. **2:**68-73, 1950
Slocum, D. B.	An atlas of amputations	St. Louis, 1949, The C. V. Mosby Co.
	Amputations of the fingers and the hand	Clin. Orthop. **15:**35-59, 1959
Swanson, A. B.	Levels of amputation of fingers and hand—considerations for treatment	Surg. Clin. N. Amer. **44:**1115-1126, 1964
	The Krukenberg procedure in the juvenile amputee	J. Bone Joint Surg. **46-A:**1540-1548, 1964
Thompson, R. G., Hanger, H. B., Fryer, C. M., and Wilson, A. B., Jr.	Above-the-knee amputations and prosthetics	American Academy of Orthopaedic Surgeons Instructional Courses, J. Bone Joint Surg. **47-A:**619-630, 1965
Zanoli, R.	Krukenberg-Putti amputation-plasty	J. Bone Joint Surg. **39-B:**230-232, 1957

Prostheses

American Academy of Orthopaedic Surgeons, Office of the Surgeon General of the U. S. Army, and the Veterans Administration	Orthopaedic appliances atlas, vol. 2: artificial limbs	Ann Arbor, 1960, J. W. Edwards
Anderson, M. H., Bechtol, C. O., and Sollars, R. E.	Clinical prosthetics for physicians and therapists	Springfield, Ill., 1959, Charles C Thomas, Publisher
Bechtol, C. O., and Compere, C. L.	The above-knee suction-socket artificial leg	American Academy of Orthopaedic Surgeons Instructional Course Lectures, Ann Arbor, 1952, J. W. Edwards, vol. 9, pp. 247-249
Bunnell, S.	The management of the nonfunctional hand—reconstruction vs. prosthesis	Artif. Limbs **4:**76-102, 1957
Klopsteg, P. E., and Wilson, P. D.	Human limbs and their substitutes (facsimile reprint of 1954 edition)	New York, 1969, Hafner Publishing Co., Inc.

Chapter 9: Amputations, prostheses, braces—cont'd

Prostheses—cont'd

Lambert, C. N.	Upper-extremity prostheses in juvenile amputees	J. Bone Joint Surg. **38-A:**421-426, 1956
McKenzie, D. S.	The prosthetic management of congenital deformities of the extremities	J. Bone Joint Surg. **39-B:**233-247, 1957
	The clinical application of externally powered artificial arms	J. Bone Joint Surg. **47-B:**399-410, 1965
Murphy, E. F., and Wilson, A. B.	Anatomical and physiological considerations in below-knee prosthetics	Artif. Limbs **6:**4-15, 1962
Nickel, V. L., Savill, D. L., Karchak, A., Jun., and Allen, J. R.	Synthetically powered orthotic systems	J. Bone Joint Surg. **47-B:**458-464, 1965
Tosberg, W. A.	Upper and lower extremity prostheses	Springfield, Ill., 1962, Charles C Thomas, Publisher
Wilson, A. B.	Limb prosthetics—1970	Artif. Limbs **14:**1-52, 1970

Braces

Alldredge, R. H., and Snow, B. M.	Principles of ambulatory bracing of the lower extremity	American Academy of Orthopaedic Surgeons Instructional Course Lectures, Ann Arbor, 1953, J. W. Edwards, vol. 10, pp. 293-298
American Academy of Orthopaedic Surgeons, Office of the Surgeon General of the U. S. Army, and the Veterans Administration	Orthopaedic appliances atlas, vol. 2: braces, splints, shoe alterations	Ann Arbor, 1952, J. W. Edwards
Anderson, M. H.	Functional bracing of the upper extremities	Springfield, Ill., 1958, Charles C Thomas, Publisher
Bloomberg, M. H.	Orthopedic braces: rationale, classification, and prescription	Philadelphia, 1964, J. B. Lippincott Co.
Bunnell, S.	Splinting the hand	American Academy of Orthopaedic Surgeons Instructional Course Lectures, Ann Arbor, 1952, J. W. Edwards, vol. 9, pp. 233-243
Deaver, G. G., and Brittis, A. L.	Braces, crutches, wheelchairs	New York, 1953, Institute of Physical Medicine and Rehabilitation
Jebsen, R. H.	Use and abuse of ambulation aids	J.A.M.A. **199:**63-68, 1967
Jordan, H. H.	Orthopaedic appliances: the principles and practice of brace construction, ed. 2	Springfield, Ill., 1963, Charles C Thomas, Publisher
Perry, J., and Hislot, H. J.	Principles of lower extremity bracing	New York, 1967, American Physical Therapy Association
Schottstaedt, E. R., and Robinson, G. B.	Functional bracing of the arm	J. Bone Joint Surg. **38-A:**477-499, 841-856, 1956

Chapter 9: Amputations, prostheses, braces—cont'd

Braces—cont'd

Street, D. M.	Paraplegic bracing	American Academy of Orthopaedic Surgeons Instructional Course Lectures, Ann Arbor, 1957, J. W. Edwards, vol. 14, pp. 336-341
Thomas, A.	Braces for the spine and trunk	American Academy of Orthopaedic Surgeons Instructional Course Lectures, Ann Arbor, 1953, J. W. Edwards, vol. 10, pp. 298-302

Chapter 10

Affections of the spine and thorax

Bingold, A. C.	Congenital kyphosis	J. Bone Joint Surg. **35-B:**579-583, 1953
Blount, W. P., Schmidt, A. C., Keever, E. D., and Leonard, E. T.	The Milwaukee brace in the operative treatment of scoliosis	J. Bone Joint Surg. **40-A:**511-525, 1958
Chin, E. F.	Surgery of funnel chest and congenital sternal prominence	Brit. J. Surg. **44:**360-376, 1957
Cobb, J. R.	Technique, after-treatment, and results of spine fusion for scoliosis	American Academy of Orthopaedic Surgeons Instructional Course Lectures, Ann Arbor, 1952, J. W. Edwards, vol. 9, pp. 65-70
Collis, D. K., and Ponseti, I. V.	Long-term follow-up of patients with idiopathic scoliosis not treated surgically	J. Bone Joint Surg. **51-A:**425-445, 1969
Compere, E. L., Johnson, W. E., and Coventry, M. B.	Vertebra plana (Calvé's disease) due to eosinophilic granuloma	J. Bone Joint Surg. **36-A:**969-980, 1954
Epstein, B. S.	The spine, a radiological text and atlas, ed. 3	Philadelphia, 1969, Lea & Febiger
Eyring, E. J., Peterson, C. A., and Bjornson, D. R.	Intervertebral-disc calcification in childhood	J. Bone Joint Surg. **46-A:**1432-1441, 1964
Flinchum, D.	Rib resection in the treatment of scoliosis	Southern Med. J. **56:**1378-1380, 1963
Goldstein, L.	Treatment of idiopathic scoliosis by Harrington instrumentation and fusion with fresh autogenous iliac bone grafts, results in eighty patients	J. Bone Joint Surg. **51-A:**209-222, 1969
Harrington, P. R.	Treatment of scoliosis, correction and internal fixation by spine instrumentation	J. Bone Joint Surg. **44-A:**591-610, 1962
	Management of scoliosis by spine instrumentation, evaluation of more than 200 cases	Southern Med. J. **56:**1367-1377, 1963
James, J. I. P.	Paralytic scoliosis	J. Bone Joint Surg. **38-B:**660-685, 1956

Chapter 10: Affections of the spine and thorax—cont'd

James, J. I. P., Lloyd-Roberts, G. C., and Pilcher, M. F.	Infantile structural scoliosis	J. Bone Joint Surg. **41-B:**719-735, 1959
James, J. I. P., Zorab, P. A., and Wynne-Davies, R.	Scoliosis	Edinburgh, 1967, E. & S. Livingstone, Ltd.
Keegan, J. J.	Alterations of the lumbar curve related to posture and seating	J. Bone Joint Surg. **35-A:**589-603, 1953
Keith, A.	Man's posture: its evolution and disorders	Brit. Med. J. **1:**451-454, 1923
Kilfoyle, R. M., Foley, J. J., and Norton, P. L.	Spine and pelvic deformity in childhood and adolescent paraplegia: a study of 104 cases	J. Bone Joint Surg. **47-A:**659-682, 1965
Lester, C. W.	Pigeon breast, funnel chest, and other congenital deformities of the chest	J.A.M.A. **156:**1063-1067, 1954
McKenzie, K. G., and Dewar, F. P.	Scoliosis with paraplegia	J. Bone Joint Surg. **31-B:**162-174, 1949
Moe, J. H.	The management of paralytic scoliosis	Southern Med. J. **50:**67-81, 1957
	A critical analysis of methods of fusion for scoliosis	J. Bone Joint Surg. **40-A:**529-554, 1958
	The Milwaukee brace in the treatment of scoliosis	Clin. Orthop. **77:**18-31, 1971
Moe, J. H., and Gustilo, R. B.	Treatment of scoliosis, results in 196 patients treated by cast correction and fusion	J. Bone Joint Surg. **46-A:**293-312, 1964
Moe, J. H., and Kettleson, D. N.	Idiopathic scoliosis: Analysis of curve patterns and the preliminary results of Milwaukee-brace treatment in one hundred sixty-nine patients	J. Bone Joint Surg. **52-A:**1509-1533, 1970
Morgan, T. H., and Scott, J. C.	Treatment of infantile idiopathic scoliosis	J. Bone Joint Surg. **38-B:**450-457, 1956
Nassim, R., and Burrows, H. J., editors	Modern trends in diseases of the vertebral column	New York, 1959, Paul B. Hoeber, Inc.
Outland, T., and Snedden, H. E.	Juvenile dorsal kyphosis	Clin. Orthop. **5:**155-163, 1955
Ponseti, I. V., and Friedman, B.	Prognosis in idiopathic scoliosis	J. Bone Joint Surg. **32-A:**381-395, 1950
	Changes in the scoliotic spine after fusion	J. Bone Joint Surg. **32-A:**751-766, 1950
Risser, J. C.	The iliac apophysis: an invaluable sign in the management of scoliosis	Clin. Orthop. **11:**111-119, 1958
	Scoliosis: past and present	American Academy of Orthopaedic Surgeons Instructional Courses, J. Bone Joint Surg. **46-A:**167-199, 1964
Risser, J. C., and Norquist, D. M.	A follow-up study of the treatment of scoliosis	J. Bone Joint Surg. **40-A:**555-569, 1958

Chapter 10: Affections of the spine and thorax—cont'd

Roaf, R.	Paralytic scoliosis	J. Bone Joint Surg. **38-B:**640-659, 1956
Schmorl, G., and Junghanns, H.	The human spine in health and disease	New York, 1959, Grune & Stratton, Inc.
Scott, J. C.	Scoliosis and neurofibromatosis	J. Bone Joint Surg. **47-B:**240-246, 1965
Shands, A. R., Jr., and Eisberg, H. B.	The incidence of scoliosis in the state of Delaware, a study of 50,000 minifilms of the chest made during a survey for tuberculosis	J. Bone Joint Surg. **37-A:**1243-1249, 1955
Sutherland, I. D.	Funnel chest	J. Bone Joint Surg. **40-B:**244-251, 1958
Tambornino, J. M. Armbrust, E. N., and Moe, J. H.	Harrington instrumentation in correction of scoliosis, a comparison with cast correction	J. Bone Joint Surg. **46-A:**313-323, 1964
Weston, W. J., and Goodson, G. M.	Vertebra plana (Calvé)	J. Bone Joint Surg. **41-B:**477-485, 1959
Williams, H. J., and Pugh, D. G.	Vertebral epiphysitis: comparison of the clinical and roentgenologic findings	Amer. J. Roentgen. **90:**1236-1247, 1963
Winter, R. B., Moe, J. H., and Eilers, V.	Congenital scoliosis: a study of 234 patients treated and untreated. Parts I and II	J. Bone Joint Surg. **50-A:**1-47, 1968

Chapter 11

Affections of the low back

Abbott, K. H., and Retter, R. H.	Protrusions of thoracic intervertebral disks	Neurology **6:**1-10, 1956
Adkins, E. W. O.	Spondylolisthesis	J. Bone Joint Surg. **37-B:**48-62, 1955
Aitken, A. P.	Rupture of the intervertebral disc in industry: further observations on the end results	Amer. J. Surg. **84:**261-267, 1952
Armstrong, J. R.	Lumbar disc lesions: pathogenesis and treatment of low back pain and sciatica, ed. 3	Baltimore, 1965, The Williams & Wilkins Co.
Barr, J. S.	Low-back and sciatic pain: results of treatment	J. Bone Joint Surg. **33-A:**633-649, 1951
Barr, J. S., Hampton, A. O., and Mixter, W. J.	Pain low in the back and "sciatica," due to lesions of the intervertebral disks	J.A.M.A. **109:**1265-1270, 1937
Bosworth, D. M.	Surgery of the spine	American Academy of Orthopaedic Surgeons Instructional Course Lectures, Ann Arbor, 1957, J. W. Edwards, vol. 14, pp. 39-55
Bosworth, D. M., Fielding, J. W., Demarest, L., and Bonaquist, M.	Spondylolisthesis: a critical review of a consecutive series of cases treated by arthrodesis	J. Bone Joint Surg. **37-A:**767-786, 1955

Chapter 11: Affections of the low back—cont'd

Brady, L. P., Parker, L. B., and Vaughen, J.	An evaluation of the electromyogram in the diagnosis of the lumbar-disc lesion	J. Bone Joint Surg. **51-A:**539-547, 1969
Brav, E. A.	An analysis of orthopedic causes of low back and sciatic pain	Amer. J. Surg. **87:**235-240, 1954
Bucy, P. C.	Neuroanatomical and neurosurgical aspects of herniated intervertebral discs	American Academy of Orthopaedic Surgeons Instructional Course Lectures, St. Louis, 1961, The C. V. Mosby Co., vol. 18, pp. 21-34
Cain, J. P., Jr.	Low-back pain: evaluation of disability	Arch. Industr. Health **19:**593-595, 1959
Chrisman, O. D., Mittnacht, A., and Snook, G. A.	A study of the results following rotatory manipulation in the lumbar intervertebral-disc syndrome	J. Bone Joint Surg. **46-A:**517-524, 1964
Cloward, R. B.	Vertebral body fusion for ruptured lumbar discs: a roentgenographic study	Amer. J. Surg. **90:**969-976, 1955
Colonna, P. C.	Spondylolisthesis: analysis of 201 cases	J.A.M.A. **154:**398-402, 1954
Colonna, P. C., and Friedenberg, Z. B.	Disc syndrome: results of conservative care of patients with positive myelograms	J. Bone Joint Surg. **31-A:**614-618, 1949
Compere, E. L.	Origin, anatomy, physiology, and pathology of the intervertebral disc	American Academy of Orthopaedic Surgeons Instructional Course Lectures, St. Louis, 1961, The C. V. Mosby Co., vol. 18, pp. 15-20
Coventry, M. B., Ghormley, R. K., and Kernohan, J. W.	The intervertebral disc, its microscopic anatomy and pathology	J. Bone Joint Surg. **27:**105-112, 233-247, 460-474, 1945
Craig, W. M., Svien, H. J., Dodge, H. W., Jr., and Camp, J. D.	Intraspinal lesions masquerading as protruded lumbar intervertebral disks	J.A.M.A. **149:**250-253, 1952
da Roza, A. C.	Primary intraspinal tumours: their clinical presentation and diagnosis	J. Bone Joint Surg. **46-B:**8-15, 1964
Deyerle, W. M., and May, V. R., Jr.	Sciatic tension test	Southern Med. J. **49:**999-1005, 1956
Eyring, E. J.	The biochemistry and physiology of the intervertebral disk	Clin. Orthop. **67:**16-28, 1969
Farfan, H. F., Cossette, J. W., Robertson, G. H., Wells, R. V., and Kraus, H.	The effects of torsion on the lumbar intervertebral joints: the role of torsion in the production of disc degeneration	J. Bone Joint Surg. **52-A:**468-497, 1970
Feffer, H. L., and Adams, J. P.	Sacro-iliac changes associated with dysfunction of the spine	Southern Med. J. **51:**986-993, 1958
Filtzer, D. L., and Bahnson, H. T.	Low back pain due to arterial obstruction	J. Bone Joint Surg. **41-B:**244-247, 1959
Ford, L. T., editor	Symposium: Complications of lumbar-disc surgery, prevention and treatment	American Academy of Orthopaedic Surgeons Instructional Courses, J. Bone Joint Surg. **50-A:**382-428, 1968

Chapter 11: Affections of the low back—cont'd

Ford, L. T., and Key, J. A.	An evaluation of myelography in the diagnosis of intervertebral-disc lesions in the low back	J. Bone Joint Surg. **32-A:**257-266, 1950
Friberg, S.	Studies on spondylolisthesis	Acta Chir. Scand., supp. 55, 1939
Friedman, J., and Goldner, M. Z.	Discography in evaluation of lumbar disk lesions	Radiology **65:**653-662, 1955
Ghormley, R. K.	The problem of multiple operations on the back	American Academy of Orthopaedic Surgeons Instructional Course Lectures, Ann Arbor, 1957, J. W. Edwards, vol. 14, pp. 56-63
Greenwood, J., Jr., McGuire, T. H., and Kimbell, F.	A study of the causes of failure in the herniated intervertebral disc operation: an analysis of 67 reoperated cases	J. Neurosurg. **9:**15-20, 1952
Harris, R. I., and Macnab, I.	Structural changes in the lumbar intervertebral discs, their relationship to low back pain and sciatica	J. Bone Joint Surg. **36-B:**304-322, 1954
Henderson, E. D.	Results of the surgical treatment of spondylolisthesis	J. Bone Joint Surg. **48-A:**619-642, 1966
Howorth, B.	The painful coccyx	Clin. Orthop. **14:**145-161, 1959
	Low backache and sciatica: results of surgical treatment	J. Bone Joint Surg. **46-A:**1485-1519, 1964
Isley, J. K., Jr., and Baylin, G. J.	Prognosis in osteitis condensans ilii	Radiology **72:**234-237, 1959
Krusen, E. M., and Ford, D. E.	Compensation factor in low back injuries	J.A.M.A. **166:**1128-1133, 1958
Lansche, W. E., and Ford, L. T.	Correlation of the myelogram with clinical and operative findings in lumbar disc lesions	J. Bone Joint Surg. **42-A:**193-206, 1960
Laurent, L. E.	Spondylolisthesis: a study of 53 cases treated by spine fusion and 32 cases treated by laminectomy	Acta Orthop. Scand., supp. 35, 1958
Love, J. G., and Rivers, M. H.	Spinal cord tumors simulating protruded intervertebral disks	J.A.M.A. **179:**878-881, 1962
McBride, E. D.	The conservative treatment of backache	Amer. Surg. **18:**504-512, 1952
McCracken, W. J.	Low back disability	Canad. Med. Ass. J. **80:**331-336, 1959
Mensor, M. C.	Non-operative treatment, including manipulation for lumbar intervertebral disc syndrome	J. Bone Joint Surg. **37-A:**925-936, 1955
Millikan, C. H.	The problem of evaluating treatment of protruded lumbar intervertebral disk: observations of results of conservative and surgical treatment in 429 cases	J.A.M.A. **155:**1141-1143, 1954
Moreton, R. D.	Spondylolysis	J.A.M.A. **195:**671-674, 1966

Chapter 11: Affections of the low back—cont'd

Morgan, F. P., and King, T.	Primary instability of lumbar vertebrae as a common cause of low back pain	J. Bone Joint Surg. **39-B:**6-22, 1957
Morris, J. M., Lucas, D. B., and Bresler, B.	Role of the trunk in stability of the spine	J. Bone Joint Surg. **43-A:**327-351, 1961
Mutch, J., and Walmsley, R.	The aetiology of cleft vertebral arch in spondylolisthesis	Lancet **1:**74-77, 1956
Newman, P. H.	Sprung back	J. Bone Joint Surg. **34-B:**30-37, 1952
Newman, P. H., and Stone, K. H.	The etiology of spondylolisthesis	J. Bone Joint Surg. **45-B:**39-59, 1963
O'Connell, J. E. A.	Involvement of the spinal cord by intervertebral disk protrusions	Brit. J. Surg. **43:**225-247, 1955
Pietra, A. D.	Stabilizing spinal fusion following herniated disk removal without fusion	J.A.M.A. **157:**701-702, 1955
Pyper, J. B.	Excision of the coccyx for coccydynia	J. Bone Joint Surg. **39-B:**733-737, 1957
Raney, R. B.	Isthmus defects of the fifth lumbar vertebra	Southern Med. J. **38:**166-176, 1945
Reynolds, F. C., McGinnis, A. E., and Morgan, H. C.	Surgery in the treatment of low-back pain and sciatica	J. Bone Joint Surg. **41-A:**223-235, 1959
Rojkó, A., and Farkas, K.	Osteitis condensans ossis ilii	Acta Orthop. Scand. **29:**108-120, 1959
Runge, C. F.	Roentgenographic examination of the lumbosacral spine in routine pre-employment examinations	J. Bone Joint Surg. **36-A:**75-84, 1954
Schneider, C. C.	Diagnosis, treatment, and rehabilitation of the industrial low back cripple	American Academy of Orthopaedic Surgeons Instructional Course Lectures, St. Louis, 1959, The C. V. Mosby Co., vol. 16, pp. 173-183
	Trends in disability evaluation, with particular reference to the low back	American Academy of Orthopaedic Surgeons Instructional Course Lectures, Ann. Arbor, 1956, J. W. Edwards, vol. 13, pp. 293-298
Sell, L. S.	Misdiagnosis and mismanagement of early intervertebral disk lesions	J.A.M.A. **150:**987-990, 1952
Shapiro, R.	Myelography, ed. 2	Chicago, 1968, Year Book Medical Publishers, Inc.
Shaw, E. G., and Taylor, J. G.	The results of lumbo-sacral fusion for low back pain	J. Bone Joint Surg. **38-B:**485-497, 1956
Shutkin, N. M.	Syndrome of the degenerated intervertebral disc	Amer. J. Surg. **84:**162-171, 1952
Splithoff, C. A.	Lumbosacral junction, roentgenographic comparison of patients with and without backaches	J.A.M.A. **152:**1610-1613, 1953

Chapter 11: Affections of the low back—cont'd

Stinchfield, F. E., and Sinton, W. A.	Criteria for spine fusion with use of "H" bone graft following disc removal: results in 100 cases	Arch. Surg. **65**:542-550, 1952
Sullivan, C. R., Bickel, W. H., and Svien, H. J.	Infections of vertebral interspaces after operations on intervertebral disks	J.A.M.A. **166**:1973-1977, 1958
Sullivan, J. E.	Backache due to visceral lesions of the chest and the abdomen	Clin. Orthop. **26**:67-73, 1963
Taylor, T. K. F., and Akeson, W. H.	Intervertebral disc prolapse: a review of morphologic and biochemic knowledge concerning the nature of prolapse	Clin. Orthop. **76**:54-79, 1971
Thiele, G. H.	Coccygodynia, the mechanism of its production and its relationship to anorectal disease	Amer. J. Surg. **79**:110-116, 1950
Toumey, J. W., Poppen, J. L., and Hurley, M. T.	Cauda equina tumors as a cause of the low-back syndrome	J. Bone Joint Surg. **32-A**:249-256, 1950
Von Werssowetz, O. F.	Back braces and supports	Clin. Orthop. 5:169-183, 1955
White, A. W. M.	Low back pain in men receiving workmen's compensation	Canad. Med. Ass. J. **101**:61-67, 1969
Williams, P. C.	The conservative management of lesions of the lumbosacral spine	American Academy of Orthopaedic Surgeons Instructional Course Lectures, Ann Arbor, 1953, J. W. Edwards, vol. 10, pp. 90-121
	The lumbosacral spine: emphasizing conservative management	New York, 1965, Blakiston Division, McGraw-Hill Book Co.
Willis, T. A.	Man's back	Springfield, Ill., 1953, Charles C Thomas, Publisher
Wiltse, L. L.	The etiology of spondylolisthesis	J. Bone Joint Surg. **44-A**:539-560, 1962
Young, H. H.	Non-neurological lesions simulating protruded intervertebral disk	J.A.M.A. **148**:1101-1105, 1952
Young, H. H., and Love, J. G.	End results of removal of protruded lumbar intervertebral discs with and without fusion	American Academy of Orthopaedic Surgeons Instructional Course Lectures, St. Louis, 1959, The C. V. Mosby Co., vol. 16, pp. 213-216

Chapter 12
Affections of the hip

Arterial supply of the femoral head

Howe, W. W., Jr., Lacey, T., II, and Schwartz, R. P.	A study of the gross anatomy of the arteries supplying the proximal portion of the femur and the acetabulum	J. Bone Joint Surg. **32-A**:856-866, 1950

Chapter 12: Affections of the hip—cont'd

Arterial supply of the femoral head—cont'd

Trueta, J.	The normal vascular anatomy of the human femoral head during growth	J. Bone Joint Surg. **39-B:**358-394, 1957
Trueta, J., and Harrison, M. H. M.	The normal vascular anatomy of the femoral head in adult man	J. Bone Joint Surg. **35-B:**442-461, 1953
Tucker, F. R.	Arterial supply to the femoral head and its clinical importance	J. Bone Joint Surg. **31-B:**82-93, 1949
Wertheimer, L. G., and Lopes, S.	Arterial supply of the femoral head: a combined angiographic and histologic study	J. Bone Joint Surg. **53-A:**545-556, 1971

Avascular necrosis of the femoral head

Bohr, H., and Larsen, E. H.	On necrosis of the femoral head after fracture of the neck of the femur	J. Bone Joint Surg. **47-B:**330-338, 1965
Chung, S. M. K., and Ralston, E. L.	Necrosis of the femoral head associated with sickle-cell anemia and its genetic variants	J. Bone Joint Surg. **51-A:**33-58, 1969
Committee of Surgery Study Sections, National Institutes of Health, Reynolds, F. C., Chairman	Proceedings of the conference on aseptic necrosis of the femoral head	St. Louis, 1964
Cruess, R. L., Blennerhassett, J., MacDonald, F. R., MacLean, L. D., and Dossetor, J.	Aseptic necrosis following renal transplantation	J. Bone Joint Surg. **50-A:**1577-1590, 1968
d'Aubigné, R. M., Postel, M., Mazabraud, A., Massias, P., and Gueguen, J.	Idiopathic necrosis of the femoral head in adults	J. Bone Joint Surg. **47-B:**612-633, 1965
Dubois, E. L., and Cozen, L.	Avascular (aseptic) bone necrosis associated with systemic lupus erythematosus	J.A.M.A. **174:**966-971, 1960
Martel, W., and Sitterley, B. H.	Roentgenologic manifestations of osteonecrosis	Amer. J. Roentgen. **106:**509-522, 1969
Patterson, R. J., Bickel, W. H., and Dahlin, D. C.	Idiopathic avascular necrosis of the head of the femur: a study of 52 cases	J. Bone Joint Surg. **46-A:**267-282, 1964
Sutton, R. D., Benedek, T. G., and Edwards, G. A.	Aseptic bone necrosis and corticosteroid therapy	Arch. Intern. Med. **112:**594-602, 1963

Coxa plana

Carpenter, E. B., and Powell, D. O.	Osteochondrosis of capital epiphysis of femur (Legg-Calvé-Perthes disease)	J.A.M.A. **172:**525-527, 1960
Chung, S. M. K., and Moe, J. H.	Legg-Calvé-Perthes disease: clinical-radiographic correlations	Clin. Orthop. **41:**116-124, 1965

Chapter 12: Affections of the hip—cont'd

Coxa plana—cont'd

Danielsson, L. G., and Hernborg, J.	Late results of Perthes' disease	Acta Orthop. Scand. **36**:70-81, 1965
Eaton, G. O.	Long-term results of treatment in coxa plana	J. Bone Joint Surg. **49-A**:1031-1042, 1967
Edgren, W.	Coxa plana	Acta Orthop. Scand., supp. 84, 1965
Evans, D. L.	Legg-Calvé-Perthes' disease: a study of late results	J. Bone Joint Surg. **40-B**:168-181, 1958
Evans, D. L., and Lloyd-Roberts, G. C.	Treatment in Legg-Calvé-Perthes' disease	J. Bone Joint Surg. **40-B**:182-189, 1958
Gill, A. B.	Legg-Perthes disease of the hip: its early roentgenographic manifestations and its cyclical course	J. Bone Joint Surg. **22**:1013-1047, 1940
Goff, C. W.	Legg-Calvé-Perthes syndrome and related osteochondroses of youth	Springfield, Ill., 1954, Charles C Thomas, Publisher
Herndon, C. H., and Heyman, C. H.	Legg-Perthes disease: an evaluation of treatment by traction and ischial weight-bearing brace	J. Bone Joint Surg. **34-A**:25-46, 1952
Legg, A. T.	An obscure affection of the hip-joint	Boston Med. Surg. J. **162**:202-204, 1910
Pedersen, H. E., and McCarrol, H. R.	Treatment in Legg-Perthes disease	J. Bone Joint Surg. **33-A**:591-600, 1951
Petrie, J. G., and Bitenc, I.	The abduction weight-bearing treatment in Legg-Perthes' disease	J. Bone Joint Surg. **53-B**:54-62, 1971
Ponseti, I. V.	Legg-Perthes disease: observations on pathological changes in two cases	J. Bone Joint Surg. **38-A**:739-750, 1956
Ralston, E. L.	Legg-Calvé-Perthes disease—factors in healing	J. Bone Joint Surg. **43-A**:249-260, 1961
Sanders, J. A., and MacEwen, G. D.	A long-term follow-up on coxa plana at the Alfred I. Dupont Institute	Southern Med. J. **62**:1042-1047, 1969
Shephard, E.	Multiple epiphysial dysplasia	J. Bone Joint Surg. **38-B**:458-467, 1956
Stamp, W. G., Canales, G., and Odell, R. T.	Late results in osteochondrosis of capital epiphysis of femur (Legg-Calvé-Perthes disease)	J.A.M.A. **169**:1443-1446, 1959
Wansbrough, R. M., Carrie, A. W., Walker, N. F., and Ruckerbauer, G.	Coxa plana, its genetic aspects and results of treatment with the long Taylor walking caliper	J. Bone Joint Surg. **41-A**:135-146, 1959

Congenital coxa vara

Amstutz, H. C., and Wilson, P. D., Jr.	Dysgenesis of the proximal femur (coxa vara) and its surgical management	J. Bone Joint Surg. **44-A**:1-24, 1962
Babb, F. S., Ghormley, R. K., and Chatterton, C. C.	Congenital coxa vara	J. Bone Joint Surg. **31-A**:115-131, 1949

Chapter 12: Affections of the hip—cont'd

Congenital coxa vara—cont'd

Finby, N., Jacobson, H. G., and Poppel, M. H.	Idiopathic coxa vara in childhood	Radiology **67:**10-16, 1956
Le Mesurier, A. B.	Developmental coxa vara	J. Bone Joint Surg. **30-B:**595-605, 1948

Slipping of the capital femoral epiphysis

Billing, L., and Severin, E.	Slipping epiphysis of the hip	Acta Radiol., supp. 174, 1959
Cleveland, M., Bosworth, D. M., Daly, J. N., and Hess, W. E.	Study of displaced capital femoral epiphyses	J. Bone Joint Surg. **33-A:**955-967, 1951
DePalma, A. F., Danyo, J. J., and Stose, W. G.	Slipping of the upper femoral epiphysis	Clin. Orthop. **37:**167-183, 1964
Durbin, F. C.	Treatment of slipped upper femoral epiphysis	J. Bone Joint Surg. **42-B:**289-302, 1960
Fahey, J. J., and O'-Brien, E. T.	Acute slipped capital femoral epiphysis	J. Bone Joint Surg. **47-A:**1105-1127, 1965
Herndon, C. H., Heyman, C. H., and Bell, D. M.	Treatment of slipped capital femoral epiphysis by epiphyseodesis and osteoplasty of the femoral neck	J. Bone Joint Surg. **45-A:**999-1012, 1963
Heyman, C. H., Herndon, C. H., and Strong, J. M.	Slipped femoral epiphysis with severe displacement	J. Bone Joint Surg. **39-A:**293-303, 1957
Howorth, B., editor	Symposium: Slipped capital femoral epiphysis	Clin. Orthop. **48:**5-164, 1966
Klein, A., Joplin, R. J., and Reidy, J. A.	Treatment of slipped capital femoral epiphysis	J.A.M.A. **136:**445-451, 1948
Maurer, R. C., and Larsen, I. J.	Acute necrosis of cartilage in slipped capital femoral epiphysis	J. Bone Joint Surg. **52-A:**39-50, 1970
Meyer, L. C., Stelling, F. H., and Wiese, F.	Slipped capital femoral epiphysis	Southern Med. J. **50:**453-459, 1957
Mozan, L. C., and Thompson, R. G.	Treatment of the minimally slipped capital femoral epiphysis	J.A.M.A. **204:**674-678, 1968
Ponseti, I. V., and McClintock, R.	The pathology of slipping of the upper femoral epiphysis	J. Bone Joint Surg. **38-A:**71-83, 1956
Salenius, P., and Kivilaakso, R.	Results of treatment of slipped upper femoral epiphysis: a survey of 99 treated hips	Acta Orthop. Scand., supp. 114, 1968
Wilson, P. D., Jacobs, B., and Schecter, L.	Slipped capital femoral epiphysis: an end-result study	J. Bone Joint Surg. **47-A:**1128-1145, 1965

Other affections of the hip

Adams, J. P.	Coxa magna	Southern Med. J. **49:**604-607, 1956
Bryson, A. F.	Treatment of pathological dislocation of the hip joint after suppurative arthritis in infants	J. Bone Joint Surg. **30-B:**449-453, 1948
Dickinson, A. M.	Bilateral snapping hip	Amer. J. Surg. **6:**97-101, 1929

Chapter 12: Affections of the hip—cont'd

Other affections of the hip—cont'd

Ferguson, A. B., and Howorth, M. B.	Coxa magna, a condition of the hip related to coxa plana	J.A.M.A. **104:**808-812, 1935
Finder, J. G.	Iliopectineal bursitis	Arch. Surg. **36:**519-538, 1938
Friedenberg, Z. B.	Protrusio acetabuli	Amer. J. Surg. **85:**764-770, 1953
Gilmour, J.	Adolescent deformities of the acetabulum; an investigation into the nature of protrusio acetabuli	Brit. J. Surg. **26:**670-699, 1939
Gordon, E. J.	Trochanteric bursitis and tendinitis	Clin. Orthop. **20:**193-202, 1961
Hooper, J. C., and Jones, E. W.	Primary protrusion of the acetabulum	J. Bone Joint Surg. **53-B:**23-29, 1971
Jones, G. B.	Paralytic dislocation of the hip	J. Bone Joint Surg. **44-B:**573-587, 1962
Nachemson, A., and Scheller, S.	A clinical and radiological follow-up study of transient synovitis of the hip	Acta Orthop. Scand. **40:**479-500, 1969
Sharrard, W. J. W.	Posterior iliopsoas transplantation in the treatment of paralytic dislocation of the hip	J. Bone Joint Surg. **46-B:**426-444, 1964
Simril, W. A.	Roentgen manifestations of hip disease in children	American Academy of Orthopaedic Surgeons Instructional Course Lectures, St. Louis, 1961, The C. V. Mosby Co., vol. 18, pp. 187-206
Smith-Petersen, M. N.	A new supra-articular subperiosteal approach to the hip joint	Amer. J. Orthop. Surg. **15:**592-595, 1917
Spear, I. M., and Lipscomb, P. R.	Noninfectious trochanteric bursitis and peritendinitis	Surg. Clin. N. Amer. **32:**1217-1224, 1952
Tachdjian, M. O., and Minear, W. L.	Hip dislocation in cerebral palsy	J. Bone Joint Surg. **38-A:**1358-1364, 1956

Chapter 13

Affections of the knee

Internal derangements

Abbott, L. C., Saunders, J. B. DeC. M., Bost, F. C., and Anderson, C. E.	Injuries to the ligaments of the knee joint	J. Bone Joint Surg. **26:**503-521, 1944
Becton, J. L., and Young, H. H.	Cysts of semilunar cartilage of the knee	Arch. Surg. **90:**708-712, 1965
Bonnin, J. G.	Cysts of the semilunar cartilages of the knee-joint	Brit. J. Surg. **40:**558-565, 1953
Brantigan, O. C., and Voshell, A. F.	The mechanics of the ligaments and menisci of the knee joint	J. Bone Joint Surg. **23:**44-66, 1941
Brewer, B. J.	Injuries to the knee	American Academy of Orthopaedic Surgeons Instructional Course Lectures, St. Louis, 1959, The C. V. Mosby Co., vol. 16, pp. 29-34

Chapter 13: Affections of the knee—cont'd

Internal derangements—cont'd

Butt, W. P., and McIntyre, J. L.	Double-contrast arthrography of the knee	Radiology **92**:487-499, 1969
Campbell, C. J., and Ranawat, C. S.	Osteochondritis dissecans: the question of etiology	J. Trauma **6**:201-221, 1966
DeLorme, T. L.	Restoration of muscle power by heavy-resistance exercises	J. Bone Joint Surg. **27**:645-667, 1945
DePalma, A. F.	Diseases of the knee: management in medicine and surgery	Philadelphia, 1954, J. B. Lippincott Co.
du Toit, G. T.	Internal derangement of the knee	American Academy of Orthopaedic Surgeons Instructional Course Lectures, Ann Arbor, 1955, J. W. Edwards, vol. 12, pp. 9-34
Green, W. T., and Banks, H. H.	Osteochondritis dissecans in children	J. Bone Joint Surg. **35-A**:26-47, 1953
Haldeman, K. O.	Internal derangements of the knee	American Academy of Orthopaedic Surgeons Instructional Course Lectures, St. Louis, 1959, The C. V. Mosby Co., vol. 16, pp. 161-169
Helfet, A. J.	Mechanism of derangements of the medial semilunar cartilage and their management	J. Bone Joint Surg. **41-B**:319-336, 1959
	The management of internal derangements of the knee	Philadelphia, 1963, J. B. Lippincott Co.
	Diagnosis and management of internal derangements of the knee joint	American Academy of Orthopaedic Surgeons Instructional Course Lectures, St. Louis, 1970, The C. V. Mosby Co., vol. 19, pp. 63-77
Howorth, B.	Injuries of the menisci of the knee	Western J. Surg. **72**:203-215, 1964
Hughston, J. C.	Acute knee injuries in athletes	Clin. Orthop. **23**:114-133, 1962
Jackson, J. P.	Degenerative changes in the knee after meniscectomy	Brit. Med. J. **2**:525-527, 1968
Jaffe, H. L., Lichtenstein, L., and Sutro, C. J.	Pigmented villonodular synovitis, bursitis and tenosynovitis	Arch. Path. **31**:731-765, 1941
Jones, K. G.	Reconstruction of the anterior cruciate ligament	J. Bone Joint Surg. **45-A**:925-932, 1963
Kaplan, E. B.	Discoid lateral meniscus of the knee joint	J. Bone Joint Surg. **39-A**:77-87, 1957
	Injuries and afflictions of the menisci of the knee	American Academy of Orthopaedic Surgeons Instructional Course Lectures, St. Louis, 1959, The C. V. Mosby Co., vol. 16, pp. 153-160
King, D.	The function of semilunar cartilages	J. Bone Joint Surg. **18**:1069-1076, 1936
Lannin, D. R.	Rehabilitation of knee meniscus injury with associated malacia of the patella	J.A.M.A. **171**:1662-1664, 1959

Chapter 13: Affections of the knee—cont'd

Internal derangements—cont'd

Larmon, W. A.	Pigmented villonodular synovitis	Med. Clin. N. Amer. **49**:141-150, 1965
Lewin, P.	The knee and related structures: injuries, deformities, diseases, disabilities	Philadelphia, 1952, Lea & Febiger
McMaster, P. E.	Pigmented villonodular synovitis with invasion of bone: report of six cases	J. Bone Joint Surg. **42-A**:1170-1183, 1960
Meyers, M. H., and McKeever, F. M.	Fracture of the intercondylar eminence of the tibia	J. Bone Joint Surg. **41-A**:209-222, 1959
Murdoch, G.	Errors of diagnosis revealed at meniscectomy	J. Bone Joint Surg. **39-B**:502-507, 1957
Murphy, F. P., Dahlin, D. C., and Sullivan, C. R.	Articular synovial chondromatosis	J. Bone Joint Surg. **44-A**:77-86, 1962
Mussey, R. D., Jr., and Henderson, M. S.	Osteochondromatosis	J. Bone Joint Surg. **31-A**:619-627, 1949
Nathan, P. A., and Cole, S. C.	Discoid meniscus, a clinical and pathologic study	Clin. Orthop. **64**:107-113, 1969
O'Donoghue, D. H.	An analysis of end results of surgical treatment of major injuries to the ligaments of the knee	J. Bone Joint Surg. **37-A**:1-13, 1955
	Surgical repair of knee ligament injuries	American Academy of Orthopaedic Surgeons Instructional Course Lectures, Ann Arbor, 1958, J. W. Edwards, vol. 15, pp. 105-115
	Injuries to the menisci of the knee	Curr. Pract. Orthop. Surg. **1**:44-64, 1963
Smillie, I. S.	Observations on the regeneration of the semilunar cartilages in man	Brit. J. Surg. **31**:398-401, 1944
	Osteochondritis dissecans	Baltimore, 1960, The Williams & Wilkins Co.
	Injuries of the knee joint, ed. 4	Edinburgh, 1970, E. & S. Livingstone, Ltd.
Slocum, D. B., and Larson, R. L.	Rotatory instability of the knee	J. Bone Joint Surg. **50-A**:211-225, 1968

Other affections of the knee

Bentley, G.	Chondromalacia patellae	J. Bone Joint Surg. **52-A**:221-232, 1970
Blount, W. P.	Tibia vara, osteochondrosis deformans tibiae	J. Bone Joint Surg. **19**:1-29, 1937
	Unequal leg length	American Academy of Orthopaedic Surgeons Instructional Course Lectures, St. Louis, 1960, The C. V. Mosby Co., vol. 17, pp. 218-245

Chapter 13: Affections of the knee—cont'd

Other affections of the knee—cont'd

Bowker, H. H., and Thompson, E. B.	Surgical treatment of recurrent dislocation of the patella: a study of 48 cases	J. Bone Joint Surg. **46-A:**1451-1461, 1964
Burleson, R. J., Bickel, W. H., and Dahlin, D. C.	Popliteal cyst: a clinicopathological survey	J. Bone Joint Surg. **38-A:**1265-1274, 1956
Cave, E. F., and Rowe, C. R.	The patella, its importance in derangement of the knee	J. Bone Joint Surg. **32-A:**542-553, 1950
Cohen, B., and Wilkinson, R. W.	The Osgood-Schlatter lesion	Amer. J. Surg. **95:**731-742, 1958
Ehrenborg, G.	The Osgood-Schlatter lesion: a clinical and experimental study	Acta Chir. Scand., supp. 288, 1962
Green, W. T., and Anderson, M.	Epiphyseal arrest for the correction of discrepancies in length of the lower extremities	J. Bone Joint Surg. **39-A:**853-872, 1957
Gristina, A. G., and Wilson, P. D.	Popliteal cysts in adults and children: a review of 90 cases	Arch. Surg. **88:**357-363, 1964
Heywood, A. W. B.	Recurrent dislocation of the patella	J. Bone Joint Surg. **43-B:**508-517, 1961
Hoffman, B. K.	Cystic lesions of the popliteal space	Surg. Gynec. Obstet. **116:**551-558, 1963
Holt, J. F., Latourette, H. B., and Watson, E. H.	Physiologic bowing of the legs in young children	J.A.M.A. **154:**390-394, 1954
Hughston, J. C.	Subluxation of the patella	American Academy of Orthopaedic Surgeons Instructional Courses, J. Bone Joint Surg. **50-A:**1003-1026, 1968
Langenskiöld, A., and Riska, E. B.	Tibia vara (osteochondrosis deformans tibiae)	J. Bone Joint Surg. **46-A:**1405-1420, 1964
Morley, A. J. M.	Knock-knee in children	Brit. Med. J. **2:**976-979, 1957
Nachlas, I. W.	The Pellegrini-Stieda para-articular calcification	Clin. Orthop. **3:**121-127, 1954
Osgood, R. B.	Lesions of the tibial tubercle occurring during adolescence	Boston Med. Surg. J. **148:**114-117, 1903
Rapp, I. H., and Lazerte, G.	Clinical pathological correlation in Osgood-Schlatter's disease	Southern Med. J. **51:**909-912, 1958
Scuderi, C.	Ruptures of the quadriceps tendon: study of twenty tendon ruptures	Amer. J. Surg. **95:**626-635, 1958
Sherman, M. S.	Physiologic bowing of the legs	Southern Med. J. **53:**830-836, 1960
Sofield, H. A., Blair, S. J., and Millar, E. A.	Leg-lengthening	J. Bone Joint Surg. **40-A:**311-322, 1958
Voshell, A. F., and Brantigan, O. C.	Bursitis in the region of the tibial collateral ligament	J. Bone Joint Surg. **26:**793-798, 1944
Weiner, A. D., and Ghormley, R. K.	Periodic benign synovitis, idiopathic intermittent hydrarthrosis	J. Bone Joint Surg. **38-A:**1039-1055, 1956
West, F. E., and Soto-Hall, R.	Recurrent dislocation of the patella in the adult	J. Bone Joint Surg. **40-A:**386-394, 1958

Chapter 13: Affections of the knee—cont'd
Other affections of the knee—cont'd

Wiles, P., Andrews, P. S., and Devas, M. B.	Chondromalacia of the patella	J. Bone Joint Surg. **38-B**:95-113, 1956
Wilson, P. D., Eyre-Brook, A. L., and Francis, J. D.	A clinical and anatomical study of the semimembranosus bursa in relation to popliteal cyst	J. Bone Joint Surg. **20**:963-984, 1938

Chapter 14
Affections of the ankle and foot

Andreasen, E.	Calcaneo-navicular coalition, late results of resection	Acta Orthop. Scand. **39**:424-432, 1968
Arner, O., and Lindholm, A.	Subcutaneous rupture of the Achilles tendon: a study of 92 cases	Acta Chir. Scand., supp. 239, 1959
Bernstein, A., and Stone, J. R.	March fracture	J. Bone Joint Surg. **26**:743-750, 1944
Bingold, A. C., and Collins, D. H.	Hallux rigidus	J. Bone Joint Surg. **32-B**:214-222, 1950
Blockey, N. J.	Peroneal spastic flat foot	J. Bone Joint Surg. **37-B**:191-202, 1955
Bonney, G., and Macnab, I.	Hallux valgus and hallux rigidus: a critical survey of operative results	J. Bone Joint Surg. **34-B**:366-385, 1952
Burman, M.	Stenosing tendovaginitis of the foot and ankle	Arch. Surg. **67**:686-698, 1953
Carr, C. R., and Boyd, B. M.	Correctional osteotomy for metatarsus primus varus and hallux valgus	J. Bone Joint Surg. **50-A**:1353-1367, 1968
Chandler, F. A.	Children's feet, normal and presenting common abnormalities	Amer. J. Dis. Child. **63**:1136-1146, 1942
Cleveland, M., and Winant, E. M.	An end-result study of the Keller operation	J. Bone Joint Surg. **32-A**:163-175, 1950
Cole, W. H.	The treatment of claw-foot	J. Bone Joint Surg. **22**:895-908, 1940
Conway, J. J., and Cowell, H. R.	Tarsal coalition: clinical significance and roentgenographic demonstration	Radiology **92**:799-811, 1969
Crego, C. H., Jr., and Ford, L. T.	An end-result study of various operative procedures for correcting flat feet in children	J. Bone Joint Surg. **34-A**:183-195, 1952
Dickson, F. D., and Diveley, R. L.	Functional disorders of the foot: their diagnosis and treatment, ed. 2	Philadelphia, 1944, J. B. Lippincott Co.
Freiberg, A. H.	The so-called infractions of the second metatarsal bone	J. Bone Joint Surg. **8**:257-261, 1926
Freiberg, J. A.	The diagnosis and treatment of common painful conditions of the foot	American Academy of Orthopaedic Surgeons Instructional Course Lectures, Ann Arbor, 1957, J. W. Edwards, vol. 14, pp. 238-247

Chapter 14: Affections of the ankle and foot—cont'd

Friedman, B., and Smith, E. E.	Foot problems in infants and children, rotational deviations of lower extremities	J. Pediat. **46:**573-580, 1955
Froimson, A. I.	Tennis leg	J.A.M.A. **209:**415-416, 1969
Giannestras, N. J.	Foot disorders: medical and surgical management	Philadelphia, 1967, Lea & Febiger
Gillies, H., and Chalmers, J.	The management of fresh ruptures of the tendo achillis	J. Bone Joint Surg. **52-A:**337-343, 1970
Haines, R. W., and McDougall, A.	The anatomy of hallux valgus	J. Bone Joint Surg. **36-B:**272-293, 1954
Hammond, G.	The operative treatment of hallux valgus and metatarsus primus varus	Surg. Clin. N. Amer. **32:**733-745, 1952
Harris, R. I.	Peroneal spastic flatfoot	American Academy of Orthopaedic Surgeons Instructional Course Lectures, Ann Arbor, 1958, J. W. Edwards, vol. 15, pp. 116-134
Harris, R. I., and Beath, T.	Hypermobile flat-foot with short tendo achillis	J. Bone Joint Surg. **30-A:**116-140, 1948
	The short first metatarsal, its incidence and clinical significance	J. Bone Joint Surg. **31-A:**553-565, 1949
Hauser, E. D. W.	Diseases of the foot, ed. 2	Philadelphia, 1950, W. B. Saunders Co.
Hughes, E. S. R.	Painful heels in children	Surg. Gynec. Obstet. **86:**64-68, 1948
Hutter, C. G., Jr., and Scott, W.	Tibial torsion	J. Bone Joint Surg. **31-A:**511-518, 1949
Inman, V. T., and DuVries, H. L.	Surgery of the foot	St. Louis, 1971, The C. V. Mosby Co.
Joplin, R. J.	Some common foot disorders amenable to surgery	American Academy of Orthopaedic Surgeons Instructional Course Lectures, Ann Arbor, 1958, J. W. Edwards, vol. 15, pp. 144-158
Keck, S. W., and Kelly, P. J.	Bursitis of the posterior part of the heel: evaluation of surgical treatment of 18 patients	J. Bone Joint Surg. **47-A:**267-273, 1965
Keith, A.	The history of the human foot and its bearing on orthopaedic practice	J. Bone Joint Surg. **11:**10-32, 1929
Kelikian, H.	Hallux valgus, allied deformities of the forefoot and metatarsalgia	Philadelphia, 1965, W. B. Saunders Co.
Kendrick, J. I.	Treatment of calcaneonavicular bar	J.A.M.A. **172:**1242-1244, 1960
Kidner, F. C.	The prehallux (accessory scaphoid) in its relation to flatfoot	J. Bone Joint Surg. **11:**831-837, 1929
Kite, J. H.	Torsion of the lower extremities in small children	J. Bone Joint Surg. **36-A:**511-520, 1954
	Morton's toe neuroma	Southern Med. J. **59:**20-25, 1966

Chapter 14: Affections of the ankle and foot—cont'd

Kohler, A.	Typical disease of the second metatarsophalangeal joint	Amer. J. Roentgen. **10:**705-710, 1923
Kroening, P. M., and Shelton, M. L.	Stress fractures	Amer. J. Roentgen. **89:**1281-1286, 1963
Lawrence, G. H., Cave, E. F., and O'Connor, H.	Injury to the Achilles tendon, experience at the Massachusetts General Hospital, 1900-1954	Amer. J. Surg. **89:**795-802, 1955
Lewin, P.	The foot and ankle: their injuries, diseases, deformities and disabilities, ed. 4	Philadelphia, 1959, Lea & Febiger
McBride, E. D.	Hallux valgus bunion deformity	American Academy of Orthopaedic Surgeons Instructional Course Lectures, Ann Arbor, 1952, J. W. Edwards, vol. 9, pp. 334-346
McDougall, A.	The os trigonum	J. Bone Joint Surg. **37-B:**257-265, 1955
Milgram, J. E.	Office measures for relief of the painful foot	American Academy of Orthopaedic Surgeons Instructional Courses, J. Bone Joint Surg. **46-A:**1095-1116, 1964
Mitchell, G. P.	Spasmodic flatfoot	Clin. Orthop. **70:**73-78, 1970
Morton, D. J.	The human foot: its evolution, physiology and functional disorders	New York, 1935, Columbia University Press
Nissen, K. I.	Plantar digital neuritis, Morton's metatarsalgia	J. Bone Joint Surg. **30-B:**84-94, 1948
O'Donoghue, D. H.	Impingement exostoses of the talus and tibia	J. Bone Joint Surg. **39-A:**835-852, 1957
O'Rahilly, R.	A survey of carpal and tarsal anomalies	J. Bone Joint Surg. **35-A:**626-642, 1953
Schwartz, R. P., and Heath, A. L.	Conservative treatment of functional disorders of the feet in the adolescent and adult	J. Bone Joint Surg. **31-A:**501-510, 1949
Shands, A. R., Jr.	The accessory bones of the foot	Southern Med. Surg. **93:**326-334, 1931
Shands, A. R., Jr., and Wentz, I. J.	Congenital anomalies, accessory bones, and osteochondritis in the feet of 850 children	Surg. Clin. N. Amer. **33:**1643-1666, 1953
Steindler, A.	The pathomechanics of the static disabilities of foot and ankle	American Academy of Orthopaedic Surgeons Instructional Course Lectures, Ann Arbor, 1952, J. W. Edwards, vol. 9, pp. 327-334
Wang, C. C., Lowrey, C. W., and Severance, R. L.	Fatigue fracture of the pelvis and the lower extremity	New Eng. J. Med. **260:**958-962, 1959
Waugh, W.	The ossification and vascularisation of the tarsal navicular and their relation to Köhler's disease	J. Bone Joint Surg. **40-B:**765-777, 1958
Webster, F. S., and Roberts, W. M.	Tarsal anomalies and peroneal spastic flatfoot	J.A.M.A. **146:**1099-1104, 1951

Chapter 14: Affections of the ankle and foot—cont'd

Wickstrom, J., and Williams, R. A.	Shoe corrections and orthopaedic foot supports	Clin. Orthop. **70:**30-42, 1970
Williams, A. A.	Tenosynovitis of the tendo achillis	Brit. Med. J. **2:**377-378, 1941
Zadek, I., and Gold, A. M.	The accessory tarsal scaphoid	J. Bone Joint Surg. **30-A:**957-968, 1948

Chapter 15

Affections of the neck, shoulder, and jaw

Torticollis

Coventry, M. B., and Harris, L. E.	Congenital muscular torticollis in infancy	J. Bone Joint Surg. **41-A:**815-822, 1959
Hulbert, K. F.	Congenital torticollis	J. Bone Joint Surg. **32-B:**50-59, 1950
Kiesewetter, W. B., Nelson, P. K., Palladino, V. S., and Koop, C. E.	Neonatal torticollis	J.A.M.A. **157:**1281-1285, 1955
MacDonald, D.	Sternomastoid tumour and muscular torticollis	J. Bone Joint Surg. **51-B:**432-443, 1969
Middleton, D. S.	The pathology of congenital torticollis	Brit. J. Surg. **18:**188-204, 1930
Sorensen, B. F., and Hamby, W. B.	Spasmodic torticollis: results in 71 surgically treated patients	J.A.M.A. **194:**706-708, 1965

Cervical root syndrome and thoracic outlet syndromes

Brannon, E. W.	Cervical rib syndrome, an analysis of 19 cases and 24 operations	J. Bone Joint Surg. **45-A:**977-998, 1963
Brodsky, A., and Gol, A.	Costoclavicular syndrome	Southern Med. J. **63:**50-58, 1970
Clarke, E.	Cervical myelopathy, a common neurological disorder	Lancet **1:**171-176, 1955
Edmonson, R. L.	Neurovascular compression syndromes of the upper extremity	Southern Med. J. **58:**754-759, 1965
Friedenberg, Z. B., Broder, H. A., Edeiken, J. E., and Spencer, H. N.	Degenerative disk disease of cervical spine: clinical and roentgenographic study	J.A.M.A. **174:**375-380, 1960
Gage, M., and Parnell, H.	Scalenus anticus syndrome	Amer. J. Surg. **73:**252-268, 1947
Hadley, L. A.	The covertebral articulations and cervical foramen encroachment	J. Bone Joint Surg. **39-A:**910-920, 1957
Hirsch, C., Schajowicz, F., and Galante, J.	Structural changes in cervical spine: a study on autopsy specimens in different age groups	Acta Orthop. Scand., supp. 109, 1967
Holt, S., and Yates, P. O.	Cervical spondylosis and nerve root lesions: incidence at routine necropsy	J. Bone Joint Surg. **48-B:**407-423, 1966

Chapter 15: Affections of the neck, shoulder, and jaw—cont'd

Cervical root syndrome and thoracic outlet syndromes—cont'd

Jackson, R.	The cervical syndrome	American Academy of Orthopaedic Surgeons Instructional Course Lectures, Ann Arbor, 1953, J. W. Edwards, vol. 10, pp. 65-90
McGowan, J. M.	Cervical rib: the rôle of the clavicle in occlusion of the subclavian artery	Ann. Surg. **124**:71-89, 1946
Odom, G. L., Finney, W., and Woodhall, B.	Cervical disk lesions	J.A.M.A. **166**:23-28, 1958
Raaf, J.	Surgery for cervical rib and scalenus anticus syndrome	J.A.M.A. **157**:219-223, 1955
Rosati, L. M., and Lord, J. W.	Neurovascular compression syndromes of the shoulder girdle	New York, 1961, Grune & Stratton, Inc.
Scoville, W. B.	Types of cervical disc lesions and their surgical approaches	J.A.M.A. **196**:479-481, 1966
Smith, G. W., and Robinson, R. A.	The treatment of certain cervical-spine disorders by anterior removal of the intervertebral disc and interbody fusion	J. Bone Joint Surg. **40-A**:607-624, 1958
Steindler, A.	The cervical pain syndrome	American Academy of Orthopaedic Surgeons Instructional Course Lectures, Ann Arbor, 1957, J. W. Edwards, vol. 14, pp. 1-10
Telford, E. D., and Mottershead, S.	Pressure at the cervico-brachial junction, an operative and anatomical study	J. Bone Joint Surg. **30-B**:249-265, 1948

Affections of the shoulder

Badgley, C. E.	Sports injuries of the shoulder girdle	J.A.M.A. **172**:444-448, 1960
Bankart, A. S. B.	An operation for recurrent dislocation (subluxation) of the sternoclavicular joint	Brit. J. Surg. **26**:320-323, 1938
	The pathology and treatment of recurrent dislocation of the shoulder joint	Brit. J. Surg. **26**:23-29, 1938
Bateman, J. E.	The shoulder and environs	St. Louis, 1955, The C. V. Mosby Co.
	The diagnosis and treatment of ruptures of rotator cuff	Surg. Clin. N. Amer. **43**:1523-1530, 1963
Bost, F. C., and Inman, V. T.	The pathological changes in recurrent dislocation of the shoulder: a report of Bankart's operative procedure	J. Bone Joint Surg. **24**:595-613, 1942
Bosworth, B. M.	Calcium deposits in the shoulder and subacromial bursitis: a survey of 12,122 shoulders	J.A.M.A. **116**:2477-2482, 1941
Brav, E. A.	Recurrent dislocation of the shoulder: ten years' experience with the Putti-Platt reconstruction procedure	Amer. J. Surg. **100**:423-430, 1960

Chapter 15: Affections of the neck, shoulder, and jaw—cont'd

Affections of the shoulder—cont'd

Caldwell, G. A., and Unkauf, B. M.	Results of treatment of subacromial bursitis in 340 cases	Ann. Surg. **132:**432-442, 1950
Codman, E. A.	The shoulder: rupture of the supraspinatus tendon and other lesions in or about the subacromial bursa	Boston, 1934, Thomas Todd Co.
Crenshaw, A. H., and Kilgore, W. E.	Surgical treatment of bicipital tenosynovitis	J. Bone Joint Surg. **48-A:**1496-1502, 1966
DePalma, A. F.	Surgery of the shoulder	Philadelphia, 1950, J. B. Lippincott Co.
	Bicipital tenosynovitis	Surg. Clin. N. Amer. **33:**1693-1702, 1953
Dickson, J. A., Humphries, A. W., and O'Dell, H. W.	Recurrent dislocation of the shoulder	Baltimore, 1953, The Williams & Wilkins Co.
Friedmann, E.	Rupture of distal biceps brachii tendon: report on 13 cases	J.A.M.A. **184:**60-63, 1963
Gilcreest, E. L.	The common syndrome of rupture, dislocation and elongation of the long head of the biceps brachii, an analysis of 100 cases	Surg. Gynec. Obstet. **58:**322-340, 1934
Godsil, R. D., Jr., and Linscheid, R. L.	Intratendinous defects of the rotator cuff	Clin. Orthop. **69:**181-188, 1970
Haggart, G. E., Dignam, R. J., and Sullivan, T. S.	Management of the "frozen" shoulder	J.A.M.A. **161:**1219-1222, 1956
Hitchcock, H. H., and Bechtol, C. O.	Painful shoulder, observations on role of the tendon of the long head of the biceps brachii in its causation	J. Bone Joint Surg. **30-A:**263-273, 1948
Howorth, M. B.	Calcification of the tendon cuff of the shoulder	Surg. Gynec. Obstet. **80:**337-345, 1945
	Tendinitis of the shoulder	J. Trauma **2:**502-529, 1962
Inman, V. T., Saunders, J. B. deC. M., and Abbott, L. C.	Observations on the function of the shoulder joint	J. Bone Joint Surg. **26:**1-30, 1944
Lazcano, M., Anzel, S. H., and Kelly, P. J.	Complete dislocation and subluxation of the acromioclavicular joint, end result in seventy-three cases	J. Bone Joint Surg. **43-A:**379-391, 1961
Lloyd-Roberts, G. C., and French, P. R.	Periarthritis of the shoulder, a study of the disease and its treatment	Brit. Med. J. **1:**1569-1571, 1959
MacAusland, W. R.	Recurrent anterior dislocation of the shoulder	Amer. J. Surg. **91:**323-331, 1956
McLaughlin, H. L.	Lesions of the musculotendinous cuff of the shoulder:	
	I. The exposure and treatment of tears with retraction	J. Bone Joint Surg. **26:**31-51, 1944

Chapter 15: Affections of the neck, shoulder, and jaw—cont'd

Affections of the shoulder—cont'd

McLaughlin, H. L.—cont'd	II. Differential diagnosis of rupture	J.A.M.A. **128**:563-568, 1945
	III. Observations on the pathology, course and treatment of calcific deposits	Ann. Surg. **124**:354-362, 1946
McLaughlin, H. L., and Asherman, E. G.	Lesions of the musculotendinous cuff of the shoulder: IV. Some observations based upon the results of surgical repair	J. Bone Joint Surg. **33-A**:76-86, 1951
Moseley, H. F.	Ruptures of the rotator cuff	Springfield, Ill., 1952, Charles C Thomas, Publisher
	Recurrent dislocation of the shoulder	Edinburgh, 1961, E. & S. Livingstone, Ltd.
	Shoulder lesions, ed. 3	Baltimore, 1969, The Williams & Wilkins Co.
Neviaser, J. S.	Arthrography of the shoulder joint, study of the findings in adhesive capsulitis of the shoulder	J. Bone Joint Surg. **44-A**:1321-1330, 1962
	The treatment of old unreduced dislocations of the shoulder	Surg. Clin. N. Amer. **43**:1671-1678, 1963
Nicola, T.	Recurrent dislocation of the shoulder	Amer. J. Surg. **86**:85-91, 1953
Norwich, I.	Calcification of the supraspinatus tendon: infiltration therapy with local anesthesia and multiple needling	Surg. Gynec. Obstet. **86**:183-191, 1948
O'Donoghue, D. H.	Injuries to the shoulder girdle	American Academy of Orthopaedic Surgeons Instructional Course Lectures, St. Louis, 1960, The C. V. Mosby Co., vol. 17, pp. 392-405
Osmond-Clarke, H.	Habitual dislocation of the shoulder, the Putti-Platt operation	J. Bone Joint Surg. **30-B**:19-25, 1948
Rowe, C. R.	Prognosis in dislocations of the shoulder	J. Bone Joint Surg. **38-A**:957-977, 1956
Samilson, R. L., et al.	Shoulder arthrography	J.A.M.A. **175**:773-778, 1961
Steinbrocker, O., Spitzer, N., and Friedman, H. H.	The shoulder-hand syndrome in reflex dystrophy of the upper extremity	Ann. Intern. Med. **29**:22-52, 1948
Steindler, A.	Interpretation of pain in the shoulder	American Academy of Orthopaedic Surgeons Instructional Course Lectures, Ann Arbor, 1958, J. W. Edwards, vol. 15, pp. 159-171

Affections of the jaw

Burman, M., and Sinberg, S. E.	Condylar movement in the study of internal derange-	J. Bone Joint Surg. **28**:351-373, 1946

Chapter 15: Affections of the neck, shoulder, and jaw—cont'd

Affections of the jaw—cont'd

Burman, M., and Sinberg, S. E.—cont'd	ment of the temporomandibular joint	
Gerry, R. G., and Rowan, R. L.	Temporomandibular joint disease, abnormal mandibular function as basis	Arch. Surg. **69:**635-645, 1954
Kazanjian, V. H.	Ankylosis of the temporomandibular joint	Surg. Gynec. Obstet. **67:**333-348, 1938
Sarnat, B. G., editor	The temporomandibular joint, ed. 2	Springfield, Ill., 1964, Charles C Thomas, Publisher
Schwartz, L.	Disorders of the temporomandibular joint	Philadelphia, 1959, W. B. Saunders Co.
Silver, C. M., and Simon, S. D.	Operative treatment for recurrent dislocation of the temporomandibular joint	J. Bone Joint Surg. **43-A:**211-218, 1961
	Meniscus injuries of the temporomandibular joint	J. Bone Joint Surg. **45-A:**113-124, 1963

Chapter 16

Affections of the elbow, wrist, and hand

Affections of the elbow

Ackerman, L. V.	Extra-osseous localized non-neoplastic bone and cartilage formation (so-called myositis ossificans)	J. Bone Joint Surg. **40-A:**279-298, 1958
Conacher, C.	Volkmann's ischaemic contracture of the forearm	Med. J. Aust. **2:**383-386, 1954
Dobbie, R. P.	Avulsion of the lower biceps brachii tendon, analysis of fifty-one previously unreported cases	Amer. J. Surg. **51:**662-683, 1941
Eichler, G. R., and Lipscomb, P. R.	The changing treatment of Volkmann's ischemic contractures from 1955 to 1965 at the Mayo Clinic	Clin. Orthop. **50:**215-223, 1967
Garden, R. S.	Tennis elbow	J. Bone Joint Surg. **43-B:**100-106, 1961
Goldie, I.	Epicondylitis lateralis humeri	Acta Chir. Scand., supp. 339, 1964
Hart, G. M.	Subluxation of the head of the radius in young children	J.A.M.A. **169:**1734-1736, 1959
Laurent, L. E., and Lindström, B. L.	Osteochondrosis of the capitulum humeri (Panner's disease)	Acta Orthop. Scand. **26:**111-119, 1956
Lipscomb, P. R.	The etiology and prevention of Volkmann's ischaemic contracture	Surg. Gynec. Obstet. **103:**353-361, 1956
Seddon, H. J.	Volkmann's contracture: treatment by excision of the infarct	J. Bone Joint Surg. **38-B:**152-174, 1956
	Volkmann's ischaemia	Brit. Med. J. **1:**1587-1592, 1964

Chapter 16: Affections of the elbow, wrist, and hand—cont'd

Affections of the elbow—cont'd

Smith, F. M.	Surgery of the elbow	Springfield, Ill., 1954, Charles C Thomas, Publisher
Spencer, G. E., Jr., and Herndon, C. H.	Surgical treatment of epicondylitis	J. Bone Joint Surg. **35-A**:421-424, 1953

Affections of the wrist and hand

Anton, J. I., Reitz, G. B., and Spiegel, M. B.	Madelung's deformity	Ann. Surg. **108**:411-439, 1938
Barnes, W. E., Larsen, R. D., and Posch, J. L.	Review of ganglia of the hand and wrist with analysis of surgical treatment	Plast. Reconstr. Surg. **34**:570-578, 1964
Bickel, W. H., Kimbrough, R. F., and Dahlin, D. C.	Tuberculous tenosynovitis	J.A.M.A. **151**:31-35, 1953
Boyes, J. H.	Dupuytren's contracture, notes on the age at onset and the relationship to handedness	Amer. J. Surg. **88**:147-154, 1954
	Bunnell's surgery of the hand, ed. 5	Philadelphia, 1970, J. B. Lippincott Co.
Boyes, J. H., Wilson, J. N., and Smith, J. W.	Flexor-tendon ruptures in the forearm and hand	J. Bone Joint Surg. **42-A**:637-646, 1960
Britt, L. P.	Principles of hand rehabilitation	Southern Med. J. **47**:205-209, 1954
Buchman, J.	Traumatic osteoporosis of the carpal bones	Ann. Surg. **87**:892-910, 1928
Butler, E. D., Hamill, J. P., Seipel, R. S., and de Lorimier, A. A.	Tumors of the hand: a 10-year survey and report of 437 cases	Amer. J. Surg. **100**:293-302, 1960
Carroll, R. E., Sinton, W., and Garcia, A.	Acute calcium deposits in the hand	J.A.M.A. **157**:422-426, 1955
Conway, H.	Dupuytren's contracture	Amer. J. Surg. **87**:101-119, 1954
Davis, J. E.	On surgery of Dupuytren's contracture	Plast. Reconstr. Surg. **36**:277-314, 1965
Dwight, T.	A clinical atlas: variations of the bones of the hands and feet	Philadelphia, 1907, J. B. Lippincott Co.
Fahey, J. J., and Bollinger, J. A.	Trigger-finger in adults and children	J. Bone Joint Surg. **36-A**:1200-1218, 1954
Felman, A. H., and Kirkpatrick, J. A.	Madelung's deformity: observations in 17 patients	Radiology **93**:1037-1042, 1969
Flynn, J. E., editor	Hand surgery	Baltimore, 1966, The Williams & Wilkins Co.
Gillespie, H. S.	Excision of the lunate bone in Kienbock's disease	J. Bone Joint Surg. **43-B**:245-249, 1961
Goldner, J. L.	Deformities of the hand incidental to pathological changes of the extensor and intrinsic muscle mechanisms	J. Bone Joint Surg. **35-A**:115-131, 1953
Harris, C., Jr., and Riordan, D. C.	Intrinsic contracture in the hand and its surgical treatment	J. Bone Joint Surg. **36-A**:10-20, 1954

Chapter 16: Affections of the elbow, wrist, and hand—cont'd

Affections of the wrist and hand—cont'd

Holm, C. L., and Embick, R. P.	Anatomical considerations in the primary treatment of tendon injuries of the hand	J. Bone Joint Surg. **41-A**:599-608, 1959
Howard, L. D., Jr.	Contracture of the thumb web	J. Bone Joint Surg. **32-A**:267-273, 1950
Hueston, J. T.	Dupuytren's contracture	Baltimore, 1963, The Williams & Wilkins Co.
Kanavel, A. B.	Infections of the hand, ed. 7	Philadelphia, 1939, Lea & Febiger
Kaplan, E. B.	Anatomy, injuries and treatment of the extensor apparatus of the hand and digits	Clin. Orthop. **13**:24-41, 1959
	Functional and surgical anatomy of the hand, ed. 2	Philadelphia, 1965, J. B. Lippincott Co.
Kelly, A. P., Jr.	Primary tendon repairs: a study of 789 consecutive tendon severances	J. Bone Joint Surg. **41-A**:581-598, 1959
Lamphier, T. A., Long, N. G., and Dennehy, T.	DeQuervain's disease, an analysis of 52 cases	Ann. Surg. **138**:832-841, 1953
Lapidus, P. W., and Fenton, R.	Stenosing tenovaginitis at the wrist and fingers: report of 423 cases in 369 patients with 354 operations	Arch. Surg. **64**:475-487, 1952
Larsen, R. D., and Posch, J. L.	Dupuytren's contracture	J. Bone Joint Surg. **40-A**:773-792, 1958
Littler, J. W.	The severed flexor tendon	Surg. Clin. N. Amer. **39**:435-447, 1959
Loomis, L. K.	Variations of stenosing tenosynovitis at the radial styloid process	J. Bone Joint Surg. **33-A**:340-346, 1951
Luck, J. V.	Dupuytren's contracture	J. Bone Joint Surg. **41-A**:635-664, 1959
Moorhead, J. J.	Trauma and Dupuytren's contracture	Amer. J. Surg. **85**:352-358, 1953
Nichols, H. M.	Manual of hand injuries, ed. 2	Chicago, 1955, Year Book Medical Publishers, Inc.
O'Rahilly, R.	A survey of carpal and tarsal anomalies	J. Bone Joint Surg. **35-A**:626-642, 1953
Peacock, E. E., Jr., and Hartrampf, C. R.	The repair of flexor tendons in the hand	Int. Abstr. Surg. **113**:411-432, 1961
Pedersen, H. E., and Day, A. J.	Dupuytren's disease of the foot	J.A.M.A. **154**:33-35, 1954
Piver, J. D., and Raney, R. B.	DeQuervain's tendovaginitis	Amer. J. Surg. **83**:691-694, 1952
Posch, J. L.	Tumors of the hand	J. Bone Joint Surg. **38-A**:517-540, 1956
Pulvertaft, R. G.	Tendon grafts for flexor tendon injuries in the fingers and thumb	J. Bone Joint Surg. **38-B**:175-194, 1956
Rhode, C. M.	Treatment of hand infections	Amer. Surg. **27**:85-115, 1961

Chapter 16: Affections of the elbow, wrist, and hand—cont'd

Affections of the wrist and hand—cont'd

Riddell, D. M.	Spontaneous rupture of the extensor pollicis longus	J. Bone Joint Surg. **45-B:**506-510, 1963
Riordan, D. C.	Dupuytren's contracture	Southern Med. J. **54:**1391-1394, 1961
Salter, R. B., and Zaltz, C.	Anatomic investigations of the mechanism of injury and pathologic anatomy of "pulled elbow" in young children	Clin. Orthop. **77:**134-143, 1971
Stark, H. H., Boyes, J. H., and Wilson, J. N.	Mallet finger	J. Bone Joint Surg. **44-A:**1061-1068, 1962
Strandell, G.	Peritendinitis calcarea in the hand	Acta Chir. Scand. **125:**42-51, 1963
Straub, L. R., and Wilson, E. H., Jr.	Spontaneous rupture of extensor tendons in the hand associated with rheumatoid arthritis	J. Bone Joint Surg. **38-A:**1208-1217, 1956

Index

B

E

F

I

J

P

Q

R

T